Lecture Notes in Bioinformatics   5724

Edited by S. Istrail, P. Pevzner, and M. Waterman

Editorial Board: A. Apostolico   S. Brunak   M. Gelfand
T. Lengauer   S. Miyano   G. Myers   M.-F. Sagot   D. Sankoff
R. Shamir   T. Speed   M. Vingron   W. Wong

Subseries of Lecture Notes in Computer Science

T0189943

Steven L. Salzberg Tandy Warnow (Eds.)

Algorithms in Bioinformatics

9th International Workshop, WABI 2009
Philadelphia, PA, USA, September 12-13, 2009
Proceedings

 Springer

Volume Editors

Steven L. Salzberg
Center for Bioinformatics and Computational Biology
and Department of Computer Science
University of Maryland
College Park, MD, USA
E-mail: salzberg@umiacs.umd.edu

Tandy Warnow
Department of Computer Sciences
The University of Texas at Austin
Austin, TX, USA
E-mail: tandy@cs.utexas.edu

Library of Congress Control Number: 2009933693

CR Subject Classification (1998): J.3, I.5, F.2.2, E.1, G.2, F.2

LNCS Sublibrary: SL 8 – Bioinformatics

ISSN   0302-9743

ISBN 978-3-642-04240-9 Springer Berlin Heidelberg New York

Typesetting: Camera-ready by author, data conversion by Scientific Publishing Services, Chennai, India
Printed on acid-free paper   SPIN: 12749041   06/3180   5 4 3 2 1 0

Preface

These proceedings contain papers from the 2009 Workshop on Algorithms in Bioinformatics (WABI), held at the University of Pennsylvania in Philadelphia, Pennsylvania during September 12–13, 2009. WABI 2009 was the ninth annual conference in this series, which focuses on novel algorithms that address important problems in genomics, molecular biology, and evolution. The conference emphasizes research that describes computationally efficient algorithms and data structures that have been implemented and tested in simulations and on real data. WABI is sponsored by the European Association for Theoretical Computer Science (EATCS) and the International Society for Computational Biology (ISCB). WABI 2009 was supported by the Penn Genome Frontiers Institute and the Penn Center for Bioinformatics at the University of Pennsylvania.

For the 2009 conference, 90 full papers were submitted for review by the Program Committee, and from this strong field of submissions, 34 papers were chosen for presentation at the conference and publication in the proceedings. The final program covered a wide range of topics including gene interaction networks, molecular phylogeny, RNA and protein structure, and genome evolution. Extended versions of selected papers will appear in a special issue of the journal *Algorithms for Molecular Biology*, published by BiomedCentral. We thank all the members of the Program Committee for their detailed and timely reviews, and we also thank the authors of all the papers, whose contributions made the workshop possible. Thanks also to the 2009 Keynote speakers, Shelley Berger of the Wistar Institute (Philadelphia, Pennsylvania) and Elchanan Mossel of the Weizmann Institute of Science (Rehovot, Israel). Above all, we want to extend our appreciation to the Organizing Committee at the University of Pennsylvania – Junhyong Kim (Chair), Stephen Fisher, and Li-San Wang – who handled all of the local arrangements, the conference website, and many other details that made the conference a success.

September 2009

Steven L. Salzberg
Tandy Warnow

Organization

Program Chairs

Steven Salzberg	University of Maryland, College Park, USA
Tandy Warnow	University of Texas, Austin, USA

Program Committee

Elizabeth Allman	University of Alaska, Fairbanks, USA
Tanya Y. Berger-Wolf	University of Illinois, Chicago, USA
Paola Bonizzoni	UniversitáDegli Studi di Milano-Bicocca, Italy
Dan Brown	University of Waterloo, Canada
Rita Casadio	Universitá Degli Studi di Milano-Bicocca, Italy
Mark Clement	Brigham Young University, USA
Maxime Crochemore	Université de Marne-la-Vallé, Carte, France
Miklós Csűrös	Université de Montréal, Canada
Aaron Darling	University of California, Davis, USA
Nadia El-Mabrouk	Université de Montréal, Canada
Eleazar Eskin	University of California, Los Angeles, USA
Liliana Florea	University of Maryland, College Park, USA
Ganesh Ganapathy	Duke University, Durham, USA
Raffaele Giancarlo	Universitá degli Studi di Palermo, Italy
Concettina Guerra	Universitá di Padova, Italy
Roderic Guigó	University of Barcelona, Spain
Sridhar Hannenhalli	University of Pennsylania, Philadelphia, USA
Daniel Huson	University of Tübingen, Germany
Shane T. Jensen	University of Pennsylvania, Philadelphia, USA
Carl Kingsford	University of Maryland, College Park, USA
Ming Li	University of Waterloo, Canada
C. Randal Linder	University of Texas, Austin, USA
Stefano Lonardi	University of California, Riverside, USA
Lusheng Lusheng-Wang	City University of Hong Kong, Hong Kong
Jian Ma	University of California, Santa Cruz, USA
Erick Matsen	University of California, Berkeley, USA
István Miklós	Rényi Institute, Budapest, Hungary
Bernard Moret	Swiss Federal Institute of Technology, Lausanne, Switzerland
Burkhard Morgenstern	University of Göttingen, Germany
Elchanan Mossel	University of California, Berkeley, USA

Luay K. Nakhleh	Rice University, Houston, USA
Macha Nikolski	Université Bordeaux, France
Laxmi Parida	IBM T.J. Watson Research Center, Yorktown Heights, USA
Kunsoo Park	Seoul National University, Korea
Ron Pinter	Israel Institute of Technology, Haifa, Israel
Cinzia Pizzi	University of Padova, Italy
David Posada	University of Vigo, Spain
Sven Rahmann	Technische Universität Dortmund, Germany
Ben Raphael	Brown University, Providence, USA
Knut Reinert	Freie Universität Berlin, Germany
Allen Rodrigo	University of Auckland, New Zealand
David Sankoff	University of Ottawa, Canada
Mona Singh	Princeton University, USA
Saurabh Sinha	University of Illinois, Urbana, USA
Steven Skiena	State University of New York, Stonybrook, USA
Peter F. Stadler	Universität Leipzig, Germany
Alexandros Stamatakis	Technische Universität München, Germany
Jens Stoye	Universität Bielefeld, Germany
Jijun Tang	University of South Carolina, Columbia, USA
Glenn Tesler	University of California, San Diego, USA
Olga Troyanskaya	Princeton University, USA
Alfonso Valencia	Spanish National Cancer Research Centre, Madrid, Spain
Chris Workman	Technical University of Denmark, Lyngby, Denmark

Local Organization

Junhyong Kim (Chair)	University of Pennsylvania, Philadelphia, USA
Stephen Fisher	University of Pennsylvania, Philadelphia, USA
Li-San Wang	University of Pennsylvania, Philadelphia, USA

Table of Contents

Minimum Factorization Agreement of Spliced ESTs

Paola Bonizzoni[1], Gianluca Della Vedova[2], Riccardo Dondi[3], Yuri Pirola[1],
and Raffaella Rizzi[1]

[1] DISCo, Univ. Milano-Bicocca
bonizzoni@disco.unimib.it, pirola@disco.unimib.it, rizzi@disco.unimib.it
[2] Dip. Statistica, Univ. Milano-Bicocca
gianluca.dellavedova@unimib.it
[3] Dip. Scienze dei Linguaggi, della Comunicazione e degli Studi Culturali, Univ.
Bergamo
riccardo.dondi@unibg.it

Abstract. Producing *spliced EST* sequences is a fundamental task in
the computational problem of reconstructing splice and transcript variants, a crucial step in the alternative splicing investigation. Now, given
an EST sequence, there can be several spliced EST sequences associated
to it, since the original EST sequences may have different alignments
against wide genomic regions.

In this paper we address a crucial issue arising from the above step:
given a collection C of different spliced EST sequences that are associated
to an initial set S of EST sequences, how can we extract a subset C' of C
such that each EST sequence in S has a putative spliced EST in C' and
C' agree on a common alignment region to the genome or gene structure?

We introduce a new computational problem that models the above
issue, and at the same time is also relevant in some more general settings,
called *Minimum Factorization Agreement* (MFA). We investigate some
algorithmic solutions of the MFA problem and their applicability to real
data sets. We show that algorithms solving the MFA problem are able to
find efficiently the correct spliced EST associated to an EST even when
the splicing of sequences is obtained by a rough alignment process. Then
we show that the MFA method could be used in producing or analyzing
spliced EST libraries under various biological criteria.

1 Introduction

RNA splicing is the process that takes place in all eukaryotic cells when a pre-
mRNA matures to become a messenger RNA (mRNA.) The maturation process
includes splicing, that is the removal of intronic regions and the concatenation of
exonic tracts to form a functional mature mRNA. The splicing process requires
a large ribonucleoproteic complex, the spliceosome, which recognizes specific
conserved sites at the exon/intron boundaries (donor, acceptor and branch site)
to carry out the intron excision and exon ligations. However, several alternative
splicing sites can be recognized, in a tightly regulated way, that may produce

S.L. Salzberg and T. Warnow (Eds.): WABI 2009, LNBI 5724, pp. 1–12, 2009.

alternative transcripts (and proteins) from the same gene. This process, known as alternative splicing (AS) is the major determinant for the expansion of the transcriptome and proteome in eukaryotic organisms. AS has recently emerged as a key mechanism to modulate gene expression in a tissue-specific manner and is also involved in the onset and progression of serious pathologies including cancer [5].

The investigation of AS can be carried on by developing computational tools that use EST data available from UNIGENE or produced by sequencing processes (see [3] for a survey of the computational methods.) Indeed an EST (Expressed Sequence Tag) is a short fragment of complementary DNA, produced in vitro by making a mirror copy of an mRNA. Then, except for a few sequencing errors, an EST sequence should be almost identical to a portion of the gene sequence, while a cluster of ESTs for a given gene should provide the information to identify all transcribed regions for a specific gene. The large amount of available EST sequences represents a very valuable source to predict alternative splicing, mainly when used in combination with the related genomic sequence [12,10]. Indeed, they play a crucial role in many computational approaches to investigate and reconstruct the human transcriptome [20,9,18,4]. A challenging computational problem that can be faced when using ESTs data is that of predicting transcript isoforms (mRNA) of which ESTs are fragments [11]. The reconstruction of isoforms can be formulated as a new computational assembly problem [2].

This problem requires as input data a set of ESTs that have been factorized into their constitutive exons, called *spliced ESTs*. Spliced ESTs are mainly produced by computational tools that predict AS [11,15,19,13,8]. Recently the ASPIc tool implements an assembly algorithm to reconstruct isoforms from spliced ESTs: such data is available in a database [7,6]. New algorithmic issues are emerging in this field since the traditional computational approaches to AS are unable to manage instances consisting of short transcripts (short-reads) produced by new sequencing technologies [14].

Alternatively, spliced ESTs are computed by using alignment algorithms of EST sequences against genomic sequences, which is complicated by the fact that genomic sequences may contain repeated substrings; moreover sequencing errors make the exact location of splice sites ambiguous.

For the previously discussed reasons, software tools that produce spliced ESTs data may give more than one factorization of a single EST sequence into its genomic exons, even when there exists a unique way to locate splice sites on the EST. This fact may occur for example when short exons, or a short prefix and suffix of the EST may have different alignments to the genomic sequence, since they may correspond to partial exons (a suffix, or prefix of an exon, respectively.)

In this paper we investigate the following crucial issue arising during the process of predicting AS from spliced ESTs; given some spliced ESTs associated to a set of ESTs, how can we extract a spliced EST for each EST sequence that explains the same gene structure?

We first formalize this problem by defining the *Minimum Factorization Agreement* (MFA) problem over an instance consisting of a set C of factorized sequences, called *compositions*, i.e. sequences that are the concatenation of factors in a finite set F. More precisely, the MFA problem over a set S of sequences asks for a minimum subset $F' \subseteq F$ such that each sequence in S has associated a composition in C using only factors in F'.

In this formalization an EST is a sequence $s \in S$ to which we associate a set of sequences over a finite alphabet F of genomic exons (or prefix and suffices of exons) of the gene, called spliced ESTs for s. Each spliced EST for s represents a possible alignment of substrings of the sequence s to the genome, each substring being aligned to a putative exon, i.e. a factor of F in our formalization.

Then, the MFA problem formalizes the idea of computing a minimum cardinality set of genomic exons that can explain at least one spliced EST associated to each sequence s in S. In fact the MFA problem, as we will show in the paper, allows to face the more general problem of reconciling a set of EST sequences data w.r.t. different types of biological information. Indeed, the alphabet F by which a sequence is factorized could represent splice sites or splice patterns instead of exons and thus the MFA problem asks for the minimum set of factors (common splice sites or significant patterns) on which all input sequences agree.

We first show that the MFA problem is NP-hard even over simple instances. Successively we provide an exact algorithm for the problem based on a precomputation step where all factors are extracted. Though the worst-case time complexity algorithm is exponential in the number of factors, the algorithm is efficient over practical instances, as the precomputation step quickly and effectively trims the set of possible factors. Such claim is assessed experimentally, showing different applications of the MFA method to retrieve common information from a set of spliced ESTs associated to an initial set of ESTs.

Two experiments complete our paper. In the first experiment we extract a minimum cardinality set of genomic exons that explains the whole set of spliced ESTs produced by the tool Gmap [17] and by another procedure for EST alignment that we have implemented. A second experimental analysis is devoted to the study of the application of the MFA approach to find common alignment regions of spliced ESTs produced by our rough factorization method. The last experiment shows the applicability of the method when the number of input factors is quite large.

2   The MFA Problem

In this section the *Minimum Factorization Agreement* problem for a set C of factorized sequences is proposed. The problem formalizes the basic issue stated in the introduction and investigated in the paper.

Let $F = \{f_1, f_2, \ldots, f_{|F|}\}$ be an ordered finite set of *factors*, and let S be a set of sequences. We can assume to have an explicit enumeration of the sequences in S, that is $S = \{s_1, s_2, \ldots, s_{|S|}\}$. A *composition* c is a sequence over alphabet F where each factor appears at most once in c and c respects the ordering of F. Equivalently c can be thought of as a subsequence of F. The *factor set* of a composition c is the set of factors belonging to c.

Informally, each sequence in S represents an EST sequence, while a composition corresponds to a spliced EST for a sequence $s \in S$, as we are not interested in the actual nucleotide-level sequences and we assume that the set of factors has already been determined. Since each sequence s has associated a certain number of spliced ESTs, each composition is colored by some sequences in S, that is for each composition we state from which sequences the composition is derived. Notice that the same composition (spliced EST) can be derived from more than one sequence (original EST.) In the following we will assume that each composition c is actually colored by some sequences in S, and we denote by $\lambda(c)$ such set of sequences. Conversely, given a sequence $s \in S$, we denote by $C(s)$ the set of compositions that are colored by s. By a slight abuse of language, given a subset $S_1 \subseteq S$, we denote by $C(S_1)$ the set $\bigcup_{s \in S_1} C(s)$. Therefore $C(S)$ consists of all compositions.

Given a subset $F' \subseteq F$ of factors and the set $C(S)$, then F' is a *factorization agreement set for* $C(S)$ iff for each sequence $s \in S$, there exists a composition $c \in C(s)$ such that the factor set of c is entirely contained in F'.

Our main computational problem, *Minimum Factorization Agreement Set (MFA)* is on instances made of a set F of factors, a set S of sequences and a set C of (colored) compositions. The problem asks for a minimum cardinality subset $F' \subseteq F$ such that F' is a factorization agreement set for $C(S)$.

First we will prove, via an L-reduction, that MFA shares all known inapproximability results of Set Cover [1,16]. We recall that an instance of Set Cover consists of a universe set U and a collection $\mathcal{C}$ of subsets of U. Moreover the goal of the problem is to find a minimum cardinality subset $\mathcal{C}_1$ of $\mathcal{C}$ such that the union of all sets in $\mathcal{C}_1$ is exactly U.

We provide an alternative formulation of the MFA problem based on binary matrices. Given an instance $(F, S, C(S))$ of the MFA problem, we construct a binary matrix M where the columns (respectively rows) are in a 1-to-1 correspondence with the factors in F (resp. with the compositions in $C(S)$) – in the following we identify each column (resp. row) with a factor (resp. a composition.) Moreover each row is colored and the set of colors is the set S of input sequences – notice that each row can be more than one color. The entry $M[\alpha, f]$ (that is the entry of the row identified by the factor f and the composition α) is equal to 1 if and only if the composition α contains the factor f.

Given a composition α, we denote by $1(\alpha)$ and $0(\alpha)$ respectively the set of columns of M containing 1 and 0 in row α. The MFA problem then asks for a minimum cardinality subset $F_1 \subseteq F$ such that, for each possible color c, there is a composition α with $1(\alpha) \subseteq F_1$ and the row α of M is colored by c (notice that this row can also be different colors.)

The matrix formulation leads immediately to a reduction from Set Cover. In fact let $(U, \mathcal{C})$ be an instance of Set Cover (we recall that U is the universe set and $\mathcal{C}$ is a collection of subsets of U.) Then let U be the color set, while the factor set and the composition set are both equal to $\mathcal{C}$. The entries of the matrix M are equal to the entries of the identity matrix and the set of colors of a row (identified by a subset $\mathcal{C}_1 \in \mathcal{C}$) is $\mathcal{C}_1$ itself. Notice that any agreement factor set

Table 1. Example of instance of MFA on factors $F = \{A, B, C, D\}$

Sequence	Compositions
1	A, B, D
2	A, C
3	B, C

Colors	Factors	A B C D
1, 2	A	1 0 0 0
1, 3	B	0 1 0 0
2, 3	C	0 0 1 0
1	D	0 0 0 1

F is also a set cover on $(U, \mathcal{C})$. An illustrative example where the instance of Set Cover is over universe $\{1, 2, 3\}$ and $\mathcal{C}$ consists of four subsets $A = \{1, 2\}$, $B = \{1, 3\}$, $C = \{2, 3\}$, $D = \{1\}$, is represented as Table 1.

It is immediate to notice that the covers of $(U, \mathcal{C})$ are also factorizations agreement sets of M and vice versa. Moreover the reduction produces instances of MFA where all compositions consists of only one factor. Also, the sizes of approximate and optimal solutions of the instances of Set Cover and MFA are the same, hence all known inapproximability results for Set Cover (namely there exists a constant $c > 0$ such that no polynomial-time approximation algorithm can achieve a guaranteed approximation ratio $c \log n$, unless $\mathbf{P} \neq \mathbf{NP}$ [16]) hold for such a restriction.

3   An Algorithm for Solving the MFA

In this section we study the matrix formulation of the MFA problem and we present an algorithm that is of practical interest. The first step of the algorithm consists of reducing the matrix, that is removing some rows and/or columns, while keeping at least an optimal solution.

More precisely, we provide some rules that must be enforced, if possible, on the matrix M. The reduction process is iterated until no such rule applies any more. At the end we will obtain a reduced matrix M_1 together with a set S_F of factors that must be added to any feasible solution of M_1 to obtain a solution of M. Let us list the five rules.

1. If there exist two rows r_1, r_2 such that $M[r_1, c] \geq M[r_2, c]$ for each column c of M and $\lambda(r_1) \cap \lambda(r_2) \neq \emptyset$, then remove all sequences in $\lambda(r_1) \cap \lambda(r_2)$ from the colors of r_1. If $\lambda(r_1) = \emptyset$ then remove r_1 from M. (Rationale: all feasible solutions that include all factors in r_1 also include all factors in r_2.)
2. If there exists a column c containing only 1s, then add c to S_F and remove c from M. (Rationale: c must necessarily belong to all solutions of M.)
3. If there exists a column c and a color i such that all rows of M that are colored i have value 1 in c, then add c to S_F and remove c from M. (Rationale: since at least one of the compositions colored by i have to be factorized by a solution, c must necessarily belong to all solutions of M.) Notice that this rule is more general than Rule 1, which is nonetheless explicitly stated since it is computationally easier to enforce.

4. If there exists a row r containing only 0s and r is colored C, then remove from M all colors in $\lambda(r)$. Remove from M all rows r such that $\lambda(r)$ becomes empty. (Rationale: S_F contains all factors of a composition of r, therefore r can be removed from M without enlarging the size of a feasible solution. Moreover, if $\lambda(r) \neq \emptyset$ and the row r has only zeroes for each color $x \in \lambda(r)$, then S_F contains all factors used in a composition colored by x.)
5. If there exists a column c containing only 0s, then remove c from M. (Rationale: no actual composition contains c, which therefore is not an interesting factor any more.)

Consequently, in the following we can assume that our algorithms deal with reduced matrices.

3.1   A Naïve Algorithm

A simple approach for solving exactly the MFA problem consists of computing all possible subsets $F_1 \subseteq F$ in non-decreasing order of cardinality of such subsets, and checking if F_1 is a factorization agreement set. As we are interested in a minimum-cardinality agreement set, the algorithm halts as soon as a factorization agreement set is found.

The analysis of the time complexity is straightforward. In fact, given a subset F_1 of factors, deciding if F_1 is a factorization agreement set for S requires time $O(|F||C(S)|)$. Moreover, since the factor subsets are enumerated in non-decreasing order of cardinality and denoting by opt the cardinality of a minimum composition agreement set, the algorithm stops after considering $\sum_{k=1}^{opt} \binom{|C(S)|}{k}$ subsets. Thus, the overall time complexity is $O\left(\sum_{k=1}^{opt} \left(\binom{|C(S)|}{k}|F||C(S)|\right)\right)$ which is clearly $O\left(2^{|C(S)|}|F||C(S)|\right)$.

Even if the time is bounded only by an exponential function, two considerations are in order. First of all, on real data the reduction process usually results in a matrix much smaller than the original, therefore lessening the impact of the exponential factor (which is relative to the reduced matrix.) Moreover, the implementation of such simple method actually exploits some features of real-world architectures, such as bit-parallelism and data locality (in fact any reduced matrix with at most 100 factors and 10^5 compositions resides in the cache memory of a mainstream PC.)

3.2   A Refined Algorithm

In this section we describe an algorithm which is more refined and somewhat more complex than the one of the previous section. While we expect the refined algorithm to outperform the naïve one on hard and/or large instances, we have currently been unable to find some real-world instances where those improvements have resulted in appreciable decreases in the running times.

The algorithm is based on the idea of building a binary matrix $FM(\cdot,\cdot)$, where each element $FM(X,s)$ is equal to 1 if and only if the set of factors $X \subseteq F$ contains a composition of the sequence $s \in S$. The following Lemma leads immediately to a recursive definition of $FM(\cdot,\cdot)$.

Lemma 1. *Let X be a subset of F and let s be a sequence in S. Then $FM(X,s) = 1$ iff X is a minimal factor set of a composition of s or there exists $x \in X$ with $FM(X - \{x\}, s) = 1$.*

Proof. Assume initially that $FM(X, s) = 1$ and that X is not a minimal set that is the factor set of a composition of s. Then, by definition of minimality, there exists $x \in X$ such that $X - \{x\}$ is composition of s. By definition of $FM(\cdot, \cdot)$, $FM(X - \{x\}, s) = 1$. The other direction is trivial.   □

If we have an algorithm for checking if a subset $X \subseteq F$ is a minimal factor set of a composition of a sequence $s \in S$, then we can describe a simple exponential time, yet effective, algorithm for the MFA problem, whose correctness is an immediate consequence of Lemma 1. Our algorithm, whose pseudocode is represented in Algorithm 1, is called MFA2 and is bit-parallel. In fact, we will use the array $Fact(\cdot)$ instead of the matrix $FM(\cdot, \cdot)$, where $Fact$ is indexed by a subset X of F and each element of $Fact$ is actually the vector corresponding to an entire row of $FM(\cdot, \cdot)$. Therefore, $Fact(X)$ is the characteristic binary vector of the sequences which have a composition whose factor set is contained in X.

We will also use another array $MinimalFA(\cdot)$, which is also indexed by a subset X of F and each $MinimalFA(X)$ is the characteristic binary vector of the sequences which have a composition whose factor set is exactly X (it is immediate to notice that it is equivalent to the fact that X is a minimal set such that there exists a composition whose factor set is X, as the algorithm works on a reduced matrix.) We recall that a sequence can color more than one composition.

Algorithm 1. MFA2(F, S)

Data: An instance $(F, S, C(S))$ of MFA
1   Compute $MinimalFA$;
2   $Fact(\emptyset) \leftarrow \bar{0}$;
3   **foreach** $F_1 \subseteq F$ *in non-decreasing order of cardinality,* $F \neq \emptyset$ **do**
4      $Fact(F_1) \leftarrow \bigvee_{x \in F_1} Fact(F_1 - \{x\}) \bigvee MinimalFA(F_1)$;
5   **if** $Fact(F_1) \leftarrow \bar{1}$ **then**
6   **return** F_1

It is paramount for the efficiency of the MFA2 algorithm that we are able to compute and query efficiently $MinimalFA(\cdot)$. Since almost all the vectors in $MinimalFA(\cdot)$ are equal to $\bar{0}$, it is more efficient to actually use a hash for storing $MinimalFA(\cdot)$, assuming that an entry not found in the hash is equal to $\bar{0}$. It is immediate to notice that the time required by Alg. 2 is $O(|S||C(S)||F|)$, assuming that we have chosen an implementation of hash tables where each operation can be performed in time logarithmic in the number of elements, and noticing that $|C(S)| \leq 2^{|F|}$.

The analysis of Alg. 1 is based on the simple observation that all bit-level operations can be performed in $O(|S|)$ time. Moreover, since the subsets of F

Algorithm 2. Build $MinimalFA$

   Data: An instance $(F, S, C(S))$ of MFA
1 $MinimalFA \leftarrow$ empty hash;
2 **foreach** $c \in C(s)$ **do**
3   **foreach** $s \in S$ **do**
4   **if** c *is a composition of* s **then**
5   Pose to 1 the bit of $MinimalFA(f)$ corresponding to sequence s;
6 **return** $MinimalFA(\cdot)$

---

are generated by non-decreasing cardinality, at most $O(|F|^{opt})$ such subsets are considered, where *opt* is the size of the optimal solution. The overall time complexity is therefore $O(|S||F|^{opt})$. The time for reducing the original matrix is clearly polynomial in the size of the matrix, as enforcing each rule results in the removal of at least a row or a column. The most trivial implementation of the reduction rules has an $O(|S||F|^3|C(S)|^3)$ overall time complexity.

4   Experimental Analysis

In this section we report the results of two main different experiments that validate our proposal to use the MFA approach to analyze and compare EST sequence data. Both experiments have been performed on the same type of instances, as detailed in Table 2.

Each instance is related to a gene and its associated UNIGENE cluster; in fact the instance consists of compositions of EST sequences obtained by merging two groups of spliced ESTs. The first group of spliced ESTs, denoted by G, is generated by applying Gmap [17] to the UNIGENE cluster.

Gmap, just as other tools, chooses a single high-quality EST alignment to the gene in the data we considered, thus for each EST in the cluster only one spliced EST, i.e. a composition, has been reported (see last row of Table 2).

To obtain the second group of spliced ESTs, denoted by A, we have used an alignment procedure that we have designed, called BASICALIGN, which produces some possible alternative alignments to the genomic region of the EST sequences of each cluster without applying specific biological criteria, such as canonical splice sites or intron length constraints used in most alternative splicing predictions to select or refine spliced EST [3]. In our experiment the main use of the alignments produced by BASICALIGN is to introduce some noise.

The goal of the first experiment is to validate that the MFA method is able to extract a factorization agreement corresponding to a solution that is both biologically validated and corresponding to the factorization induced by the Gmap compositions. To evaluate how the MFA approach tackles the above issue, we executed it on instances consisting of the set $G \cup A$. Table 3 reports the results for some genes. For all genes, the number of factors for the compositions of G given in Table 2 turned out to be equal to the number of factors computed by MFA on set $G \cup A$, thus showing that our approach is resistant to the perturbations produced by adding to G spliced ESTs of lower quality.

Indeed, it must be pointed out that even though BASICALIGN can quite efficiently produce good quality alignments , usually it generates a larger set of factors (or candidate genomic exons) for a cluster of EST data, therefore providing an interesting data set to test the time complexity of our MFA algorithms, whose time requirements can be substantial for some of the generated data. On the other hand, whenever the number of factors is quite large, the MFA approach can be used to locate efficiently the correct exonic regions where a factorization agreement for the EST sequences can be found.

The goal of the second experiment is to test the efficiency of the method over large real data sets arising from several compositions of the same EST cluster. Moreover, it mainly shows the capability of the MFA approach to infer a factorization agreement for large input data. The idea is to cluster together very similar occurrences of each EST in the genomic sequences and to extract a representative factor from each cluster so that the resulting instance is smaller than the original one. The reduction process has the additional advantage that spurious occurrences of an EST in the genomic sequence does not negatively affect the running time or the overall quality of the solution computed by our MFA procedure.

The experiment consists of two steps. In the first step we reduce the instance that will be given to the MFA procedure. More precisely, we compound aligned factors of $G \cup A$ (reported in Table 2) into equivalence classes w.r.t. an equivalence relation between factors representing a common exonic region over the genome: such a relation holds between two exons of each spliced EST that are reasonably similar or variants of the same exon. Since our definition of MFA allows arbitrary factor sets, the factor set passed to our MFA procedure consists of the classes of the equivalence relation. Moreover, the compositions are updated so that each original factor f is replaced with the equivalence class to which f belongs. After such a process, we remove all copies but one of the duplicate compositions (two compositions are duplicate if they are equal factor-wise and are the same color.) Notice that by keeping track of the equivalence classes and the modifications

Table 2. Description of the instance for experiments 1 and 2

Gene	ABCB10	DENND2D	RSBN1	HNRPR
Number of compositions in G	44	159	29	754
Number of compositions in A	40	268	48	1576
Number of factors in G	17	12	11	22
Number of factors in A	47	92	28	449
Average number of compositions per sequence on $G \cup A$	1.91	2.68	2.65	3.09
Max number of compositions per sequence on $G \cup A$	5	11	2	21
Min number of compositions per sequence on $G \cup A$	1	1	1	1

Table 3. Summary of the results for experiment 1, which regards spliced ESTs generated by Gmap and spliced ESTs generated by BASICALIGN on the UNIGENE clusters associated to some genes

Gene	ABCB10	DENND2D	RSBN1	HNRPR
Number of factors	63	103	39	470
Number of columns removed in the reduction step	13	9	3	15
Percentage of columns removed in the reduction step	20.6%	8.7%	7.7%	3.19%
Number of factors computed by MFA	17	12	11	22
Number of compositions explained by the factors computed by MFA	44	159	29	781
Average number of compositions explained by the factors computed by MFA	1	1	1	1.04
Running time (μs)	6,629	22,533	1,583,162	2,113,154

Table 4. Summary of the results for experiment 2

Gene	ABCB10	DENND2D	RSBN1	HNRPR
Number of factors	17	9	8	37
Number of factors computed by MFA	11	8	7	11
Number of columns removed in the reduction step	11	8	7	11
Percentage of columns removed in the reduction step	66.4%	88.9%	87.5%	29.73%

performed, we can output the solution on the original instance corresponding to the computed solution on the reduced instance.

Subsequently, the MFA procedure is executed on the new data set where each factor represents an equivalence class and thus the number of input factors is usually largely reduced (for example, 449 factors are grouped into 37 factors in HRNPR gene as reported in Table 4).

In both experiments the algorithm was quite fast and was able to filter out erroneous EST alignments consisting of factors aligned outside the region of the putative gene. Moreover, Table 4 shows that the number of factors in the second experiment is quite similar to the number of factors of G. Clearly, intron retention or exons variants are represented by the same exon after the first reduction step of the original instance specified in Table 2. Remember that it is easy to obtain the factorization agreement computed in the first experiment by finding the exons that are in the same equivalence class.

Tables 3 and 4 report some results on the performance of the algorithm. The second experiment points out that the reduction phase of the input matrix greatly reduces the number of input factors, thus producing a reduced matrix over which even the naïve algorithm performs well. Summarizing the comparison of the experiments over the same instances both in terms of efficiency and accuracy, we can conclude that the two-step MFA method used in the second experiment can be applied for computing factorization agreements even over larger instances than the one reported in the Tables.

In all experiments we used a naïve version of the MFA algorithm, since its running times have turned out to be adequate, as shown in the results. In fact, the reduction step succeeds in drastically reducing the the number of factors.

5 Conclusions and Open Problems

In this paper we have given a novel formulation of the computational problem of comparing spliced EST sequences in order to extract a minimum set of ordered factors (exons) that can explain a unique way of splicing each single EST from an EST data set. We have shown that the problem is hard to approximate and we have proposed two algorithms that exhibit a nice behavior in practice. Nonetheless, both the theoretical and the experimental analysis in only at a preliminary stage, and we plan to further investigate the problem on both levels.

In our opinion the MFA approach could be used in other frameworks where it is required to find a minimum set of alignment regions that can explain one alignment of a sequence among multiple ones, such as for in example, in transcript assembly from short-reads. Moreover, the MFA approach could be useful in comparing spliced EST data from different databases.

Acknowledgments

We thank Marcello Varisco for his sharp implementations of the algorithms proposed in the paper. PB, GDV, YP and RR have been partially supported by the FAR 2008 grant "Computational models for phylogenetic analysis of gene variations". PB has been partially supported by the MIUR PRIN grant "Mathematical aspects and emerging applications of automata and formal languages".

References

1. Ausiello, G., Crescenzi, P., Gambosi, V., Kann, G., Marchetti-Spaccamela, A., Protasi, M.: Complexity and Approximation: Combinatorial optimization problems and their approximability properties. Springer, Heidelberg (1999)
2. Bonizzoni, P., Mauri, G., Pesole, G., Picardi, E., Pirola, Y., Rizzi, R.: Detecting alternative gene structures from spliced ests: A computational approach. Journal of Computational Biology 16(1), 43–66 (2009) PMID: 19119993
3. Bonizzoni, P., Rizzi, R., Pesole, G.: Computational methods for alternative splicing prediction. Briefings in Functional Genomics and Proteomics Advance 5(1), 46–51 (2006)
4. Brett, D., Hanke, J., Lehmann, G., Haase, S., Delbruck, S., Krueger, S., Reich, J., Bork, P.: EST comparison indicates 38% of human mRNAs contain possible alternative splice forms. FEBS Letters 474(1), 83–86 (2000)
5. Caceres, J., Kornblihtt, A.: Alternative splicing: multiple control mechanisms and involvement in human disease. Trends Genet. 18(4), 186–193 (2002)

6. Castrignanò, T., D'Antonio, M., Anselmo, A., Carrabino, D., Meo, A.D.D., D'Erchia, A.M., Licciulli, F., Mangiulli, M., Mignone, F., Pavesi, G., Picardi, E., Riva, A., Rizzi, R., Bonizzoni, P., Pesole, G.: Aspicdb: A database resource for alternative splicing analysis. Bioinformatics 24(10), 1300–1304 (2008)
7. Castrignanò, T., Rizzi, R., Talamo, I.G., Meo, P.D.D., Anselmo, A., Bonizzoni, P., Pesole, G.: Aspic: a web resource for alternative splicing prediction and transcript isoforms characterization. Nucleic Acids Research 34, 440–443 (2006)
8. Eyras, E., Caccamo, M., Curwen, V., Clamp, M.: ESTGenes: alternative splicing from ESTs in Ensembl. Genome Res. 14, 976–987 (2004)
9. Galante, P., Sakabe, N., Kirschbaum-Slager, N., de Souza, S.: Detection and evaluation of intron retention events in the human transcriptome. RNA 10(5), 757–765 (2004)
10. Gupta, S., Zink, D., Korn, B., Vingron, M., Haas, S.: Genome wide identification and classification of alternative splicing based on EST data. Bioinformatics 20(16), 2579–2585 (2004)
11. Heber, S., Alekseyev, M., Sze, S., Tang, H., Pevzner, P.: Splicing graphs and EST assembly problem. Bioinformatics 18(suppl. 1), S181–S188 (2002)
12. Kan, Z., Rouchka, E.C., Gish, W.R., States, D.J.: Gene structure prediction and alternative splicing analysis using genomically aligned ESTs. Genome Res. 11(5), 889–900 (2001)
13. Kim, N., Shin, S., LeeSanghyuk: Ecgene: genome-based est clustering and gene modeling for alternative splicing. Genome Research 15(4), 5 (2005)
14. Lacroix, V., Sammeth, M., Guigó, R., Bergeron, A.: Exact transcriptome reconstruction from short sequence reads. In: Crandall, K.A., Lagergren, J. (eds.) WABI 2008. LNCS (LNBI), vol. 5251, pp. 50–63. Springer, Heidelberg (2008)
15. Leipzig, J., Pevzner, P., Heber, S.: The Alternative Splicing Gallery (ASG): bridging the gap between genome and transcriptome. Nucleic Acids Res. 32(13), 3977–3983 (2004)
16. Raz, R., Safra, S.: A sub-constant error-probability low-degree test, and a sub-constant error-probability PCP characterization of NP. In: STOC, pp. 475–484 (1997)
17. Wu, T.D., Watanabe, C.K.: Gmap: a genomic mapping and alignment program for mRNA and est sequence. Bioinformatics 21(9), 1859–1875 (2005)
18. Xie, H., Zhu, W., Wasserman, A., Grebinskiy, V., Olson, A., Mintz, L.: Computational analysis of alternative splicing using EST tissue information. Genomics 80(3), 326–330 (2002)
19. Xing, Y., Resch, A., Lee, C.: The multiassembly problem: reconstructing multiple transcript isoforms from EST fragment mixtures. Genome Res. 14(3), 426–441 (2004)
20. Xu, Q., Modrek, B., Lee, C.: Genome-wide detection of tissue-specific alternative splicing in the human transcriptome. Nucleic Acids Res. 30(17), 3754–3766 (2002)

Annotating Fragmentation Patterns

Sebastian Böcker[1,2], Florian Rasche[1], and Tamara Steijger[1]

[1] Lehrstuhl für Bioinformatik, Friedrich-Schiller-Universität Jena,
Ernst-Abbe-Platz 2, 07743 Jena, Germany
{sebastian.boecker,florian.rasche,tamara.steijger}@uni-jena.de
[2] Jena Centre for Bioinformatics, Jena, Germany

Abstract. Mass spectrometry is one of the key technologies in metabolomics for the identification and quantification of molecules in small concentrations. For identification, these molecules are fragmented by, e.g., tandem mass spectrometry, and masses and abundances of the resulting fragments are measured. Recently, methods for *de novo* interpretation of tandem mass spectra and the automated inference of fragmentation patterns have been developed. If the correct structural formula is known, then peaks in the fragmentation pattern can be annotated by substructures of the underlying compound. To determine the structure of these fragments manually is tedious and time-consuming. Hence, there is a need for automated identification of the generated fragments.

In this work, we consider the problem of annotating fragmentation patterns. Our input are fragmentation trees, representing tandem mass spectra where each peak has been assigned a molecular formula, and fragmentation dependencies are known. Given a fixed structural formula and any fragment molecular formula, we search for all structural fragments that satisfy elemental multiplicities. Ultimately, we search for a fragmentation pattern annotation with minimum total cleavage costs. We discuss several algorithmic approaches for this problem, including a randomized and a tree decomposition-based algorithm. We find that even though the problem of identifying structural fragments is NP-hard, instances based on molecular structures can be efficiently solved with a classical branch-and-bound algorithm.

1 Introduction

Mass spectrometry in combination with liquid or gas chromatography (LC-MS, GC-MS) is the most widely used high-throughput technique to analyze metabolites. Since the manual interpretation and annotation of such mass spectra is time-consuming, automated methods for this task are needed. Established methods for metabolite identification rely on comparison with a database of reference spectra. A major challenge is that most of the metabolites remain unknown: Current estimates are up to 20 000 metabolites for any given higher eukaryote [8]. Even for model organisms, only a tiny part of these metabolites has been identified. Recently, different techniques for the *de novo* interpretation of mass spectra have been proposed [12, 3, 4], but these techniques are limited to the

S.L. Salzberg and T. Warnow (Eds.): WABI 2009, LNBI 5724, pp. 13–24, 2009.
© Springer-Verlag Berlin Heidelberg 2009

determination of molecular formulas. But only if the structure of a molecule is known, one can consider the molecule fully identified, as this structure determines chemical properties and, hence, the function of a metabolite, as well as possible interactions with proteins.

In mass spectrometry, fragmentation spectra are used to gather information about the structure of a molecule. In this work, we annotate fragment peaks in a spectrum of a *known* compound with molecular structures. Firstly, this can verify hits in a metabolite database, which are based, say, on molecular formula alone. Secondly, structurally annotated fragmentation spectra of reference compounds may help us to deduce structural information about an unknown compound. This might be achieved by comparing the fragmentation pattern of the unknown and the reference compound. An automated method for fragmentation pattern alignment has been proposed recently [5].

There exist several tools for the annotation of fragmentation spectra (Mass Frontier, MS Fragmenter), but these are usually based on predefined fragmentation rules. Rule-based systems will err if the fragmentation of a molecule differs from what has been known so far. Also, rule-based approaches usually show unsatisfactory prediction accuracy for compounds above 300 Da, as too many fragmentation pathways can explain any fragment mass. Rule-based systems fail especially at the interpretation of tandem mass spectra from LC-MS experiments, as fragmentation principles for these spectra are only partly understood.

Heinonen *et al.* [9] propose a method to annotate a tandem mass spectrum with structural fragments of a known compound: For each peak, candidate fragments are generated and ranked according to the costs of cleaving them out of the molecular structure. For single-step fragmentation, each fragment is cleaved directly from the parent molecule, whereas multistep fragmentation allows fragments to be further fragmented. Unfortunately, their Integer Linear Program for multistep fragmentation may require several days to process even a medium-sized metabolite of about 350 Da.

We propose an approach for the automated structural annotation of tandem mass spectra, combining the fragmentation model from [9] with the *de novo* interpretation of Böcker and Rasche [4]: All peaks in the tandem mass spectrum are annotated with molecular formulas, and a hypothetical fragmentation tree represents dependencies between these fragments. Here, we identify fragments of the molecular structure that fit molecular formulas in the fragmentation tree, such that fragmentation costs are minimized. This leads to the EDGE-WEIGHTED GRAPH MOTIF problem which, unfortunately, is NP-hard. We present randomized, branch-and-bound, and heuristic algorithms for its solution. The randomized algorithm is fixed-parameter tractable [11], and guarantees to find the exact solution with high probability. We limit our branch-and-bound search to a fixed maximal number of bonds that can break. Finally, we present a heuristic based on tree decomposition and dynamic programming.

Our ultimate goal is to assign substructures to all nodes of the fragmentation tree, such that *total* fragmentation costs in the tree are minimized. We propose a branch-and-bound heuristic that recursively follows the TOP-p substructures along the fragmentation tree. We also correct errors in the underlying fragmentation tree, that stem from fragments inserted "too deep" in the tree. We have validated our methods using Orbitrap and orthogonal time-of-flight MS data. We find that each fragmentation step requires to break only a small number of bonds in the molecular structure. Despite the above hardness result, we can process molecules of masses up to 500 Da in a matter of seconds. Finally, our method allows us to validate the fragmentation tree computation from [4].

2 Preliminaries

Fragmentation spectra of molecules can be measured with different experimental setups. Here, we concentrate on collision-induced dissociation (CID), where ions collide with a neutral gas. The overall size of fragments can be adjusted using the collision energy. If the sample is separated by mass twice, once before and once after fragmentation, one speaks of tandem mass spectrometry. For peptide and glycan fragmentation, the experimental setup is chosen such that the fragmentation happens mainly at the peptide and glycosidic bonds. In contrast, fragmentation of metabolites is rather unpredictable. In the following, we will not use prior information about metabolite fragmentation such as fragmentation rules, with the exception of edge weights that represent fragmentation energies.

Böcker and Rasche [4] propose an approach for the determination of molecular formulas from tandem mass spectra. They construct a fragmentation graph where vertices represent all molecular formulas that can explain a peak in the fragmentation spectrum, and edges represent possible fragment relationships. Finding a fragmentation tree in this graph leads to the NP-hard MAXIMUM COLORFUL SUBTREE problem. Using the exact algorithms from [4], fragmentation trees often resemble trees constructed manually by an expert, and can be computed in seconds. Vertices in the fragmentation tree are labeled by the hypothetical molecular formula of the corresponding fragment. See Fig. 1 for an example. We will use such vertex-labeled fragmentation trees as input for our approach. The method may insert fragments too low in the fragmentation tree, because the molecular formulas of fragments allow to do so, increasing the score of the fragmentation tree. There no way to tackle this methodical problem unless the molecular structure of the compound is known.

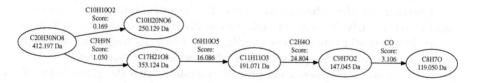

Fig. 1. The fragmentation tree of hexosyloxycinnamoyl choline as calculated by exact algorithms from [4]. Nodes are labeled with the molecular formula of the fragment and the mass of the peak. Edge labels consist of the neutral loss and a score that represents the likelihood that the corresponding fragmentation reaction is real.

3   Formal Problem Definition

We use two different models for the fragmentation process: One-step fragmentation and multistep fragmentation [9]. Note, that single-step fragmentation [9] is a special case of one-step fragmentation, where all fragments are cleaved from the parent molecule.

Let Σ be the alphabet of elements in our molecules, such as $\Sigma = \{C, H, N, O, P, S\}$. A *molecular structure* M consists of a simple, undirected, connected graph where all vertices are labeled with elements from Σ, and edges are weighted by positive weights $w(e) > 0$. The elements of Σ will be called *colors* in this context. The *molecular formula* indicates how many vertices of each color are present in a molecular structure, e.g., $C_{20}H_{30}NO_8$. For *one-step fragmentation*, we are given a molecular structure M and a molecular formula f over Σ, and we try to find a connected subgraph of M that can be cleaved out with minimum costs, that is, minimum sum of energies for all cleaved bonds, and that has colors corresponding to f.

EDGE-WEIGHTED GRAPH MOTIF PROBLEM. Given a vertex-colored edge-weighted graph $G = (V, E)$ and a multiset of colors C of size k, find a connected subgraph $H = (U, F)$ of G such that the multiset of colors of U equals C, and H has minimum weight $w(H) := \sum_{\{u,v\} \in E, u \in U, v \in V \setminus U} w(\{u, v\})$.

This problem is a generalization of the GRAPH MOTIF problem, where no edge weights exist, and one asks whether any such subgraph exists. This problem is NP hard even for bipartite graphs of bounded degree and two colors [7]. Betzler *et al.* [2] developed a randomized FPT algorithm for the GRAPH MOTIF problem using color-coding [1]: Solving the problem with error probability ϵ takes $O(|\log \epsilon| \cdot 4.32^k \cdot k^2 \cdot |E|)$ time.

However, fragmentation pathways can consist of consecutive fragmentation steps [9], where fragments can be cleaved from other fragments. Fragmentation pathways can be represented by *fragmentation trees*, directed trees where the root corresponds to the parent molecule and each edge represents a fragmentation step. Each node n is labeled with a molecular formula that has to be a sub-formula of the molecular formula attached to n's parent.

For the *multistep fragmentation* model, we are given a molecular structure M and a fragmentation tree T. We want to assign sub-structures to the nodes of the fragmentation tree that match their molecular formulas, such that the total cost of cutting out the substructures, over all edges of the fragmentation tree, is minimized. Clearly, it does not suffice to search for the optimal graph motif in every fragmentation step independently, since following fragmentation steps may be cleaved from a suboptimal substructure with lower total costs.

In order to find a fragmentation process consistent with the given fragmentation tree, we use a search tree. Since it does not suffice to take the fragment with minimum costs in every fragmentation step, our heuristic allows the user to specify a number p so that in every step, the best p fragments are considered. For each such fragment, we build up the search tree recursively, and accept the fragment that results in lowest total costs. Finally, we check whether moving up a node in a fragmentation tree by one level, will decrease the total cost of fragmentation. To do so, we compare the total costs of cleaving fragment f and all subsequent fragments from its parent, with the total costs of cleaving them from its grandfather.

We have noted that using methods from [4], some fragments may be attached too deep in the fragmentation tree. Following this line of thought, we can decide that the fragmentation tree cannot be trusted and should be re-computed in our optimization, comparable to the problem considered in [9]. In order to find the multistep fragmentation with the lowest total costs, we have to build up a fragmentation graph representing all possible fragmentation pathways, we omit the details. For small molecules, this can be solved using exhaustive enumeration [9]. But finding the fragmentation process with lowest costs leads to the MINIMUM COLORFUL SUBTREE problem which, again, is known to be NP-hard [4]. In addition, building the fragmentation graph requires to solve a huge number of EDGE-WEIGHTED GRAPH MOTIF instances. In view of this seemingly inevitable complexity, we refrain from attacking this more general problem.

4   Algorithms

In the following, we describe three algorithms to solve the EDGE-WEIGHTED GRAPH MOTIF problem. For the multistep fragmentation search tree, our algorithms also have to calculate suboptimal solutions.

4.1   Random Separation

Cai *et al.* [6] proposed a randomized technique called *random separation* based on color-coding [1]. The key idea of random separation is to partition the vertices by coloring them randomly with two different colors. Then, connected components are identified and appropriate components are tested for optimality. Random separation can be used to solve a wide range of fixed-parameter tractable problems, especially when the input graph has bounded degree. This is the case for molecular structures, where vertex degrees are bounded by the valences of elements.

We now apply random separation to the EDGE-WEIGHTED GRAPH MOTIF problem. Let k be the cardinality of the color multiset C. We search for a substructure $H = (U, F)$ that minimizes $w(H)$, where $|U| = k$. Let $N(U)$ denote the neighborhood of U in G. Given a graph $G = (V, E)$ and a random separation of G that partitions V into V_1 and V_2, there is a $2^{-(k+|N(U)|)+1}$ chance that U is entirely in V_1 and its neighborhood $N(U)$ is entirely in V_2 or vice versa. We use depth-first search to identify the connected components in V_1 and V_2. Simultaneously, colors are counted and costs for the partition are calculated. If the colors of a connected component correspond to the colors of the given multiset C and the costs are smaller than the costs of the best solution so far, the connected component is stored. In order to find the optimal solution with error probability ϵ, the procedure has to be repeated $\lceil |\log \epsilon| / |\log(1 - 2^{-(k+kd)+1})| \rceil$ times, where d is the maximum vertex degree in G.

We now analyze the worst-case running time of this approach: Coloring takes $O(|V|)$ time. Depth-first search has a running time of $O(|V| + |E|)$ but since molecular structures have bounded degree, the overall running time of one trial is $O(|V|)$. Accordingly, the overall running time of the random separation algorithm is $O(|\log \epsilon| \, 2^{(k+kd)} \cdot |V|)$. Recall that d is bounded in molecular structures. Also note that the term kd is due to the neighborhood of U in G. In our experiments, we observe that one-step fragmentation usually requires only few bonds to break. In this case, we can substitute the worst case estimation kd with maximal number b of bonds breaking in a fragmentation step. In our implementation, b is an input parameter that can be used to reduce the number of trials and, hence, to decrease running time. Obviously, b has to be chosen large enough to guarantee that, with high probability, the optimal solution is found.

4.2   Branch-and-Bound

The second algorithm is a classical branch-and-bound algorithm. It branches over edge sets that might break during a fragmentation step. Given an edge set, its deletion might separate G into a set of connected components. Similar to the random separation approach, depth-first search is used to identify components that might be selected as a solution. If a solution has been found, its costs are used as an upper bound for pruning. The user can specify the maximum number of bonds b that may break during one single fragmentation step. We then try to cut out a solution with exactly $b' = 1, \ldots, b$ edges.

Since the costs of a solution correspond to the sum of weights of deleted edges, it is not necessary to iterate over all possible edge sets. To efficiently traverse the search tree, we use an edge set iterator that avoids edge sets with too high costs. Edges are sorted in increasing order with respect to their weight. Now, we can easily iterate over all edge sets of a fixed cardinality such that the edge sets have increasing costs. Thus, as soon as a solution with b' edges has been found, or the costs exceed that of the best solution found so far, all following edge sets of the same cardinality will have higher costs and can be omitted.

Sorting edges costs $O(|E| \log |E|)$ time. Running time of the depth-first search is $O(|V|)$, as explained for random separation. The branch-and-bound algorithm iterates over $O(|V|^b)$ edge sets. This results in an overall running time of $O(|E| \log |E| + |V|^b)$. Unfortunately, running time is exponentially in b. But if the number of bonds that break in one single fragmentation step is small and bounded, b can be assumed as a constant and hence, the algorithm can be executed in polynomial time.

4.3   Tree Decomposition-Based Algorithm

Fellows *et al.* [7] show that the GRAPH MOTIF problem is already NP-hard on trees with maximum vertex degree three, assuming an unbounded alphabet. If, however, the alphabet is bounded, then the GRAPH MOTIF problem can be solved in polynomial time for graphs with bounded treewidth [7]. Unfortunately, running times of this algorithm are too high for applications. Similarly, there is a polynomial time algorithm to solve the EDGE-WEIGHTED GRAPH MOTIF problem on trees, we defer the details to the full paper. We now adapt ideas from this algorithm for a heuristic algorithm, that uses the concept of tree decomposition to solve the EDGE-WEIGHTED GRAPH MOTIF problem on arbitrary graphs.

A *tree decomposition* of a graph $G = (V, E)$ is a pair $\langle \{X_i \mid i \in I\}, T \rangle$ where each X_i is a subset of V, called a *bag*, and T is a tree containing the elements of I as nodes, satisfying: a) $\bigcup_{i \in I} X_i = V$; b) for every edge $\{u, v\} \in E$, there is an $i \in I$ such that $\{u, v\} \subseteq X_i$; and c) for all $i, j, h \in I$, if j lies on the path between i and h in T then $X_i \cap X_h \subseteq X_j$. Let $\omega := \max\{|X_i| \mid i \in I\}$ be the maximal size of a bag in the tree decomposition. The *width* of the tree decomposition equals $\omega - 1$. The *treewidth* of G is the minimum number $\omega - 1$ such that G has a tree decomposition of width $\omega - 1$.

To simplify the description and analysis of our algorithm, we use nice tree decompositions in the remainder of this paper. Here, we assume the tree T to be arbitrarily rooted. A tree decomposition is a *nice* tree decomposition if every node of the tree has at most two children, and there exist only three types of nodes in the tree: A *join node* i has two children j and h such that $X_i = X_j = X_h$. An *introduce node* i has only one child j, and $|X_i| = |X_j| + 1$ as well as $X_j \subset X_i$ holds. A *forget node* i has only one child j, and $|X_i| = |X_j| - 1$ as well as $X_i \subset X_j$ holds. Using methods from [10], we can transform a tree decomposition with m nodes into a nice tree decomposition with the same width and $O(m)$ nodes in linear time. Finally, for each leaf X_i with $|X_i| > 1$ we can insert additional introduce nodes under X_i, such that the new leaf contains only a single vertex.

Assume that a nice tree decomposition $\langle \{X_i \mid i \in I\}, T \rangle$ of width $\omega - 1$ and $O(m)$ nodes of the molecule graph $M = (V, E)$ is given. In the following, we describe a dynamic programming heuristic to solve the EDGE-WEIGHTED GRAPH MOTIF problem using the nice tree decomposition of the molecule graph. Let Y_i be the vertices in V that are contained in the bags of the subtree below node i. Let $c(v)$ be the color of a vertex $v \in V$ in the molecule graph, and let $c(U)$ be the *multiset* of colors for a vertex set $U \subseteq V$. We define costs for $U \subseteq V$ and color multiset C as $costs(U, C) := w(U)$ if $c(U) = C$, and $costs(U, C) := \infty$ otherwise. For each node i of the tree decomposition, we want to calculate the costs $W_i(C, U)$ for building up a fragment using the multiset of colors C and vertices $U \subseteq X_i$. These values are stored in a table and calculated using dynamic programming. Our algorithm starts at the leaves of the tree decomposition. For each leaf i with $X_i = \{v\}$ we initialize

$$W_i(\{c(v)\}, \{v\}) = w(\{v\}).$$

During the bottom-up traversal, the algorithm distinguishes if i is an introduce node, a forget node, or a join node. We now give recurrences to compute the matrix $W_i(C, U)$ where $U \subseteq X_i$ and $c(U) \subseteq C$. For readability, we omit the condition that all solutions need to be connected from the recurrences. This has to be checked in addition for all introduce and join nodes.

Forget nodes. These nodes result in a simple recurrence, so we treat them first: Let i be the parent node of j and choose v with $X_j \setminus X_i = \{v\}$. Then,

$$W_i(C, U) = \min\{W_j(C, U), W_j(C, U \cup \{v\}), costs(U, C)\}.$$

Join nodes. Let i be the parent node of j and h, where $X_i = X_j = X_h$. We want to compute $W_i(C, U)$. To do so, we iterate over all $U_1, U_2 \subseteq U$ such that $U_1 \cup U_2 = U$. Let $C' = C \setminus c(U)$ be the multiset of remaining colors. We then iterate over all bipartitions C_1', C_2' of C', that is, $C_1' \cup C_2' = C'$ and $C_1' \cap C_2' = \emptyset$. Let $C_1 := c(U_1) \cup C_1'$ and $C_2 := c(U_2) \cup C_2'$. We now access the values $W_j(C_1, U_1)$ and $W_h(C_2, U_2)$ that represent minimal cost for the respective instances. Using our traceback matrix, we backtrace through the tree decomposition and reconstruct the sets of vertices $V_1 \subseteq Y_j$, $V_2 \subseteq Y_h$ used in the corresponding optimal solutions with weights $W_j(C_1, U_1)$ and $W_h(C_2, U_2)$. If $V_1 \cap V_2 \not\subseteq U$ then we stop, as the partial solutions overlap outside of the current bag X_i. Otherwise, we can compute $costs(V_1 \cup V_2, C)$. We take the minimum of all these values as $W_i(C, U)$.

Introduce nodes. Let i be the parent node of j and choose v with $X_i \setminus X_j = \{v\}$. We distinguish two different situations: If v is the last vertex of a cycle in M, so that v *closes a cycle*, then we have to compute $W_i(C, U)$ using a traceback through the tree decomposition, combining two optimal solutions. This is analogous to the case of a join node, we defer the details to full paper. Otherwise, we set

$$W_i(C, U) = \begin{cases} \min\{W_j(C \setminus \{c(v)\}, U \setminus \{v\}), costs(U, C)\} & \text{if } v \in U, \\ \min\{W_j(C, U), costs(U, C)\} & \text{otherwise.} \end{cases}$$

The minimum costs to cleave a fragment from M are $\min_{i, U \subseteq X_i} W_i(C, U)$. The corresponding fragment can be computed by a traceback.

We can close a cycle either by introducing the last vertex of it, or by joining two parts of a cycle. In this case, we cannot guarantee that the above recurrences result in an optimal solution: Optimal partial solutions in a cycle might overlap and be discarded, whereas suboptimal, non-overlapping partial solutions may be used to build up an optimal solution but are not stored, as we limit calculations to optimal solutions at all times. In order to check for connectivity the algorithm needs to perform tracebacks through the matrices W_i. To achieve feasible running times, use of an explicit traceback matrix was inevitable. The drawback is that the tree decomposition heuristic has only limited support for finding suboptimal solutions, severely limiting its use for multistep fragmentation.

The algorithm needs $O(m \cdot |\Sigma|^k \cdot 2^\omega)$ memory in the worst case. Forget nodes and introduce nodes that do not close a cycle can be calculated in $O(\omega \cdot d)$ while join nodes and the remaining introduce nodes need $O(m \cdot k \cdot 2^k \cdot 3^\omega)$. Running time of the algorithm is $O(m \cdot |\Sigma|^k \cdot 2^\omega \cdot (m \cdot k \cdot 2^k \cdot 3^\omega + \omega \cdot d))$, we defer the slightly technical details of the running time analysis to the full paper.

The molecular formulas in X_i are restricted by the available elements of f and further by the elements of the vertices in Y_i, so the number of molecular formulas that have to be checked for each node i is significantly smaller than what worst-case running times suggest. We use a hash map to store and access only those entries of the matrices $W_i(C, U)$ that are smaller than infinity. This reduces both running times and memory of our approach in applications.

5 Experimental Results

We implemented our algorithms in Java 1.5. Running times were measured on an Intel Core 2 Duo processor, 2.5 GHz with 3 GB memory. To compute the tree decomposition of the molecule graphs, we used the method QuickBB in the library LibTW implemented by van Dijk *et al.* (http://www.treewidth.com). We implemented a method to transform the computed tree decomposition into a nice tree decomposition. For the random separation algorithm, we use an error probability $\epsilon = 0.1\%$, so that the optimal solution will be found with a probability of 99.9%. In the multistep fragmentation evaluation, we set $p = 5$, thus, keeping the five best substructures in each fragmentation step.

As test data we used 35 fragmentation trees calculated from CID mass spectra measured on an API QStar Pulsar Hybrid instrument (Applied Biosystems) and 8 trees from an LTQ Orbitrap XL instrument (Thermo Fisher Scientific) using PQD fragmentation. The test compounds consisted of biogenic amino acids, complex choline derivatives, and commercially available pharmacological agents. QStar data is publicly available from http://msbi.ipb-halle.de/MassBank.

Since hydrogen atoms are often subject to rearrangements, we do not include them in our calculations. We do, however, support some minor structural rearrangements such as hydroxyl group rearrangements. These occur frequently as a result of cyclizations. We model this using pseudo-edges. Our model is not biochemically correct but enables us to reconstruct fragmentation trees with minor rearrangements, e.g., the fragmentation tree of arginine where a hydroxyl group is rearranged because of cyclization.

Detailed information about running times of the multistep heuristic using the different approaches can be found in Table 1. One can see that the branch-and-bound

Table 1. The average running times of the algorithms: neighborhood for random separation has been estimated with $b = 5$, branch-and-bound allowed $b = 3$ (BB-3) and $b = 5$ (BB-5) bonds to break, running time of tree decomposition-based algorithm includes computing the tree decomposition itself. Multistep fragmentation considered the 5 best fragments in every step.

Mass (Da)	#comp.	multistep heuristic			
		RS	BB-3	BB-5	TD
< 100	1	< 1 s	< 1 s	< 1 s	< 1 s
100–200	17	23.2 s	0.1 s	0.2 s	1.1 s
200–300	16	50.7 min	0.7 s	6.0 s	13.3 s
300–400	3	4.8 h	6.7 s	55.7 s	49.2 s
400–500	5	> 1 day	0.5 s	5.6 s	1.7 min
> 500	1	> 1 week	10.2 min	4.6 h	2.0 h

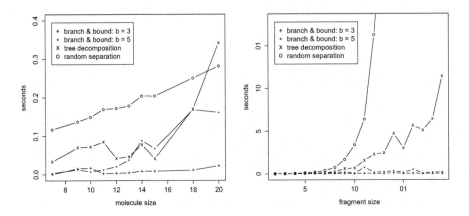

Fig. 2. Running time comparison of the three algorithms: The left diagram shows the average running time depending on the molecule size $|M|$ given a fixed fragment size of six. In the right diagram the average running time for several molecule sizes in dependence on the fragment's size is displayed.

algorithms outperforms the two more sophisticated algorithms. The running time of the tree decomposition algorithm grows exponentially, but all test instances could be calculated in a feasible amount of time. The random separation algorithm, however, performs fast for small instances, but requires several days for molecules > 400 Da.

Fig. 2 shows how the running times of the algorithms depend on the size of the molecular structure M and on the size of the fragments. It illustrates that the running time of the branch-and-bound algorithm mainly depends on the size of M, particularly for larger b, whereas that of the tree decomposition algorithm depends both on fragment size and size of the molecular structure. Finally, running time of the random separation algorithm depends mainly on fragment size.

For all instances that finished computation using the random separation algorithm, we reach identical total costs as for the branch-and-bound algorithm. Annotations differ only marginally in the sequence of cleavages. The annotations found by the branch-and-bound algorithm with $b = 3$ and $b = 5$ also have identical costs. This supports our assumption that instances based on molecular graphs do not resemble the full complexity of the EDGE-WEIGHTED GRAPH MOTIF problem.

The heuristic algorithm failed to annotate 3 fragmentation trees consistently, and returned a clearly suboptimal solution in 11 cases due to its limited support for suboptimal partial solutions. Often, a suboptimal fragment that split up a ring had to be chosen early in the fragmentation tree in order to reduce subsequent costs.

Our annotations turned out to be valuable to validate the fragmentation trees proposed by methods in [4]. Our analysis of the annotated fragmentation trees identified peaks in several fragmentation trees that were annotated with a molecular formula but probably are noise peaks. These peaks have a low intensity and are also scored very low. In our analysis, we were unable to assign a fragment to the molecular formula. For example, the 250 Da node in Figure 1 was identified as noise peak. The score of the corresponding fragmentation step is very low compared to the others, and a fragment with formula $C_{10}H_{20}NO_6$ cannot be cleaved from the structure of hexosyloxycinnamoyl choline without major rearrangements. We also identified an intense peak that could not be annotated with any fragment. Consultation with an expert resulted in the conclusion that the spectrum was contaminated.

Furthermore, we identified three nodes in two fragmentation trees that had been inserted too low into the fragmentation tree, and pulling them up one level resulted in a fragmentation pattern with significantly decreased total costs. In two other fragmentation trees, we identified nodes where pulling-up results in slightly reduced costs. A closer look at the fragmentation patterns revealed that in these cases, two competitive paths might co-occur.

6   Conclusion

We have presented a branch-and-bound heuristic for the multistep fragmentation problem that aims at annotating fragmentation cascades. As a sub-task, this heuristic repeatedly needs to solve the NP-hard EDGE-WEIGHTED GRAPH MOTIF problem. We proposed a randomized FPT-algorithm, an exact branch-and-bound algorithm, as well as a heuristic to solve the problem. Our experimental results reveal that despite its theoretical complexity, real world instances of the problem can be solved quickly, as only few bonds break in each fragmentation step.

We find that the branch-and-bound search outperforms the two more involved algorithms. In case a large number of bonds has to be broken, this might be an indication of either structural rearrangements, or errors in the fragmentation tree. Thus, in those cases an experimentalist should manually annotate the fragmentation pattern. Our method was able to correct a few errors in the fragmentation trees computed in [4], making alignments of those trees, as proposed in [5], more reliable.

In the future, we want to improve the support for structural rearrangements, e.g., by letting the user introduce pseudo-edges in the molecule which represent expected rearrangements. Furthermore, the cost function can be improved by taking into consideration aromatic bonds, and by updating the valences after each fragmentation step. Additionally, we plan to integrate fragmentation rules into our model through the modification of edge weights. To improve the annotation reliability when using the multistep fragmentation model, intermediate concepts, such as two-step fragmentation, might be helpful. Since our method is "model-free" it can, in principle, also be applied to other fragmentation techniques.

Acknowledgments. We thank the department of Stress and Developmental Biology at the Leibniz Institute of Plant Biochemistry in Halle, Germany and Aleš Svatoš from the Max Planck Institute for Chemical Ecology in Jena, Germany for supplying us with the metabolite mass spectra, and Christoph Böttcher for the manual identification of fragmentation trees.

References

1. Alon, N., Yuster, R., Zwick, U.: Color-coding. J. ACM 42(2), 844–856 (1995)
2. Betzler, N., Fellows, M.R., Komusiewicz, C., Niedermeier, R.: Parameterized algorithms and hardness results for some graph motif problems. In: Ferragina, P., Landau, G.M. (eds.) CPM 2008. LNCS, vol. 5029, pp. 31–43. Springer, Heidelberg (2008)
3. Böcker, S., Letzel, M., Lipták, Z., Pervukhin, A.: SIRIUS: Decomposing isotope patterns for metabolite identification. Bioinformatics 25(2), 218–224 (2009)
4. Böcker, S., Rasche, F.: Towards de novo identification of metabolites by analyzing tandem mass spectra. Bioinformatics 24, I49–I55 (2008); Proc. of European Conference on Computational Biology (ECCB 2008)
5. Böcker, S., Zichner, T., Rasche, F.: Automated classification of unknown biocompounds using tandem MS. In: Poster Proc. of Conference of the American Society for Mass Spectrometry (ASMS 2009), p. W690 (2009)
6. Cai, L., Chan, S.M., Chan, S.O.: Random separation: a new method for solving fixed-cardinality optimization problems. In: Bodlaender, H.L., Langston, M.A. (eds.) IWPEC 2006. LNCS, vol. 4169, pp. 239–250. Springer, Heidelberg (2006)
7. Fellows, M., Fertin, G., Hermelin, D., Vialette, S.: Sharp tractability borderlines for finding connected motifs in vertex-colored graphs. In: Arge, L., Cachin, C., Jurdziński, T., Tarlecki, A. (eds.) ICALP 2007. LNCS, vol. 4596, pp. 340–351. Springer, Heidelberg (2007)
8. Fernie, A.R., Trethewey, R.N., Krotzky, A.J., Willmitzer, L.: Metabolite profiling: from diagnostics to systems biology. Nat. Rev. Mol. Cell Biol. 5(9), 763–769 (2004)
9. Heinonen, M., Rantanen, A., Mielikäinen, T., Kokkonen, J., Kiuru, J., Ketola, R.A., Rousu, J.: FiD: a software for ab initio structural identification of product ions from tandem mass spectrometric data. Rapid Commun. Mass Spectrom. 22(19), 3043–3052 (2008)
10. Kloks, T.: Treewidth, Computation and Approximation. Springer, Heidelberg (1994)
11. Niedermeier, R.: Invitation to Fixed-Parameter Algorithms. Oxford University Press, Oxford (2006)
12. Rogers, S., Scheltema, R.A., Girolami, M., Breitling, R.: Probabilistic assignment of formulas to mass peaks in metabolomics experiments. Bioinformatics 25(4), 512–518 (2009)

biRNA: Fast RNA-RNA Binding Sites Prediction

Hamidreza Chitsaz[1], Rolf Backofen[2], and S. Cenk Sahinalp[1]

[1] School of Computing Science, Simon Fraser University, 8888 University Drive, Burnaby, British Columbia, Canada
{hrc4,cenk}@cs.sfu.ca
[2] Institut für Informatik, Albert-Ludwigs-Universität, Georges-Koehler-Allee, Freiburg, Germany
backofen@informatik.uni-freiburg.de

Abstract. We present biRNA, a novel algorithm for prediction of binding sites between two RNAs based on minimization of binding free energy. Similar to RNAup approach [30], we assume the binding free energy is the sum of accessibility and the interaction free energies. Our algorithm maintains tractability and speed and also has two important advantages over previous similar approaches: (1) biRNA is able to predict multiple simultaneous binding sites and (2) it computes a more accurate interaction free energy by considering both intramolecular and intermolecular base pairing. Moreover, biRNA can handle crossing interactions as well as hairpins interacting in a zigzag fashion. To deal with simultaneous accessibility of binding sites, our algorithm models their joint probability of being unpaired. Since computing the exact joint probability distribution is intractable, we approximate the joint probability by a polynomially representable graphical model namely a Chow-Liu tree-structured Markov Random Field. Experimental results show that biRNA outperforms RNAup and also support the accuracy of our approach. Our proposed Bayesian approximation of the Boltzmann joint probability distribution provides a powerful, novel framework that can also be utilized in other applications.

1 Introduction

Following the recent discovery of RNA interference (RNAi), the post transcriptional silencing of gene expression via interactions between mRNAs and their regulatory RNAs, RNA-RNA interaction has moved from a side topic to a central research topic. Recent studies have shown that a large fraction of the genome gives rise to RNA transcripts that do not code for proteins [40]. Several of these non-coding RNAs (ncRNAs) regulate gene expression post-transcriptionally through base pairing (and establishing a joint structure) with a target mRNA, as per the eukaryotic miRNAs and small interfering RNAs (siRNAs), antisense RNAs or bacterial small regulatory RNAs (sRNAs) [16]. In addition to such endogenous regulatory ncRNAs, antisense oligonucleotides have been used as exogenous inhibitors of gene expression; antisense technology is now commonly used as a research tool as well as for therapeutic purposes. Furthermore, synthetic nucleic acids systems have been engineered to self assemble into complex structures performing various dynamic mechanical motions.

S.L. Salzberg and T. Warnow (Eds.): WABI 2009, LNBI 5724, pp. 25–36, 2009.

A key tool for all the above advances is a fast and accurate computational method for predicting RNA-RNA interactions. Existing comprehensive methods for analyzing binding thermodynamics of nucleic acids are computationally expensive and prohibitively slow for real applications [1,9]. Other existing methods suffer from a low specificity, possibly because several of these methods consider restricted versions of the problem (e.g. simplified energy functions or restricted types of interactions) - this is mostly for computational reasons. In this paper we present an algorithm to predict the binding sites of two interacting RNA strands. Our most important goal in this work is tractability as well as high specificity. While our algorithm considers the most general type of interactions, it is still practically tractable by making simplifying assumptions on the energy function. These assumptions are however natural and adopted by many other groups as well [6,22,30,41].

Our Contribution

We give an algorithm to predict the binding sites of two interacting RNAs and also the interaction secondary structure constrained by the predicted binding sites. As opposed to previous approaches that are able to predict only one binding site [6,30,42], our algorithm predicts multiple simultaneous binding sites. We define a binding site to be a subsequence which interacts with exactly one binding site in the other strand. Crossing interactions (external pseudoknots) and zigzags (see [1] for exact definition) are particularly allowed. To the best of our knowledge, this allows for the most general type of interactions considered in the literature. Although intramolecular pseudoknots are not considered in the current work, they can be incorporated into our framework at the expense of additional computational complexity.

Following the RNAup approach [30], we assume the total interaction free energy is the sum of two terms: (1) the free energy needed to make binding sites accessible in each molecule and (2) the free energy released as a result of intermolecular bonds formed between the interacting binding site pairs. Based on that energy model, our algorithm is essentially composed of three consecutive steps: (1) building a tree-structured Markov Random Field (MRF) to approximate accessibility of a collection of potential binding sites, (2) computing pairwise interaction free energies for potential binding sites of one strand against those of the other strand, and (3) finding a minimum free energy matching of binding sites. Unlike RNAup that computes only the hybridization partition function for step (2), our algorithm computes the full interaction partition function [9]. Therefore, our algorithm not only considers multiple simultaneous binding sites but also computes a more accurate free energy of binding.

The time complexity of the first two steps is $O(n^3r + m^3s + nmw^4)$ in which n and m denote the lengths of sequences, w denotes maximum binding site length, and $r \leq nw$ and $s \leq mw$ are the number of potential sites heuristically selected out of the $O(nw)$ and $O(mw)$ possible ones. More importantly, the space complexity of the first two steps is $O(r^2 + s^2 + nmw^2)$. The third step requires a nontrivial optimization namely minimum energy bipartite matching of two tree-structured Markov Random Fields, a topic on which we are currently working. In this paper, we implement an exhaustive search for the third step. Therefore, the running time of the third step is currently $O(r^\kappa s^\kappa)$ where κ is the maximum number of simultaneous binding sites.

Related Work

Since the initial works of Tinoco et al. [43], Nussinov et al. [31], and Waterman and Smith [46] several computational methods have emerged to study the secondary structure thermodynamics of a single nucleic acid molecule. Those initial works laid the foundation of modern computational methods by adopting a divide and conquer strategy. That view, which originally exhibited itself in the form of a simple base pair counting energy function, has evolved into Nearest Neighbor Thermodynamic model which has become the standard energy model for a nucleic acid secondary structure [26]. The standard energy model is based on the assumption that stacking base pairs and loop entropies contribute additively to the free energy of a nucleic acid secondary structure. Based on additivity of the energy, efficient dynamic programming algorithms for predicting the minimum free energy secondary structure [31,37,46,47] and computing the partition function of a single strand [14,27] have been developed.

Some previous attempts to analyze the thermodynamics of multiple interacting nucleic acids concatenate input sequences in some order and consider them as a single strand. For example, pairfold [2] and RNAcofold from Vienna package concatenate the two input sequences into a single strand and predict its minimum free energy structure. Dirks et al. present a method, as a part of NUPack, that concatenates the input sequences in some order, carefully considering symmetry and sequence multiplicities, and computes the partition function for the whole ensemble of complex species [13]. However, concatenating the sequences is not accurate at all as even if pseudoknots are considered, some useful interactions are excluded while many physically impossible interactions are included. Several other methods, such as RNAhybrid [35], UNAFold [12,24], and RNAduplex from Vienna package avoid intramolecular base-pairing in either strand and compute minimum free energy hybridization secondary structure. Those approaches naturally work only for simple cases involving typically very short strands.

Alternatively, a number of studies aimed to take a more fundamental stance and compute the minimum free energy structure of two interacting strands under energy models with growing complexity. For instance, Pervouchine devised a dynamic programming algorithm to maximize the number of base pairs among interacting strands [33]. A followup work by Kato et al. proposed a grammar based approach to RNA-RNA interaction prediction [18]. An approximation algorithm for RNA-RNA interaction prediction is given by Mneimneh [28]. More generally, Alkan et al. [1] studied the joint secondary structure prediction problem under three different models: (1) base pair counting, (2) stacked pair energy model, and (3) loop energy model. Alkan et al. proved that the general RNA-RNA interaction prediction under all three energy models is an NP-hard problem. Therefore, they suggested some natural constraints on the topology of possible joint secondary structures, which are satisfied by all examples of complex RNA-RNA interactions in the literature. The resulting algorithm computes the minimum free energy secondary structure among all possible joint secondary structures that do not contain (internal) pseudoknots, crossing interactions (i.e. external pseudoknots), and *zigzags* (see [1] for the exact definition). In our previous work [9], we gave an algorithm piRNA to compute the partition function, base pair probabilities, and minimum free energy structure over the type of interactions that Alkan et al. considered. We extended the standard energy model for a single RNA to an energy model for the joint secondary structure of interacting strands by considering new types of (joint) structural components. Although piRNA outperforms existing alternatives, its $O(n^4m^2 + n^2m^4)$ time and $O(n^2m^2)$ space complexity make it prohibitive for many practical, particularly high-throughput, applications.

A third set of methods predict the secondary structure of each individual RNA independently, and predict the (most likely) hybridization between accessible regions of the two molecules. More sophisticated methods in this category view interaction as a multi step process [6,30,45]: (1) unfolding of the two molecules to expose bases needed for hybridization, (2) the hybridization at the binding site, and (3) restructuring of the complex to a new minimum free energy conformation. Some approaches in this set, such as IntaRNA [6] and RNAup [30], assume that binding happens at only one location, which is not the case for some known interacting RNAs such as OxyS-fhlA [3] and CopA-CopT [20,21]. Those programs are able to predict only one binding site, so in this paper, we consider multiple simultaneous binding sites.

2   Preliminaries

Our algorithm is based on the assumption that binding is practically a stepwise process, a view that has been proposed by others as well [1,6,22,30,41]. In real world, each nucleic acid molecule has a secondary structure before taking part in any interaction. To form an interaction, as the first step the individual secondary structures are deformed so that the binding sites in both molecules become unpaired. As the second step, pairwise matching between the binding sites takes place. Each step is associated with an energy the sum of which gives the free energy of binding ΔG. Specifically, denote the energy difference that is needed for unpairing *all the binding sites* in **R** and **S** by ED_u^R and ED_u^S respectively, and denote by ΔG_b^{RS} the free energy that is released as a result of binding. Similar to [30],

$$\Delta G = ED_u^R + ED_u^S + \Delta G_b^{RS}. \tag{1}$$

This assumption is intuitively plausible because each molecule needs to expose its interacting parts before the actual binding happens; moreover, these two steps are assumed to be independent from one another. Note that previous approaches such as IntaRNA [6], RNAup [30], and RNAplex [42] consider only one binding site in each molecule, which makes the problem easier, whereas we consider multiple binding sites. It is sometimes argued that nature does not usually favor too highly entangled structures [30]. Our algorithm easily accommodates an upper bound on the number of potential binding sites, which is another advantage of our approach.

To reduce the complexity, we assume that the length of a potential binding site is not more than a window size w in this work. This is a reasonable assumption, which has also been made in similar approaches [30], as most known RNA-RNA interactions such as OxyS-fhlA and CopA-CopT do not exhibit lengthy binding sites [3,20,21]. We call a subsequence of length not more than w a *site*.

3   Algorithm

Based on the assumption above, our program biRNA finds a combination of binding sites that minimizes ΔG. Let $\mathcal{V}^R$ and $\mathcal{V}^S$ denote the set of potential binding sites, which is the collection of subsequences of length not more than w in our case, of **R** and **S** respectively. biRNA is composed of five consecutive steps:

(I) For every site $W = [i, j]$ in $\mathcal{V}^R$ or $\mathcal{V}^S$, compute the probability $P_u^R(W)$ or $P_u^S(W)$ that W is unpaired.

(II) For every pair of sites W_1 and W_2, compute the joint probabilities $P_u^R(W_1, W_2)$ and $P_u^S(W_1, W_2)$ that W_1 and W_2 are simultaneously unpaired.

(III) Build tree-structured Markov Random Fields (MRF) $\mathcal{T}^R = (\mathcal{V}^R, \mathcal{E}^R)$ and $\mathcal{T}^S = (\mathcal{V}^S, \mathcal{E}^S)$ to approximate the joint probability distribution of multiple unpaired sites. Denote the $\mathcal{T}$-approximated joint probability of unpaired sites $W_1, W_2, \ldots, W_k$ by $P_u^*(W_1, W_2, \ldots, W_k)$.

(IV) Compute $Q_{W^R W^S}^I$, the interaction partition function restricted to subsequences W^R and W^S, for every $W^R \in \mathcal{V}^R$ and $W^S \in \mathcal{V}^S$.

(V) Find a non-overlapping matching $M = \{(W_1^R, W_1^S), (W_2^R, W_2^S), \ldots, (W_k^R, W_k^S)\}$ that minimizes $\Delta G(M) = ED_u^R(M) + ED_u^S(M) + \Delta G_b^{RS}(M)$, in which

$$ED_u^R(M) = -RT \log P_u^{R*}(W_1^R, W_2^R, \ldots, W_k^R) \tag{2}$$

$$ED_u^S(M) = -RT \log P_u^{S*}(W_1^S, W_2^S, \ldots, W_k^S) \tag{3}$$

$$\Delta G_b^{RS}(M) = -RT \sum_{1 \leq i \leq k} \log(Q_{W_i^R W_i^S}^I - Q_{W_i^R} Q_{W_i^S}). \tag{4}$$

Above, R is the universal gas constant and T is temperature. To demonstrate (2) and (3), let for instance $P_u^R(W_1^R, W_2^R, \ldots, W_k^R)$ be the exact probability that the sites are unpaired. In that case, $ED_u^R(W_1^R, W_2^R, \ldots, W_k^R) = \Delta G^R(W_1^R, W_2^R, \ldots, W_k^R) - \Delta G^R$, and

$$\Delta G^R(W_1^R, W_2^R, \ldots, W_k^R) - \Delta G^R = -RT \log Q_R(W_1^R, W_2^R, \ldots, W_k^R) + RT \log Q_R$$

$$= -RT \log \frac{Q_R(W_1^R, W_2^R, \ldots, W_k^R)}{Q_R} = -RT \log P_u^R(W_1^R, W_2^R, \ldots, W_k^R), \tag{5}$$

in which Q_R is the partition function of **R** and $Q_R(W_1^R, W_2^R, \ldots, W_k^R)$ is the partition function of those structures in which $W_1^R, W_2^R, \ldots, W_k^R$ are unpaired.

In the following, we describe each step in more details. In Section 3.1 we explain (I) and (II) above. Section 3.2 is dedicated to (III) and also inference in tree-structured Markov Random Fields namely computing P_u^*. In Section 3.3 we describe (IV). Finally, (V) is presented in Section 3.4.

3.1 Accessibility of Site Pairs

As part of RNAup, Mückstein et al. present an efficient algorithm for computing the probability of an unpaired subsequence [30]. Their algorithm computes the probability of being unpaired for all subsequences in $O(n^3)$ time in which n is the sequence length. Based on RNAup algorithm, we present an $O(n^4 w)$ time and $O(n^2)$ space complexity algorithm to compute the joint probabilities of all unpaired site pairs. For every site, our algorithm uses constrained McCaskill's [27] and constrained RNAup algorithm to compute the conditional probabilities of all other unpaired subsequences. There are $O(nw)$ sites for each of which the algorithm takes $O(n^3)$ time. For triple joint probabilities, the same method is applicable but the running time will be multiplied by another factor of $O(nw)$. Therefore, we only compute pairwise probabilities and approximate the whole joint probability distribution by graphical models.

3.2   Simultaneous Accessibility of Multiple Sites

To deal with simultaneous accessibility of binding sites, we must model their joint probability of being unpaired. One way is to compute the exact joint probability distribution by using constrained McCaskill's algorithm [22]. For every collection of sites, the algorithm has polynomial time complexity, however, since there are exponential number of different collections, this naïve approach is intractable. In this paper, we approximate the joint probability by a polynomially representable graphical model namely a Markov Random Field. Graphical models, including Bayesian Networks and Markov Random Fields, are powerful tools for approximating joint probabilities. They generally have enough expressive power, which intuitively explains why the inference problem for general graphical models is NP-hard [11]. Fortunately, there is an efficient inference algorithm for tree-structured models [32]. In this work, we build a Chow-Liu tree, which is a tree-structured Markov Random Field, to approximate the exact joint probability distribution [10].

To describe the Chow-Liu algorithm, let G be the complete weighted graph on $\mathcal{V}$, the set of potential binding sites, in which the weight of an edge between W_1 and W_2 is $I(W_1, W_2)$, the mutual information given by

$$I(W_1, W_2) = \sum_{\substack{x_1 \in \{W_1, \sim W_1\} \\ x_2 \in \{W_2, \sim W_2\}}} P(x_1, x_2) \log \left(\frac{P(x_1, x_2)}{P(x_1)P(x_2)} \right). \tag{6}$$

Above, $P(W_1, \sim W_2)$ is for instance the joint probability that W_1 is unpaired and W_2 is not unpaired. In Section 3.1, we explained how to compute the joint probabilities of all site pairs. The following equations calculate all the necessary terms from $P(W_1, W_2)$: $P(W_1, \sim W_2) = P(W_1) - P(W_1, W_2)$, $P(\sim W_1, W_2) = P(W_2) - P(W_1, W_2)$, $P(\sim W_1, \sim W_2) = 1 - P(W_1) - P(W_2) + P(W_1, W_2)$. The Chow-Liu tree $\mathcal{T}$ is the best tree-structured approximation for a joint probability distribution, in the sense that $\mathcal{T}$ has the maximum mutual information with the joint probability distribution [10]. Chow and Liu proved that $\mathcal{T}$ is the maximum spanning tree of G. To compute $\mathcal{T}$, we use a standard maximum spanning tree algorithm such as Chazelle's algorithm [7]. We refer the reader to [17] or [32] for a detailed description of inference algorithm in $\mathcal{T}$. In summary, $P_u^*(W_1, W_2, \ldots, W_k)$ is computed by marginalizing over $\mathcal{V} \setminus \{W_1, W_2, \ldots, W_k\}$ the joint probability distribution defined by $\mathcal{T}$. Inference in $\mathcal{T}$ can be done in $O(|\mathcal{V}|)$ time [17].

3.3   Free Energy of Interaction

The local free energy of interaction for a pair of sites W^R and W^S is $-RT \log(Q_{W^R W^S}^I - Q_{W^R} Q_{W^S})$ in which Q^I is the interaction partition function restricted to W^R and W^S [9] and Q is McCaskill's partition function restricted to W. Note that a simple version of Q^I would calculate only the hybridization partition function between W^R and W^S (see

[30]); however, this would exclude any intramolecular structure in the binding sites. For that reason, we use our approach for Q^I which considers intermolecular as well as intramolecular structures. Our modified algorithm is the dynamic programming in [9] that starts with $l_R = 1, l_S = 1$ and incrementally computes all the recursive quantities up to $l_R = w, l_S = w$. Therefore, the windowed version of our interaction partition function algorithm has $O(nmw^4)$ time and $O(nmw^2)$ space complexity, in which n and m are the lengths of $\mathbf{R}$ and $\mathbf{S}$ respectively. Finally, for a non-overlapping matching $M = \{(W_1^R, W_1^S), (W_2^R, W_2^S), \dots, (W_k^R, W_k^S)\}$ the free energy of interaction is $\Delta G_b^{RS}(M) = -RT \sum_{1 \leq i \leq k} \log(Q_{W_i^R W_i^S}^I - Q_{W_i^R} Q_{W_i^S})$, where $Q_{W_i^R W_i^S}^I - Q_{W_i^R} Q_{W_i^S}$ is the partition function for those structures that constitute at least one intermolecular bond. Note that this is based on the simplifying assumption in Section 3.

3.4 Binding Sites Matching

Having built the machinery to compute ΔG for a matching of binding sites, we would like to find a matching that minimizes ΔG. To clarify the importance and difficulty of the problem, suppose the binding sites were independent so that $P_u(W_1, W_2, \dots, W_k) = P_u(W_1)P_u(W_2)\cdots P_u(W_k)$. In that case, the problem would reduce to finding a minimum weight bipartite matching with $\text{weight}(W_i^R, W_j^S) = -RT \log[P_u^R(W_i^R)P_u^S(W_j^S)(Q_{W_i^R W_j^S}^I - Q_{W_i^R} Q_{W_j^S})]$. There are efficient algorithms for minimum weight bipartite matching, but the issue is that the independence assumption is too crude of an approximation. Therefore, we propose the following problem, which has not been solved to our knowledge:

Minimum Weight Chow-Liu Trees Matching Problem

Given a pair of Chow-Liu trees $T^R = (V^R, E^R)$ and $T^S = (V^S, E^S)$, compute a non-perfect matching M between the nodes of T^R and T^S that minimizes $\Delta G(M)$.
Input: Chow-Liu trees T^R and T^S.
Output: A matching $M = \{(W_1^R, W_1^S), (W_2^R, W_2^S), \dots, (W_k^R, W_k^S)\} \subset V^R \times V^S$.

The complexity of minimum weight Chow-Liu trees matching problem is currently unknown. We are working on the problem, and we hope to either prove its hardness or to give a polynomial algorithm; we incline toward the latter. In this paper, we implement an exhaustive search on single, pair, and triple sites.

3.5 Complexity Analysis

Let n denote the length of $\mathbf{R}$, m denote the length of $\mathbf{S}$, and w denote the window length. Step (I) of the algorithm takes $O(n^3 + m^3)$ time and $O(n^2 + m^2)$ space. If we consider all site pairs, then (II) takes $O(n^4 w + m^4 w)$ time and $O(n^2 w^2 + m^2 w^2)$ space to store the joint probabilities. It is often reasonable to filter potential binding sites, for example based on the probability of being unpaired or the interaction partition function with another site in the other molecule. Suppose r potential sites out of nw possible ones and s sites out of mw ones are selected. In that case, (II) takes $O(n^3 r + m^3 s)$

time and $O(r^2 + s^2)$ space. Step (III) takes $O(r^2\alpha(r^2, r) + s^2\alpha(s^2, s))$ time where α is the classical functional inverse of the Ackermann function [7]. The function α grows extremely slowly, so that for all practical purposes it may be considered a constant. Step (IV) takes $O(nmw^4)$ time and $O(nmw^2)$ space. In this paper, we implement an exhaustive search for (V). Therefore, its running time is currently $O(r^\kappa s^\kappa)$ where κ is the maximum number of simultaneous binding sites. Therefore, the algorithm takes $O(n^3 r + m^3 s + nmw^4 + r^\kappa s^\kappa)$ time and $O(r^2 + s^2 + nmw^2)$ space. Note that $r \leq nw$ and $s \leq mw$, so that the algorithm has $O(n^4 w + m^4 w + n^2 m^2 w^4)$ time and $O(n^2 w^2 + m^2 w^2)$ space complexity without heuristic filtering, considering maximum two simultaneous binding sites. We are working on (V) and hope to either find an efficient algorithm or to prove the problem's hardness.

3.6 Interaction Structure Prediction

Once the binding sites are predicted, a constrained minimum free energy structure prediction algorithm predicts the secondary structure of each RNA. For each binding site pair, a modified version of our partition function algorithm in [9] yields the interaction structure. Our partition function algorithm is modified for structure prediction by replacing summation with minimization, multiplication with summation, and exponentiation with the identity function.

4 Results

To evaluate the performance of biRNA, we used the program to predict the binding site(s) of 21 bacterial sRNA-mRNA interactions studied in the literature. Since all these sRNAs bind their target in close proximity to the ribosome binding site at the Shine-Dalgarno (SD) sequence, we restricted the program to a window of maximum 250 bases around the start codon of the gene. We compared our results with those obtained by RNAup on the same sequences with the same window size $w = 20$. The results are summarized in Table 1. RNAup and biRNA have generally very close performance for the cases with one binding site. However, biRNA outperforms RNAup in some single binding site cases such as GcvB-gltI. That is because biRNA uses the interaction partition function as opposed to the hybridization partition function used by RNAup. The interaction partition function is more accurate than the hybridization partition function as the interaction partition function accounts for both intermolecular and intramolecular base pairing [9]. Also, biRNA significantly outperforms RNAup for OxyS-fhlA and CopA-CopT which constitute more than one binding site. We noticed that our predicted energies are generally lower than those predicted by RNAup which may be due to different energy parameters. We used our piRNA energy parameters [9] which in turn are based on UNAFold v3.6 parameters [24].

We implemented biRNA in C++ and used OpenMP to parallelize it on Shared-Memory multiProcessor/core (SMP) platforms. Our experiments were run on a Sun Fire X4600 Server with 8 dual AMD Opteron CPUs and 64GB of RAM. The sequences were 71-253 nt long (see the supplementary materials for sequences) and the running time of biRNA with full features was from about 10 minutes to slightly more

Table 1. Binding sites reported in the literature and predicted by biRNA and RNAup with window size $w = 20$. ΔG is in kcal/m. Two RNAs interact in opposite direction, hence, sites in the second RNA are presented in reverse order. See the supplementary materials for sequences.

Pair		Binding Site(s) Literature		biRNA Site(s)		$-\Delta G$	RNAup Site		$-\Delta G$	Ref.
GcvB	gltI	[66,77]	[44,31]	(64,81)	(44,26)	11.5	(75,93)	(38,19)	18.7	[39]
GcvB	argT	[75,91]	[104,89]	(71,90)	(108,90)	13.1	(72,91)	(107,89)	20.2	[39]
GcvB	dppA	[65,90]	[150,133]	(62,81)	(153,135)	14.7	(62,81)	(153,135)	23.5	[39]
GcvB	livJ	[63,87]	[82,59]	(66,84)	(73,54)	13.1	(71,90)	(67,49)	14.9	[39]
GcvB	livK	[68,77]	[177,165]	(67,86)	(175,156)	12.2	(67,86)	(175,157)	19.0	[39]
GcvB	oppA	[65,90]	[179,155]	(67,86)	(176,158)	9.3	(67,86)	(176,158)	15.3	[39]
GcvB	STM4351	[70,79]	[52,44]	(69,77)	(52,44)	9.6	(69,87)	(52,33)	17.7	[39]
MicA	lamB	[8,36]	[148,122]	(8,26)	(148,131)	6.1	(8,27)	(148,129)	12.9	[5]
MicA	ompA	[8,24]	[128,113]	(8,24)	(128,113)	14.0	(8,24)	(128,113)	19.4	[34]
DsrA	rpoS	[8,36]	[38,10]	(21,40)	(25,7)	9.4	(13,32)	(33,14)	16.3	[36]
RprA	rpoS	[33,62]	[39,16]	(40,51)	(32,22)	4.3	(33,51)	(39,22)	10.7	[23]
IstR	tisA	[65,87]	[79,57]	(66,85)	(78,59)	18.1	(66,85)	(78,59)	29.0	[44]
MicC	ompC	[1,30]	[139,93]	(1,16)	(119,104)	18.5	(1,16)	(119,104)	18.7	[8]
MicF	ompF	[1,33]	[125,100]	(14,30)	(118,99)	8.0	(17,33)	(116,100)	14.7	[38]
RyhB	sdhD	[9,50]	[128,89]	(22,41)	(116,98)	15.8	(22,41)	(116,98)	21.5	[25]
RyhB	sodB	[38,46]	[60,52]	(38,46)	(64,48)	9.7	(38,57)	(60,45)	10.3	[15]
SgrS	ptsG	[157,187]	[107,76]	(174,187)	(89,76)	14.5	(168,187)	(95,76)	22.9	[19]
Spot42	galK	[1,61]	[126,52]	(1,8)	(128,119)	20.5	(27,46)	(84,68)	14.6	[29]
				(25,37)	(86,73)					
				(46,60)	(64,53)					
lncRNA$_{54}$	repZ	[16,42]	[54,28]	(19,38)	(51,32)	35.3	(19,38)	(51,32)	37.5	[4]
OxyS	fhlA	[22,30]	[95,87]	(23,30)	(94,87)	7.9	-	-	10.3	[3]
		[98,104]	[45,39]	(96,104)	(48,39)		(96,104)	(48,39)		
CopA	CopT	[22,33]	[70,59]	(22,31)	(70,61)	25.9	-	-	23.9	[20]
		[48,56]	[44,36]	(49,57)	(43,35)		(49,67)	(43,24)		
		[62,67]	[29,24]	(58,67)	(33,24)		-	-		

than one hour per sRNA-mRNA pair. The biRNA software and webserver are available at http://compbio.cs.sfu.ca/taverna/

5 Conclusions and Future Work

In this paper, we presented biRNA, a new thermodynamic framework for prediction of binding sites between two RNAs based on minimization of binding free energy. Similar to RNAup approach, we assume the binding free energy is the sum of the energy needed to unpair all the binding sites and the interaction free energy released as a result of binding.

Our algorithm is able to predict multiple binding sites which is an important advantage over previous approaches. More importantly, our algorithm can handle crossing interactions as well as zigzags (hairpins interacting in a zigzag fashion, see [1]). To assess the performance of biRNA, we compared its predictions with those of RNAup for 21 bacterial sRNA-mRNA pairs studied in the literature. The results are presented in

Table 1. As it was expected, biRNA outperforms RNAup for those RNA pairs that have multiple binding sites such as OxyS-fhlA and CopA-CopT. Moreover, biRNA performs slightly better than RNAup for those pairs that have only one binding site because biRNA accounts for intramolecular as well as intermolecular base pairing in the binding sites.

To deal with simultaneous accessibility of binding sites, our algorithm models their joint probability of being unpaired. Since computing the exact joint probability distribution is intractable, we approximate the joint probability by a polynomially representable graphical model namely a tree-structured Markov Random Field computed by the Chow-Liu algorithm [10]. Calculating a joint probability in the Chow-Liu tree is performed by efficient marginalization algorithms [32]. Eventually, two Chow-Liu trees, pertaining to the two input RNAs, are matched to find the minimum binding free energy matching. To the best of our knowledge, the complexity of minimum weight Chow-Liu trees matching problem is currently unknown. We are working on the problem, and we hope to either prove its hardness or give a polynomial algorithm. In this paper, we implemented an exhaustive search on the set of all collections of single, pair, and triple sites.

Our proposed Bayesian approximation of the Boltzmann joint probability distribution provides a novel, powerful framework which can also be utilized in individual and joint RNA secondary structure prediction algorithms. As graphical models allow for models with increasing complexity, our proposed Bayesian framework may inspire more accurate but tractable RNA-RNA interaction prediction algorithms in future work.

Acknowledgement. H. Chitsaz received funding from Combating Infectious Diseases (BCID) initiative. S.C. Sahinalp was supported by Michael Smith Foundation for Health Research Career Award. R. Backofen received funding from the German Research Foundation (DFG grant BA 2168/2-1 SPP 1258), and from the German Federal Ministry of Education and Research (BMBF grant 0313921 FRISYS).

References

1. Alkan, C., Karakoc, E., Nadeau, J.H., Cenk Sahinalp, S., Zhang, K.: RNA-RNA interaction prediction and antisense RNA target search. Journal of Computational Biology 13(2), 267–282 (2006)
2. Andronescu, M., Zhang, Z.C., Condon, A.: Secondary structure prediction of interacting RNA molecules. J. Mol. Biol. 345, 987–1001 (2005)
3. Argaman, L., Altuvia, S.: fhlA repression by OxyS RNA: kissing complex formation at two sites results in a stable antisense-target RNA complex. J. Mol. Biol. 300, 1101–1112 (2000)
4. Asano, K., Mizobuchi, K.: Structural analysis of late intermediate complex formed between plasmid ColIb-P9 Inc RNA and its target RNA. How does a single antisense RNA repress translation of two genes at different rates? J. Biol. Chem. 275, 1269–1274 (2000)
5. Bossi, L., Figueroa-Bossi, N.: A small RNA downregulates LamB maltoporin in Salmonella. Mol. Microbiol. 65, 799–810 (2007)
6. Busch, A., Richter, A.S., Backofen, R.: IntaRNA: Efficient prediction of bacterial sRNA targets incorporating target site accessibility and seed regions. Bioinformatics 24(24), 2849–2856 (2008)
7. Chazelle, B.: A minimum spanning tree algorithm with inverse-Ackermann type complexity. J. ACM 47(6), 1028–1047 (2000)

8. Chen, S., Zhang, A., Blyn, L.B., Storz, G.: MicC, a second small-RNA regulator of Omp protein expression in Escherichia coli. J. Bacteriol. 186, 6689–6697 (2004)
9. Chitsaz, H., Salari, R., Sahinalp, S.C., Backofen, R.: A partition function algorithm for interacting nucleic acid strands. Bioinformatics 25(12), i365–i373 (2009)
10. Chow, C., Liu, C.: Approximating discrete probability distributions with dependence trees. IEEE Transactions on Information Theory 14(3), 462–467 (1968)
11. Cooper, G.F.: The computational complexity of probabilistic inference using bayesian belief networks (research note). Artif. Intell. 42(2-3), 393–405 (1990)
12. Dimitrov, R.A., Zuker, M.: Prediction of hybridization and melting for double-stranded nucleic acids. Biophysical Journal 87, 215–226 (2004)
13. Dirks, R.M., Bois, J.S., Schaeffer, J.M., Winfree, E., Pierce, N.A.: Thermodynamic analysis of interacting nucleic acid strands. SIAM Review 49(1), 65–88 (2007)
14. Dirks, R.M., Pierce, N.A.: A partition function algorithm for nucleic acid secondary structure including pseudoknots. Journal of Computational Chemistry 24(13), 1664–1677 (2003)
15. Geissmann, T.A., Touati, D.: Hfq, a new chaperoning role: binding to messenger RNA determines access for small RNA regulator. EMBO J. 23, 396–405 (2004)
16. Gottesman, S.: Micros for microbes: non-coding regulatory RNAs in bacteria. Trends in Genetics 21(7), 399–404 (2005)
17. Jordan, M.I., Weiss, Y.: Graphical models: probabilistic inference. In: Arbib, M. (ed.) Handbook of Neural Networks and Brain Theory. MIT Press, Cambridge (2002)
18. Kato, Y., Akutsu, T., Seki, H.: A grammatical approach to RNA-RNA interaction prediction. Pattern Recognition 42(4), 531–538 (2009)
19. Kawamoto, H., Koide, Y., Morita, T., Aiba, H.: Base-pairing requirement for RNA silencing by a bacterial small RNA and acceleration of duplex formation by Hfq. Mol. Microbiol. 61, 1013–1022 (2006)
20. Kolb, F.A., Engdahl, H.M., Slagter-Jger, J.G., Ehresmann, B., Ehresmann, C., Westhof, E., Wagner, E.G., Romby, P.: Progression of a loop-loop complex to a four-way junction is crucial for the activity of a regulatory antisense RNA. EMBO J. 19, 5905–5915 (2000)
21. Kolb, F.A., Malmgren, C., Westhof, E., Ehresmann, C., Ehresmann, B., Wagner, E.G., Romby, P.: An unusual structure formed by antisense-target RNA binding involves an extended kissing complex with a four-way junction and a side-by-side helical alignment. RNA 6, 311–324 (2000)
22. Lu, Z.J., Mathews, D.H.: Efficient siRNA selection using hybridization thermodynamics. Nucleic Acids Res. 36, 640–647 (2008)
23. Majdalani, N., Hernandez, D., Gottesman, S.: Regulation and mode of action of the second small RNA activator of RpoS translation, RprA. Mol. Microbiol. 46, 813–826 (2002)
24. Markham, N.R., Zuker, M.: UNAFold: software for nucleic acid folding and hybridization. Methods Mol. Biol. 453, 3–31 (2008)
25. Massé, E., Gottesman, S.: A small RNA regulates the expression of genes involved in iron metabolism in Escherichia coli. Proc. Natl. Acad. Sci. U.S.A. 99, 4620–4625 (2002)
26. Mathews, D.H., Sabina, J., Zuker, M., Turner, D.H.: Expanded sequence dependence of thermodynamic parameters improves prediction of RNA secondary structure. J. Mol. Biol. 288, 911–940 (1999)
27. McCaskill, J.S.: The equilibrium partition function and base pair binding probabilities for RNA secondary structure. Biopolymers 29, 1105–1119 (1990)
28. Mneimneh, S.: On the approximation of optimal structures for RNA-RNA interaction. IEEE/ACM Transactions on Computational Biology and Bioinformatics (to appear)
29. Møller, T., Franch, T., Udesen, C., Gerdes, K., Valentin-Hansen, P.: Spot 42 RNA mediates discoordinate expression of the E. coli galactose operon. Genes Dev. 16, 1696–1706 (2002)

30. Mückstein, U., Tafer, H., Bernhart, S.H., Hernandez-Rosales, M., Vogel, J., Stadler, P.F., Hofacker, I.L.: Translational control by RNA-RNA interaction: Improved computation of RNA-RNA binding thermodynamics. In: Elloumi, M., Küng, J., Linial, M., Murphy, R.F., Schneider, K., Toma, C. (eds.) BIRD. Communications in Computer and Information Science, vol. 13, pp. 114–127. Springer, Heidelberg (2008)
31. Nussinov, R., Piecznik, G., Grigg, J.R., Kleitman, D.J.: Algorithms for loop matchings. SIAM Journal on Applied Mathematics 35, 68–82 (1978)
32. Pearl, J.: Probabilistic Reasoning in Intelligent Systems: Networks of Plausible Inference. Morgan Kaufmann, San Francisco (1988)
33. Pervouchine, D.D.: IRIS: intermolecular RNA interaction search. Genome Inform. 15, 92–101 (2004)
34. Rasmussen, A.A., Eriksen, M., Gilany, K., Udesen, C., Franch, T., Petersen, C., Valentin-Hansen, P.: Regulation of ompA mRNA stability: the role of a small regulatory RNA in growth phase-dependent control. Mol. Microbiol. 58, 1421–1429 (2005)
35. Rehmsmeier, M., Steffen, P., Hochsmann, M., Giegerich, R.: Fast and effective prediction of microRNA/target duplexes. RNA 10, 1507–1517 (2004)
36. Repoila, F., Majdalani, N., Gottesman, S.: Small non-coding RNAs, co-ordinators of adaptation processes in Escherichia coli: the RpoS paradigm. Mol. Microbiol. 48, 855–861 (2003)
37. Rivas, E., Eddy, S.R.: A dynamic programming algorithm for RNA structure prediction including pseudoknots. J. Mol. Biol. 285, 2053–2068 (1999)
38. Schmidt, M., Zheng, P., Delihas, N.: Secondary structures of Escherichia coli antisense micF RNA, the 5'-end of the target ompF mRNA, and the RNA/RNA duplex. Biochemistry 34, 3621–3631 (1995)
39. Sharma, C.M., Darfeuille, F., Plantinga, T.H., Vogel, J.: A small RNA regulates multiple ABC transporter mRNAs by targeting C/A-rich elements inside and upstream of ribosome-binding sites. Genes Dev. 21, 2804–2817 (2007)
40. Storz, G.: An expanding universe of noncoding RNAs. Science 296(5571), 1260–1263 (2002)
41. Tafer, H., Ameres, S.L., Obernosterer, G., Gebeshuber, C.A., Schroeder, R., Martinez, J., Hofacker, I.L.: The impact of target site accessibility on the design of effective siRNAs. Nat. Biotechnol. 26, 578–583 (2008)
42. Tafer, H., Hofacker, I.L.: RNAplex: a fast tool for RNA-RNA interaction search. Bioinformatics 24(22), 2657–2663 (2008)
43. Tinoco, I., Uhlenbeck, O.C., Levine, M.D.: Estimation of secondary structure in ribonucleic acids. Nature 230, 362–367 (1971)
44. Vogel, J., Argaman, L., Wagner, E.G., Altuvia, S.: The small RNA IstR inhibits synthesis of an SOS-induced toxic peptide. Curr. Biol. 14, 2271–2276 (2004)
45. Walton, S.P., Stephanopoulos, G.N., Yarmush, M.L., Roth, C.M.: Thermodynamic and kinetic characterization of antisense oligodeoxynucleotide binding to a structured mRNA. Biophys. J. 82, 366–377 (2002)
46. Waterman, M.S., Smith, T.F.: RNA secondary structure: A complete mathematical analysis. Math. Biosc. 42, 257–266 (1978)
47. Zuker, M., Stiegler, P.: Optimal computer folding of large RNA sequences using thermodynamics and auxiliary information. Nucleic Acids Research 9(1), 133–148 (1981)

Quantifying Systemic Evolutionary Changes by Color Coding Confidence-Scored PPI Networks

Phuong Dao, Alexander Schönhuth, Fereydoun Hormozdiari, Iman Hajirasouliha,
S. Cenk Sahinalp, and Martin Ester

School of Computing Science, Simon Fraser University, 8888 University Drive
Burnaby, BC, V5A 1S6, Canada

Abstract. A current major challenge in systems biology is to compute statistics on *biomolecular network motifs*, since this can reveal significant systemic differences between organisms. We extend the "color coding" technique to weighted edge networks and apply it to PPI networks where edges are weighted by probabilistic confidence scores, as provided by the *STRING* database. This is a substantial improvement over the previously available studies on, still heavily noisy, binary-edge-weight data. Following up on such a study, we compute the expected number of occurrences of non-induced subtrees with $k \leq 9$ vertices. Beyond the previously reported differences between unicellular and multicellular organisms, we reveal major differences between prokaryotes and unicellular eukaryotes. This establishes, for the first time on a statistically sound data basis, that evolutionary distance can be monitored in terms of elevated systemic arrangements.

Keywords: Biomolecular Network Motifs, Color Coding, Evolutionary Systems Biology, Protein-Protein Interaction Networks.

1 Introduction

A current major issue in evolutionary systems biology is to reliably quantify both organismic complexity and evolutionary diversity from a systemic point of view. While currently available biomolecular networks provide a data basis, the assessment of network similarity has remained both biologically and computationally challenging. Since currently available network data is still incomplete, simple edge statistics, for example, do not apply. Moreover, recent research has revealed that many biomolecular networks share global topological features which are robust relative to missing edges, which rules out many more straightforward approaches to the topic (see e.g. [13] for a related study on global features such as degree distribution, k-hop reachability, betweenness and closeness). On the more sophisticated end of the scale of such approaches would be attempts to perform and appropriately score alignments of the collection of all systemic subunits of two organisms. However, the development of workable scoring schemes in combination with related algorithms comes with a variety of obvious, yet unresolved, both biological and computational issues. Clearly, any such scoring schemes would already establish some form of condensed, systemic evolutionary truth by themselves.

This explains why recent approaches focused on monitoring differences between biomolecular networks in terms of *local structures*, which likely reflect biological

S.L. Salzberg and T. Warnow (Eds.): WABI 2009, LNBI 5724, pp. 37–48, 2009.

arrangements such as functional subunits. A seminal study which reported that statistically overrepresented graphlets, i.e. small subnetworks, are likely to encode pathway fragments and/or similar functional cellular building blocks [17] sparked more general interest in the topic. In the meantime, to discover and to count *biomolecular network motifs* has become a thriving area of research which poses some intriguing algorithmic problems. As is summarized in the comprehensive review [6], such approaches are supported by various arguments.

In this paper, following up on a series of earlier studies, we will focus on physical protein-protein interaction (PPI) network data. Related studies focused on determining the number of all possible "induced" subgraphs in a PPI network, which already is a very challenging task. Przulj et al. [19] devised algorithms with which to count induced PPI subgraphs on up to $k = 5$ vertices. Recently developed techniques improved on this by counting induced subgraphs of size up to $k = 6$ [13] and $k = 7$ [11]. However, the running time of these techniques all increase exponentially with k. To count subgraphs of size $k \geq 8$ required novel algorithmic tools. A substantial advance was subsequently provided in [1] which introduced the "color coding" technique for counting non-induced occurrences of subgraph topologies in the form of bounded treewidth subgraphs, which includes trees as the most obvious special case. Counting non-induced occurrences of network motifs is not only challenging but also quite desirable since non-induced patterns are often correlated to induced occurrences of denser patterns. Trees in particular can be perceived as the backbones of induced dense and connected subgraphs and there is abundant evidence that dense and connected (induced) subgraphs reflect functional cellular building blocks (e.g. [26]). See also [7] for a successful approach to count all patterns of density at least 0.85 in several PPI networks as well as in synthetic networks from popular generative random models. See also [7] for a successful approach to count all patterns of density at least 0.85 in several PPI networks as well as in synthetic networks from popular generative random models.

While these studies successfully revealed differences between PPI networks of uni- and multicellular organisms, a binary edge has remained a notoriously noisy datum. However, none of the studies considered PPI networks with weighted edges where edge weights reflect the confidence that the interactions are of cellular relevance instead of being experimental artifacts. Weighted network data have recently become available and have already been employed for other purposes (see e.g. [22] and the references therein for a list of weighted network data sources). One of the main reasons for the lack of network motif studies on such data might be that to exhaustively mine biomolecular networks with probabilistic edge weights poses novel computational challenges.

In this paper, we show how to apply the "color coding" technique to networks with arbitrary edge weights and two different scoring schemes for weighted subgraphs. Edge weights are supposed to reflect our confidence in the interactions, as provided by the *STRING* database, and we will apply a scoring scheme which reflects our expectation[1] in entire subgraphs to be present or not. *STRING* is a major resource for assessments of protein interactions and/or associations predicted by large-scale experimental data (in the broad sense, including e.g. literature data) of various types (see [15] for the latest

---

[1] Expectation is meant to be in the very sense of probability theory, by interpreting confidence scores as probabilities.

issue of STRING and the references therein for earlier versions). Here, we only employ experimentally determined physical protein interactions in order to both follow up on the recent discussions and to avoid inconsistencies inherent to using several types of protein interactions at once. Clearly, statistics on such networks will establish substantial improvements over studies on binary network data in terms of statistical significance and robustness.

We compute the expected number of non-induced occurrences (E-values) of tree motifs G' ("treelets") with k vertices in a network G with n vertices in time polynomial in n, provided $k = O(\log n)$. Note that, in binary edge weight graphs, computation of the number of expected occurrences and counting occurrences is equivalent when interpreting an edge to be an interaction occurring with probability 1. This provides the basis on which we can benchmark our results against previous studies. We use our algorithm to obtain normalized treelet distributions, that is the sum of the weights of non-induced occurrences of different tree topologies of size $k = 8, 9^2$ normalized by the total weight of all non-induced trees of size $8, 9$ for weighted PPI networks. We analyze the prokaryotic, unicellular organisms (*E.coli*, *H.pylori*), *B. subtilis* and *T. pallidum*, which are all quite similar, the eukaryotic unicellular organism *S.cerevisiae* (Yeast), and a multicellular organism (*C.elegans*). Beyond the previously reported similarities among the prokaryotic organisms, we were able to also reveal strong differences between Yeast and the prokaryotes. As before, statistics on C.elegans are still different from all other ones. As a last point, we demonstrate that our weighted treelet distributions are *robust* relative to reasonable amounts of network sparsification as suggested by [12].

To summarize, we have presented a novel randomized approximation algorithm to count the weight of non-induced occurrences of a tree T with k vertices in a weighted-edge network G with n vertices in time polynomial with n, provided $k = O(\log n)$ for a given error probability and an approximation ratio. We prove that resulting weighted treelet distributions are robust and sensitive measures of PPI network similarity. Our experiments then confirm, for the first time on a statistically reliable data basis, that uni- and multicellular organisms are different on an elevated systemic cellular level. Moreover, for the first time, we report such differences also between pro- and eukaryotes.

Related Work

Flum and Grohe [10] showed that the problem of counting the number of paths of length k in a network is $\#W[1]$-complete. Thus it is unlikely that one can count the number of paths of length k efficiently even for small k. The most recent approaches such as [5,23] offer a running time of $O(n^{k/2+O(1)})$. As mentioned before, [19,11,13] describe practical approaches to counting all induced subgraphs of at most $k = 5, 6$ and $k = 7$ vertices in a PPI network.

Approximate counting techniques have been devised which were predominantly based on the color coding technique as introduced by Alon et al. [3]. Color coding is based on assigning random colors to the vertices of an input graph and, subsequently, counting only "colorful" occurrences, that is, subgraphs where each vertex has a different color. This can usually be done in polynomial time. Iterating over different color

---

[2] We recall that there are 23 resp. 47 different tree topologies on 8 resp. 9 nodes, see e.g. [18].

assignments sufficiently many, but still a polynomial number of times yields statistically reliable counts. In the seminal study [3], color coding was used to detect (but not to count) simple paths, trees and bounded treewidth subgraphs in unlabelled graphs.

Scott et al. [20], Shlomi et al. [21] and Huffner et al. [14] designed algorithms for querying paths within a PPI network. More recently, Dost et al. [9] have extended these algorithms in the QNet software to allow searching for trees and bounded treewidth graphs. Arvind and Raman [4] use the color coding approach to count the number of subgraphs in a given graph G which are isomorphic to a *bounded treewidth graph* H. The framework which they use is based on approximate counting via sampling [16]. However, even when $k = O(\log n)$, the running time of this algorithm is *super-polynomial* with n, and thus is not practical. Alon and Gutner [2] derandomized the color coding technique by the construction of balanced families of hash functions. Recently, Alon et al. [1] presented a randomized approximation algorithm that, given an additive error ϵ and error probability δ, with success probability $1 - \delta$, outputs a number within ϵ times the number of non-induced occurrences of a tree T of k vertices in a graph G of n vertices running in time $O(|E| \cdot 2^{O(k)} \cdot \log(1/\delta) \cdot \frac{1}{\epsilon^2})$. Note that if $k = O(\log n)$ and ϵ, δ are fixed, this results in a polynomial time algorithm.

Note that all the previous works tried either to count exactly the total weight of non-induced occurrences of a pattern in a given graph or to approximate the occurrences where the weights of all edges are 1. The exact counting methods, due to parameterized complexity as mentioned earlier, give exponential running time even for paths of $k = O(\log n)$ vertices. We will give approximate based counting methods that offer polynomial running time given that patterns are trees of $k = O(\log n)$ vertices, a fixed approximation factor and an error probability.

2 Methods

In the following, let $G = (V, E)$ be a graph on n vertices and $w : E \rightarrow \mathbb{R}$ be an edge-weight function. Let T be a tree on k vertices where, in the following, $k = O(\log n)$. We define $\mathcal{S}(G, T)$ to be the set of non-induced subgraphs of G which are isomorphic to T and let $E(H)$ to be the edges of such a subgraph $H \in \mathcal{S}$. We extend w to weight functions on the members of $\mathcal{S}(G, T)$ by either defining

$$w(H) = \prod_{e \in E(H)} w(e) \quad \text{or} \quad w(H) = \sum_{e \in E(H)} w(e) \tag{1}$$

Note that if $w(e)$ is interpreted as the probability that e is indeed present in G then, assuming independence between the edges, $w(H)$ of the first case is just the probability that H is present in G. In the following, we will focus on the first case. Proofs for the second choice of $w(H)$ can be easily obtained, mutatis mutandis, after having replaced multiplication by addition in the definition of $w(H)$. Finally, let $w(G, T) = \sum_{H \in \mathcal{S}(G,T)} w(H)$ be the total weight of non-induced occurrences of T in G. We would like to provide reliable estimates $\hat{w}(G, T)$ on $w(G, T)$. Note that $w(G, T)$ is the number of expected occurrences of T in G due to the linearity of expectation.

Consider Fig. 1. Here, T is a star-like tree on 4 vertices. There are two subgraphs H and H' in G which are isomorphic to T; therefore, $w(G, T) = w(H) + w(H')$. In the

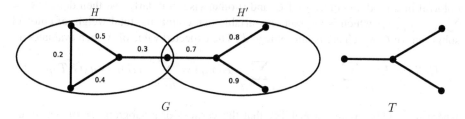

Fig. 1. An example of counting the total weight of non-induced subgraphs in a given network G which are isomorphic to a query tree T

case that the weight of a subgraph in G is calculated as the product of the weights of its edges, we have $w(G,T) = w(H) + w(H') = 0.5 \times 0.4 \times 0.3 + 0.7 \times 0.8 \times 0.9$. In the other case, we have $w(G,T) = w(H) + w(H') = (0.5 + 0.4 + 0.3) + (0.7 + 0.8 + 0.9)$.

In the following, in order to estimate $w(G,T)$ by color coding, we will randomly assign k colors to the vertices of G where k is the size of T. Therefore, we introduce the notations $[k] = \{1, ..., k\}$ for the set of k colors and $S(G, T, [k])$ for the set of all non-induced subgraphs of G which are *colorful* in terms of $[k]$, that is occurrences of T where each vertex has been assigned to a different color.

The following algorithm APPROXWEIGHTEDOCCUR, when given an approximation factor ϵ and an error probability δ, computes an estimate $\hat{w}(G,T)$ of $w(G,T)$ efficiently in n and k, given that $k = O(\log n)$ such that with probability $1 - 2\delta$, $\hat{w}(G,T)$ lies in the range $[(1 - \epsilon)w(G,T), (1 + \epsilon)w(G,T)]$.

Algorithm 1. APPROXWEIGHTEDOCCURR (G, T, ϵ, δ)

$G = (V, E)$, $k \leftarrow |V(T)|$, $t \leftarrow \log(1/\delta)$, $p \leftarrow k!/k^k$, $s \leftarrow 4/(\epsilon^2 p)$
for $i = 1$ to t **do**
    $Y_i \leftarrow 0$
    for $j = 1$ to s **do**
        Color each vertex of G independently and uniformly at random with one of k colors
        $X \leftarrow$ total weight of colorful subgraphs of G which are isomorphic to T
        $Y_i \leftarrow Y_i + X$
    end for
    $Y_i \leftarrow Y_i/s$
end for
$Z \leftarrow$ median of $Y_1 \ldots Y_t$
Return Z/p as the estimate $\hat{w}(G,T)$ of $w(G,T)$

The following lemmata give rise to a theorem that supports our claims from above.

Lemma 1. *The algorithm* APPROXWEIGHTEDOCCURR *(G, T, ϵ, δ) returns $\hat{w}(G,T)$ such that with probability at least $1 - 2\delta$ we have $(1 - \epsilon)w(G,T) \leq \hat{w}(G,T) \leq (1 + \epsilon)w(G,T)$*

Proof. The proof proceeds quite similar to that of [1], after having replaced x_F there by x_H here where x_H is the indicator random variable whose value is $w(H)$ if H is colorful in a random coloring of G and 0 otherwise. Similarly, we then define $X = \sum_{H \in \mathcal{S}(G,T)} x_H$ which is the random variable that counts the total weight of colorful subgraphs of G which are isomorphic to T. The expected value of X then evaluates as

$$E(X) = E\left(\sum_{H \in \mathcal{S}(G,T)} x_H \right) = \sum_{H \in \mathcal{S}(G,T)} E(x_H) = \sum_{H \in \mathcal{S}} w(H)p = w(G,T)p \qquad (2)$$

where $p = k!/k^k$ is the probability that the vertices of a subgraph H of size k are assigned to different colors. To obtain a bound on the variance $Var(X)$ of X, one observes that $Var(x_H) = E(x_H^2) - E^2(x_H) \le E(x_H^2) = [w(H)]^2 p$. Moreover, the probability that both H and H' are colorful is at most p which implies

$$Cov(x_H, x_{H'}) = E(x_H x_{H'}) - E(x_H)E(x_{H'}) \le E(x_H x_{H'}) \le w(H)w(H')p.$$

Therefore, in analogy to [1], the variance of X satisfies $Var(X) = (\sum_{H \in \mathcal{S}} w(H))^2 p = w^2(G,T)p$. Since Y_i is the average of s independent copies of random variable X, we have $E(Y) = E(X) = w(G,T)p$ and $Var(Y_i) = Var(X)/s \le w^2(G,T)p/s$. Again in analogy to [1], we obtain $P(|Y_i - w(G,T)p| \ge \epsilon w(G,T)p) \le \frac{1}{4}$.

Thus, with constant error probability, Y_i/p is an ϵ-approximation of $w(G,T)$. To obtain error probability $1 - 2\delta$, we compute t independent samples of Y_i (using the first for loop) and replace Y_i/p by Z/p where Z is the median of Y_i's. The probability that Z is less than $(1 - \epsilon)w(G,T)p$ is the probability that at least half of the copies of Y_i computed are less than Z, which is at most $\binom{t}{t/2} 4^{-t} \le 2^{-t}$. Similarly we can estimate the probability that Z is bigger than $(1 + \epsilon)w(G,T)p$. Therefore, if $t = \log(1/\delta)$ then with probability $1 - 2\delta$ the value of $\hat{o}$ will lie in $[(1 - \epsilon)w(G,T), (1 + \epsilon)w(G,T)]$. $\diamond$

We still need to argue that given the graph G where each vertex is colored with one of k colors, we can compute the total weight of all non-induced colorful occurrences $w(G,T,[k])$ of T in G which refers to the variable X in the second for loop efficiently.

Lemma 2. *Given a graph G where each vertex has one of k colors, we can estimate $w(G,T,[k])$ in time $O(|E| \cdot 2^{O(k)})$.*

Proof. We pick a vertex ρ of T and consider T_ρ to be a rooted version of the query tree T with designated root ρ. We will compute $w(G, T_\rho, [k])$ recursively in terms of subtrees of T_ρ; so let $T'_{\rho'}$ be any subtree T' of T with designated root ρ'. Let $C \subset [k]$ and $\mathcal{S}(G, T'_{\rho'}, v, C)$ be the set of all non-induced occurrences of $T'_{\rho'}$ in G which are rooted at v and colorful with colors from C and $w(v, T'_{\rho'}, v, C) = \sum_{H \in \mathcal{S}(G, T'_{\rho'}, v, C)} w(H)$ to be the total weight of all such occurrences. We observe that

$$w(G, T, [k]) = \frac{1}{q} \sum_{v \in G} w(G, T_\rho, v, [k]) \qquad (3)$$

where q is equal to one plus the number of vertices ϱ in T for which there is an automorphism that ρ is mapped to ϱ. For example, if T in Figure 2 is rooted at ρ'', q is equal to 3. The key observation is that we can compute $w(G, T_\rho, v, [k])$ or the total weight of colorful non-induced subtrees rooted at v in G which are isomorphic to T_ρ in terms of

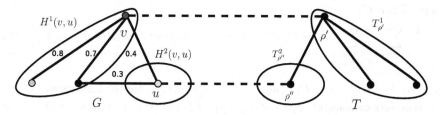

Fig. 2. Counting the total weight of colorful non-induced occurrences of T in G by counting the total weight of colorful non-induced occurrences of subtrees $T_{\rho'}$ and $T_{\rho''}$ in G

total weight of colorful non-induced occurrences of subtrees of T_ρ in G. Let $T'_{\rho'}$ be an arbitrary rooted subtree of T_ρ. We decompose $T'_{\rho'}$ into two smaller subtrees and count total weight colorful non-induced occurrences of these subtrees in G as follows. We choose a child ρ'' of ρ' and, by removing the edge between ρ' and ρ'' to decompose $T'_{\rho'}$ into two rooted subtrees $T^1_{\rho'}$ that do not contain ρ'' and $T^2_{\rho''}$ that do not contain ρ' for example Figure 2. Analogously, for every neighbor u of v in G, we denote a colorful copy of $T^1_{\rho'}$ at v by $H^1(v, u)$ and a colorful copy of $T^2_{\rho''}$ at u by $H^2(v, u)$. To obtain a copy H of $T'_{\rho'}$ in G by combining $H^1(v, u)$ and $H^2(v, u)$, $H^1(v, u)$ and $H^2(v, u)$ must be colorful for color sets $C_1(v, u), C_2(v, u)$ such that $C_1(v, u) \cap C_2(v, u) = \emptyset, C_1 \cup C_2 = C$ where the cardinality of C is the number of vertices of $T'_{\rho'}$. Finally, independent of the choice of u, we have

$$w(H) = w(H^1(v, u))w(H^2(v, u))w(vu) \qquad (4)$$

To initialize the base case of single-vertex trees $T'_{\rho'}$, we set $w(G, T'_{\rho'}, v, \{i\}) = 1$ if the color of v is i; otherwise 0. In general, we have

$$w(G, T'_{\rho'}, v, C) = \frac{1}{d} \sum_{u \in N(v)} \sum_{\substack{C_1 \cap C_2 = \emptyset \\ C_1 \cup C_2 = C}} \sum_{\substack{H^1 \in S(G, T^1_{\rho'}, v, C_1) \\ H^2 \in S(G, T^2_{\rho''}, u, C_2)}} w(H^1(v, u))w(H^2(v, u))w(vu)$$

$$= \frac{1}{d} \sum_{u \in N(v)} \sum_{\substack{C_1 \cap C_2 = \emptyset \\ C_1 \cup C_2 = C}} w(G, T^1_{\rho'}, v, C_1)w(vu)w(G, T^2_{\rho''}, u, C_2).$$

$$(5)$$

Note that $w(H)$ will, as in the summands, appear exactly d times across the different suitable choices of color sets C_1, C_2. For example, H in Fig. 2, rooted at v, is a colorful copy of a star-like rooted tree T with three leaves. There are three different ways by which one can decompose H into a path of length 2, denoted by H_1 and a single node H_2, meaning that $d = 3$ in this case. Now observe that the weight of H $w(H)$ will appear three times as a summand in the above summation scheme, according to the three different decompositions of H. The proof for the case of additive weight schemes proceeds mutatis mutandis, after having replaced multiplication by addition and adjusted the base case ($w(G, T'_{\rho'}, v, \{i\} = 0$ regardless of the color of v). Let $N(v)$ be the set of neighbors of v, we have

$$w(G, T'_{\rho'}, v, C)$$

$$= \frac{1}{d} \sum_{u \in N(v)} \sum_{\substack{C_1 \cap C_2 = \emptyset \\ C_1 \cup C_2 = C}} n_2 w(G, T^1_{\rho'}, v, C_1) + n_1 n_2 w(vu) + n_1 w(G, T^2_{\rho''}, u, C_2) \quad (6)$$

where $n_1 = |\mathcal{S}(G, T^1_{\rho'}, v, C_1)|, n_2 = |\mathcal{S}(G, T^2_{\rho''}, u, C_2)|$ are the cardinalities of the respective sets of colorful copies of $T^1_{\rho'}$ resp. $T^2_{\rho''}$ rooted at v resp. u which can be computed efficiently [1] and parallely with $w(G, T^1_{\rho'}, v, C_1)$ and $w(G, T^2_{\rho''}, u, C_2)$.

Note that each $w(G, T'_{\rho'}, v, C)$ can be computed in $O(deg(v) \cdot 2^{O(k)})$ time where $deg(v)$ is the degree of v. Thus, the computation of total weight of colorful non-induced occurrences of T in G is in $O(|E|2^{O(k)})$ time.   ◇

Theorem 1. *The algorithm* APPROXWEIGHTEDOCCURR *(G, T, ϵ, δ) estimates the total weight of non-induced occurrences of a tree T in G with additive error ϵ and with probability at least $1 - 2\delta$ and runs in time $O(|E| \cdot 2^{O(k)} \cdot \log(1/\delta) \cdot \frac{1}{\epsilon^2})$ where $|E|$ is the number of edges in the input network.*

Proof. Now we only need to consider its running time. Notice that we need to repeat the color coding step and counting step $s \cdot t$ times and each iteration runs in time $O(|E| \cdot 2^{O(k)})$ where $|E|$ is the number of edges in the input network. Thus, since $p = k^k/k! = O(e^k) = O(2^{O(k)})$, the asymptotic running time of our algorithm evaluates as

$$O(s \cdot t \cdot |E| \cdot 2^{O(k)}) = O(|E| \cdot 2^{O(k)} \cdot \log(1/\delta) \cdot \frac{1}{\epsilon^2 p}) = O(|E| \cdot 2^{O(k)} \log(1/\delta) \cdot \frac{1}{\epsilon^2}). \quad (7)$$

3   Results

3.1   Data and Implementation

Weighted PPI Networks. We downloaded PPI networks with confidence scores from the *STRING* database, version 8.0 [15] for the prokaryotic, unicellular organisms *E.coli*, *H.pylori*, *B.subtilis*, *T. pallidum*, the eukaryotic, unicellular organism *S.cerevisiae* (Yeast) and the eukaryotic, multicellular organism *C.elegans*. Edge weights exclusively reflect confidence in experimentally determined physical interactions to be functional cellular entities and not only experimental artifacts (see [25,24] for detailed information). See Table 1 for some basic statistics about these networks.

Query Tree Topologies. There are 23 and 47 possible tree topologies with 8 and 9 nodes respectively. We obtained the list of treelets from the Combinatorial Object

Table 1. Number of vertices, edges, in the studied PPI networks

	E.coli	H.pylori	B. subtilis	T. pallidum	S.cerevisiae	C.elegans
Vertices	2482	1019	939	398	5913	5353
Edges	22476	9274	9184	4198	207075	43100

Server [18]. See our supplementary website [8] for diagrams of the respective tree topologies.

Implementation / Choice of Parameters. We implemented our algorithm APPROX-WEIGHTOCCUR with the multiplicative weight scheme such that the weight of a query tree can be interpreted as its probability to be present in the network, as aforementioned. We set $\epsilon = 0.01$ as the approximation ratio and $\delta = 0.001$ as the error probability. Then we computed expected numbers of occurrences of all query trees of size 8 and 9 for each of the networks described above. By normalizing the occurrences of the different query trees of size 8 resp.9 over the 23 resp. 47 different query trees, we obtained size 8 resp. size 9 treelet distributions which we refer to as *normalized weighted treelet distributions*. The idea behind normalizing expected occurrences is for comparing PPI networks with different number of nodes and edges and to increase robustness with respect to missing data which still is a considerable issue in PPI network studies. We will demonstrate the robustness by an approved series of experiments [12] in the subsequent subsection 3.3. Experiments were performed on a Sun Fire X4600 Server with 64GB RAM and 8 dual AMD Opteron CPUs with 2.6 Ghz speed each.

3.2   Comparison of PPI Networks

In order to be able to appropriately benchmark our results against previous findings we considered the same organisms that were examined in [1]. We also considered the two prokaryotic organisms *B.subtilis* (a Gram-negative bacterium commonly found in soil) and *T.pallidum* (a Gram-negative pathogen giving rise to congenital syphilis). The corresponding weighted treelet distributions are displayed in Fig. 3. The upper row of figures shows that the treelet distributions of the prokaryotic organisms are all similar. This is quite amazing since the weighted PPI networks have been determined in experiments which were independent of one another and without the integration of cross-species associations [15]. As can be seen in the middle row of Fig. 3, the treelet distributions of the Yeast PPI network is quite different from the ones of the prokaryotic organisms, which had not been observed in the boolean networks used in [1]. Still, there are obvious differences between the unicellular organisms and *C.elegans*, the multicellular model organism under consideration. It might be interesting to note that the greatest differences occur for the expected numbers of occurrences of tree topologies 23 resp. 47, which are the stars with 8 resp. 9 nodes. As a last point, note that global features such as degree and clustering coefficient distributions of these networks do not differ much (see the supplementary materials [8] for respective results).

3.3   Robustness Analysis

In order to assess the reliability of the normalized weighted treelet distributions as a measure of weighted PPI network similarity one needs to ensure that they are robust w.r.t. small alterations to the network. This is motivated by that there still might be some insecurity about the amount and the quality of currently available PPI data. In this section, we evaluate the robustness of normalized weighted treelet distributions meaning that minor changes in the weighted PPI networks do not result in drastic changes in their normalized weighted treelet distributions.

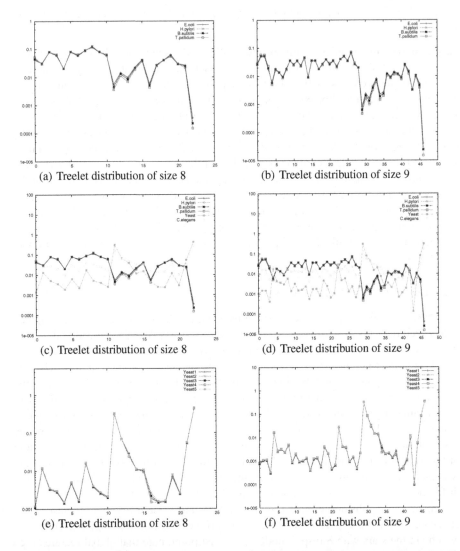

Fig. 3. Normalized weighted treelet distributions of the prokaryotes *H.pylori, E.coli, B.subtilis, T. pallidum* PPI networks (top row) and of the prokaryotes *H.pylori, E.coli, B.subtilis, T. pallidum, S.cervisiae* (Yeast) and *C.elegans* PPI networks (middle row). and of five networks (a) size 8, (b) size 9 generated from the *S.cervisiae* Yeast PPI network as outlined in ssec. 3.3 (bottom row).

Therefore, we used the random sparsification method which was proposed in [12] and was applied in earlier studies [1]. The method iteratively sparsifies networks by removing vertices and edges in a sampling procedure and specifically addresses the pecularities of experimentally generating PPI networks. It is based on two parameters, the bait sampling probability α_b and the edge sampling probability α_e which refer to sampling vertices and edges. As in [1], we set $\alpha_b = 0.7$ and $\alpha_e = 0.7$ and shrank the weighted PPI network of Yeast to five smaller networks accordingly. A comparison of the normalized weighted treelet distributions of the shrunken networks is displayed in the bottom row of Fig. 3. As can be seen, the normalized weighted treelet distributions are very similar to one another which confirms that normalized treelet distributions are robust w.r.t. experimentally induced noise and/or missing data.

4   Conclusions

To quantify organismic complexity and evolutionary diversity from a systemic point of view poses challenging biological and computational problems. Here, we have investigated *normalized weighted treelet distributions*, based on exploration of PPI network whose edges are assigned to confidence scores, which can be retrieved from the STRING database as an appropriate measure. As a theoretical novelty, we have extended the color coding technique to weighted networks. As a result, we were able to reveal differences between uni- and multicellular as well as pro- and eukaryotic organisms. Systemic differences based on local features in PPI networks between pro- and eukaryotes had not been reported before. In sum, our study reveals novel systemic differences and confirms previously reported ones on a substantially more reliable data.

References

1. Alon, N., Dao, P., Hajirasouliha, I., Hormozdiari, F., Sahinalp, S.C.: Biomolecular network motif counting and discovery by color coding. Bioinformatics 24(Supp. 1), i241–i249 (2008)
2. Alon, N., Gutner, S.: Balanced families of perfect hash functions and their applications. In: Arge, L., Cachin, C., Jurdziński, T., Tarlecki, A. (eds.) ICALP 2007. LNCS, vol. 4596, pp. 435–446. Springer, Heidelberg (2007)
3. Alon, N., Yuster, R., Zwick, U.: Color-coding. J. ACM 42(4), 844–856 (1995)
4. Arvind, V., Raman, V.: Approximation algorithms for some parameterized counting problems. In: Bose, P., Morin, P. (eds.) ISAAC 2002. LNCS, vol. 2518, pp. 453–464. Springer, Heidelberg (2002)
5. Bjorklund, A., Husfeldt, T., Kaski, P., Koivisto, M.: The fast intersection transform with applications to counting paths (2008), http://arxiv.org/abs/0809.2489
6. Ciriello, G., Guerra, C.: A review on models and algorithms for motif discovery in protein-protein interaction networks. Briefings in Functional Genomics and Proteomics (2008)
7. Colak, R., Hormozdiari, F., Moser, F., Schönhuth, A., Holman, J., Sahinalp, S.C., Ester, M.: Dense graphlet statistics of protein interaction and random networks. In: Proceedings of the Pacific Symposium on Biocomputing, vol. 14, pp. 178–189 (2009)
8. Dao, P., Schoenhuth, A., Hormozdiari, F., Hajirasouliha, I., Sahinalp, S.C., Ester, M.: Quantifying systemic evolutionary changes by color coding confidence-scored ppi networks. Supplementary Materials (2009),
   http://www.cs.sfu.ca/~pdao/personal/weightedmotifsup.pdf
9. Dost, B., Shlomi, T., Gupta, N., Ruppin, E., Bafna, V., Sharan, R.: QNet: A tool for querying protein interaction networks. In: Speed, T., Huang, H. (eds.) RECOMB 2007. LNCS (LNBI), vol. 4453, pp. 1–15. Springer, Heidelberg (2007)
10. Flum, J., Grohe, M.: The parameterized complexity of counting problems. SIAM J. Comput. 33, 892–922 (2004)

11. Grochow, J.A., Kellis, M.: Network motif discovery using subgraph enumeration and symmetry-breaking. In: Speed, T., Huang, H. (eds.) RECOMB 2007. LNCS (LNBI), vol. 4453, pp. 92–106. Springer, Heidelberg (2007)

12. Han, J., Dupuy, D., Bertin, N., Cusick, M., Vidal, M.: Effect of sampling on topology predictions of protein-protein interaction networks. Nature Biotech. 23, 839–844 (2005)

13. Hormozdiari, F., Berenbrink, P., Przulj, N., Sahinalp, S.C.: Not all scale-free networks are born equal: The role of the seed graph in ppi network evolution. PLoS Comput. Biol. 3(7) (2007)

14. Huffner, F., Wernicke, S., Zichner, T.: Algorithm engineering for color coding with applications to signaling pathways. Algorithmica 52(2), 114–132 (2008)

15. Jensen, L.J., Kuhn, M., Stark, M., Chaffron, S., Creevey, C., Muller, J., Doerke, T., Julien, P., Roth, A., Simonovic, M., Bork, P., von Mering, C.: String 8—a global view on proteins and their functional interactions in 630 organisms. Nucleic Acids Research 37(Database issue), D412–D416 (2009)

16. Karp, R.M., Luby, M.: Monte-carlo algorithms for enumeration and reliability problems. In: FOCS, pp. 56–64 (1983)

17. Milo, R., Shen-Orr, S., Itzkovitz, S., Kashtan, N., Chklovskii, D., Alon, U.: Network motifs: simple building blocks of complex networks. Science 298(5594), 824–827 (2002)

18. University of Victoria. Combinatorial object server (since 1995),
    http://www.theory.csc.uvic.ca/~cos

19. Przulj, N., Corneil, D.G., Jurisica, I.: Modeling interactome: scale-free or geometric? Bioinformatics 20(18), 3508–3515 (2004)

20. Scott, J., Ideker, T., Karp, R.M., Sharan, R.: Efficient algorithms for detecting signaling pathways in protein interaction networks. J. Comput. Biol. 13(2), 133–144 (2006)

21. Shlomi, T., Segal, D., Ruppin, E., Sharan, R.: Qpath: a method for querying pathways in a protein-protein interaction network. BMC Bioinformatics 7, 199 (2006)

22. Ulitsky, I., Shamir, R.: Identifying functional modules using expression profiles and confidence-scored protein interactions. Bioinformatics (2009),
    doi:10.1093/bioinformatics/btp118

23. Vassilevska, V., Wiliams, R.: Finding, minimizing and counting weighted subgraphs. In: Proceedings of the Symposium of the Theory of Computing, STOC (to appear, 2009)

24. von Mering, C., Jensen, L.J.: News about the string and stitch databases (2008),
    http://string-stitch.blogspot.com/

25. von Mering, C., Jensen, L.J., Snel, B., Hooper, S.D., Krupp, M., Foglierini, M., Jouffre, N., Huynen, M.A., Bork, P.: String: known and predicted protein-protein associations, integrated and transferred across organisms. Nucleic Acids Research 33(Database issue), D433–D437(2005)

26. Zhu, X., Gerstein, M., Snyder, M.: Getting connected: analysis and principles of biological networks. Genes and Development 21, 1010–1024 (2007)

PMFastR: A New Approach to Multiple RNA Structure Alignment

Daniel DeBlasio[1], Jocelyne Bruand[2], and Shaojie Zhang[1]

[1] University of Central Florida, Orlando, FL 32816, USA
{deblasio,shzhang}@eecs.ucf.edu
[2] University of California, San Diego, La Jolla, CA 92093, USA
jbruand@ucsd.edu

Abstract. Multiple RNA structure alignment is particularly challenging because covarying mutations make sequence information alone insufficient. Many existing tools for multiple RNA alignments first generate pairwise RNA structure alignments and then build the multiple alignment using only the sequence information. Here we present PMFastR, an algorithm which iteratively uses a sequence-structure alignment procedure to build a multiple RNA structure alignment. PMFastR has low memory consumption allowing for the alignment of large sequences such as 16S and 23S rRNA. The algorithm also provides a method to utilize a multi-core environment. Finally, we present results on benchmark data sets from BRAliBase, which shows PMFastR outperforms other state-of-the-art programs. Furthermore, we regenerate 607 Rfam seed alignments and show that our automated process creates similar multiple alignments to the manually-curated Rfam seed alignments.

Keywords: multiple RNA alignment, RNA sequence-structure alignment, iterative alignment.

1 Introduction

A high quality multiple alignment of RNA sequences is crucial for RNA homology search [29], structure analysis [6] and discovery [20,26]. Even though methods for multiple alignments of DNA and protein sequences are well studied [4,7,19,23], multiple RNA structure alignment is still an open problem.

The problem of aligning two RNA sequences with consideration of the structural information comes as an extension of the RNA structure prediction studies [13,17,31]. When aligning two RNA sequences, we can consider three problems and their associated methods depending on the availability of the structural information for these RNA sequences [1,14]. Without considering the pseudo-knots in the RNAs, all of these problems can be solved in polynomial time. First, it is possible to align two RNA sequences without any structural information and to predict their common secondary structure [21]; this is the RNA *sequence-sequence* alignment problem. Another potential input is a structural profile or other structural information for one of the sequences but not for the second one; this is the *sequence-structure* alignment problem. Several methods have

S.L. Salzberg and T. Warnow (Eds.): WABI 2009, LNBI 5724, pp. 49–61, 2009.

been developed to address this issue and are used for finding RNA homologs. FastR [30] and PFastR [29] use a guided trees and dynamic programming to globally align a sequence or profile to a given target sequence. CMSEARCH [27] and RSEARCH [16] use covariance models to supplement dynamic programming for their alignment procedure. Finally, the structural information can be given for both RNA sequences, allowing us to find common motifs between two RNA structures [14]; this is the RNA *structure-structure* alignment problem.

Much work has already been done on the multiple RNA structure alignment problem. Most of these RNA multiple alignment methods (PMmulti [10], MARNA [22], Stemloc [12], STRAL [3], and LARA [1]) use pairwise RNA alignment (sequence-sequence or structure-structure) for all sequences and then combine these alignments into a multiple alignment using T-Coffee [19] or other progressive strategies. For instance, LARA can take in the structure or predict the structure of the input sequences. The program then uses a connection graph and integer linear programming to create pairwise alignments. The score of these pairwise alignments are then fed into T-Coffee to generate a multiple alignment. However, in the case of sequence-sequence alignment, these methods predict the RNA structure or pairing probabilities from scratch at the expense of RNA structure accuracy. On the other hand, structure-structure alignment is not feasible on very long RNA sequences, such as 16S rRNA and 23S rRNA. Here, we use the RNA sequence-structure alignment strategy to build a multiple alignment. Eddy and Durban [6] have used this strategy in the past by using a covariance model but their algorithm requires an initial multiple alignment as input. In contrast, the algorithm presented here requires only one sequence with structure and a database of other unaligned sequences as input.

Databases of multiple RNA alignments such as Rfam [9] maintain very high quality alignments which are obtained by integration of multiple sources and manual curation. We use these databases for baseline comparison as a mean to assay how well our algorithm is performing. We show in this paper that the proposed algorithm can produce comparable results without manually curation of the alignments.

In this paper, we present the Profile based Multiple Fast RNA Alignment (PMFastR) algorithm. An overview of this program is presented in Figure 1. PMFastR does a multiple structure-based alignment from scratch while using a relatively small amount of memory and can be used to make a multiple structure alignment of long RNA sequences such as 16S rRNA or 23S rRNA sequences. The input is one sequence with secondary structure information and a group of sequences without such information. Our algorithm consists of three major steps. The first step is a structure-sequence alignment of an RNA sequence from the database with the original structure. This outputs an aligned profile of two sequences with structure information. We can then align the next element from the database to the output profile and obtain a new alignment profile. This can be repeated iteratively until all of the elements of the input dataset are aligned. Finally, we run the CMBuild refinement program to improve the unpaired gap regions in the alignment. In the Methods section, we present the algorithm itself

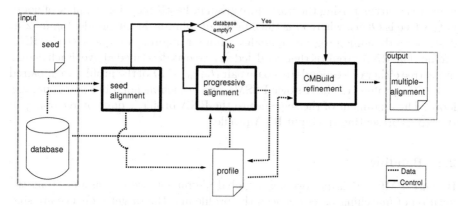

Fig. 1. Overview of PMFastR. The three major steps are highlighted in the solid square boxes, while the input and output of the program are represented in dashed boxes. After generation of a seed alignment, the algorithm progressively aligns the rest of the input data set. Finally, we run the CMBuild refinement program to improve the unpaired gap regions in the alignment.

and details on all of the major improvements over FastR. In the Results section, we run several benchmarking tests on PMFastR with other multiple RNA alignment tools using BRAliBase (version 2.1) [28] as test data sets, which are extracted from Rfam database. We also regenerate all Rfam hand-curated seed alignments (version 8.1) [9] automatically with comparable results.

2   Methods

PMFastR is based on the methods used in FastR [30] and PFastR [29] alignment procedures, which were designed for RNA sequence-structure alignment. This section first briefly describes the sequence-structure alignment algorithm itself, then the improvements made upon it. These improvements are banding, multithreading, and multiple alignment. By using banding, the algorithm has a reduced running time and space consumption by several orders of magnitude. Multithreading allows the algorithm to utilize multiple cores when available to further improve wall time. Finally, by using a progressive profile alignment, the algorithm is able to produce multiple sequence alignments.

2.1   The Alignment Procedure

We make the assumption that all base-pairs are non-crossing, and let M be the set of all such base-pairs. Thus, for each base-pair $(i, j) \in M$, there is a unique *parent* base pair (i', j') such that $i' < i < j < j'$, and there is no base-pair (i'', j'') such that $i < i'' < i'$ or $j' < j'' < j$. The alignment can be done by recursing on a tree which is representing the RNA profile with the secondary structural information. Since this tree can have high degree and not all columns of the profile participate in it, we binarize it using the procedure given in FastR [30]. The number of nodes in this tree is $O(m)$, where m is the number of columns in the profile. Figure 2(a) describes a dynamic programming algorithm for aligning a sequence to an RNA profile. The RNA profile is then modeled as a tree as described above. Each node v in the tree either corresponds to a base pair $(l_v, r_v) \in M$ of the profile, an unpaired base of the profile (and has its own PSSM), or a branching point in a pair of parallel loops. The alignment of the sequence to the RNA profile is done by recursing on the tree representing the input RNA profile.

2.2 Banding

Because PMFastR is performing a global alignment, we can assume that the location of matching bases between the profile and the target sequence are similar. In particular, we assume that it is within some banding constant of the original location. Once this assumption has been established, the search space is limited to these bounds. This allows for a reduction in running time since we do not examine locations with a very low likelihood of matching. The memory consumption is also reduced since we only need to store the alignment values for the locations within the banding region. A banding variable, *band*, is defined and can be adjusted depending on the type and length of sequence being examined, as well as the precision of the result desired. Thus, for any node v in the tree, the algorithm only needs to examine the locations in the query where the corresponding base-pair or unpaired base might exist. For example, let v be a node in the binarized tree as described above where v represents the base-pair at (l_v, r_v). The algorithm looks for the corresponding base-pair in the query and only examines the potential pairing sites (i, j) in the query where $l_v - band \leq i \leq l_v + band$ and $r_v - band \leq j \leq r_v + band$.

This banding entails that any potential pairings outside of the bounds are never assigned a score. Since the banding constant is given at the beginning of the algorithm, we only need to allocate the space to store results within the banding bounds. If there is a reference to a location outside those bounds the initialization value is returned. The running time and space complexity of this algorithm is then reduced from $\sim O(n^2 * m)$ to $\sim O(band^2 * m)$, where n is the length of the target sequence and m is the number of nodes in M' which are bounded by the length of the profile. Figures 2(b) and 2(c) show the methods used for mapping into the new bounded array. We can see that these procedures incur only a small amount of overhead, thus the effect on running time is not significant.

A problem arises when we work with sequences that are not of similar length. To overcome this we can make adjustments to the analysis parameters to allow for these alignments to still be computed. The first solution is to adjust the banding parameter to search a larger space when the difference in size is high. In particular, the banding value should always be more than one and a half times the difference in length. Additionally, any columns in the profile that do not meet a certain quality criteria (percentage of sequences represented) are removed before alignment, and are added back when the profile is output. By adjusting the quality criteria we can also effect the length of the profile used for alignment.

procedure PAln
(*M is the set of base-pairs in RNA profile R. M' is the augmented set. *)
for all nodes $v \in M'$,
        all intervals (i, j), $l_v - band \leq i \leq l_v + band$ and $r_v - band \leq j \leq r_v + band$
    if $v \in M$

$$value = \max \begin{cases} mapRetrieve(\text{child}(v), i+1, j-1) + \delta(l_v, r_v, t[i], t[j]), \\ mapRetrieve(v, i, j-1) + \gamma('-', t[j]), \\ mapRetrieve(v, i+1, j) + \gamma('-', t[i]), \\ mapRetrieve(\text{child}(v), i+1, j) + \gamma(l_v, t[i]) + \gamma(r_v, '-'), \\ mapRetrieve(\text{child}(v), i, j-1) + \gamma(l_v, '-') + \gamma(r_v, t[j]), \\ mapRetrieve(\text{child}(v), i, j) + \gamma(l_v, '-') + \gamma(r_v, '-'), \end{cases}$$

    else if $v \in M' - M$, and v has one child

$$value = \max \begin{cases} mapRetrieve(\text{child}(v), i, j-1) + \gamma(r_v, t[j]), \\ mapRetrieve(\text{child}(v), i, j) + \gamma(r_v, '-'), \\ mapRetrieve(v, i, j-1) + \gamma('-', t[j]), \\ mapRetrieve(v, i+1, j) + \gamma('-', t[i]), \end{cases}$$

    else if $v \in M' - M$, and v has two children
        $value = \max_{i \leq k \leq j} \{$
                $mapRetrieve(\text{left\_child}(v), i, k-1) +$
                $mapRetrieve(\text{right\_child}(v), k, j)$
        $\}$
    end if
    $mapSet(v, i, j, value)$
end for

(a)

procedure mapRetrieve(v, i, j)
(*i and j are the global position
in the table assuming that banding is not used. *)
if i & j are within the *band* of l_v and r_v
    $i_t = i - l_v + band$
    $j_t = j - r_v + band$
    return A$[i_t, j_t, v]$
else
    return initialization value for (i, j, v)
end if

(b)

procedure mapSet$(v, i, j, value)$
if i & j are within the *band* of l_v and r_v
    $i_t = i - l_v + band$
    $j_t = j - r_v + band$
    A$[i_t, j_t, v] = value$
end if

(c)

Fig. 2. (a) An algorithm for aligning an RNA profile R with m columns against a database string t of length n. The query consensus structure M was *Binarized* to obtain M'. Each node v in the tree corresponds to a base-pair $(l_v, r_v) \in M'$. (b) and (c) The mapping functions that make the transition to a memory-saving banding array. It is assumed that the array A is of size $n * n * m$ while the new banding array is of size $band * band * m$.

```
procedure processNode(v)
(*v is the current node to be processed. *)
if   v has one or more children
        spawn thread on processNode(left_child(v))
end if
if   v has two children
        spawn thread on processNode(right_child(v))
end if
wait for all (if any) child threads to complete
signal that this node is ready for processing
wait until resources are freed to process
execute PAln(v)
signal that the next node can be processed
```

Fig. 3. The multithreaded design elements of the improved algorithm. Because the threads are queued, we can control the amount of simultaneous computation.

2.3   Multithread Design

To improve the feasibility on large sequences, parallelization is used to improve the wall time of PMFastR. The intuition comes from the fact that, at any given time, only one node is processed from the tree. Each node only depends on its two children, which in turn depend only on their children, and so on. This means that the two children do not depend on each other and can be processed simultaneously and independently.

By altering the procedure in Figure 2(a) to take the node v as input, we can run each node independently. Another procedure is used to manage the threading: given a node, it runs the alignment on its two children (if any exist), then runs the alignment on the input node. If each of the children is spawned as a new thread, then this can be carried out by the processing environment in parallel.

This improvement allows the majority of the processing time to be run in parallel. Figure 3 shows the node processing procedure. A *signal* and *wait* queue is used to have processes wait for resources allocated. This allows a restriction on the amount of processor being used in a shared environment by giving control to the user of how many active process threads are allowed to be in the "running state" at any given point.

2.4   Multiple Alignment

We implemented a single pass progressive profile multiple alignment. The algorithm first aligns one input sequence with structure and the rest of the given sequences without structure. Note that we can also give a profile rather than a single sequence. The first step involves running the profile alignment algorithm described above on the input profile and a single sequence from the set, which outputs an alignment between the sequence and the profile. We then use this alignment to create a new profile composed of the original profile, with additional columns for the gaps inserted in the alignment, followed by the aligned (gapped) sequence.

Since the profile is constructed in a single pass, the order in which the sequences are added becomes very important. While this can be done in an ad hoc manner and adequate results are achieved, we found that having some guidance greatly improved the output quality. Hence, we retrieve the alignment of the sequences from ClustalW [18] and use the ClustalW guided tree to direct the order of the input for PMFastR.

Figure 4 shows the complete alignment procedure. This procedure assumes that the input is a single sequence with structure and a set of sequences without structure. Note the differences between the sequence-structure alignment and the profile-sequence alignment; a PSSM is used to help the alignment in the profile alignment, whereas this step is excluded from the sequence-structure alignment. If the input is a profile instead of a single sequence, the third and fourth lines of the algorithm are be skipped. Finally, we also use the refinement option of CMBuild [5] to refine the reinserted unpaired columns.

---

procedure multipleAlignment
(*Here S is an array of sequence file names without extension. *)
run ClustalW to get the alignment tree for the non-structured sequences
order the input sequences $\{S_1, S_2, S_3, ...S_k\}$ as ordered above
run sequence-structure alignment on S_0 with structure and S_1 without
output the alignment to a profile file p.
for S_i as the rest of the sequences S_2 to S_k
    compact p remove unpaired columns with less than cut_off sequences
           present
    run the profile-sequence alignment on p and S_i
    reinsert the unpaired columns removed above
    output this alignment to p
end for
execute CMBuild –refine on p and *output* the multiple alignment with structure
        annotation.

Fig. 4. The multiple alignment procedure. The profile is built progressively using the structure from a seed sequence. All subsequent sequences are then aligned to this profile.

3   Experimental Results

We evaluated the performance of PMFastR based on running time, quality of the alignments and memory consumption. We first looked at the improvement in the running time of the multi-threaded version of PMFastR in function of the number of simultaneous processes. We then compared the performance of PMFastR to five other multiple alignment tools [1]. We also generated multiple alignments for every RNA family from the Rfam 8.1 database [9] and evaluated their quality by comparing them to the manually-curated seed alignments. Furthermore, we assessed the improvements in memory consumption of the PMFastR in contrast to that of its predecessor FastR; these results are shown in the supplementary data.

3.1 Multithreading

Multithreading can be used to improve the running time of PMFastR. We tested the improvement in performance from multithreading by running our algorithm on 100 alignments on a multi-core system while varying the number of active threads. In Table 1, we show the total wall time needed to complete the jobs on four nodes of a high performance cluster with eight processors per node. The speedup was calculated as the original runtime over the improved runtime. It can be seen that for more than eight processes, the performance increase is minimal. This is due to the dominance of communication overhead as the load on each thread diminishes.

Table 1. Multithreaded restriction results

Number of Processes	Total Time	Average Time Per Alignment	Speedup
1	7:10:46	4:18	1
2	4:10:38	2:25	1.78
4	3:24:00	2:02	2.11
8	2:59:13	1:48	2.40
16	2:57:14	1:46	2.43
32	2:56:51	1:46	2.44

3.2 Alignments on BRAliBase Data Sets and Rfam Families

We chose the widely used benchmark set BRAliBase 2.1 [28], which is based on seed alignments from the Rfam 7.0 database, as our test set. We compared PMFasR to four other structure-based RNA alignment tools (LARA, FOLDALIGNM [25], MARNA, and STRAL) and one purely sequence-based multiple alignment tool (MAFFT [15]).

In order to compare the alignments generated by PMFastR with the alignments generated by these other tools, we used four measures: Compalign, Sum-of-Pairs Score (SPS), Structure Conservation Index (SCI) and Structure Conservation Rate (SCR). Sum-of-Pairs Score (SPS) is the fraction of pairs aligned in both the reference alignments and the predicted alignments and has been used in many alignment benchmarks [8,24]. It is similar to Compalign which has been used in LARA's benchmarking [1]. For SPS and Compalign, a value of 1 is achieved when the reference alignment is the same as the test alignment. On the other hand, Structure Conservation Rate (SCR) calculates the fraction of the conserved base pairs in the alignments. It resembles the Structure Conservation Index (SCI) [8,26] with

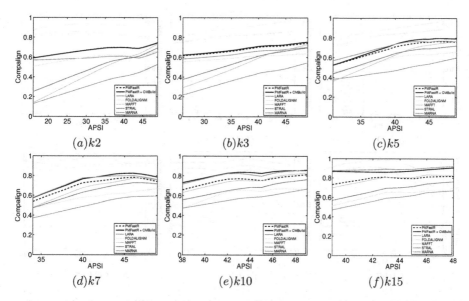

Fig. 5. Compalign Benchmarking. Each alignment was run though Compalign, this score is shown on the y-axis. The x-axis shows the average pairwise sequence identity (APSI) for each dataset. The data used were sets of (a)2, (b)3, (c)5, (d)7, (e)10, and (f)15 sequences per dataset.

the differences that SCR rewards the compensatory mutations and uses reference structure from the benchmarking alignment, while SCI first uses RNAalifold [11] to predict the consensus structures and then calculates the structural conservation index with compensatory mutations. Since we already have the structural information of the benchmarking data set, SCR is a better way to check the paired regions alignments. Since other programs (LARA, FOLDALIGNM, MARNA, and STRAL) do not output the structural information, we have to use RNAalifold to predict the consensus structure for SCR.

To remain consistent with Bauer et al. [1], Figure 5 and Figure 6 show Compalign and SCI results for benchmarking data set. We can see that PMFastR with CMBuild refinement outperforms the other programs in these tests. The results for LARA, FOLDALIGNM, MAFFT, MARNA, and STRAL were obtained from the supplementary data of Bauer et al. [1]. Due to space restrictions, the full benchmarking results, including the SPS and SCR test results, are available on the supplementary website.

To further test our program, we use PMFastR with CMBuild to recreate all seed alignments from Rfam 8.1 database starting from only one sequence with its annotated structure. There are 607 families in the Rfam database. The detailed alignment results of all RFam families are available in the supplementary data. Figure 7 shows that alignments predicted by PMFastR with CMBuild are comparable with the manually curated seed alignments from the Rfam database. Moreover, we generated a multiple structure alignment for the 567 16S rRNA

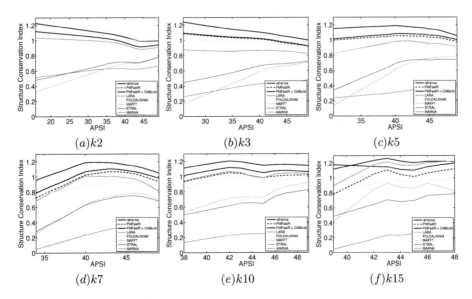

Fig. 6. SCI Benchmarking. Structure Conservation Index (SCI) score is obtained for each alignment, which is shown on the y-axis. The x-axis shows the average pairwise sequence identity (APSI) for each dataset. The data used were sets of (a)2, (b)3, (c)5, (d)7, (e)10, and (f)15 sequences per dataset.

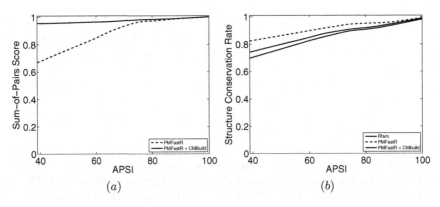

Fig. 7. Comparison of alignments generated by PMFastR on RFam families with manually curated seed alignments by two measures, SPS (a) and SCR (b)

from CRW database [2] with average pairwise sequence identity of 48%. The alignment took ~40 hours, and had an average maximum memory consumption per alignment of 3.63 Gb. To our knowledge, this is the first automated multiple alignment of 16S RNA sequences which takes into account secondary structure information. This alignment is available on our supplementary data website.

4   Conclusion

We presented an algorithm which aligns RNA sequences using both sequence and structure information. The algorithm only requires one sequence to have structure information and is able to align all other input sequences without human interaction. Because we are able to drastically reduce the memory consumption as compared to previous work, our algorithm is able to run on very long RNA sequences, such as 16S and 23S rRNA.

We propose three major ideas which improve the performance of PMFastR and which can be applied to other work. Banding allows a significant reduction in both run time and space consumption of our alignment algorithm. In fact, we are able to reduce the space consumption from cubic to linear when comparing to the predecessor, FastR. Moreover, reordering the inner loops of this algorithm allows us to run it in a multi-threaded manner, thus drastically reducing the wall time. These modifications do not alter the quality of the algorithm's results, and we show that PMFastR with CMBuild refinement does better than other state-of-the-art RNA multiple alignment tools and creates comparable alignments to the hand-curated ones in the Rfam database. All results, as well as the application source code, can be found on the project web page at http://genome.ucf.edu/PMFastR.

Acknowledgements

This work is supported in part by the UCF STOKES High Performance Computing Cluster project.

References

1. Bauer, M., Klau, G., Reinert, K.: Accurate multiple sequence-structure alignment of RNA sequences using combinatorial optimization. BMC Bioinformatics 8(1), 271 (2007)
2. Cannone, J., Subramanian, S., Schnare, M., Collett, J., D'Souza, L., Du, Y., Feng, B., Lin, N., Madabusi, L., Muller, K., Pande, N., Shang, Z., Yu, N., Gutell, R.: The Comparative RNA Web (CRW) Site: an online database of comparative sequence and structure information for ribosomal, intron, and other RNAs. BMC Bioinformatics 3(1), 2 (2002)
3. Dalli, D., Wilm, A., Mainz, I., Steger, G.: STRAL: progressive alignment of non-coding RNA using base pairing probability vectors in quadratic time. Bioinformatics 22, 1593–1599 (2006)
4. Do, C.B., Mahabhashyam, M.S., Brudno, M., Batzoglou, S.: ProbCons: Probabilistic consistency-based multiple sequence alignment. Genome Res. 15, 330–340 (2005)
5. Eddy, S.R.: Infernal package, http://infernal.janelia.org/
6. Eddy, S.R., Durbin, R.: RNA sequence analysis using covariance models. Nucleic Acids Res. 22, 2079–2088 (1994)
7. Edgar, R.: MUSCLE: multiple sequence alignment with high accuracy and high throughput. Nucleic Acids Res. 32, 1792–1797 (2004)

8. Gardner, P.P., Wilm, A., Washietl, S.: A benchmark of multiple sequence alignment programs upon structural RNAs. Nucleic Acids Res. 33, 2433–2439 (2005)

9. Griffiths-Jones, S., Moxon, S., Marshall, M., Khanna, A., Eddy, S.R., Bateman, A.: Rfam: annotating non-coding RNAs in complete genomes. Nucleic Acids Res. 33, D121–D124 (2005)

10. Hofacker, I.L., Bernhart, S.H., Stadler, P.F.: Alignment of RNA base pairing probability matrices. Bioinformatics 20, 2222–2227 (2004)

11. Hofacker, I.L., Fekete, M., Stadler, P.F.: Secondary structure prediction for aligned RNA sequences. J. Mol. Biol. 319, 1059–1066 (2002)

12. Holmes, I.: Accelerated probabilistic inference of RNA structure evolution. BMC Bioinformatics 6, 73 (2005)

13. Jaeger, J.A., Turner, D.H., Zuker, M.: Improved predictions of secondary structures for RNA. Proc. Natl. Acad. Sci. U.S.A. 86, 7706–7710 (1989)

14. Jiang, T., Lin, G., Ma, B., Zhang, K.: A general edit distance between RNA structures. Journal of Computational Biology 9, 2002 (2002)

15. Katoh, K., Kuma, K., Toh, H., Miyata, T.: MAFFT version 5: improvement in accuracy of multiple sequence alignment. Nucleic Acids Res. 33, 511–518 (2005)

16. Klein, R.J., Eddy, S.R.: RSEARCH: finding homologs of single structured RNA sequences. BMC Bioinformatics 4, 44 (2003)

17. Knudsen, B., Hein, J.: Pfold: RNA secondary structure prediction using stochastic context-free grammars. Nucleic Acids Res. 31, 3423–3428 (2003)

18. Larkin, M., Blackshields, G., Brown, N., Chenna, R., McGettigan, P., McWilliam, H., Valentin, F., Wallace, I., Wilm, A., Lopez, R., Thompson, J., Gibson, T., Higgins, D.: Clustal W and Clustal X version 2.0. Bioinformatics 23(21), 2947–2948 (2007)

19. Notredame, C., Higgins, D.G., Heringa, J.: T-Coffee: A novel method for fast and accurate multiple sequence alignment. J. Mol. Biol. 302, 205–217 (2000)

20. Rivas, E., Eddy, S.R.: Noncoding RNA gene detection using comparative sequence analysis. BMC Bioinformatics 2, 8 (2001)

21. Sankoff, D.: Simulations solution of the RNA folding, alignment and protosequence problems. SIAM J. Appl. Math. 45(5), 810–825 (1985)

22. Siebert, S., Backofen, R.: MARNA: multiple alignment and consensus structure prediction of RNAs based on sequence structure comparisons. Bioinformatics 21, 3352–3359 (2005)

23. Thompson, J.D., Higgins, D.G., Gibson, T.J.: CLUSTAL W: improving the sensitivity of progressive multiple sequence alignment through sequence weighting, position-specific gap penalties and weight matrix choice. Nucleic Acids Res. 22, 4673–4680 (1994)

24. Thompson, J.D., Plewniak, F., Poch, O.: A comprehensive comparison of multiple sequence alignment programs. Nucleic Acids Res. 27, 2682–2690 (1999)

25. Torarinsson, E., Havgaard, J.H., Gorodkin, J.: Multiple structural alignment and clustering of RNA sequences. Bioinformatics 23, 926–932 (2007)

26. Washietl, S., Hofacker, I.L., Lukasser, M., Httenhofer, A., Stadler, P.F.: Mapping of conserved RNA secondary structures predicts thousands of functional noncoding RNAs in the human genome. Nat. Biotechnol. 23, 1383–1390 (2005)

27. Weinberg, Z., Ruzzo, W.L.: Exploiting conserved structure for faster annotation of non-coding RNAs without loss of accuracy. Bioinformatics 20(suppl. 1), i334–i341 (2004)

28. Wilm, A., Mainz, I., Steger, G.: An enhanced RNA alignment benchmark for sequence alignment programs. Algorithms Mol. Biol. 1, 19 (2006)

29. Zhang, S., Borovok, I., Aharonowitz, Y., Sharan, R., Bafna, V.: A sequence-based filtering method for ncRNA identification and its application to searching for riboswitch elements. Bioinformatics 22(14), e557–e565 (2006)
30. Zhang, S., Haas, B., Eskin, E., Bafna, V.: Searching genomes for noncoding RNA using FastR. IEEE/ACM Transactions on Computational Biology and Bioinformatics 2(4), 366–379 (2005)
31. Zuker, M., Sankoff, D.: RNA secondary structures and their prediction. Bulletin of Mathematical Biology 46(4), 591–621 (1984)

On the Upper Bound of the Prediction Accuracy of Residue Contacts in Proteins with Correlated Mutations: The Case Study of the Similarity Matrices

Pietro Di Lena[1], Piero Fariselli[2], Luciano Margara[1], Marco Vassura[1], and Rita Casadio[2]

[1] Department of Computer Science, University of Bologna, Italy
{dilena,margara,vassura}@cs.unibo.it
[2] Biocomputing Group, Department of Biology, University of Bologna, Italy
{piero,casadio}@biocomp.unibo.it

Abstract. Correlated mutations in proteins are believed to occur in order to preserve the protein functional folding through evolution. Their values can be deduced from sequence and/or structural alignments and are indicative of residue contacts in the protein three-dimensional structure. A correlation among pairs of residues is routinely evaluated with the Pearson correlation coefficient and the MCLACHLAN similarity matrix. In this paper, we describe an optimization procedure that maximizes the correlation between the Pearson coefficient and the protein residue contacts with respect to different similarity matrices, including random. Our results indicate that there is a large number of equivalent matrices that perform similarly to MCLACHLAN. We also obtain that the upper limit to the accuracy achievable in the prediction of the protein residue contacts is independent of the optimized similarity matrix. This suggests that poor scoring may be due to the choice of the linear correlation function in evaluating correlated mutations.

1 Introduction

A large-scale statistical analysis indicates that through evolution side chain mutations tend to preserve more the protein structure than its sequence [11]. As a consequence, residue substitutions, when occurring, must be compensated in spatially close neighbors by other mutations (two residues in the same spatial neighborhood are termed in *contact*). This basic idea has been exploited first by searching for pairs of residues that might have co-evolved and then by inferring that these pairs are indeed close in the three dimensional space of the protein [8,15]. The Pearson coefficient (or its variants) is routinely adopted to score the assumed linear correlation between co-evolving residues [5,8,13,15]. For a residue pair in the protein sequence the Pearson coefficient measures the degree of correlation between pairwise substitutions. A set of similar sequences compiled in a multiple alignment and a similarity matrix to weigh residue substitutions in each position are necessary to compute each pairwise correlation.

S.L. Salzberg and T. Warnow (Eds.): WABI 2009, LNBI 5724, pp. 62–72, 2009.
© Springer-Verlag Berlin Heidelberg 2009

Scores for residue substitutions are routinely provided by the MCLACHLAN similarity matrix [12], in which substitution values are assigned on the basis of physico-chemical similarity between side chains. If two residues a, b have a high level of physical similarity then the substitution $a \rightarrow b$ ($b \rightarrow a$) is rewarded with a positive score, otherwise a negative score is assigned.

To the best of our knowledge there is no justification to the use of the MCLACHLAN matrix instead of other substitution matrices, such as BLO-SUM62 [7] or PAM250 [14]. Recently MCLACHLAN was used in conjunction with a second matrix derived from wet contacts (side chains in contact with water molecules co-crystallized within the protein) and the results suggest that water may play a role in contact predictions [16].

The main motivation of this work is to explore the space of all the possible similarity matrices to maximize the correlation between residue contacts and correlated mutation values, as calculated with the Pearson coefficient. Under this condition, we show that the problem of computing the optimal similarity matrix for correlated mutations can be defined as a minimization problem of some particular error function. We performed several experimental tests by optimizing the similarity matrices starting from different initial solutions. For each optimized matrix, the accuracy of contact prediction is similar to that obtained with MCLACHLAN. We obtain that the space of the error function seems to have a huge number of solutions that perform equally well. This result provides a justification for adopting MCLACHLAN and, at the same time, defines an upper limit to the accuracy achievable in predicting residue contacts.

2   Preliminaries

2.1   Contact Maps and Evaluation Criteria for Contact Prediction

The map of contacts (*contact map*) of a protein P is a two-dimensional approximation of the three-dimensional structure of P. There are several definitions of residue contacts in literature. We adopt the *physical* representation, which defines two residues to be in contact when the minimum distance between every possible pair of atoms in the two residues (one in the first residue and one in the second residue) is less than or equal to 5 Angstrom (Å) [10]. For a given protein P, its contact map M^P is a symmetric square matrix, which can be defined as

$$M_{i,j}^P = \begin{cases} 1 & \text{if the } i-\text{th and the } j-\text{th residues of } P \text{ are in contact} \\ -1 & \text{otherwise} \end{cases}$$

for $1 \leq i, j \leq L$, where L is the length of the protein P. The presence of a contact between two residues is a structural constraint for the correct fold. If we were able to predict just few suitable contacts of a target protein then we could use such predictions as an intermediate step for the reconstruction of the tertiary structure [18] or as a tool for the validation of novel structure predictions computed by other methods [4].

The research in contact prediction is assessed every two years in CASP experiments [9] following the EVAcon indications [6]. The most important evaluation measure is the accuracy of prediction, which is defined as

(number of correctly predicted contacts) / (predicted contacts).

Contact predictors assign a contact score to each pair of residues. Then the accuracy is calculated by sorting scores in decreasing order and by applying the accuracy definition to the first L, $L/2$, $L/5$, $L/10$ most probable contacts. To have a better measure of accuracy, three different classes of residue pairs are considered, depending on their sequence separation (number of residues included in their interleaving segment): long range (sequence separation ≥ 24), medium range (≥ 12) and short range (≥ 6). Residue contacts whose sequence separation is below 6 do not provide information on the protein folding and are not evaluated. The difficulty in contact prediction increases at increasing sequence separation length. Long range contacts are the most informative about the protein structure since they impose the strongest constraints to the protein fold. Thus, the performances of a contact predictor are usually ranked in terms of the prediction accuracy for long range contacts.

2.2   Correlated Mutations and Contact Prediction

Correlated mutations have been intensively exploited for residue contact predictions [5,13,15]. All implementations compute the correlated mutations for substitution pairs with the Pearson correlation coefficient (or its variants). In the following we describe in detail the standard implementation of this method [5].

Consider a protein P and denote with A^P the matrix representing the multiple sequence alignment for a family of proteins similar to P. The first column of A^P contains the residue chain of P , the remaining columns of A^P correspond to the aligned protein chains and every row i of A^P corresponds to the observed substitutions for the i-th residue of P (according to the alignment). In detail, for every position i, we have the sequence of amino acid substitutions

$$A^P_{i,0} \to A^P_{i,1}, A^P_{i,0} \to A^P_{i,2}, \ldots, A^P_{i,1} \to A^P_{i,2}, A^P_{i,1} \to, A^P_{i,3}, \ldots$$

which correspond to the vector of scores

$$s^i = (\mathcal{S}_{A^P_{i,0}, A^P_{i,1}}, \mathcal{S}_{A^P_{i,0}, A^P_{i,2}}, \ldots, \mathcal{S}_{A^P_{i,1}, A^P_{i,2}}, \mathcal{S}_{A^P_{i,1}, A^P_{i,3}}, \ldots)$$

where $\mathcal{S}$ is some amino acid similarity matrix (i.e. $\mathcal{S}_{A^P_{i,0}, A^P_{i,1}}$ is the score provided by the similarity matrix $\mathcal{S}$ for the substitution $A^P_{i,0} \to A^P_{i,1}$). Scores are not computed for substitutions which involve gaps. MCLACHLAN is the similarity matrix adopted to compute the substitution scores for correlation coefficients. For every pair of positions i, j we compute the substitution score vectors s^i, s^j as described above, taking care to maintain a correspondence between substitution pairs and entries of s^i and s^j, i.e. if for some k, s^i_k is the score of the substitution

$A_{ip}^P \rightarrow A_{iq}^P$ then s_k^j must be the score of the substitution $A_{jp}^P \rightarrow A_{jq}^P$. The Pearson correlation coefficient for positions i, j of P is then defined by

$$C_{ij}^P = \frac{1}{N} \sum_{k=1}^{N} \frac{(s_k^i - \bar{s}^i)(s_k^j - \bar{s}^j)}{\sigma^i \sigma^j} \qquad (1)$$

where N is the length of the substitution score vectors s^i, s^j and

$$\bar{s}^i = \frac{1}{N} \sum_{k=1}^{N} s_k^i \quad (average \ of \ s^i)$$

$$\sigma^i = \sqrt{\frac{\sum_{k=1}^{N}(s_k^i - \bar{s}^i)^2}{N-1}} \quad (standard \ deviation \ of \ s^i)$$

The coefficient $-1 \leq C_{ij}^P \leq 1$ is a real number, which quantifies the degree of linear correlation for the (observed) mutations of the i-th residue with respect to the mutations of the j-th residue of P. If $C_{ij}^P \sim 1$ then the mutations at sites i, j are positively correlated, if $C_{ij}^P \sim -1$ they are negatively correlated. When $C_{ij}^P \sim 0$ mutations are uncorrelated.

Referring to the contact prediction problem, it is generally believed that the values of the correlated mutations provide an indication of the contacts between residues in proteins: the higher is the coefficient C_{ij}^P, the higher is the probability that residues i, j are in contact. Perfectly conserved positions (i.e. positions which correspond to residues conserved in the alignment) and positions containing more than 10% gaps in the alignment are not considered since they are uninformative for this analysis.

3   Method

The values of the correlation coefficient provided by Equation (1) depend on the similarity matrix S used to compute the substitution scores. The aim of our paper is to search in the space of substitution matrices and to find the one which maximizes the correlation between the Pearson coefficient and the residue contacts. This can be done in a very effective way by transforming the problem to find the *optimal* similarity matrix for correlated mutations into a minimization problem.

Considering some similarity matrix S and some *training set* T of proteins, the *error* function with respect to T and S is defined as

$$E = - \sum_{P \in T} \sum_{i \geq 1} \sum_{j > i+6} C_{ij}^P M_{i,j}^P \qquad (2)$$

where M^P is the contact map of protein P (see Section 2.1) and C_{ij}^P is the Pearson correlation coefficient for residues i, j of P computed with similarity matrix S (see Section 2.2). Note that we defined the error function for only residue pairs with sequence separation ≥ 6, since this is the minimum sequence separation for which the contact prediction is meaningful (see Section 2.1). Moreover, note that, the error function E is sound with respect to the correlation between residue contacts and correlated mutation coefficients: if residues i, j are not in contact (i.e. $M^P_{i,j} = -1$) then E increases of C^P_{ij} while if i, j are in contact (i.e. $M^P_{i,j} = 1$) then E increases of $-C^P_{ij}$. The similarity matrix that minimizes E maximizes the correlation between residue contacts and correlated mutation coefficients (for the set T). Starting from an initial similarity matrix $\mathcal{S} = \mathcal{S}^{(0)}$, we can compute a new similarity matrix which minimizes E by using the standard gradient descent method [17]. The similarity matrix $\mathcal{S}^{(k)}$ at step $k > 0$ is computed by

$$\mathcal{S}^{(k)}_{l,m} = \mathcal{S}^{(k-1)}_{l,m} - \lambda \frac{\partial E}{\partial \mathcal{S}^{(k-1)}_{l,m}}$$

where λ is the learning rate of the minimization procedure. The partial derivative of the correlation coefficient C^P_{ij} with respect to $\mathcal{S}_{l,m}$ (i.e. with respect to the substitution $l \to m$) is defined by

$$\frac{\partial C^P_{ij}}{\partial \mathcal{S}_{l,m}} = \frac{1}{N\sigma^i\sigma^j} \left[\sum_{k=1}^{N} \delta(\mathcal{S}_{l,m}, s^i_k) \left(\xi^j_k - \frac{c_{ij}}{(\sigma^i)^2} \xi^i_k \right) + \sum_{k=1}^{N} \delta(\mathcal{S}_{l,m}, s^j_k) \left(\xi^i_k - \frac{c_{ij}}{(\sigma^j)^2} \xi^j_k \right) \right]$$

where

$$c_{ij} = \frac{1}{N} \sum_{k=1}^{N} (s^i_k - \overline{s}^i)(s^j_k - \overline{s}^j) = \frac{1}{N} \sum_{k=1}^{N} s^i_k s^j_k - \overline{s}^i \overline{s}^j$$

$$\xi^i_k = (s^i_k - \overline{s}^i)$$

$$\xi^j_k = (s^j_k - \overline{s}^j)$$

and

$$\delta(\mathcal{S}_{l,m}, s^i_k) = \begin{cases} 1 & \text{if } s^i_k \text{ corresponds to the substitution } l \to m \\ 0 & \text{otherwise} \end{cases}$$

The minimization procedure halts when the *distance* between two consecutive similarity matrices $\mathcal{S}^{(k-1)}, \mathcal{S}^{(k)}$ is sufficiently small. For our test we adopted the following condition:

$$\sum_{i=1}^{20} \sum_{j=i}^{20} \frac{|\mathcal{S}^{(k)}_{i,j} - \mathcal{S}^{(k-1)}_{i,j}|}{210} \leq 10^{-5}.$$

The number of contacts is quite sparse with respect to the number of non-contacts and the minimization procedure will tend to favor non-contacts, leading to a similarity matrix which minimizes the correlation coefficients for almost all pairs of contacts. We overcome the problem in the following way. At every iteration step and for every protein $P \in T$, we compute the error function balancing the number of negative and positive examples. This is done by taking all the available n contacts (with sequence separation ≥ 6) and adding n randomly chosen (distinct) residue pairs not in contact. By this the number of contacts and non-contacts is balanced at every iteration step and randomness assures that most of the non-contact pairs are taken into account during the minimization process.

3.1 Data Sets

We selected from PDB [3] all protein chains whose structures have resolution < 2.5 Å. We excluded from this set all proteins whose structures contain holes in the coordinates and removed sequence redundancy with BLAST [1] in order to obtain a set of protein chains with sequence similarity lower than 25%. For each protein P in this set we performed BLAST search against the non-redundant sequence dataset to compile a multiple sequence alignment with respect to each query protein P. From this set we removed all protein chains for which BLAST returned less than 100 aligned sequences (the correlation coefficients are no reliable when the set of aligned sequences is small [5,13]). We ended up with a set of 900 protein chains with the correspondent multiple sequence alignments. This set has been randomly partitioned in four disjoint sets (to which will refer as set1, set2, set3, set4) of the same cardinality, well balanced with respect to the SCOP structural classification [2] and protein chain lengths. We performed tests in cross validation: when set$i$ is considered as the test set, the training set consists of $\cup_{j \neq i}$set$j$. Thus, for each experiment, we performed four distinct tests.

3.2 Similarity Matrices

In order to obtain meaningful tests, it is useful to consider several different initial similarity matrices. This is necessary since the gradient descent method can be trapped in local minima.

	MCLACHLAN	BLOSUM62	PAM250	RANDOM1	RANDOM2
MCLACHLAN		0.886	0.829	0.015	0.029
BLOSUM62	0.886		0.841	0.025	-0.036
PAM250	0.829	0.841		-0.002	-0.032
RANDOM1	0.015	0.025	-0.002		0.039
RANDOM2	0.029	-0.036	-0.032	0.039	

Fig. 1. Pearson correlation coefficients between similarity matrices. Darker colors correspond to higher correlation.

Table 1. Accuracies (%) of contact prediction (with correlated mutations) for sequence separation ≥ 24 (see Section 2.1). The predictions have been computed by using the substitution matrices in the first column. The accuracies shown are the average accuracies $\pm$ the standard deviation over the accuracies obtained on set1, set2, set3, set4 (see Section 3.1).

Similarity Matrix	L	L/2	L/5	L/10
MCLACHLAN	7.2 ± 0.2	9.1 ± 0.3	12.2 ± 0.5	15.2 ± 1.0
BLOSUM62	7.4 ± 0.2	9.3 ± 0.2	12.3 ± 0.5	15.3 ± 1.2
PAM250	6.7 ± 0.2	8.3 ± 0.2	10.9 ± 0.4	13.6 ± 0.6
RANDOM1	4.4 ± 0.1	5.4 ± 0.2	7.3 ± 0.6	8.7 ± 0.7
RANDOM2	4.9 ± 0.3	6.0 ± 0.3	7.8 ± 0.5	9.2 ± 0.7

We considered five distinct initial similarity matrices for our test: MCLACHLAN[12], BLOSUM62 [7], PAM250 [14] and two symmetric randomly-generated matrices, RANDOM1 and RANDOM2, whose entries are rational numbers in the interval $[-1, 1]$. We can use the Pearson correlation coefficient (Equation (1)) to evaluate the correlation between the substitution matrices. This can be done by defining the vectors s^i, s^j to contain the entries of the similarity matrices $i, j \in$ {MCLACHLAN,BLOSUM62,PAM250,RANDOM1,RANDOM2}). The correlation coefficients are shown in Fig. 1. We can notice that MCLACHLAN, BLOSUM62, PAM250 are highly correlated rather independently of the property/s from which they were derived. As expected, theres no (positive or negative) correlation between the two random matrices (RANDOM1,RANDOM2) and those derived from different properties.

4   Results

The general scheme of our tests is the following. We start with an *initial* similarity matrix S and compute the *optimized* similarity matrix S' by using the minimization procedure described in Section 3. Then we compare the accuracies of prediction obtained by using MCLACHLAN and S'.

For sake of clarity in Table 1 we show the accuracy obtained when predicting contacts with the similarity matrices discussed above (without minimization). For every matrix the average accuracies have been computed separately for each test set (set1, set2, set3, set4) (Table 1). We show only the accuracies obtained for sequence separation ≥ 24, since they are the most valuable with respect to the contact prediction problem (see Section 2.1). The accuracies obtained by using MCLACHLAN and BLOSUM62 are comparable. PAM250 has slightly worst performances. This is in agreement with our previous observation (Fig.1) that such matrices are highly correlated. The random similarity matrices have poor performances. Nevertheless, the accuracies obtained with RANDOMs are still better than those we would obtain with a random prediction of contacts ($\sim 2\%$) [6]. MCLACHLAN in Table 1 represents the *standard method* [5,13] and can be taken as benchmark of the current state of the art evaluated following EVAcon [6].

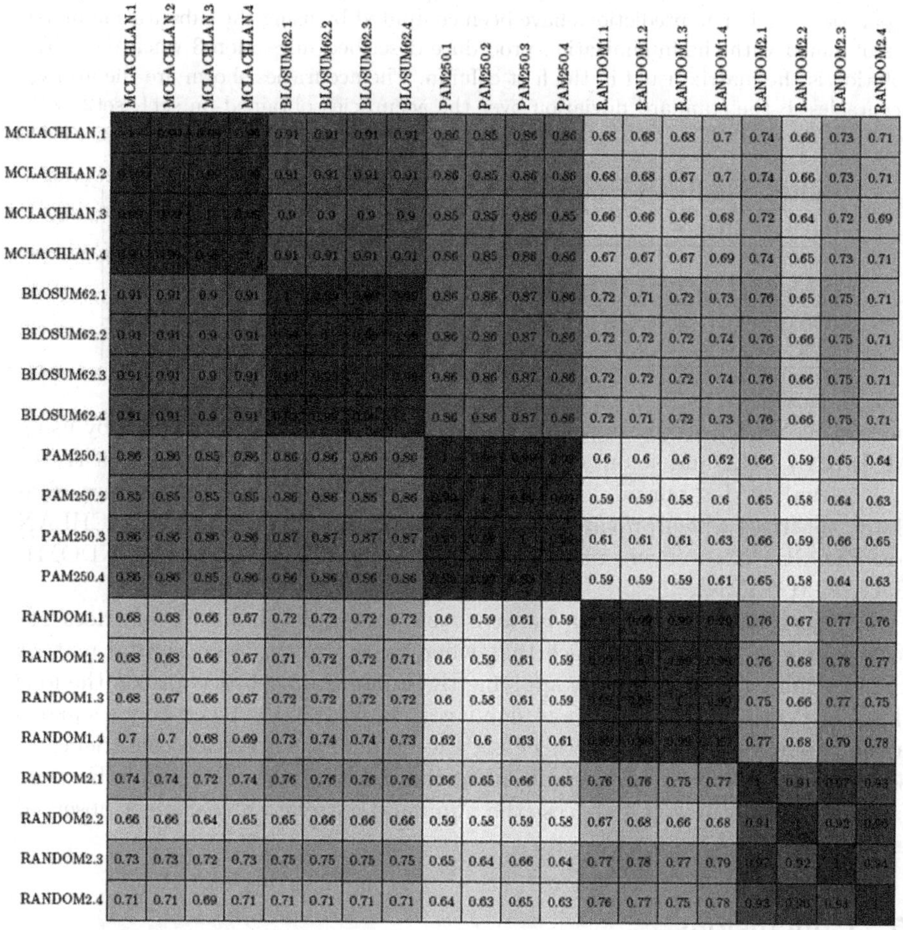

Fig. 2. Pearson correlation coefficients between the similarity matrices computed with the minimizaton procedure of Section 3. Darker colors correspond to higher level of correlation.

In Table 2 we show the accuracies obtained by using the similarity matrices computed with the minimization procedure described in Section 3. The initial similarity matrix for the minimization procedure is listed in the first column of Table 2. As specified in Section 3.2, when the test set is set$i$ (for $i \in \{1,2,3,4\}$), the training set is defined by $\cup_{j \neq i}$set$j$ (for $j \in \{1,2,3,4\}$). Thus for each test we obtained four distinct optimized similarity matrices. We will refer to such matrices in the following way: the optimized matrix obtained by using the MCLACHLAN as initial solution and by training on $\cup_{i \in \{2,3,4\}}$set$i$ is denoted as MCLACHLAN.1, the one obtained by training on $\cup_{i \in \{1,3,4\}}$set$i$ will be denoted as MCLACHLAN.2 and so on. In this way, the contact predictions for set1

Table 2. Accuracies (%) of contact prediction (with correlated mutations) for sequence separation ≥ 24. The predictions have been computed by using the substitution matrices obtained with the minimization procedure described in Section 3 when the initial solution is the matrix listed in the first column. The accuracies shown are the average accuracies $\pm$ the standard deviation over the accuracies obtained on set1, set2, set3, set4 (see Section 3.1).

Initial Similarity Matrix	L	L/2	L/5	L/10
MCLACHLAN	7.4 ± 0.2	9.5 ± 0.3	12.6 ± 0.7	15.4 ± 0.8
BLOSUM62	7.5 ± 0.2	9.5 ± 0.3	12.5 ± 0.6	15.4 ± 0.8
PAM250	7.2 ± 0.1	9.1 ± 0.2	12.2 ± 0.5	15.0 ± 0.8
RANDOM1	7.7 ± 0.2	9.8 ± 0.4	13.0 ± 0.4	15.8 ± 0.8
RANDOM2	7.6 ± 0.2	9.8 ± 0.3	13.0 ± 0.5	15.8 ± 0.6

have been obtained by using MCLACHLAN.1, the predictions for set2 by using MCLACHLAN.2 and so on. The procedure is the same for the other matrices.

It is evident that our minimization procedure produces in all tested cases similarity matrices which provide almost the same performances of MCLACHLAN and BLOSUM62, even when the initial solution is randomly chosen (RANDOM1, RANDOM2) (Table 2).

The correlation coefficients between the computed similarity matrices are shown in Fig. 2. We can notice that when we start with some initial matrix, the four matrices, obtained by applying the minimization procedure on the four different training sets, are always highly correlated (i.e. the black 4 by 4 squares along the main diagonal). In this case, the minimum correlation coefficient observed is in fact ~ 0.91. It is also evident that there is some correlation between every pair of optimized matrices (the minimum correlation coefficient observed is ~ 0.6).

5 Conclusions

Correlated mutations are indicative of residue contacts in proteins. Pearson correlation coefficient or some variants are routinely used to compute correlated mutations from protein sequence alignments. Standard implementations exploit the MCLACHLAN similarity matrix to provide a measure of residue substitutions. So far, the choice of the MCLACHLAN matrix was not justified in literature and there was no evidence of its optimality with respect to the contact prediction problem. In this paper we show that other classical substitution matrices achieve similar performances in contact prediction. Furthermore, we cast the problem of computing the optimal similarity matrix as a minimization problem for the error function (2). Our results show that for each optimized matrix, the contact prediction accuracy achieves a performance comparable with that obtained with MCLACHLAN, indicating that the space of the error function has a huge number of equivalent minima.

We can conclude that the space of similarity matrices for correlated mutation contains a huge number of equivalent matrices, i.e. matrices which provide the same performances in terms of contact prediction accuracy. With our optimization procedure an upper limiting value of about 16% accuracy is found when predicting long range residue contacts (at sequence separation $\geq$ 24). The value is independent of the optimized similarity matrix adopted to compute correlated mutations in proteins and represents the-state-of-the-art performance at sequence separation $\geq$ 24. This finding suggests that a possible improvement in the accuracy value of contact prediction may be obtained by exploiting higher order correlation functions between co-evolving pairs of residues.

References

1. Altschul, S.F., Madden, T.L., Schäffer, A.A., Zhang, J., Zhang, Z., Miller, W., Lipman, D.J.: Gapped BLAST and PSI-BLAST: a new generation of protein database search programs. Nucleic Acids Res. 25(17), 3389–3402 (1997)
2. Andreeva, A., Howorth, D., Brenner, S.E., Hubbard, T.J., Chothia, C., Murzin, A.G.: SCOP database in 2004: refinements integrate structure and sequence family data. Nucleic Acids Res. 32(Database issue), D226–D229 (2004)
3. Berman, H.M., Westbrook, J., Feng, Z., Gilliland, G., Bhat, T.N., Weissig, H., Shindyalov, I.N., Bourne, P.E.: The Protein Data Bank. Nucleic Acids Res. 28(1), 235–242 (2000)
4. Das, R., Baker, D.: Macromolecular modeling with rosetta. Annu. Rev. Biochem. 77, 363–382 (2008)
5. Göbel, U., Sander, C., Schneider, R., Valencia, A.: Correlated mutations and residue contacts in proteins. Proteins 18(4), 309–317 (1994)
6. Graña, O., Eyrich, V.A., Pazos, F., Rost, B., Valencia, A.: EVAcon: a protein contact prediction evaluation service. Nucleic Acids Res. 33(Web Server issue), W347–W351 (2005)
7. Henikoff, S., Henikoff, J.G.: Amino acid substitution matrices from protein blocks. Proc. Natl. Acad. Sci. U S A 89(22), 10915–10919 (1992)
8. Horner, D.S., Pirovano, W., Pesole, G.: Correlated substitution analysis and the prediction of amino acid structural contacts. Brief Bioinform. 9(1), 46–56 (2008); Epub. (November 13, 2007)
9. Izarzugaza, J.M., Graña, O., Tress, M.L., Valencia, A., Clarke, N.D.: Assessment of intramolecular contact predictions for CASP7. Proteins 69(suppl. 8), 152–158 (2007)
10. Hinds, D.A., Levitt, M.: A lattice model for protein structure prediction at low resolution. Proc. Natl. Acad. Sci. U S A 89(7), 2536–2540 (1992)
11. Lesk, A.: Introduction to Bioinformatics. Oxford University Press, Oxford (2006)
12. McLachlan, A.D.: Tests for comparing related amino-acid sequences. Cytochrome c and cytochrome c 551. J. Mol. Biol. 61(2), 409–424 (1971)
13. Olmea, O., Valencia, A.: Improving contact predictions by the combination of correlated mutations and other sources of sequence information. Fold Des. 2(3), S25–S32 (1997)
14. Dayhoff, M.O., Schwartz, R.M., Orcutt, B.C.: A model of evolutionary change in proteins. Atlas of Protein Sequence and Structure 5(3), 345–352 (1978); Dayhoff, M.O. (ed.)

15. Pollock, D.D., Taylor, W.R.: Effectiveness of correlation analysis in identifying protein residues undergoing correlated evolution. Protein Eng. 10(6), 647–657 (1997)
16. Samsonov, S.A., Teyra, J., Anders, G., Pisabarro, M.T.: Analysis of the impact of solvent on contacts prediction in proteins. BMC Struct. Biol. 9(1), 22 (2009)
17. Snyman, J.A.: Practical mathematical optimization: an introduction to basic optimization theory and classical and new gradient-based algorithms. Springer, New York (2005)
18. Vassura, M., Margara, L., Di Lena, P., Medri, F., Fariselli, P., Casadio, R.: Reconstruction of 3D structures from protein contact maps. IEEE/ACM Trans. Comput. Biol. Bioinform. 5(3), 357–367 (2008)

Constructing Majority-Rule Supertrees*

Jianrong Dong[1], David Fernández-Baca[1], and F.R. McMorris[2]

[1] Department of Computer Science, Iowa State University, Ames, IA 50011, USA
[2] Department of Applied Mathematics, Illinois Institute of Technology, Chicago, IL 60616, USA

Abstract. Supertree methods combine the phylogenetic information from multiple partially-overlapping trees into a larger phylogenetic tree called a supertree. Several supertree construction methods have been proposed to date, but most of these are not designed with any specific properties in mind. Recently, Cotton and Wilkinson proposed extensions of the majority-rule consensus tree method to the supertree setting that inherit many of the appealing properties of the former. Here we study a variant of one of their methods, called majority-rule (+) supertrees. After proving that a key underlying problem for constructing majority-rule (+) supertrees is NP-hard, we develop a polynomial-size integer linear programming formulation of the problem. We then report on a preliminary computational study of our approach. The results indicate that our method is computationally feasible for moderately large inputs. Perhaps more significantly, our results suggest that the majority-rule (+) approach produces biologically meaningful results.

1 Introduction

A supertree method begins with a collection of phylogenetic trees with possibly different leaf (taxon) sets, and assembles them into a larger phylogenetic tree, a *supertree*, whose taxon set is the union of the taxon sets of the input trees. Interest in supertrees was sparked by Gordon's paper [18]. Since then, particularly during the past decade, there has been a flurry of activity with many supertree methods proposed and studied from the points-of-view of algorithm design, theory, and biology. The appeal of supertree synthesis is that it can provide a high-level perspective that is harder to attain from individual trees. A recent example of the use of this approach is the species-level phylogeny of nearly all extant Mammalia constructed by Bininda-Emonds [6] from over 2,500 partial estimates. Several of the known supertree methods are reviewed in the book edited by Bininda-Emonds [5] — more recent papers with good bibliographies include [32,26]. There is still much debate about what specific properties should (can), or should not (can not), be satisfied by supertree methods. Indeed, it is often a challenging problem to rigorously determine the properties of a supertree method that gives seemingly good results on data, but is heuristic.

The well-studied consensus tree problem can be viewed as the special case of the supertree problem where the input trees have identical leaf sets. Consensus methods

---

* This work was supported in part by the National Science Foundation under grants DEB-0334832 and DEB-0829674.

S.L. Salzberg and T. Warnow (Eds.): WABI 2009, LNBI 5724, pp. 73–84, 2009.
© Springer-Verlag Berlin Heidelberg 2009

in systematics date back to [1]; since then, many consensus methods have been designed. A good survey of these methods, their properties, and their interrelationships is given by Bryant [8], while the axiomatic approach and the motivation from the social sciences is found in Day and McMorris' book [12]. One of the most widely used methods is the majority-rule consensus tree [3,22], which is the tree that contains the splits displayed by the majority of the input trees. Not only does this tree always exist, but it is also unique, can be efficiently constructed [2], and has the property of being a *median tree* relative to the symmetric-difference distance (also known as the Robinson-Foulds distance [23,27]). That is, the majority-rule consensus tree is a tree whose total Robinson-Foulds distance to the input trees is minimum.

The appealing qualities of the majority-rule consensus method have made it attractive to try to extend the method to the supertree setting, while retaining as many of its good characteristics as possible. Cotton and Wilkinson [10] were able to define two such extensions (despite some doubts about whether such an extension was possible [17]) and at least two additional ones have been studied since [14]. Here we study one of the latter variants, called *graft-refine majority-rule (+) supertrees* in [14], and here simply referred to as *majority-rule (+) supertrees*. These supertrees satisfy certain desirable properties with respect to what information from the input trees, in the form of splits, is displayed by them (see Section 2). The key idea in this method is to expand the input trees by grafting leaves onto them to produce trees over the same leaf set. The expansion is done so as to minimize the distance from the expanded trees to their median relative to the Robinson-Foulds distance. The supertree returned is the strict consensus of the median trees with minimum distance to the expanded input trees; these median trees are called *optimal candidate supertrees*.

After showing that computing an optimal candidate supertree is NP-hard, we develop a characterization of these supertrees that allows us to formulate the problem as a polynomial-size integer linear program (ILP). We then describe an implementation that enables us to solve moderately large problems exactly. We show that, in practice, the majority-rule (+) supertree can be constructed quickly once an optimal candidate supertree has been identified. Furthermore, we observe that the supertrees produced are similar to biologically reasonable trees, adding further justification to the majority-rule (+) approach.

We should mention that the supertree method most commonly used in practice is matrix representation with parsimony (MRP) [4,25]. MRP first encodes the input trees as incomplete binary characters, and then builds a maximum-parsimony tree for this data. The popularity of MRP is perhaps due to the widespread acceptance of the philosophy underlying parsimony approaches and the availability of excellent parsimony software (e.g., [31]). However, while parsimony is relatively easy to justify in the original tree-building problem (in which homoplasy represents additional assumptions of evolutionary changes) a justification for its use as a supertree construction method is not quite as obvious. Perhaps the main criticism of MRP, as well as other tree construction methods, is that it can produce unsupported groups [16,24]. The provable properties of majority-rule (+) supertrees [10,14] prevent such anomalies.

There has been previous work on ILP in phylogenetics, much of it dealing with parsimony or its relative, compatibility [7,19,20,21,29]. Our work uses some of these ideas (especially those of [21]), but the context and the objective function are quite different. In particular, the need to handle all possible expansions of the input trees necessitates the introduction of new techniques.

2 Preliminaries

2.1 Basic Definitions and Notation

Our terminology largely follows [28]. A *phylogenetic tree* is an unrooted leaf-labeled tree where every internal node has degree at least three. We will use "tree" and "phylogenetic tree" interchangeably. The leaf set of a tree T is denoted by $\mathcal{L}(T)$.

A *profile* is a tuple of trees $P = (t_1, \ldots, t_k)$. Each t_i in P is called an *input tree*. Let $\mathcal{L}(P) = \bigcup_{i \in K} \mathcal{L}(t_i)$, where K denotes the set $\{1, \ldots, k\}$. An input tree t_i is *plenary* if $\mathcal{L}(t_i) = \mathcal{L}(P)$. Tree T is a *supertree* for profile P if $\mathcal{L}(T) = \mathcal{L}(P)$.

A *split* is a bipartition of a set. We write $A|B$ to denote the split whose parts are A and B. The order here does not matter, so $A|B$ is the same as $B|A$. Split $A|B$ is *nontrivial* if each of A and B has at least two elements. Split $A|B$ *extends* another split $C|D$ if $A \supseteq C$ and $B \supseteq D$, or $A \supseteq D$ and $B \supseteq C$.

Phylogenetic tree T *displays* split $A|B$ if there is an edge in T whose removal gives trees T_1 and T_2 such that $A \subseteq \mathcal{L}(T_1)$ and $B \subseteq \mathcal{L}(T_2)$. A split $A|B$ is *full* with respect to a tree T if $A \cup B = \mathcal{L}(T)$; $A|B$ is *partial* with respect to T if $A \cup B \subset \mathcal{L}(T)$. Split $A|B$ is *plenary* with respect to a profile P if $A \cup B = \mathcal{L}(P)$.

The set of all nontrivial full splits of T is denoted $Spl(T)$. It is well known that the full splits of T uniquely identify T [28, p. 44]. Let $S \subseteq \mathcal{L}(T)$. The *restriction of T to S*, denoted $T|S$, is the phylogenetic tree with leaf set S such that

$$Spl(T|S) = \{A \cap S | B \cap S : A|B \in Spl(T) \text{ and } |A \cap S|, |B \cap S| > 1\}.$$

Let T' be a phylogenetic tree such that $S = \mathcal{L}(T') \subseteq \mathcal{L}(T)$. Then, T *displays* T' if $Spl(T') \subseteq Spl(T|S)$.

A set of splits is *compatible* if there is a tree T that displays them all. Tree T is *compatible with* a set of splits $\mathcal{X}$ if there is a tree T' that displays T and $\mathcal{X}$.

Let T_1 and T_2 be two phylogenetic trees over the same leaf set. The *symmetric-difference distance*, also known as *Robinson-Foulds distance* [27], between T_1 and T_2, denoted $d(T_1, T_2)$, is defined as

$$d(T_1, T_2) = |(Spl(T_1) \setminus Spl(T_2)) \cup (Spl(T_2) \setminus Spl(T_1))|. \tag{1}$$

The *majority splits* in a profile $P = (t_1, \ldots, t_k)$ are the splits displayed by more than $\frac{k}{2}$ of the input trees. A *majority plenary split* is a plenary split that is also a majority split. Similarly, a *majority partial split* is a partial split that is also a majority split.

Rooted phylogenetic trees can be viewed as a special case of unrooted trees. That is, we can view a profile of rooted trees as unrooted trees, all of which have a common taxon called the root. Thus, in a split in a rooted tree, one of the two parts must contain the root; the part that does not contain the root is called a *cluster* (or *clade*, or *mono-phyletic group*). All of the above concepts (e,g., compatibility and distance), as well as those introduced in the rest of this paper, directly apply to rooted trees. However, we shall not elaborate on this here.

To close this section, we examine the *consensus problem*, the special case of the supertree problem where the profile $P = (T_1, \ldots, T_k)$ consists of trees that have the same leaf set. The *strict consensus* of P is the tree that displays exactly the plenary splits present in every tree in the profile. The *majority-rule consensus tree* of P, denoted $Maj(P)$, is the tree that displays all the majority plenary splits in P [22]. For any phylogeny T with $\mathcal{L}(T) = \mathcal{L}(P)$, define the *distance* from T to P as $dist(T, P) = \sum_{i \in K} d(T, T_i)$, where d denotes the symmetric-difference distance. Any T with leaf set $\mathcal{L}(P)$ that minimizes $dist(T, P)$ is called a *median tree* for P. It is known that $Maj(P)$ is a median tree for P; indeed, it follows from [3] that $Maj(P)$ is the strict consensus of the median trees for P. The (median) *score* of P is defined as $s(P) = \min_{T:\mathcal{L}(T)=\mathcal{L}(P)} dist(T, P)$. Thus, $s(P) = dist(Maj(P), P)$.

2.2   Majority-Rule (+) Supertrees

Here we describe a variant (suggested by Bill Day) of one Cotton and Wilkinson's [10] extensions of majority-rule consensus to the supertree setting.

The *span* of an input tree t, denoted by $\langle t \rangle$, is the set of all trees on $\mathcal{L}(P)$ that display t. The *span* of a profile $P = (t_1, \ldots, t_k)$, denoted $\langle P \rangle$, is the set of all k-tuples $R = (T_1, \ldots, T_k)$, where $T_i \in \langle t_i \rangle$ for each $i \in K$. Each $R \in \langle P \rangle$ is called a *representative selection* for P and $Maj(R)$ is called a *candidate supertree*. An *optimal representative selection* is a representative selection R with minimum score $s(R)$ over all $R \in \langle P \rangle$. We refer to $Maj(R)$ as the *optimal candidate supertree* associated with R. The *majority-rule (+) supertree* of profile P, denoted by $Maj^+(P)$, is the strict consensus of all the optimal candidate supertrees.

We have shown elsewhere [14] that $Maj^+(P)$ satisfies the following appealing properties (originally conjectured by Cotton and Wilkinson).

(CW1) $Maj^+(P)$ displays all of the majority plenary splits in P.
(CW2) $Maj^+(P)$ is compatible with each majority partial split in P.
(CW3) Each split in $Maj^+(P)$ is compatible with a majority of the trees in P.
(CW4) Every plenary split in $Maj^+(P)$ extends at least one input tree full split.

We should note that majority-rule (+) supertrees, as defined above, do not generalize majority-rule consensus. That is, when used in the consensus setting, $Maj^+(P)$ is not, in general, the same as $Maj(P)$. Nevertheless, we have shown [15] that majority-rule (+) consensus trees have a simple characterization that yields an efficient algorithm for computing them (see Theorem 1 of Section 3).

The majority-rule (+) supertrees we study differ from other variants in the way the span of an input tree is defined. Cotton and Wilkinson originally defined the span of a tree t as the set of all plenary *binary* trees that display t [10]. This version does not generalize majority-rule consensus and does not satisfy (CW4) [14]. In a more recent version, suggested by Wilkinson (personal communication), the span of t is the set of all plenary trees T such that $T|\mathcal{L}(t) = t$. This definition of span prohibits refinement of any original polytomies (nodes of degree at least four) in t. It can be shown that the supertree method that results from using this definition generalizes majority-rule consensus, and that it satisfies properties (CW1)–(CW4) [14]. Nonetheless, we have preferred

Day's variant for two reasons. First, we have found it computationally easier to deal with than the others. More importantly, there are reasons why a strict generalization of majority-rule consensus might not be the ideal approach for supertree construction: In practice, one often encounters profiles where different trees "specialize" in different groups of taxa, leaving other groups largely unresolved. The goal is that, jointly, the trees will produce a well-resolved supertree, with each input tree contributing its own specialized information. A strict generalization of majority rule would disallow this, since the method discards minority information. The majority-rule (+) supertrees presented here preserve this fine-grained information, unless it were substantially contradicted by the remaining trees (the sense in which this is true can be gleaned from Theorem 1 of Section 3).

3 Constructing Optimal Candidate Trees

We first consider the consensus version of the problem. Let $P = (T_1, \ldots, T_k)$ be a profile of trees over the same leaf set. Given a plenary split $X = A|B$, define

$$K_X(P) = \{i \in K : X \text{ is displayed by } T_i\}$$

and

$$K_{\overline{X}}(P) = \{i \in K : X \text{ is not compatible with } T_i\}.$$

The theorem below, proved elsewhere [15], characterizes the majority-rule (+) consensus tree of a profile and implies that this tree can be computed in polynomial time.

Theorem 1. *For any profile P, $Maj^+(P)$ is precisely the tree that displays every split X such that $|K_X(P)| > |K_{\overline{X}}(P)|$. Furthermore, $Maj^+(P)$ is an optimal candidate tree for P, as well as the strict consensus of all optimal candidate trees for P.*

On the other hand, the next result suggests that finding the majority-rule (+) supertree for a profile of trees with partially overlapping leaf sets may be hard.

Theorem 2. *There is no polynomial-time algorithm to construct an optimal candidate supertree unless P = NP.*

Proof. We show that if there is a polynomial time algorithm to compute an optimal candidate tree, then there exists a polynomial-time algorithm for the quartet compatibility problem, which is known to be NP-complete [30]. The quartet compatibility problem asks whether, given a collection Q of trees on four leaves, there exists a single tree that displays them all. If the answer is "yes", we say that Q is *compatible*.

Let Q be an instance of quartet compatibility. Construct a profile P that simply consists of the trees in Q in some arbitrary order. We claim that Q is compatible if and only if P has an optimal candidate tree with a score of zero. Suppose first that Q is compatible and that T is any tree that displays each element of Q. Then, for each tree t in P, $T \in \langle t \rangle$, because all the splits in T must be compatible with t, so any split in T that is not in t can be added to t. Hence, T is a candidate tree for P with a score of zero, and thus T is also an optimal candidate tree. Conversely, if P has a candidate tree with zero score, it can be seen that T displays all the quartets in Q; i.e., Q is compatible. $\square$

In the next sections, we show that despite the above result, moderately large majority-rule (+) supertree problems can be solved using integer linear programming. For this, we need to address a potential complication: Since the definition of $\langle t \rangle$ allows refinement of multifurcations in t, a tree $T \in \langle t \rangle$ can contain many more nontrivial splits than t; indeed, we cannot predetermine the number of additional such splits T will contain. We circumvent this potential problem by defining a restricted version of the span.

Given an input tree t in a profile P, the *restricted span* of t, denoted $\langle t \rangle_r$ is the set of all plenary trees T such that every nontrivial split in T extends a distinct nontrivial split in t. Thus, $|Spl(T)| = |Spl(t)|$. Note that T is obtained from t by filling it in each of t's splits, by adding zero or more taxa to each part, to make them plenary splits in such a way that the resulting splits are compatible. Note also that $\langle t \rangle_r \subseteq \langle t \rangle$. The *restricted span* of a profile $P = (t_1, \ldots, t_k)$, denoted $\langle P \rangle_r$ is the set of all $R = (T_1, \ldots, T_k)$ for P such that $T_i \in \langle t_i \rangle_r$ for each $i \in K$; each $R \in \langle P \rangle_r$ is called a *restricted representative selection* for P. Since $\langle P \rangle_r \subseteq \langle P \rangle$, the restricted span represents an intermediate level between the input profile and the original definition of span. The restricted span is more manageable than the original span because it does not allow any refinement of input trees. In the rest of this section, we will show how to obtain majority-rule (+) supertrees directly from the restricted span.

Before presenting the first of the two main results of this section, we need to introduce some new concepts. An optimal candidate tree T for a profile P is *minimal* if contracting any edge in T yields a tree that is not an optimal candidate tree. Let $R = (T_1, \ldots, T_k)$ and $R' = (T_1', \ldots, T_k')$ be two representative selections for a profile P. We say that R *displays* R' if T_i displays T_i' for every $i \in K$. Theorem 1 motivates the next definition. The *completion* of a representative selection $R = (T_1, \ldots, T_k)$ for a profile P is the representative selection $\widehat{R} = (\widehat{T}_1, \ldots, \widehat{T}_k)$ obtained as follows: For every $i \in K$, $\widehat{T}_i$ is the tree constructed by inserting into T_i each plenary split $X = A|B$ compatible with T_i such that $|K_X(R)| > |K_{\overline{X}}(R)|$.

Theorem 3. *Let T be a minimal optimal candidate tree for a profile P and let $R \in \langle P \rangle$ be such that $T = Maj(R)$. Consider any $G \in \langle P \rangle_r$ such that G is displayed by R. Then, R is the completion of G and $T = Maj^+(G)$.*

Proof. We begin by proving that T is an optimal candidate tree for G. Assume the contrary. Then, there exists another candidate tree T' for G such that (i) $T' = Maj(R')$ for some $R' \in \langle G \rangle$ and (ii) $s(R') < s(R)$. But then, since $\langle G \rangle \subseteq \langle P \rangle$, we have $R' \in \langle P \rangle$, and thus (ii) contradicts the optimality of T for P.

Next, we argue that T is a *minimal* optimal candidate supertree for profile G. Suppose this is not true. Then, T displays an optimal candidate tree T' for G such that $T \neq T'$. Consider any $R' \in \langle G \rangle$ such that $T' = Maj(R')$. Since T and T' are both optimal for G, $s(R) = s(R')$. Since R' displays P, we have $R' \in \langle P \rangle$. Hence, T' is also an optimal candidate tree for P. This, however, contradicts the assumption that T is a minimal candidate tree for P.

By Theorem 1, $Maj^+(G)$ is an optimal candidate tree for G, as well as the strict consensus of all optimal candidate trees for G. Therefore, $Maj^+(G)$ is the only minimal candidate tree for G. Hence $T = Maj^+(G)$.

Suppose $R = (T_1, \ldots, T_k)$ and let $\widehat{R} = (\widehat{T}_1, \ldots, \widehat{T}_k)$ be the completion of G. We claim that $\widehat{R} = R$. Assume, on the contrary, that there is some $i \in K$ such that $\widehat{T}_i \neq T_i$. That is, $\mathcal{X} \cup \mathcal{Y} \neq \emptyset$, where $\mathcal{X} = Spl(\widehat{T}_i) \setminus Spl(T_i)$ and $\mathcal{Y} = Spl(T_i) \setminus Spl(\widehat{T}_i)$. Set $\mathcal{X}$ consists of splits X such that $|K_X(G)| > |K_{\overline{X}}(G)|$ and $\mathcal{Y}$ consists of splits Y such that $|K_Y(G)| \leq |K_{\overline{Y}}(G)|$. By Theorem 1, $T = Maj^+(G)$ displays all splits X such that $|K_X(G)| > |K_{\overline{X}}(G)|$. Thus, $d(T, \widehat{T}_i) < d(T, T_i)$. As we are assuming that there is at least one such $i \in K$, we have $\sum_{i \in K} d(T, \widehat{T}_i) < \sum_{i \in K} d(T, T_i)$, contradicting the fact that T is a minimal optimal candidate tree for G. $\qquad\square$

Motivated by Theorem 3, we define the *adjusted score* of a representative selection R for a profile P, denoted $\widehat{s}(R)$, to be the score of the completion R' of R; i.e., $\widehat{s}(R) = s(R')$. Recall that $s(R') = dist(Maj(R'), R')$.

Theorem 4. *Let P be a profile. Define $\mathcal{G} = \{G \in \langle P \rangle_r : \widehat{s}(G) \text{ is minimum}\}$ and $\mathcal{S} = \{T = Maj^+(G) : G \in \mathcal{G}\}$. Then, $Maj^+(P)$ is the strict consensus of $\mathcal{S}$.*

Proof. Let $\mathcal{O}$ be the set of all optimal candidate trees for P and let $\mathcal{M}$ be the set of all minimal optimal candidate trees of P. In what follows, we show that $\mathcal{M} \subseteq \mathcal{S} \subseteq \mathcal{O}$. This immediately implies the theorem, because not only is (by definition) $Maj^+(P)$ the strict consensus of $\mathcal{O}$, but it must also be the strict consensus of $\mathcal{M}$.

Suppose $T \in \mathcal{M}$. We claim that $T \in \mathcal{S}$ and, therefore, that $\mathcal{M} \subseteq \mathcal{S}$. Let R be a representative selection for P such that $T = Maj(R)$. Let G be any restricted representative selection for P displayed by R. By Theorem 3, $T = Maj^+(G)$ and R is the completion of G. We claim that $G \in \mathcal{G}$; i.e., $\widehat{s}(G)$ is minimum. Assume, by way of contradiction, that there is another $G' \in \langle P \rangle_r$ such that $\widehat{s}(G') < \widehat{s}(G)$. Let R' be the completion of G'. Then, $s(R') = \widehat{s}(G') < \widehat{s}(G) = s(R)$, which contradicts the assumption that T is optimal. Therefore, $\widehat{s}(G)$ is minimum and $T \in \mathcal{S}$.

Suppose $T \in \mathcal{S}$. We claim that $T \in \mathcal{O}$ and, therefore, that $\mathcal{S} \subseteq \mathcal{O}$. Let $G \in \langle P \rangle_r$ be such that $T = Maj^+(G)$ and the adjusted score $\widehat{s}(G)$ is minimum. Let R be the completion of G. Assume, by way of contradiction, that $T \notin \mathcal{O}$. Then there is a $T' \in \mathcal{M}$ such that, if R' is a representative selection for P where $T' = Maj(R')$, then $s(R') < s(R)$. By Theorem 3, there is a $G' \in \langle P \rangle_r$ such that $T' = Maj^+(G')$ and $\widehat{s}(G') = s(R')$. Then $\widehat{s}(G') = s(R') < s(R) = \widehat{s}(G)$. This contradicts the assumption that $\widehat{s}(G)$ is minimum. $\qquad\square$

4 ILP Formulation

In this section we first describe an ILP formulation of the optimal candidate supertree problem based on Theorem 4. The optimum solution to this ILP is a $G \in \langle P \rangle_r$ with minimum adjusted score. For ease of exposition, we divide the variables of our ILP into three categories: *fill-in variables*, which represent the way taxa are added to the input trees to create G; *objective function variables*, which are used to express $\widehat{s}(G)$; and *auxiliary variables*, which are used to establish a connection between the fill-in and objective function variables. All variables are binary. After presenting our ILP model, we discuss how to use it to generate $Maj^+(P)$.

Fill-in variables. At the core of our ILP formulation is a matrix representation of the input trees similar to that used in MRP [4,25]. Let $P = (t_1, \ldots, t_k)$ be a profile where $|\mathcal{L}(P)| = n$. Assume input tree t_i has m_i nontrivial splits, which are assumed to be ordered in some fixed but arbitrary way. A *matrix representation* of t_i is a $n \times m_i$ matrix $M(t_i)$ whose columns are in one to one correspondence with the nontrivial splits of t_i. Suppose column j of $M(t_i)$ corresponds to split $A|B$ in t_i and let x be a taxon in $\mathcal{L}(P)$. Then, $M_{x,j}(t_i) = 1$ if $x \in A$, $M_{x,j}(t_i) = 0$ if $x \in B$, and $M_{x,j}(t_i) = ?$ otherwise[1]. Let $m = \sum_{i \in K} m_i$. A *matrix representation* of P, denoted $M(P)$, is a $n \times m$ matrix $M(P)$ obtained by concatenating matrices $M(t_1), M(t_2), \ldots, M(t_k)$.

A *fill-in* of the character matrix $M(P)$ is a matrix representation for a restricted representative selection G for P. Note that $M(G)$ has no question marks and that if $M_{xj}(P) \in \{0, 1\}$, then $M_{xj}(G) = M_{xj}(P)$. To represent fill-ins, the ILP associates a variable with each question-mark in $M(P)$. Setting one of these variables to 0 or 1 represents an assignment of the corresponding taxon to one of the two sides of a split.

Objective function variables. The objective is to minimize $\widehat{s}(G)$ over all $G \in \langle P \rangle_r$, where each G is represented by a fill-in of $M(P)$. By definition, $\widehat{s}(G) = dist(Maj^+(G), R)$, where $R = (\widehat{T}_1, \ldots, \widehat{T}_k)$ is the completion of $G = (T_1, \ldots, T_k)$. We do not, however, construct $Maj^+(G)$ and R explicitly. Instead, we proceed indirectly, using the fact that, by Theorems 1 and 3, all splits in $Maj^+(G)$ and R are already in G. Indeed, those theorems and the definition of Robinson-Foulds distance (Equation 1) imply that

$$dist(Maj^+(G), R) = \sum_{i \in K} |Spl(Maj^+(G)) \setminus Spl(\widehat{T}_i)| + \sum_{i \in K} |Spl(\widehat{T}_i) \setminus Spl(Maj^+(G))|.$$
(2)

The next result, which follows Theorems 1 and 3, allows us to count directly from G the contribution of each split $X \in Spl(Maj^+(G)) \cup Spl(\widehat{T}_i)$ to $d(Maj^+(G), \widehat{T}_i)$.

Lemma 1. *Let P be a profile and suppose $G \in \langle P \rangle_r$. Then, for each $i \in K$,*

(i) $X \in Spl(Maj^+(G)) \setminus Spl(\widehat{T}_i)$ if and only if $|K_X(G)| > |K_{\overline{X}}(G)|$ and $i \in K_{\overline{X}}(G)$.
(ii) $X \in Spl(\widehat{T}_i) \setminus Spl(Maj^+(G))$ if and only if $|K_X(G)| \leq |K_{\overline{X}}(G)|$ and $i \in K_X(G)$.

Suppose we have a fill-in for $M(P)$ that corresponds to some $G = (T_1, \ldots, T_k) \in \langle P \rangle_r$. Our ILP has two kinds of objective function variables. The first group of variables are denoted $w_1, \ldots, w_m$, where w_j corresponds to the jth column of $M(G)$. Suppose this column corresponds to split X in tree T_i; thus, $i \in K_X(G)$. Our ILP has constraints such that $w_j = 1$ if and only if $|K_X(G)| > |K_{\overline{X}}(G)|$. Thus, $w_j = 0$ means that $|K_X(G)| \leq |K_{\overline{X}}(G)|$, which, together with Lemma 1 (ii), implies that $\sum_{j=1}^{m} (1 - w_j) = \sum_{i \in K} |Spl(\widehat{T}_i) \setminus Spl(Maj^+(G))|$.

The second group of variables are denoted z_{ij}, $1 \leq i \leq k$, $1 \leq j \leq m$. Suppose column j of $M(P)$ corresponds to split X. Our ILP has constraints such that $z_{ij} = 1$ if and only if $w_j = 1$ (i.e., $|K_X(G)| > |K_{\overline{X}}(G)|$), $i \in K_{\overline{X}}(G)$, and $i = \min\{\ell : \ell \in K_{\overline{X}}(G)\}$. Thus, by Lemma 1 (i), $\sum_{i=1}^{k} \sum_{j=1}^{m} z_{ij} = \sum_{i \in K} |Spl(Maj^+(G)) \setminus Spl(\widehat{T}_i)|$.

---

[1] The assignment of 1 to the A side of the split and of 0 to the B side is arbitrary.

The objective function can now be expressed as

$$\text{minimize} \quad \sum_{i=1}^{k}\sum_{j=1}^{m} z_{ij} + \sum_{i=1}^{m}(1 - w_i).$$

Auxiliary variables and constraints. We now list the auxiliary variables and the constraints that connect them with each other, the fill-in variables, and the objective function variables. For lack of space, we do not write the constraints explicitly; we note, however, that several of them are similar to those used by Gusfield et al. [21].

The settings of the fill-in variables must be such that, for each input tree t_i, the resulting plenary splits associated with the tree are pairwise compatible, so that they yield a plenary tree $T_i \in \langle t_i \rangle_r$. We define variables C_{pq}, $1 \le p, q \le m$ and add constraints such that $C_{pq} = 1$ if and only if columns p and q are compatible. The constraints use the fact that splits p and q are incompatible if and only if $00, 01, 10$, and 11 all appear in some rows of columns p and q. The presence or absence of these patterns for columns p and q are indicated by the settings of variables $B_{pq}^{(ab)}$, $a, b \in \{0, 1\}$, where $B_{pq}^{(ab)} = 1$ if and only if ab appears in some row of columns p and q. The $B_{pq}^{(ab)}$'s are determined from the settings of variables $\Gamma_{rpq}^{(ab)}$, where r ranges over the rows of the filled-in matrix M. $\Gamma_{rpq}^{(ab)} = 1$ if and only if $M_{rp} = a$ and $M_{rq} = b$. There are variables E_{pq}, $1 \le p, q \le m$, where $E_{pq} = 1$ if and only if columns p and q of the filled-in matrix represent the same split. The E variables derive from the B and Γ variables. Variable D_p, $1 \le p \le m$, equals 1 if and only if column p represents the same split as some column with smaller index. The D variables are determined from the E variables.

The ILP has variables $S_{ij}^{(1)}$, $1 \le i \le m$, $1 \le j \le k$, such that $S_{ij}^{(1)} = 1$ if and only if split i is in tree j. The $S^{(1)}$ variables are determined by the E variables. There are also variables $S_{ij}^{(2)}$, where $S_{ij}^{(2)} = 1$ if and only if split i is compatible with tree j. The $S^{(2)}$ variables are determined from the C variables. The values of the w and the z variables are determined, respectively from the $S^{(1)}$ and $S^{(2)}$ variables, and from the w, $S^{(2)}$, and D variables. The resulting ILP has $O(nm^2)$ variables and $O(nm^2)$ constraints.

Building Maj$^+$(P). The ILP model just outlined allows us to find a $G \in \langle P \rangle_r$ corresponding to some optimal candidate tree T^*. To build $Maj^+(P)$ we need, in principle, the set of all such G. While there are ways to enumerate this set [11], we have found that an alternative approach works much better in practice. The key observation is that, since $Maj^+(P)$ is the strict consensus of all optimal candidate trees, each split in $Maj^+(P)$ must also be in T^*. Thus, once we have T^*, we simply need to verify which splits in T^* are in $Maj^+(P)$ and which are not. To do this, for each split $A|B$ in T^*, we put additional constraints on the original ILP requiring that the optimal tree achieve an objective value equal or smaller than that of T^* and not display split $A|B$. The resulting ILP has only $O(mn)$ more variables and constraints than the original one. If the new ILP is feasible, then $A|B \notin Spl(Maj^+(P))$; otherwise, $A|B \in Spl(Maj^+(P))$. We have found that detecting infeasibility is generally much faster than finding an optimal solution.

5   Computational Experiments

Here we report on preliminary computational tests of our ILP method on five published data sets. The *Drosophila A* data set is the example studied in [10], which was extracted from a larger Drosophila data set considered by Cotton and Page [9]. *Primates* is the smaller of the data sets from [26]. *Drosophila B* is a larger subset of the data studied in [9] than that considered in [10]. *Chordata A and B* are two extracts from a data set used in a widely-cited study by Delsuc et al. [13]. Chordata A consists of the first 6 trees with at least 35 taxa (out of 38). Chordata B consists of the first 12 trees with at least 37 taxa (out of 38). We used empirical rather than simulated data, because we felt that this would give us a more accurate picture of the effectiveness of our method. In particular, we were interested in seeing if the groupings of taxa generated by majority-rule (+) supertrees would coincide with those commonly accepted by biologists. We wrote a program to generate the ILPs from the input profiles. The ILPs were then solved using CPLEX[2] on an Intel Core 2 64 bit quad-core processor (2.83GHz) with 8 GB of main memory and a 12 MB L2 cache per processor.

The results are summarized in Table 1. Here n, m, and k are the number of taxa, total number of splits, and number of trees, respectively. U denotes the number of question marks in $M(P)$, the matrix representation of the input; N is the size of the CPLEX-generated reduced ILP. Table 1 shows the time to solve the ILP and produce an optimal candidate supertree T^* and the time to verify the splits of T^* to produce $Maj^+(P)$.

Table 1. Summary of experimental results

Data set	n	m	k	U	%U	N	Sol. (sec)	Verif. (sec)
Drosphila A	9	17	5	60	39.2	9.8 e5	0.83	1.6
Primates	33	51	3	630	37.4	9.9 e7	5.24	3.2
Drosophila B	40	55	4	1133	51.5	1.25 e9	362	19
Chordata A	38	290	6	330	3	1.40 e8	120	258
Chordata B	38	411	12	306	2	1.05 e8	986	1784

Comparison with published results. For Drosophila A we obtained exactly the same tree reported in [10]. For Primates, the output is exactly the same as [26], which was produced by PhySIC method. The coincidence with PhySIC is noteworthy, since this supertree is less controversial than the MRP, Mincut, and PhySIC$_{PC}$ supertrees reported in [26]. The reason for the coincidence may lie in the fact that, while heuristic, PhySIC requires that all topological information contained in the supertree be present in an input tree or collectively implied by the input trees, which bears some similarity with properties (CW1)–(CW4) of majority (+) supertrees.

For Drosphila B, Cotton and Page [9] show four supertrees: strict consensus of gene tree parsimony (GTP), Adams consensus of GTP, strict consensus of MRP, Adams consensus of MRP. Among the 10 clusters (splits) found by our ILP, two are in all four of these supertrees, three are found in the Adams consensus of GTP and Adams consensus

---

[2] CPLEX is a trademark of ILOG, Inc.

of MRP, one is in the strict and Adams consensus of GTP, and one is found in the strict and Adams consensus of MRP. Thus, with only four input trees we were able to generate a tree that is quite similar to the published results.

For Chordata A, the 12 splits found matched published results [13] exactly. For Chordata B, the 14 splits found matched [13].

6 Discussion

We cannot, of course, claim that our ILP method will solve problems of the larger scale found in some studies such as [6]. Indeed, the solution time increases rapidly as the model size grows, although it appears that model size is not the only factor that makes instances hard. Interestingly, it seems that a problem is either solvable quickly (from seconds to at most 20 minutes), or it takes impracticably long. Nevertheless, our results suggest that the majority-rule (+) method produces biologically reasonable results (i.e., no unsupported groups), and that the method is practical for moderately large problems. For larger data sets, a combination of the ILP approach, to obtain a "backbone" phylogeny, along with heuristic approaches may be a more feasible technique.

References

1. Adams III, E.N.: Consensus techniques and the comparison of taxonomic trees. Syst. Zool. 21(4), 390–397 (1972)
2. Amenta, N., Clarke, F., St. John, K.: A linear-time majority tree algorithm. In: Benson, G., Page, R.D.M. (eds.) WABI 2003. LNCS (LNBI), vol. 2812, pp. 216–227. Springer, Heidelberg (2003)
3. Barthélemy, J.P., McMorris, F.R.: The median procedure for n-trees. J. Classif. 3, 329–334 (1986)
4. Baum, B.R.: Combining trees as a way of combining data sets for phylogenetic inference, and the desirability of combining gene trees. Taxon 41, 3–10 (1992)
5. Bininda-Emonds, O.R.P. (ed.): Phylogenetic Supertrees: Combining Information to Reveal the Tree of Life. Series on Computational Biology, vol. 4. Springer, Heidelberg (2004)
6. Bininda-Emonds, O.R.P., Cardillo, M., Jones, K.E., MacPhee, R.D.E., Beck, R.M.D., Grenyer, R., Price, S.A., Vos, R.A., Gittleman, J.L., Purvis, A.: The delayed rise of present-day mammals. Nature 446, 507–512 (2007)
7. Brown, D.G., Harrower, I.M.: Integer programming approaches to haplotype inference by pure parsimony. IEEE/ACM Trans. Comput. Biol. Bioinformatics 3(2), 141–154 (2006)
8. Bryant, D.: A classification of consensus methods for phylogenetics. In: Janowitz, M., Lapointe, F.-J., McMorris, F., Mirkin, B.B., Roberts, F. (eds.) Bioconsensus. Discrete Mathematics and Theoretical Computer Science, vol. 61, pp. 163–185. American Mathematical Society, Providence (2003)
9. Cotton, J.A., Page, R.D.M.: Tangled trees from molecular markers: reconciling conflict between phylogenies to build molecular supertrees. In: Bininda-Emonds, O.R.P. (ed.) Phylogenetic Supertrees: Combining Information to Reveal the Tree of Life. Series on Computational Biology, vol. 4, pp. 107–125. Springer, Heidelberg (2004)
10. Cotton, J.A., Wilkinson, M.: Majority-rule supertrees. Syst. Biol. 56, 445–452 (2007)
11. Danna, E., Fenelon, M., Gu, Z., Wunderling, R.: Generating multiple solutions for mixed integer programming problems. In: Fischetti, M., Williamson, D.P. (eds.) IPCO 2007. LNCS, vol. 4513, pp. 280–294. Springer, Heidelberg (2007)

12. Day, W., McMorris, F.: Axiomatic Consensus Theory in Group Choice and Biomathematics. SIAM Frontiers in Mathematics, Philadelphia, PA (2003)
13. Delsuc, F., Brinkmann, H., Chourrout, D., Philippe, H.: Tunicates and not cephalochordates are the closest living relatives of vertebrates. Nature 439, 965–968 (2006)
14. Dong, J., Fernández-Baca, D.: Properties of majority-rule supertrees. Syst. Biol. (to appear, 2009)
15. Dong, J., Fernández-Baca, D., McMorris, F.R., Powers, R.C.: A characterization of majority-rule (+) consensus trees (in preparation)
16. Goloboff, P.: Minority rule supertrees? MRP, compatibility, and minimum flip may display the least frequent groups. Cladistics 21, 282–294 (2005)
17. Goloboff, P.A., Pol, D.: Semi-strict supertrees. Cladistics 18(5), 514–525 (2005)
18. Gordon, A.D.: Consensus supertrees: The synthesis of rooted trees containing overlapping sets of labelled leaves. J. Classif. 9, 335–348 (1986)
19. Gusfield, D.: Haplotype inference by pure parsimony. In: Baeza-Yates, R., Chávez, E., Crochemore, M. (eds.) CPM 2003. LNCS, vol. 2676, pp. 144–155. Springer, Heidelberg (2003)
20. Gusfield, D.: The multi-state perfect phylogeny problem with missing and removable data: Solutions via integer-programming and chordal graph theory. In: Proc. RECOMB (2009)
21. Gusfield, D., Frid, Y., Brown, D.: Integer programming formulations and computations solving phylogenetic and population genetic problems with missing or genotypic data. In: Lin, G. (ed.) COCOON 2007. LNCS, vol. 4598, pp. 51–64. Springer, Heidelberg (2007)
22. Margush, T., McMorris, F.R.: Consensus n-trees. Bull. Math. Biol. 43, 239–244 (1981)
23. Pattengale, N.D., Gottlieb, E.J., Moret, B.M.E.: Efficiently computing the Robinson-Foulds metric. J. Comput. Biol. 14(6), 724–735 (2007)
24. Pisani, D., Wilkinson, M.: MRP, taxonomic congruence and total evidence. Syst. Biol. 51, 151–155 (2002)
25. Ragan, M.A.: Phylogenetic inference based on matrix representation of trees. Mol. Phylogenet. Evol. 1, 53–58 (1992)
26. Ranwez, V., Berry, V., Criscuolo, A., Fabre, P.-H., Guillemot, S., Scornavacca, C., Douzery, E.J.P.: PhySIC: A veto supertree method with desirable properties. Syst. Biol. 56(5), 798–817 (2007)
27. Robinson, D.F., Foulds, L.R.: Comparison of phylogenetic trees. Math. Biosci. 53, 131–147 (1981)
28. Semple, C., Steel, M.: Phylogenetics. Oxford Lecture Series in Mathematics. Oxford University Press, Oxford (2003)
29. Sridhar, S., Lam, F., Blelloch, G.E., Ravi, R., Schwartz, R.: Mixed integer linear programming for maximum-parsimony phylogeny inference. IEEE/ACM Trans. Comput. Biol. Bioinformatics 5(3), 323–331 (2008)
30. Steel, M.A.: The complexity of reconstructing trees from qualitative characters and subtrees. J. Classif. 9, 91–116 (1992)
31. Swofford, D.: PAUP*: Phylogenetic analysis using parsimony (*and other methods). Sinauer Assoc., Sunderland, Massachusetts, U.S.A. Version 4.0 beta
32. Wilkinson, M., Cotton, J.A., Lapointe, F.-J., Pisani, D.: Properties of supertree methods in the consensus setting. Syst. Biol. 56, 330–337 (2007)

SCJ: A Variant of Breakpoint Distance for Which Sorting, Genome Median and Genome Halving Problems Are Easy

Pedro Feijão[1] and João Meidanis[1,2]

[1] Institute of Computing, University of Campinas, Brazil
[2] Scylla Bioinformatics, Brazil

Abstract. The breakpoint distance is one of the most straightforward genome comparison measures. Surprisingly, when it comes to define it precisely for multichromosomal genomes with both linear and circular chromosomes, there is more than one way to go about it. In this paper we study Single-Cut-or-Join (SCJ), a breakpoint-like rearrangement event for which we present linear and polynomial time algorithms that solve several genome rearrangement problems, such as median and halving. For the multichromosomal linear genome median problem, this is the first polynomial time algorithm described, since for other breakpoint distances this problem is NP-hard. These new results may be of value as a speedily computable, first approximation to distances or phylogenies based on more realistic rearrangement models.

1 Introduction

Genome rearrangement, an evolutionary event where large, continuous pieces of the genome shuffle around, has been studied since shortly after the very advent of genetics [1, 2, 3]. With the increased availability of whole genome sequences, gene order data have been used to estimate the evolutionary distance between present-day genomes and to reconstruct the gene order of ancestral genomes. The inference of evolutionary scenarios based on gene order is a hard problem, with its simplest version being the pairwise genome rearrangement problem: given two genomes, represented as sequences of conserved segments called *syntenic blocks*, find the most parsimonious sequence of rearrangement events that transforms one genome into the other. In some applications, one is interested only in the number of events of such a sequence — the *distance* between the two genomes.

Several rearrangement events, or operations, have been proposed. Early approaches considered the case where just one operation is allowed. For some operations a polynomial solution was found (e.g., for reversals [4], translocations [5], and block-interchanges [6]), while for others the complexity is still open (e.g., for transpositions [7, 8]). Later on, polynomial algorithms for combinations of operations were discovered (e.g., for block-interchanges and reversals [9]; for fissions, fusions, and transpositions [10]; for fissions, fusions, and block-interchanges [11]). Yancopoulos et al. [12] introduced a very comprehensive model, with reversals,

S.L. Salzberg and T. Warnow (Eds.): WABI 2009, LNBI 5724, pp. 85–96, 2009.
© Springer-Verlag Berlin Heidelberg 2009

transpositions, translocations, fusions, fissions, and block-interchanges modeled as compositions of the same basic operation, the double-cut-and-join (DCJ).

Different relative weights for the operations have been considered. Proposals have also differed in the number and type of allowed chromosomes (unichromosomal vs. multichromosomal genomes; linear or circular chromosomes).

When more than two genomes are considered, we have the more challenging problem of rearrangement-based phylogeny reconstruction, where we want to find a tree that minimizes the total number of rearrangement events. Early approaches were based on a breakpoint distance (e.g., BPAnalysis [13], and GRAPPA [14]). With the advances on pairwise distance algorithms, more sophisticated distances were used, with better results (e.g., reversal distance, used by MGR [15] and in an improved version of GRAPPA, and DCJ distance [16]).

Two problems are commonly used to find the gene order of ancient genomes in rearrangement-based phylogeny reconstruction: the median problem and the halving problem. These problems are NP-hard in most cases even under the simplest distances.

In this paper we propose a new way of computing breakpoint distances, based on the the *Single-Cut-or-Join* (SCJ) operation, and show that several rearrangement problems involving it are polynomial. For some problems, these will be the only polynomial results known to date. The SCJ distance is exactly twice the the breakpoint (BP) distance of Tannier et al. [17] for circular genomes, but departs from it when linear chromosomes are present, because of an alternative way of treating telomeres. In a way, SCJ is the simplest mutational event imaginable, and it may be of value as a speedily computable, first approximation to distances or phylogenies based on more realistic rearrangement models.

The rest of this paper is structured as follows. In Section 2 we present the basic definitions, including SCJ. Section 3 deals with the distance problem and compares SCJ to other distances. Sections 4 and 5 deal with genome medians and genome halving, respectively, and their generalizations. Finally, in Section 6 we present a brief discussion and future directions.

2   Representing Genomes

We will use a standard genome formulation [18, 17]. A *gene* is an oriented sequence of DNA that starts with a tail and ends with a head, called the *extremities* of the gene. The tail of a gene a is denoted by a_t, and its head by a_h. Given a set of genes $\mathcal{G}$, the extremity set is $\mathcal{E} = \{a_t : a \in \mathcal{G}\} \cup \{a_h : a \in \mathcal{G}\}$. An *adjacency* is an unordered pair of two extremities that represents the linkage between two consecutive genes in a certain orientation on a chromosome, for instance $a_h b_t$. An extremity that is not adjacent to any other extremity is called a *telomere*. A genome is represented by a set of adjacencies where the tail and head of each gene appear at most once. Telomeres will be omitted in our representation, since they are uniquely determined by the set of adjacencies and the extremity set $\mathcal{E}$.

The *graph representation* of a genome Π is a graph G_Π whose vertices are the extremities of Π and there is a grey edge connecting the extremities x and y

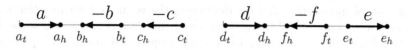

Fig. 1. Graph G_Π representing a genome with two linear chromosomes. Black directed edges represent genes, while grey edges link consecutive extremities.

when xy is an adjacency of Π or a directed black edge if x and y are head and tail of the same gene. A connected component in G_Π is a *chromosome* of Π, and it is *linear* if it is a path, and *circular* if it is a cycle. A *circular genome* is a genome whose chromosomes are all circular, and a *linear genome* is a genome whose chromosomes are all linear. A *string representation* of a genome Π, denoted by Π_S, is a set of strings corresponding to the genes of Π in the order they appear on each chromosome, with a bar over the gene if it is read from head to tail and no bar otherwise. Notice that the string representation is not unique: each chromosome can be replaced by its reverse complement.

For instance, given the set $\mathcal{G} = \{a, b, c, d, e, f\}$, and the genome $\Pi = \{a_h b_h, b_t c_h, d_h f_h, f_t e_t\}$, the graph G_Π is given in Figure 1. Notice that telomeres a_t, c_t, d_t, and e_h are omitted from the set representation without any ambiguity. A string representation of this genome is $\Pi_S = \left(a\,\overline{b}\,\overline{c}\,,\, d\,\overline{f}\,e\right)$.

In problems where gene duplicates are allowed, a gene can have any number of homologous copies within a genome. Each copy of a gene is called a *duplicated gene* and is identified by its tail and head with the addition of an integer label, from 1 to n, identifying the copy. For instance, a gene g with three copies has extremities $g_h^1, g_t^1, g_h^2, g_t^2, g_h^3$ and g_t^3. An *n-duplicate genome* is a genome where each gene has exactly n copies. An *ordinary genome* is a genome with a single copy of each gene. We can obtain n-duplicate genomes from an ordinary genome with the following operation: for an ordinary genome Π on a set $\mathcal{G}$, Π^n represents a set of n-duplicate genomes on $\mathcal{G}^n = \{a^1, a^2, \ldots, a^n : a \in \mathcal{G}\}$ such that if the adjacency xy belongs to Π, n adjacencies of the form $x^i y^j$ belong to any genome in Π^n. The assignment of labels i and j to the duplicated adjacencies is arbitrary, with the restriction that each extremity copy has to appear exactly once. For instance, for $n = 3$ a valid choice could be $\{x^1 y^2, x^2 y^1, x^3 y^3\}$. Since for each adjacency in Π we have $n!$ possible choices for adjacencies in Π^n, the number of genomes in the set Π^n is $|\Pi|^{n!}$, where $|\Pi|$ is the number of adjacencies of Π.

2.1   A New Rearrangement Operation

We will define two simple operations applied directly on the adjacencies and telomeres of a genome. A *cut* is an operation that breaks an adjacency in two telomeres (namely, its extremities), and a *join* is the reverse operation, pairing two telomeres into an adjacency. Any *cut* or *join* applied to a genome will be called a **Single-Cut-or-Join** (SCJ) operation. In this paper, we are interested in solving several rearrangement problems under the SCJ distance.

Since each genome is a set of adjacencies, standard set operations such as union, intersection and set difference can be applied to two (or more) genomes.

In the case of intersection and set difference, the result is a set of adjacencies contained in at least one of the genomes, and therefore it is also a genome. On the other hand, the set resulting from a union operation might not represent a genome since the same extremity could be present in more than one adjacency. We will use these operations throughout this paper in our algorithms, and whenever union is used, we will prove that the resulting set represents a valid genome. Set difference between sets A and B will be denoted by $A - B$.

3   Rearrangement by SCJ

The *rearrangement by SCJ problem* is stated as follows: given two genomes Π and Σ with the same set of genes $\mathcal{G}$, find a shortest sequence of SCJ operations that transforms Π into Σ. This problem is also called *genome sorting*. The length of such a sequence is called the *distance* between Π and Σ and is denoted by $d_{SCJ}(\Pi, \Sigma)$.

Since the only possible operations are to remove (cut) or include (join) an adjacency in a genome, the obvious way of transforming Π into Σ is to remove all adjacencies that belong to Π and not to Σ, and then include all adjacencies that belong to Σ and not to Π.

Lemma 1. *Consider the genomes Π and Σ, and let $\Gamma = \Pi - \Sigma$ and $\Lambda = \Sigma - \Pi$. Then, Γ and Λ can be found in linear time and they define a minimum set of SCJ operations that transform Π into Σ, where adjacencies in Γ define cuts and adjacencies in Λ define joins. Consequently, $d_{SCJ}(\Pi, \Sigma) = |\Pi - \Sigma| + |\Sigma - \Pi|$.*

Proof. Considering the effect an arbitrary cut or join on Π can have on the quantity $f_{\Sigma}(\Pi) = |\Pi - \Sigma| + |\Sigma - \Pi|$, it is straightforward to verify that $f_{\Sigma}(\Pi)$ can increase or decrease by at most 1. Hence, the original value is a lower bound on the distance. Given that the sequence of operations proposed in the statement does lead from Π to Σ along valid genomes in that number of steps, we have our lemma.   □

3.1   SCJ Distance with the Adjacency Graph

The Adjacency Graph, introduced by Bergeron et al. [18], was used to find an easy equation for the DCJ distance. The adjacency graph $AG(\Pi, \Sigma)$ is a bipartite graph whose vertices are the adjacencies and telomeres of the genomes Π and Σ and whose edges connect two vertices that have a common extremity. Therefore, vertices representing adjacencies will have degree two and telomeres will have degree one, and this graph will be a union of path and cycles.

A formula for the SCJ distance based on the cycles and paths of $AG(\Pi, \Sigma)$ can be easily found, as we will see in the next lemma. We will use the following notation: C and P represent the number of cycles and paths of $AG(\Pi, \Sigma)$, respectively, optionally followed by a subscript to indicate the number of edges (the *length*) of the cycle or path or if the length is odd or even. For instance, P_2 is the number of paths of length two, $C_{\geq 4}$ is the number of cycles with length four or more and P_{odd} is the number of paths with an odd length.

Lemma 2. *Consider two genomes Π and Σ with the same set of genes $\mathcal{G}$. Then, we have*

$$d_{SCJ}(\Pi, \Sigma) = 2[N - (C_2 + P/2)], \qquad (1)$$

where N is the number of genes, C_2 is the number of cycles of length two and P the number of paths in $AG(\Pi, \Sigma)$.

Proof. We know from the definition of SCJ distance and basic set theory that

$$d_{SCJ}(\Pi, \Sigma) = |\Pi - \Sigma| + |\Sigma - \Pi| = |\Sigma| + |\Pi| - 2|\Sigma \cap \Pi|.$$

Since the number of cycles of length two in $AG(\Pi, \Sigma)$ is the number of common adjacencies of Π and Σ, we have $|\Sigma \cap \Pi| = C_2$. For any genome Π, we know that $|\Pi| = N - t_\Pi/2$, where t_Π is the number of telomeres of Π. Since each path in $AG(\Pi, \Sigma)$ has exactly two vertices corresponding to telomeres of Π and Σ, the total number of paths in $AG(\Pi, \Sigma)$, denoted by P, is given by $P = (t_\Pi + t_\Sigma)/2$. Therefore,

$$d_{SCJ}(\Pi, \Sigma) = |\Sigma| + |\Pi| - 2|\Sigma \cap \Pi| =$$
$$= 2N - (t_\Pi + t_\Sigma)/2 - 2C_2 = 2[N - (C_2 + P/2)]. \qquad \square$$

3.2   Comparing SCJ Distance to Other Distances

Based on the adjacency graph, we have the following equation for the DCJ distance [18]:

$$d_{DCJ}(\Pi, \Sigma) = N - (C + P_{odd}/2), \qquad (2)$$

where N is the number of genes, C is the number of cycles, and P_{odd} is the number of odd paths in $AG(\Pi, \Sigma)$. For the Breakpoint (BP) distance, as defined by Tannier et al. [17], we have

$$d_{BP}(\Pi, \Sigma) = N - (C_2 + P_1/2), \qquad (3)$$

where C_2 is the number of cycles with length two and P_1 is the number of paths with length one in $AG(\Pi, \Sigma)$.

With these equations we can find relationships between SCJ, BP, and DCJ distances. First, for the BP distance, we have

$$d_{SCJ}(\Pi, \Sigma) = 2d_{BP}(\Pi, \Sigma) - P_{\geq 2}.$$

As expected, the SCJ distance is related to the BP distance, differing only by a factor of 2 and the term $P_{\geq 2}$, the number of paths with two or more edges. For circular genomes, $P = 0$ and the SCJ distance is exactly twice the BP distance. For the general case, the following sandwich formula holds:

$$d_{BP}(\Pi, \Sigma) \leq d_{SCJ}(\Pi, \Sigma) \leq 2d_{BP}(\Pi, \Sigma).$$

For the DCJ distance, a reasonable guess is that it would be one fourth of the SCJ distance, since a DCJ operation, being formed by two cuts and two joins, should correspond to four SCJ operations. This is not true, however, for two reasons.

First, a DCJ operation may correspond to four, two, or even one SCJ operation. Examples of these three cases are shown in Figure 2, with *caps* represented by the symbol ∘. In each case the target genome is $(a\ b\ c\ d)$. The figure shows a reversal, a suffix reversal, and a linear fusion, all of which have the same weight under the DCJ model, but different SCJ distances, because caps do not exist in the SCJ model. Incidentally, they have different BP distances as well.

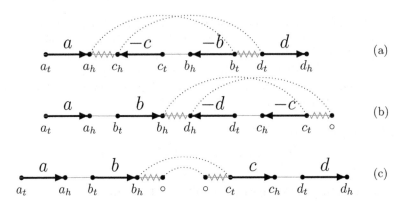

Fig. 2. Three types of single DCJ operations transforming each genome into $\Pi_S = \{a, b, c, d\}$. (a) Reversal. (b) *Suffix* Reversal. (c) Linear Fusion.

The second reason is that, when consecutive DCJ operations use common spots, the SCJ model is able to "cancel" operations, resulting in a shorter sequence. Both arguments show SCJ saving steps, which still leaves four times DCJ distance as an upper bound on SCJ distance. The complete sandwich result is

$$d_{DCJ}(\Pi, \Sigma) \le d_{SCJ}(\Pi, \Sigma) \le 4d_{DCJ}(\Pi, \Sigma).$$

4   The Genome Median Problem

The *Genome Median Problem* (GMP) is an important tool for phylogenetic reconstruction of trees with ancestral genomes based on rearrangement events. When genomes are unichromosomal this problem is NP-hard under the breakpoint, reversal and DCJ distances [19, 20]. In the multichromosomal general case, when there are no restrictions as to whether the genomes are linear or circular, Tannier et al. [17] recently showed that under the DCJ distance the problem is still NP-hard, but it becomes polynomial under the breakpoint distance (BP), the first polynomial result for the median problem. The problem can be solved in linear time for SCJ, our version of the breakpoint distance.

We show this by proposing a more general problem, *Weighted Multichromosomal Genome Median Problem* (WMGMP), where we find the genome median among any number of genomes with weights for the genomes. We will give a straightforward algorithm for this problem under the SCJ distance in the general case, from which the special case of GMP follows with unique solution, and then proceed to solve it with the additional restrictions of allowing only linear or only circular chromosomes.

4.1   Weighted Multichromosomal Genome Median Problem

This problem is stated as follows: Given n genomes $\Pi_1, \ldots, \Pi_n$ with the same set of genes $\mathcal{G}$, and nonnegative weights $w_1, \ldots, w_n$, we want to find a genome Γ that minimizes $\sum_{i=1}^{n} w_i \cdot d(\Pi_i, \Gamma)$.

We know that

$$\sum_{i=1}^{n} w_i \cdot d(\Pi_i, \Gamma) = \sum_{i=1}^{n} w_i |\Pi_i| + \sum_{i=1}^{n} w_i |\Gamma| - 2 \sum_{i=1}^{n} w_i |\Gamma \cap \Pi_i|$$

and since $\sum_{i=1}^{n} w_i |\Pi_i|$ does not depend on Γ we want to minimize

$$f(\Gamma) \equiv \sum_{i=1}^{n} w_i |\Gamma| - 2 \sum_{i=1}^{n} w_i |\Gamma \cap \Pi_i| \tag{4}$$

Now, for any adjacency d, let $\mathcal{S}_d$ be the set of indices i for which Π_i has this adjacency, that is, $\mathcal{S}_d = \{i : d \in \Pi_i\}$. To simplify the notation, we will write $f(\{d\})$ as $f(d)$. Then we have

$$f(d) = \sum_{i=1}^{n} w_i |\{d\}| - 2 \sum_{i=1}^{n} w_i |\{d\} \cap \Pi_i| = \sum_{i=1}^{n} w_i - 2 \sum_{i \in \mathcal{S}_d} w_i = \sum_{i \notin \mathcal{S}_d} w_i - \sum_{i \in \mathcal{S}_d} w_i$$

and it is easy to see that for any genome Γ we have

$$f(\Gamma) = \sum_{d \in \Gamma} f(d) \tag{5}$$

Since we want to minimize $f(\Gamma)$, a valid approach would be to choose Γ as the genome with all adjacencies d such as $f(d) < 0$. As we will see from the next lemma, this strategy is optimal.

Lemma 3. *Given n genomes $\Pi_1, \ldots, \Pi_n$ and nonnegative weights $w_1, \ldots, w_n$, the genome $\Gamma = \{d : f(d) < 0\}$, where*

$$f(d) = \sum_{i \notin \mathcal{S}_d} w_i - \sum_{i \in \mathcal{S}_d} w_i$$

and $\mathcal{S}_d = \{i : d \in \Pi_i\}$, minimizes $\sum_{i=1}^{n} w_i \cdot d(\Pi_i, \Gamma)$. Furthermore, if there is no adjacency $d \in \Pi_i$ for which $f(d) = 0$, then Γ is a unique solution.

Proof. Let xy be an adjacency such that $f(xy) < 0$. For any extremity $z \neq y$, we have $xy \in \Pi_i \Rightarrow xz \notin \Pi_i$ and $xz \in \Pi_i \Rightarrow xy \notin \Pi_i$. Therefore

$$f(xz) = \sum_{i \notin S_{xz}} w_i - \sum_{i \in S_{xz}} w_i \geq \sum_{i \in S_{xy}} w_i - \sum_{i \notin S_{xy}} w_i = -f(xy) > 0$$

This means that adjacencies d with $f(d) < 0$ do not have extremities in common and it is then possible to add all those adjacencies to form a valid genome Γ, minimizing $f(\Gamma)$ and consequently $\sum_{i=1}^{n} w_i \cdot d(\Pi_i, \Gamma)$.

To prove the uniqueness of the solution, suppose there is no adjacency d such that $f(d) = 0$. Since any adjacency d belonging to Γ satisfies $f(d) < 0$ and any other adjacency d' satisfies $f(d') > 0$, for any genome $\Gamma' \neq \Gamma$ we have $f(\Gamma') > f(\Gamma)$, confirming that Γ is a unique solution. If there is d with $f(d) = 0$, then $\Gamma' = (\Gamma \cup d)$ is a valid genome (that is, the extremities of d are telomeres in Γ), which is also a solution, and uniqueness cannot be guaranteed.   □

After solving the general case, we will restrict the problem to circular or linear genomes in the next two sections.

4.2   The Weighted Multichromosomal Circular Median Problem

In this section we will solve the WMGMP restricted to circular genomes: given n *circular* genomes $\Pi_1, \ldots, \Pi_n$ with the same set of genes $\mathcal{G}$, and nonnegative weights $w_1, \ldots, w_n$, we want to find a *circular* genome Γ which minimizes $\sum_{i=1}^{n} w_i \cdot d(\Pi_i, \Gamma)$.

It is easy to see that a genome is circular if and only if it has N adjacencies, where N is the number of genes. Basically we want to minimize the same function f defined in equation (4) with the additional constraint $|\Gamma| = N$. To solve this problem, let G be a complete graph where every extremity in the set $\mathcal{E}$ is a vertex and the weight of an edge connecting vertices x and y is $f(xy)$. Then a perfect matching on this graph corresponds to a circular genome Γ and the total weight of this matching is $f(\Gamma)$. Then, a minimum weight perfect matching can be found in polynomial time [21] and it is an optimum solution to the weighted circular median problem.

4.3   The Weighted Multichromosomal Linear Median Problem

The solution of this problem is found easily using the same strategy as in the WMGMP. Since we have no restrictions on the number of adjacencies, we find Γ as defined in Lemma 3, including only adjacencies for which $f > 0$. If Γ is linear, this is the optimum solution. If Γ has circular chromosomes, a linear median Γ' can be obtained by removing, in each circular chromosome of Γ, an adjacency xy with maximum $f(xy)$. Removing xy would allow the inclusion of new adjacencies of the forms xw and yz, but we know that $f(xy) < 0$ implies $f(xw) > 0$ and $f(yz) > 0$. Therefore, any genome Σ different from Γ' either has a circular chromosome or has $f(\Sigma) \geq f(\Gamma')$. Therefore, Γ' is an optimal solution.

This is the first polynomial result for this problem.

5   Genome Halving and Genome Aliquoting

The *Genome Halving Problem* (GHP) is motivated by whole genome duplication events in molecular evolution, postulated by Susumu Ohno in 1970 [22]. Whole genome duplication has been very controversial over the years, but recently, very strong evidence in its favor was discovered in yeast species [23]. The goal of a halving analysis is to reconstruct the ancestor of a 2-duplicate genome at the time of the doubling event.

The GHP is stated as follows: given a 2-duplicate genome Δ, find an ordinary genome Γ that minimizes $d(\Delta, \Gamma^2)$, where

$$d(\Delta, \Gamma^2) = \min_{\Sigma \in \Gamma^2} d(\Delta, \Sigma) \qquad (6)$$

If both Δ and Γ are given, computing the right hand side of Equation (6) is known as the *Double Distance* problem, which has a polynomial solution under the breakpoint distance but is NP-hard under the DCJ distance [17]. In contrast, the GHP has a polynomial solution under the DCJ distance for unichromosomal genomes [24], and for multichromosomal genomes when both linear and circular chromosomes are allowed [25].

Warren and Sankoff recently proposed a generalization of the halving problem, the *Genome Aliquoting Problem* (GAP) [26]: Given an n-duplicate genome Δ, find an ordinary genome Γ that minimizes $d(\Delta, \Gamma^n)$. In their paper, they use the DCJ distance and develop heuristics for this problem, but a polynomial time exact solution remains open. To the best of our knowledge, this problem has never been studied under any other distance. We will show that under the SCJ distance this problem can be formulated as a special case of the WMGMP.

Lemma 4. *Given an n-duplicate genome Δ, define n ordinary genomes Π_1, ..., Π_n as follows. For each ordinary adjacency xy, add it to the k first genomes $\Pi_1, \ldots, \Pi_k$, where k is the number of adjacencies of the form $x^i y^j$ in Δ. Then, for every genome Γ, we have $d(\Delta, \Gamma^n) = \sum_{i=1}^{n} d(\Gamma, \Pi_i)$.*

Proof. We have that

$$d(\Delta, \Gamma^n) = \min_{\Sigma \in \Gamma^n} d(\Delta, \Sigma) = \min_{\Sigma \in \Gamma^n} (|\Delta| + |\Sigma| - 2|\Delta \cap \Sigma|)$$

$$= |\Delta| + n|\Gamma| - 2 \max_{\Sigma \in \Gamma^n} |\Delta \cap \Sigma|$$

To maximize $|\Delta \cap \Sigma|$, let $k(xy)$ be the number of adjacencies of the form $x^i y^j$ in Δ. For each adjacency xy in Γ, we add to Σ the $k(xy)$ adjacencies of Δ plus $n - k(xy)$ arbitrarily labeled adjacencies, provided they do not collide. It is clear that this Σ maximizes $|\Delta \cap \Sigma|$ and furthermore

$$\max_{\Sigma \in \Gamma^n} |\Delta \cap \Sigma| = \sum_{xy \in \Gamma} k(xy)$$

On the other hand,

$$\sum_{i=1}^{n} d(\Gamma, \Pi_i) = n|\Gamma| + \sum_{i=1}^{n} |\Pi_i| - 2 \sum_{i=1}^{n} |\Pi_i \cap \Gamma|$$

Now $\sum_{i=1}^{n} |\Pi_i \cap \Gamma|$ is exactly $\sum_{xy \in \Gamma} k(xy)$, since any adjacency xy in Δ appears exactly $k(xy)$ times in genomes $\Pi_1, \ldots, \Pi_n$. Taking into account that $|\Delta| = \sum_{i=1}^{n} |\Pi_i|$, we have our result. □

Lemma 4 implies that the GAP is actually a special case of the WMGMP, and can be solved using the same algorithm. Another corollary is that the constrained versions of the GAP for linear or circular multichromosomal genomes are also polynomial.

5.1 Guided Genome Halving

The *Guided Genome Halving* (GGH) problem was proposed very recently, and is stated as follows: given a 2-duplicate genome Δ and an ordinary genome Γ, find an ordinary genome Π that minimizes $d(\Delta, \Pi^2) + d(\Gamma, \Pi)$. This problem is related to Genome Halving, only here an ordinary genome Γ, presumed to share a common ancestor with Π, is used to *guide* the reconstruction of the ancestral genome Π.

Under the BP distance, the GGH has a polynomial solution for general multichromosomal genomes [17] but is NP-hard when only linear chromosomes are allowed [27]. For the DCJ distance, it is NP-hard in the general case [17].

As in the Halving Problem, here we will solve a generalization of GGH, the *Guided Genome Aliquoting* problem: given an n-duplicate genome Δ and an ordinary genome Γ, find an ordinary genome Π that minimizes $d(\Delta, \Pi^n) + d(\Gamma, \Pi)$. It turns out that the version with an n-duplicate genome Δ as input is very similar to the "unguided" version with an $(n+1)$-duplicate genome.

Lemma 5. *Given an n-duplicate genome Δ and an ordinary genome Γ, let Δ' be an $(n+1)$-duplicate genome such that $\Delta' = \Delta \cup \{x^{n+1}y^{n+1} : xy \in \Gamma\}$. Then for any genome Π we have $d(\Delta', \Pi^{n+1}) = d(\Delta, \Pi^n) + d(\Gamma, \Pi)$.*

Proof. We have that

$$\min_{\Sigma \in \Pi^n} d(\Delta, \Sigma) = \min_{\Sigma \in \Pi^n} (|\Delta| + |\Sigma| - 2|\Delta \cap \Sigma|)$$
$$= |\Delta| + n|\Pi| - 2 \max_{\Sigma \in \Pi^n} |\Delta \cap \Sigma|$$

and

$$\min_{\Sigma' \in \Pi^{n+1}} d(\Delta', \Sigma') = |\Delta'| + (n+1)|\Pi| - 2 \max_{\Sigma' \in \Pi^{n+1}} |\Delta' \cap \Sigma'|.$$

Since $|\Delta'| = |\Delta| + |\Gamma|$ and $|\Gamma \cap \Pi| + \max_{\Sigma \in \Pi^n} |\Delta \cap \Sigma| = \max_{\Sigma' \in \Pi^{n+1}} |\Delta' \cap \Sigma'|$, we have our result. □

The last lemma implies that GGH is a special case of GAP, which in turn is a special case of the WMGMP. Again, the constrained linear or circular versions are also polynomial for GGH in the SCJ model.

6   Discussion and Future Directions

In this paper we show that a variant of breakpoint distance, based on the Single-Cut-or-Join (SCJ) operation, allows linear- and polynomial-time solutions to some rearrangement problems that are NP-hard under the BP distance, for instance, the multichromosomal linear versions of the genome halving, guided halving and genome median problems. In addition, the SCJ approach is able to produce a rearrangement scenario between genomes, not only a distance, which is useful for phylogeny reconstruction.

The complexity of unichromosomal median and halving remain open under the SCJ distance.

From a biological point of view, we can think of a rearrangement event as an accepted mutation, that is, a mutational event involving large, continuous genome segments that was accepted by natural selection, and therefore became fixed in a population. SCJ may model the mutation part well, but a model for the acceptance part is missing. For instance, while the mutational effort of doing a fission seems to be less than that of an inversion, the latter is more frequent as a rearrangement event, probably because it has a better chance of being accepted. This may have to do with the location and movement of origins of replication, since any free segment will need one to become fixed.

Other considerations, such as the length of segments, hotspots, presence of flanking repeats, etc. are likely to play a role in genome rearrangements, and need to be taken into account in a comprehensive model.

Although crude from the standpoint of evolutionary genomics, the new distance may serve as a fast, first-order approximation for other, better founded genomic rearrangement distances, and also for reconstructed phylogenies. We intend to pursue this line or work, applying it to real datasets and comparing the results to those obtained with other methods.

References

[1] Sturtevant, A.H., Dobzhansky, T.: Inversions in the third chromosome of wild races of *Drosophila pseudoobscura*, and their use in the study of the history of the species. PNAS 22(7), 448–450 (1936)
[2] McClintock, B.: The origin and behavior of mutable loci in maize. PNAS 36(6), 344–355 (1950)
[3] Nadeau, J.H., Taylor, B.A.: Lengths of chromosomal segments conserved since divergence of man and mouse. PNAS 81(3), 814–818 (1984)
[4] Hannenhalli, S., Pevzner, P.A.: Transforming cabbage into turnip (polynomial algorithm for sorting signed permutations by reversals). In: Proc. 27th Ann. Symp. Theory of Computing STOC 1995 (1995)
[5] Hannenhalli, S.: Polynomial-time algorithm for computing translocation distance between genomes. Discrete Appl. Math. 71(1-3), 137–151 (1996)
[6] Christie, D.A.: Sorting permutations by block-interchanges. Information Processing Letters 60, 165–169 (1996)
[7] Bafna, V., Pevzner, P.A.: Sorting by transpositions. SIAM J. Discrete Math. 11(2), 224–240 (1998)

[8] Elias, I., Hartman, T.: A 1.375-approximation algorithm for sorting by trans-
      positions. IEEE/ACM Transactions on Computational Biology and Bioinformat-
      ics 3(4), 369–379 (2006)
[9] Mira, C., Meidanis, J.: Sorting by block-interchanges and signed reversals. In:
      Proc. ITNG 2007, pp. 670–676 (2007)
[10] Dias, Z., Meidanis, J.: Genome rearrangements distance by fusion, fission, and
      transposition is easy. In: Proc. SPIRE 2001, pp. 250–253 (2001)
[11] Lu, C.L., Huang, Y.L., Wang, T.C., Chiu, H.T.: Analysis of circular genome re-
      arrangement by fusions, fissions and block-interchanges. BMC Bioinformatics 7,
      295 (2006)
[12] Yancopoulos, S., Attie, O., Friedberg, R.: Efficient sorting of genomic permuta-
      tions by translocation, inversion and block interchange. Bioinformatics 21(16),
      3340–3346 (2005)
[13] Blanchette, M., Bourque, G., Sankoff, D.: Breakpoint phylogenies. In: Genome
      Inform. Ser. Workshop Genome Inform., vol. 8, pp. 25–34 (1997)
[14] Moret, B.M., Wang, L.S., Warnow, T., Wyman, S.K.: New approaches for recon-
      structing phylogenies from gene order data. Bioinformatics 17(suppl. 1), S165–
      S173 (2001)
[15] Bourque, G., Pevzner, P.A.: Genome-scale evolution: reconstructing gene orders
      in the ancestral species. Genome Res. 12(1), 26–36 (2002)
[16] Adam, Z., Sankoff, D.: The ABCs of MGR with DCJ. Evol. Bioinform. Online 4,
      69–74 (2008)
[17] Tannier, E., Zheng, C., Sankoff, D.: Multichromosomal genome median and halv-
      ing problems. In: Crandall, K.A., Lagergren, J. (eds.) WABI 2008. LNCS (LNBI),
      vol. 5251, pp. 1–13. Springer, Heidelberg (2008)
[18] Bergeron, A., Mixtacki, J., Stoye, J.: A unifying view of genome rearrangements.
      In: Bücher, P., Moret, B.M.E. (eds.) WABI 2006. LNCS (LNBI), vol. 4175, pp.
      163–173. Springer, Heidelberg (2006)
[19] Bryant, D.: The complexity of the breakpoint median problem. Technical Report
      CRM-2579, Centre de recherches mathematiques, Université de Montréal (1998)
[20] Caprara, A.: The reversal median problem. INFORMS J. Comput. 15, 93–113
      (2003)
[21] Lovász, L., Plummer, M.D.: Matching theory. Annals of Discrete Mathematics,
      vol. 29. North-Holland, Amsterdam (1986)
[22] Ohno, S.: Evolution by gene duplication. Springer, Heidelberg (1970)
[23] Kellis, M., Birren, B.W., Lander, E.S.: Proof and evolutionary analysis of ancient
      genome duplication in the yeast saccharomyces cerevisiae. Nature 428(6983), 617–
      624 (2004)
[24] Alekseyev, M.A., Pevzner, P.A.: Colored de Bruijn graphs and the genome halving
      problem. IEEE/ACM Trans. Comput. Biol. Bioinform. 4(1), 98–107 (2007)
[25] Mixtacki, J.: Genome halving under DCJ revisited. In: Hu, X., Wang, J. (eds.)
      COCOON 2008. LNCS, vol. 5092, pp. 276–286. Springer, Heidelberg (2008)
[26] Warren, R., Sankoff, D.: Genome aliquoting with double cut and join. BMC Bioin-
      formatics 10(suppl. 1), S2 (2009)
[27] Zheng, C., Zhu, Q., Adam, Z., Sankoff, D.: Guided genome halving: hardness,
      heuristics and the history of the hemiascomycetes. Bioinformatics 24(13), i96–
      i104 (2008)

A Simple, Practical and Complete $O(\frac{n^3}{\log n})$-Time Algorithm for RNA Folding Using the *Four-Russians* Speedup

Yelena Frid and Dan Gusfield

Department of Computer Science, U.C. Davis

Abstract. The problem of computationally predicting the secondary structure (or folding) of RNA molecules was first introduced more than thirty years ago and yet continues to be an area of active research and development. The basic *RNA-folding problem* of finding a maximum cardinality, non-crossing, matching of complimentary nucleotides in an RNA sequence of length n, has an $O(n^3)$-time dynamic programming solution that is widely applied. It is known that an $o(n^3)$ worst-case time solution is possible, but the published and suggested methods are complex and have not been established to be practical. Significant practical improvements to the original dynamic programming method have been introduced, but they retain the $O(n^3)$ worst-case time bound when n is the only problem-parameter used in the bound. Surprisingly, the most widely-used, general technique to achieve a worst-case (and often practical) speed up of dynamic programming, the *Four-Russians* technique, has not been previously applied to the RNA-folding problem. This is perhaps due to technical issues in adapting the technique to RNA-folding.

In this paper, we give a simple, complete, and practical Four-Russians algorithm for the basic RNA-folding problem, achieving a worst-case time-bound of $O(n^3/\log(n))$. We show that this time-bound can also be obtained for richer nucleotide matching scoring-schemes, and that the method achieves consistent speed-ups in practice. The contribution is both theoretical and practical, since the basic RNA-folding problem is often solved multiple times in the inner-loop of more complex algorithms, and for long RNA molecules in the study of RNA virus genomes.

1 Introduction

The problem of computationally predicting the secondary structure (or folding) of RNA molecules was first introduced more than thirty years ago ([9,8,14,11]), and yet continues to be an area of active research and development, particularly due to the recent discovery of a wide variety of new types of RNA molecules and their biological importance. Additional interest in the problem comes from synthetic biology where modified RNA molecules are designed, and from the

S.L. Salzberg and T. Warnow (Eds.): WABI 2009, LNBI 5724, pp. 97–107, 2009.
© Springer-Verlag Berlin Heidelberg 2009

study of the complete genomes of RNA viruses (which can be up to 11,000 basepairs in length).

The basic *RNA-folding problem* of finding a maximum cardinality, non-crossing, matching of complimentary nucleotides in an RNA sequence of length n, is at the heart of almost all methods to computationally predict RNA secondary structure, including more complex methods that incorporate more realistic folding models, such as allowing some crossovers (pseudoknots). Correspondingly, the basic $O(n^3)$-time *dynamic-programming* solution to the RNA-folding problem remains a central tool in methods to predict RNA structure, and has been widely exposed in books and surveys on RNA folding, computational biology, and computer algorithms. Since the time of the introduction of the $O(n^3)$ dynamic-programming solution to the basic RNA-folding problem, there have been several practical heuristic speedups ([2,12]); and a complex, worst-case speedup of $O(n^3(loglogn)/(logn)^{1/2})$ time ([1]) whose practicality is unlikely and unestablished. In [2], Backofen et al. present a compelling, practical reduction in space and time using the observations of [12] that yields a worst-case improvement when additional problem parameters are included in the time-bound i.e. $O(nZ)$ were $n \leq Z \leq n^2$. The method however retains an $O(n^3)$ time-bound when only the length parameter n is used[1].

Surprisingly, the most widely-used and known, general technique to achieve a worst-case (and often practical) speed up of dynamic programming, the Four-Russians technique, has not been previously applied to the RNA-folding problem, although the general Four-Russians technique has been cited in some RNA folding papers. Two possible reasons for this are that a widely exposed version of the original dynamic-programming algorithm does not lend itself to application of the Four-Russians technique, and unlike other applications of the Four-Russians technique, in RNA folding, it does not seem possible to separate the preprocessing and the computation phases of the Four-Russians method; rather, those two phases are interleaved in our solution.

In this paper, we give a simple, complete and practical Four-Russians algorithm for the basic RNA-folding problem, achieving a worst-case time reduction from $O(n^3)$ to $O(n^3/log(n))$. We show that this time-bound can also be obtained for richer nucleotide matching scoring-schemes, and that the method achieves significant speed-ups in practice. The contribution is both theoretical and practical, since the basic RNA-folding problem is often solved multiple times in the inner-loop of more complex algorithms and for long RNA molecules in the study of RNA virus genomes.

Some of technical insights we use to make the Four-Russians technique work in the RNA-folding dynamic program come from the paper of Graham et. al. ([5])

---

[1] Backofen et al. ([2]) also comment that the general approach in ([1]) can be sped up by combining a newer paper on the all-pairs shortest path problem ([3]). That approach, if correct, would achieve a worst-case bound of $(O(\frac{n^3 * log^3(log(n))}{log^2 n}))$ which is below the $O(n^3/log n)$ bound established here. But that combined approach is highly complex, uses word tricks, is not fully exposed, and has practicality that is unestablished and not promising (in our view).

which gives a Four-Russians solution to the problem of Context-Free Language recognition[2].

2   A Formal Definition of the Basic RNA-Folding Problem

The input to the basic RNA-folding problem consists of a string K of length n over the four-letter alphabet {A,U,C,G}, and an optional integer d. Each letter in the alphabet represents an RNA *nucleotide*. Nucleotides A and U are called *complimentary* as are the nucleotides C and G. A *matching* consists of a set M of *disjoint* pairs of sites in K. If pair (i, j) is in M, then the nucleotide at site i is said to *match* the nucleotide at site j. It is also common to require a fixed minimum distance, d, between the two sites in any match. A match is a *permitted match* if the nucleotides at sites i and j are complimentary, and $|i - j| > d$.[3] A matching M is *non-crossing* or *nested* if and only if it does not contain any four sites $i < i' < j < j'$ where (i, j) and (i', j') are matches in M. Graphically, if we place the sites in K in order on a circle, and draw a straight line between the sites in each pair in M, then M is non-crossing if and only if no two such straight lines cross. Finally, a *permitted matching M* is a matching that is non-crossing, where each match in M is a permitted match. The basic RNA-folding problem is to find a permitted matching of *maximum cardinality*. In a richer variant of the problem, an n by n *integer scoring matrix* is given in the input to the problem; a match between nucleotides in sites i and j in K is given score $B(i, j)$. The problem then is to find a matching with the largest total score. Often this scoring scheme is simplified to give a constant score for each permitted A, U match, and a different constant score for each permitted C, G match.

3   The Original $O(n^3)$ Time Dynamic Programming Solution

Let $S(i, j)$ represent the score for the optimal solution that is possible for the subproblem consisting of the sites in K between i and j inclusive (where $j > i$). Then the following recurrences hold: $S(i, j) = \max\{ \underbrace{S(i + 1, j - 1) + B(i, j)}_{\text{rule a}},$

$\underbrace{S(i, j - 1)}_{\text{rule b}}, \underbrace{S(i + 1, j)}_{\text{rule c}}, \underbrace{Max_{i<k<j}S(i, k) + S(k + 1, j)}_{\text{rule d}} \}$

---

[2] Note that although it is well-known how to reduce the problem of RNA folding to the problem of *stochastic* context-free parsing ([4]), there is no known reduction to non-stochastic context-free parsing, and so it is not possible to achieve the $O(n^3/\log n)$ result by simply reducing RNA folding to context-free parsing and then applying the Four-Russians method from ([5]).

[3] We let d=1 for the remainder of the paper for simplicity of exposition, but in general, d can be any value from 1 to n.

Rule a covers all matchings that contain an (i, j) match; Rule b covers all matchings when site j is not in any match; Rule c covers all matchings when site i is not in any match; Rule d covers all matchings that can be decomposed into two non-crossing matchings in the interval $i..k$, and the interval $k + 1..j$.

In the case of Rule d, the matching is called a *bipartition*, and the interval $i..k$ is called the **head** of bipartition, and the interval $k + 1..j$ is the called the **tail** of the bipartition.

These recurrences can be evaluated in nine different ordering of the variables i,j,k [13]. A common suggestion [13,7] is to evaluate the recurrences in order of increasing distance between i and j. That is, the solution to the RNA folding problem is found for all substrings of K of length two, followed by all substrings of length three, etc. up to length n. This dynamic programming solution is widely published in textbooks, and it is easy to establish that it is correct and that it runs in $O(n^3)$ worst-case time. However, we have not found it possible to apply the Four-Russians technique using that algorithmic evaluation order, but will instead use a different evaluation order.

4   An Alternative $O(n^3)$-Time Dynamic Programming Solution

```
for j =2 to n  do
    [Independent] Calculations below don't depend on the current column j
    for i =1 to j − 1  do
        S(i,j)=max( S(i+1,j-1)+B(i,j) , S(i,j-1)) (Rules a, b )
    [Dependent] Calculations below depend on the current column j
    for i =j − 1 to 1  do
        S(i,j)=max(S(i+1,j) , S(i,j) ) (Rule c)
        for k = j − 1 to i+1 do {The loop below is called the Rule d loop}
            S(i,j)=max( S(i,j), S(i,k-1)+S(k,j) ) (Rule d)
```

The recurrences used in this algorithm are the same as before, but the order of evaluation of S(i,j) is different. It is again easy to see that this Dynamic Programming Algorithm is correct and runs in $O(n^3)$ worst-case time. We will see that this Dynamic Programming algorithm can be used in a Four-Russians speed up.

5   The Four-Russians Speedup

In the Second Dynamic Programming algorithm, each execution of the loop labeled "independent" takes $O(n)$ time, and is inside a loop that executes only $O(n)$ times, so the independent loop takes $O(n^2)$ time in total, and does not need any improvement. The cubic-time behavior of the algorithm comes from the fact that there are three nested loops, for j, i and k respectively, each incrementing $O(n)$ times when entered. The speed-up we will obtain will be due to reducing the work in the Rule d loop. Instead of incrementing k through each value from $j - 1$ down to $i + 1$, we will combine indices into groups of size q (to be determined later) so that only constant time per group will be needed. With that speed up, each execution of that Rule d loop will increment only $O(n/q)$ times when entered. However, we will also need to do some preprocessing, which takes time that increases with q. We will see that setting $q = \log_3(n)$ will yield an $O(n^3/\log n)$ overall worst-case time bound.

5.1   Speeding Up the Computation of S

We now begin to explain the speed-up idea. For now, assume that $B(i,j) = 1$ if (i,j) is a permitted match, and is 0 otherwise. First, conceptually, divide each *column* in the S matrix into groups of q rows, where q will be determined later. For this part of the exposition, suppose $j-1$ is a multiple of q, and let rows $1...q$ be in a group called Rgroup 0, rows $q+1..2q$ be in Rgroup 1 etc, so that rows $j - q...j - 1$ are Rgroup $\lfloor j - 1 \rfloor/q$. We use g as the index for the groups, so g ranges from 0 to $(\lfloor j - 1 \rfloor/q)$. See Figure 1. We will modify the Second Dynamic Program so that for each fixed i, j pair, we do not compute Rule d for each $k = j - 1$ down to $i + 1$. Rather, we will only do constant-time work for each Rgroup g that falls completely into the interval of rows from $j - 1$ down to $i+1$. For the (at most) one Rgroup that falls partially in that interval, we execute the Rule d loop as before. Over the entire algorithm, the time for those partial intervals is $O(n^2q)$, and so this detail will be ignored until the discussion of the time analysis.

Introducing vector V_g and modified Rule d loop. We first modify the Second Dynamic Program to accumulate auxiliary vectors V_g inside the Rule d loop. For a fixed j, consider an Rgroup g consisting of rows $z, z - 1, z - q + 1$, for some $z < j$, and consider the associated consecutive values $S(z, j), S(z - 1, j)...S(z - q + 1, j)$. Let V_g be the vector of *those* values in that order.

The work to accumulate V_g may seem wasted, but we will see shortly how V_g is used.

Introducing v_g. It is clear that for the simple scoring scheme of $B(i,j) = 1$ when (i,j) is a permitted match, and $B(i,j) = 0$ when (i,j) is not permitted, $S(z - 1, j)$ is either equal to S(z,j) or is one more then S(z,j). This observation holds for each consecutive pair of values in V_g. So for a single Rgroup g in column j, the change in consecutive values of V_g can be encoded by a vector of length $q - 1$, whose values are either 0 or 1. We call that vector v_g. We therefore define the function $encode{:}V_g \rightarrow v_g$ such that $v_g[i]=V_g[i\text{-}1]-V_g[i]$. Moreover, for any fixed j, immediately after all the S values have been computed for the cells in an Rgroup g, function $encode(V_g)$ can be computed and stored in $O(q)$ time, and v_g will then be available in the Rule d loop for all i smaller than the smallest row in g. Note that for any fixed j, the time needed to compute all the encode functions is just $O(n)$.

Introducing Table R and the use of v_g. We examine the action of the Second Dynamic-Programming algorithm in the Rule d loop, for fixed j and fixed $i < j - q$. For an Rgroup g in column j, let $k^*(i, g, j)$ be the index k in Rgroup g such that $S(i, k - 1) + S(k, j)$ is maximized, and let $S^*(i, g, j)$ denote the actual value $S(i, k^*(i, g, j) - 1) + S(k^*(i, g, j), j)$.

Note that the Second Dynamic-Program can find $k^*(i, g, j)$ and $S^*(i, g, j)$ during the execution of the Rule d loop, but would take $O(q)$ time to do so. However, by previously doing some preprocessing (to be described below) before column j is reached, we will reduce the work in each Rgroup g to $O(1)$ time.

To explain the idea, suppose that before column j is reached, we have *pre-computed* a table R which is indexed by i, g, v_g. Table R will have the property that, for fixed j and $i < j - q$, a *single lookup* of $R(i, g, v_g)$) will effectively return $k^*(i, g, j)$ for any g. Since $k - 1 < j$ and $k > i$, both values $S(i, k^*(i, g, j) - 1)$ and $S(k^*(i, g, j), j)$ are known when we are trying to evaluate $S(i, j)$, so we can find $S(i, k^*(i, g, j) - 1) + S(k^*(i, g, j), j)$ in $O(1)$ operations once $k^*(i, g, v_g)$ is known.

Since there are $O(j/q)$ Rgroups, it follows that for fixed j and i, by calling table R once for each Rgroup, only $O(j/q)$ work is needed in the Rule d loop. Hence, for any fixed j, letting i vary over its complete range, the work will be $O(j^2/q)$, and so the total work (over the entire algorithm) in the Rule d loop will be $O(n^3/q)$. Note that as long as $q < n$, some work has been saved, and the amount of saved work increases with increasing q. This use of the R table in the Rule d loop is summarized as follows:

Dependent section using table R.

   for $g = \lfloor (j - 1)/q \rfloor$ to $\lfloor (i + 1)/q \rfloor$ **do**
      retrieve v_g given g
      retrieve $k^*(i, g, j)$ from $R(i, g, v_g)$
      S(i,j)=max(S(i,j), $S(i, k^*(i, g, j) - 1) + S(k^*(i, g, j), j)$);

Of course, we still need to explain how R is precomputed.

5.2   Obtaining Table R

Before explaining exactly where and how table R is computed, consider the action of the Second Dynamic-Programming algorithm in the Rule d loop, for a fixed j. Let g be an Rgroup consisting of rows $z - q + 1, z - q, ..., z$, for some $z < j$. A key observation is that if one knows the single value $S(z, j)$ and the entire vector v_g, then one can determine all the values $S(z - q + 1, j)...S(z, j)$ or V_g. Each such value is exactly $S(z, j)$ plus a partial sum of the values in v_g. In more detail, for any $k \in g$, $S(k, j) = S(z, j) + \sum_{p=0}^{p=z-k-1} v_g[p]$. Let $decode(v_g)$ be a function that returns the vector V' where $V'[k] = \sum_{p=0}^{p=z-k-1} v_g[p]$.

Next, observe that if one does *not* know any of the V_g in the rows of g (e.g., the values $S(z - q + 1, j), S(z - 1, j)...S(z, j)$), but *does* know all of v_g, then, for any fixed i below the lowest row in Rgroup g (i.e., row z-q+1), one can find the value of index k in Rgroup g to maximize $S(i, k - 1) + S(k, j)$. That value of k is what we previously defined as $k^*(i, g, j)$. To verify that $k^*(i, g, j)$ can be

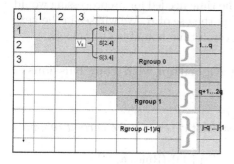

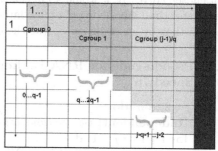

Fig. 1. Rgroups with $q=3$ **Fig. 2.** Cgroups

determined from v_g, but without knowing any S values in column j, recall that since $k - 1 < j$, $S(i, k - 1)$ is already known. We call this Fact 1.

Precomputing the R table. We now describe the preprocessing that is needed to compute table R.

Conceptually divide matrix for S into groups of *columns* of size q, i.e., the same size groups that divide each column. Columns 1 through $q - 1$ are in a group we call Cgroup 0, q through $2q - 1$ are in Cgroup 1 etc, and we again use g to index these groups.

Assume we run the Second Dynamic Program until j reaches $q - 1$. That means that all the $S(i, j)$ values have been completely and correctly computed for all columns in Cgroup 0. At that point, we compute the following:

for each binary vector v of length $q - 1$ **do**
    $V' = decode(v)$
    for each i such that $i < q - 1$ **do**
        $R(i, 0, v)$ is set to the index k in Rgroup 0 such that S(i,k-1) + V'[k] is maximized. {we let k^* denote that optimal k }

The above details the preprocessing done after all the S values in Cgroup 0 have been computed. In general, for Cgroup $g > 0$, we could do a similar preprocessing after all the entries in columns of Cgroup g have been computed. That is, $k^*(i,g,v)$ could be found and stored in R(i,g,v) for all $i < g * q$.

This describes the preprocessing that is done after the computation of the S values in each Rgroup g. With that preprocessing, the table R is available for use when computing S values in any column $j > g \times q$. Note that the preprocessing computations of table R are interleaved with the use of table R. This is different than other applications of the Four-Russians technique that we know of. Note also that the amount of preprocessing work increases with increasing q. Several additional optimizations are possible, of which one, parallel computation, is described in Section 5.5.

With this, the description of the Four-Russians speed up is complete. However, as in most applications of the Four-Russians idea, q must be chosen carefully. If not chosen correctly, it would seem that the time needed for preprocessing would be greater than any saving obtained later in the use of R. We will see in the next section that by choosing $q=log_3(n)$ the overall time bound is $O(n^3/\log(n))$.

5.3 Pseudo-code for RNA Folding Algorithm with Four Russians SpeedUp

for $j =2$ to n **do**
    [**Independent**] *Calculations below don't depend on the current column j*
    for $i =1$ to $j-1$ **do**
        S(i,j)=max(S(i+1,j-1)+B(i,j) , S(i,j-1)) (Rules a, b)
    [**Dependent**] *Calculations below depend on the current column j*
    for $i =j-1$ to 1 **do**
        for $g =\lfloor(j-1)/q\rfloor$ to $\lfloor(i+1)/q\rfloor$ **do**
            if $(i > k)$, $k \in$ Rgroup g **then** {this statement runs at most once, for the smallest g}
                find $k^*(i,g,j)$ by directly computing and comparing S(i,k-1)+S(k,j) where $k \in$ L={$g*q$ to i}[4] $|L| < q$
        else
            retrieve v_g given g
            retrieve $k^*(i,g,j)$ from $R(i,g,v_g)$
            S(i,j)=max(S(i,j), S(i,k^*(i,g,j) − 1) + S(k^*(i,g,j),j));
        if $((i-1)$ mod $q == 0)$, Compute v_g for group g and store it
    [**Table**] once Cgroup $g = \lfloor j/q\rfloor$ is complete
    for each binary vector v of length $q-1$ **do**
        $V'=decode(v)$
        for each i 1 to $i < j-1$ **do**
            $R(i,g,v)$ is set to the index k in Rgroup g such that S(i,k-1) + V'[k] is maximized.

5.4 Correctness and Time Analysis

From Fact 1, it follows that when the algorithm is in the Rule d loop, for some fixed j, and needs to use R to look up $k^*(i,g,j)$ for some g, $k^*(i,g,j) = R(i,g,v_g)$. It follows immediately, that the algorithm does correctly compute all of the S values, and fills in the complete S table. A standard traceback can be done to extract the optimal matching, in $O(n)$ time if pointers were kept when the S table built.

Time Analysis. The time analysis for any column j can be broken down into the sum of the time analyzes for the [**independent**], [**dependent**], [**table**] sections.
    For any column j the [**independent**] section of the speedup algorithm remains unchanged from the original algorithm and is $O(n)$ time. For each row

---

[4] Reversing the order of evaluation in the algorithm for i, and k would eliminate this overhead. Lowering the time for this operation from $0(q)$ to $0(1)$. However for simplicity of exposition we leave those details out.

i, the [**dependent**] section of the speedup algorithm is now broken down into n/q calls to the table R. As discussed above, the total time to compute all the encode functions is $O(n)$ per column, and this is true for all decode functions as well. Therefore in any column j, the dependent section takes $0(\frac{n^2}{q})$ time. Also, the processing of the one (possible) partial group in any column j only takes $O(n^2q)$ time. The [**table**] section sets R(i,g,v) by computing every binary vector v and then computing and storing the value k*(i,g,v). The variable i ranges from 1 to n and there are 2^{q-1} binary vectors. Hence the dependent section for any j takes $O(n * 2^{q-1})$ time.

In summary, the total time for the algorithm is $O(n^2 * q) + 0(\frac{n^3}{q}) + 0(n^2 * 2^q)$ time.

Theorem 1. (Runtime) *If* $2 < b < n$, *then the algorithm runs in* $O(n^3/\log_b(n))$ *time.*

Proof. Clearly, $O(n^2\log_b(n) + \frac{n^3}{q}) = O(n^3/\log_b(n))$.
To show that $n^2 \times 2^{\log_b(n)} = O(n^3/\log_b(n))$ for $2 < b < n$, we need to show that

$$2^{\log_b(n)} = O(n/\log_b(n)).$$

The relation holds iff

$$n^{\log_b(2)} = O(n/\log_b(n)) \text{ iff}$$
$$\log_b(n) \times n^z = O(n)$$

for $z = \log_b(2)$. $0 < z < 1$ since $b > 2$.
The above relation holds if

$$\lim_{n\to\infty} \frac{(n^z * \log_b(n))}{n} = 0$$

We simplify the above equation to $\lim_{n\to\infty} \log_b(n)/(n^{1-z})$
We find the limit by taking the derivative of top and bottom (L' Hopital's Rule)

$$\lim_{n\to\infty} \frac{\log_b(n)}{(n^{1-z})} = \lim_{n\to\infty} \frac{(\frac{1}{n\ln(b)})}{(1-z)n^{-z}} = \lim_{n\to\infty} \frac{1}{n\ln(b)*(1-z)n^{-z}} =$$

$$\lim_{n\to\infty} \frac{1}{(1-z)\ln(b)n^{1-z}} = 0 \text{ since } z < 1.$$

5.5 Parallel Computing

By exploiting the parallel nature of the computation for a specific column, one can achieve a time bound of $O(n^2)$ with vector computing. The [Independent] section computes max(S(i+1,j-1)+B(i,j), S(i,j-1)), and all three of the values are available for all i simultaneously. So for each i we could compute the maximum in parallel with an asymptotic run time of $O(1)$. The [Dependent] section encodes the entire column into vectors of size q. This could be done in parallel, but sequentially it is has $0(n)$ asymptotic run time.

Table 1. Empirical Results

$Size$	$O(n^3)$ Algorithm	$O(n^3/\log(n))$ Algorithm	ratio
1000	3.20	1.43	2.23
2000	27.10	7.62	3.55
3000	95.49	26.90	3.55
4000	241.45	55.11	4.38
5000	470.16	97.55	4.82
6000	822.79	157.16	5.24

The encoding is used to reference the R table. For each i there are q entries in the R table. The maximum for each i computed in parallel takes $O(n + q)=O(n)$ time to find. The [Tabling] section can also be done in parallel to find k^* for all possible v vectors in $O(1)$ time, entering all 2^{q-1} values into table R simultaneously. The entire algorithm then takes $O(n^2)$ time in parallel.

6   Generalized Scoring Schemes for B(i,j)

The Four Russians Speed-Up could be extended to any B(i,j) for which all possible differences between S(i,j) and S(i+1,j) do not depend on n. Let C denote the size of the set of all possible differences. The condition that C doesn't depend on n allows one to incorporate not just pairing energy information but also energy information that is dependent on the distance between matching pairs and types of loops. In fact the tabling idea currently can be applied to any scoring scheme as long as the marginal score $(S(i-1,j) - S(i,j))$is not $\Omega(n)$. In this case, the algorithm takes $O(n * (n * C^{q-1} + \frac{n^2}{q} + n)) = O(n^2 C^{q-1} + \frac{n^3}{q} + n^2)$ time. If we let $q = \log_b(n)$ with $b > C$, the asymptotic time is again $0(n^3/log(n))$. Based on the proof of Theorem 1, the base of the log must be greater then C in order to achieve the speed up. The scoring schemes in ([6,10]) have marginal scores that are not dependent on n, so the speedup method can be applied in those cases.

6.1   Empirical Results

We compare our Four-Russians algorithm to the original $O(n^3)$-time algorithm. The empirical results shown below give the average time for 50 tests of randomly generated RNA sequences and 10 downloaded sequences from genBank, for each size between 1,000bp and 6,000bp. All times had a standard deviation of .01. The purpose of these empirical results is to show that despite the additional overhead required for the Four-Russians approach, it does provide a consistent practical speed-up over the $O(n^3)$-time method, and does not just yield a theoretical result. In order to make this point, we keep all conditions for the comparisons the same, we emphasize the ratio of the running times rather than the absolute times, and we do not incorporate any additional heuristic ideas or optimizations that might be appropriate for one method but not the other. However, we are aware that there are speedup ideas for RNA folding, which do not reduce the

$O(n^3)$ bound, but provide significant practical speedups. Our empirical results are not intended to compete with those results, or to provide a finished competitive *program*, but to illustrate the practicality of the Four-Russians method, in addition to its theoretical consequences. In future work, we will incorporate all known heuristic speedups, along with the Four-Russians approach, in order to obtain an RNA folding *program* which can be directly compared to all existing methods.

Acknowledgments. This research was partially supported by NSF grants SEI-BIO 0513910, CCF-0515378, and IIS-0803564.

References

1. Akutsu, T.: Approximation and exact algorithms for RNA secondary structure prediction and recognition of stochastic context-free languages. J. Comb. Optim. 3(2-3), 321–336 (1999)
2. Backofen, R., Tsur, D., Zakov, S., Ziv-Ukelson, M.: Sparse RNA folding: Time and space efficient algorithms. In: CPM 2009 (2009)
3. Chan, T.M.: More algorithms for all-pairs shortest paths in weighted graphs. In: STOC, pp. 590–598 (2007)
4. Durbin, R., Eddy, S.R., Krogh, A., Mitchison, G.: Biological Sequence Analysis: Probabilistic Models of Proteins and Nucleic Acids. Cambridge University Press, Cambridge (1998)
5. Graham, S.L., Harrison, M., Ruzzo, W.L.: An improved context-free recognizer. ACM Trans. Program. Lang. Syst. 2(3), 415–462 (1980)
6. Hofacker, I.L., Fekete, M., Stadler, P.F.: Secondary structure prediction for aligned RNA sequences. Journal of Molecular Biology 319(5), 1059–1066 (2002)
7. Kleinberg, J., Tardos, E.: Algorithm Design. Addison-Wesley Longman Publishing Co., Inc., Boston (2005)
8. Nussinov, R., Jacobson, A.B.: Fast algorithm for predicting the secondary structure of single-stranded RNA. PNAS 77(11), 6309–6313 (1980)
9. Nussinov, R., Pieczenik, G., Griggs, J.R., Kleitman, D.J.: Algorithms for loop matchings. SIAM Journal on Applied Mathematics 35(1), 68–82 (1978)
10. Seemann, S.E., Gorodkin, J., Backofen, R.: Unifying evolutionary and thermodynamic information for RNA folding of multiple alignments. NAR (2008)
11. Waterman, M.S., Smith, T.F.: RNA secondary structure: A complete mathematical analysis. Math. Biosc. 42, 257–266 (1978)
12. Wexler, Y., Zilberstein, C.B.-Z., Ziv-Ukelson, M.: A study of accessible motifs and RNA folding complexity. Journal of Computational Biology 14(6), 856–872 (2007)
13. Zuker, M., Sankoff, D.: RNA secondary structures and their prediction. Bulletin of Mathematical Biology 46(4), 591–621 (1984)
14. Zuker, M., Stiegler, P.: Optimal computer folding of large RNA sequences using thermodynamics and auxiliary information. Nucleic Acids Research 9(1), 133–148 (1981)

Back-Translation for Discovering Distant Protein Homologies

Marta Gîrdea, Laurent Noé, and Gregory Kucherov*

INRIA Lille - Nord Europe, LIFL/CNRS, Université Lille 1, 59655 Villeneuve d'Ascq,
France

Abstract. Frameshift mutations in protein-coding DNA sequences produce a drastic change in the resulting protein sequence, which prevents
classic protein alignment methods from revealing the proteins' common
origin. Moreover, when a large number of substitutions are additionally involved in the divergence, the homology detection becomes difficult
even at the DNA level. To cope with this situation, we propose a novel
method to infer distant homology relations of two proteins, that accounts
for frameshift and point mutations that may have affected the coding sequences. We design a dynamic programming alignment algorithm over
memory-efficient graph representations of the complete set of putative
DNA sequences of each protein, with the goal of determining the two
putative DNA sequences which have the best scoring alignment under a
powerful scoring system designed to reflect the most probable evolutionary process. This allows us to uncover evolutionary information that is
not captured by traditional alignment methods, which is confirmed by
biologically significant examples.

1 Introduction

In protein-coding DNA sequences, frameshift mutations (insertions or deletions
of one or more bases) can alter the translation reading frame, affecting all the
amino acids encoded from that point forward. Thus, frameshifts produce a drastic change in the resulting protein sequence, preventing any similarity to be
visible at the amino acid level.

When the coding DNA sequence is relatively well conserved, the similarity
remains detectable at the DNA level, by DNA sequence alignment, as reported
in several papers, including [1,2,3,4].

However, the divergence often involves additional base substitutions. It has
been shown [5,6,7] that, in coding DNA, there is a base compositional bias
among codon positions, that does not apply when the translation reading frame
is changed. Hence, after a reading frame change, a coding sequence is likely to
undergo base substitutions leading to a composition that complies with this bias.
Amongst these substitutions, synonymous mutations (usually occurring on the
third position of the codon) are more likely to be accepted by natural selection,

---

* On leave in J.-V. Poncelet Lab, Moscow, Russia.

S.L. Salzberg and T. Warnow (Eds.): WABI 2009, LNBI 5724, pp. 108–120, 2009.

since they are silent with respect to the gene's product. If, in a long evolutionary time, a large number of codons in one or both sequences are affected by these changes, the sequence may be altered to such an extent that the common origin becomes difficult to observe by direct DNA comparison.

In this paper, we address the problem of finding distant protein homologies, in particular when the primary cause of the divergence is a frameshift. We achieve this by computing the best alignment of DNA sequences that encode the target proteins. This approach relies on the idea that synonymous mutations cause mismatches in the DNA alignments that can be avoided when all the sequences with the same translation are explored, instead of just the known coding DNA sequences. This allows the algorithm to search for an alignment by dealing only with non-synonymous mutations and gaps.

We designed and implemented an efficient method for aligning putative coding DNA sequences, which builds expressive alignments between hypothetical nucleotide sequences that can provide some information about the common ancestral sequence, if such a sequence exists. We perform the analysis on memory-efficient graph representations of the complete set of putative DNA sequences for each protein, described in Section 3.1. The proposed method, presented in Section 3.2, consists of a dynamic programming alignment algorithm that computes the two putative DNA sequences that have the best scoring alignment under an appropriate scoring system (Section 3.3) designed to reflect the actual evolution process from a codon-oriented perspective.

While the idea of finding protein relations by frameshifted DNA alignments is not entirely new, as we will show in Section 2 in a brief related work overview, Section 4 – presenting tests performed on artificial data – demonstrates the efficiency of our scoring system for distant sequences. Furthermore, we validate our method on several pairs of sequences known to be encoded by overlapping genes, and on some published examples of frameshifts resulting in functional proteins. We briefly present these experiments in Section 5, along with a study of a protein family whose members present high dissimilarity on a certain interval. The paper is concluded in Section 6.

2   Related Work

The idea of using knowledge about coding DNA when aligning amino acid sequences has been explored in several papers.

A *non-statistical approach* for analyzing the homology and the "genetic semi-homology" in protein sequences was presented in [8,9]. Instead of using a statistically computed scoring matrix, amino acid similarities are scored according to the complexity of the substitution process at the DNA level, depending on the number and type (transition/transversion) of nucleotide changes that are necessary for replacing one amino acid by the other. This ensures a differentiated treatment of amino acid substitutions at different positions of the protein sequence, thus avoiding possible rough approximations resulting from scoring them equally, based on a classic scoring matrix. The main drawback of this approach is that it was not designed to cope with frameshift mutations..

Regarding *frameshift mutation discovery*, many studies [1,2,3,4] preferred the plain BLAST [10,11] alignment approach: BLASTN on DNA and mRNA, or BLASTX on mRNA and proteins, applicable only when the DNA sequences are sufficiently similar. BLASTX programs, although capable of insightful results thanks to the six frame translations, have the limitation of not being able to transparently manage frameshifts that occur inside the sequence, for example by reconstructing an alignment from pieces obtained on different reading frames.

An interesting approach for *handling frameshifts at the protein level* was developed in [12]. Several substitution matrices were designed for aligning amino acids encoded on different reading frames, based on nucleotide pair matches between respective codons. This idea has the advantage of being easy to use with any classic protein alignment tool. However, it lacks flexibility in gap positioning.

On the subject of *aligning coding DNA in presence of frameshift errors*, some related ideas were presented in [13,14]. The author proposed to search for protein homologies by aligning their *sequence graphs* (data structures similar to the ones we describe in Section 3.1). The algorithm tries to align pairs of codons, possibly incomplete since gaps of size 1 or 2 can be inserted at arbitrary positions. The score for aligning two such codons is computed as the maximum substitution score of two amino acids that can be obtained by translating them. This results in a complex, time costly dynamic programming method that basically explores all the possible translations. In Section 3.2, we present an algorithm addressing the same problem, more efficient since it aligns symbols, not codons, and more flexible with respect to scoring functions. Additionally, we propose to use a scoring system relying on codon evolution rather than amino acid translations, since we believe that, in frameshift mutation scenarios, the information provided by DNA sequence dynamics is more relevant than amino acid similarities.

3   Our Approach to Distant Protein Relation Discovery

The problem of inferring homologies between distantly related proteins, whose divergence is the result of frameshifts and point mutations, is approached in this paper by determining the best pairwise alignment between two DNA sequences that encode the proteins.

Given two proteins P_A and P_B, the objective is to find a pair of DNA sequences, D_A and D_B, such that $translation(D_A) = P_A$ and $translation(D_B) = P_B$, which produce the best pairwise alignment under a given scoring system.

The alignment algorithm (described in Section 3.2) incorporates a gap penalty that limits the number of frameshifts allowed in an alignment, to comply with the observed frequency of frameshifts in a coding sequence's evolution. The scoring system (Section 3.3) is based on possible mutational patterns of the sequences. This leads to reducing the false positive rate and focusing on alignments that are more likely to be biologically significant.

3.1   Data Structures

An explicit enumeration and pairwise alignment of all the putative DNA sequences is not an option, since their number increases exponentially with the protein's length[1]. Therefore, we represent the protein's "back-translation" (set of possible source DNAs) as a directed acyclic graph, whose size depends linearly on the length of the protein, and where a path represents one putative sequence.

As illustrated in Figure 1(a), the graph is organized as a sequence of length $3n$ where n is the length of the protein sequence. At each position i in the graph, there is a group of nodes, each representing a possible nucleotide that can appear at position i in at least one of the putative coding sequences. Two nodes at consecutive positions are linked by arcs if and only if they are either consecutive nucleotides of the same codon, or they are respectively the third and the first base of two consecutive codons. No other arcs exist in the graph.

Note that in the implementation, the number of nodes is reduced by using the IUPAC nucleotide codesIf the amino acids composing a protein sequence are non-ambiguous, only 4 extra nucleotide symbols – R, Y, H and N – are necessary for their back-translation. In this condensed representation, the number of ramifications in the graph is substantially reduced, as illustrated by Figure 1. More precisely, the only amino acids with ramifications in their back-translation are amino acids R, L and S, each encoded by 6 codons with different prefixes.

3.2   Alignment Algorithm

We use a dynamic programming method, similar to the Smith-Waterman algorithm, extended to data structures described in Section 3.1 and equipped with gap related restrictions.

Given the input graphs G_A and G_B obtained by back-translating proteins P_A and P_B, the algorithm finds the best scoring local alignment between two DNA sequences comprised in the back-translation graphs (illustrated in Figure 2). The alignment is built by filling each entry $M[i, j, (\alpha_A, \alpha_B)]$ of a dynamic programming matrix M, where i and j are positions of the first and second graph respectively, and (α_A, α_B) is a pair of nodes that can be found in G_A at position i, and in G_B at position j, respectively. An example is given in Figure 3.

The dynamic programming algorithm begins with a classic local alignment initialization (0 at the top and left borders), followed by the recursion step described in equation (1). The partial alignment score from each cell $M[i, j, (\alpha_A, \alpha_B)]$ is computed as the maximum of 6 types of values:

(a) 0 (similarly to the classic Smith-Waterman algorithm, only non-negative scores are considered for local alignments).

(b) the substitution score of symbols (α_A, α_B), denoted $score(\alpha_A, \alpha_B)$, added to the score of the best partial alignment ending in $M[i-1, j-1]$, provided that the partially aligned paths contain α_A on position i and α_B on position

---

[1] With the exception of M and W, which have a single corresponding codon, all amino acids are encoded by 2, 3, 4 or 6 codons.

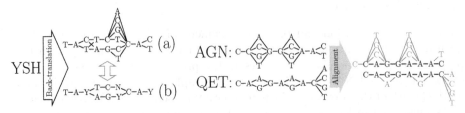

Fig. 1. Example of fully represented (a) and condensed (b) back-translation graph for the amino acid sequence YSH

Fig. 2. Alignment example. A path (corresponding to a putative DNA sequence) was chosen from each graph so that the match/mismatch ratio is maximized.

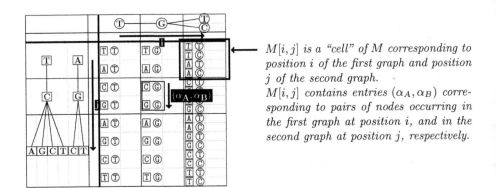

$M[i,j]$ is a "cell" of M corresponding to position i of the first graph and position j of the second graph. $M[i,j]$ contains entries (α_A, α_B) corresponding to pairs of nodes occurring in the first graph at position i, and in the second graph at position j, respectively.

Fig. 3. Example of dynamic programming matrix M

j respectively; this condition is ensured by restricting the entries of $M[i-1, j-1]$ to those labeled with symbols that precede α_A and α_B in the graphs.

(c) the cost *singleGapPenalty* of a frameshift (gap of size 1 or extension of a gap of size 1) in the first sequence, added to the score of the best partial alignment that ends in a cell $M[i, j-1, (\alpha_A, \beta_B)]$, provided that β_B precedes α_B in the second graph; this case is considered only if the number of allowed frameshifts on the current path is not exceeded, or a gap of size 1 is extended.

(d) the cost of a frameshift in the second sequence, added to a partial alignment score defined as above.

(e) the cost *tripleGapPenalty* of removing an entire codon from the first sequence, added to the score of the best partial alignment ending in a cell $M[i, j-3, (\alpha_A, \beta_B)]$.

(f) the cost of removing an entire codon from the second sequence, added to the score of the best partial alignment ending in a cell $M[i-3, j, (\beta_A, \alpha_B)]$

We adopted a non-monotonic gap penalty function, which favors insertions and deletions of full codons, and does not allow a large number of frameshifts – very rare events, usually eliminated by natural selection. As can be seen in equation (1), two particular kinds of gaps are considered: **i) frameshifts** – gaps of size 1 or 2, with high penalty, whose number in a local alignment can be limited, and **ii) codon skips** – gaps of size 3 which correspond to the insertion or deletion of a whole codon.

$$
M[i, j, (\alpha_A, \alpha_B)] =
$$
$$
max \begin{cases}
0 & \text{(a)} \\
M[i-1, j-1, (\beta_A, \beta_B)] + score(\alpha_A, \alpha_B), & \beta_k \in pred(\alpha_k); & \text{(b)} \\
(M[i, j-1, (\alpha_A, \beta_B)] + singleGapPenalty), \beta_B \in pred(\alpha_B); & \text{(c)} \\
(M[i-1, j, (\beta_A, \alpha_B)] + singleGapPenalty), \beta_A \in pred(\alpha_A); & \text{(d)} \\
(M[i, j-3, (\alpha_A, \beta_B)] + tripleGapPenalty), \ j \geq 3 & \text{(e)} \\
(M[i-3, j, (\beta_A, \alpha_B)] + tripleGapPenalty), \ i \geq 3 & \text{(f)}
\end{cases}
\quad (1)
$$

3.3   Translation-Dependent Scoring Function

In this section, we present a new translation-dependent scoring system suitable for our alignment algorithm. The scoring scheme we designed incorporates information about possible mutational patterns for coding sequences, based on a codon substitution model, with the aim of filtering out alignments between sequences that are unlikely to have common origins.

Mutation rates have been shown to vary within genomes, under the influence of several factors, including neighbor bases [15]. Consequently, a model where all base mismatches are equally penalized is oversimplified, and ignores possibly precious information about the context of the substitution.

With the aim of retracing the sequence's evolution and revealing which base substitutions are more likely to occur within a given codon, our scoring system targets pairs of triplets (α, p, a), were α is a nucleotide, p is its position in the codon, and a is the amino acid encoded by that codon, thus differentiating various contexts of a substitution. There are 99 valid triplets out of the total of 240 hypothetical combinations.

Pairwise alignment scores are computed for all possible pairs of valid triplets $(t_1, t_2) = ((\alpha_1, p_1, a_1), (\alpha_2, p_2, a_2))$ as a classic log-odds ratio:

$$
score(t_1, t_2) = \lambda \log \frac{f_{t_1 t_2}}{b_{t_1 t_2}}
\quad (2)
$$

where $f_{t_1 t_2}$ is the frequency of the $t_1 \leftrightarrow t_2$ substitution in related sequences, and $b_{t_1 t_2} = p(t_1)p(t_2)$ is the background probability.

In order to obtain the foreground probabilities $f_{t_i t_j}$, we will consider the following scenario: two proteins are encoded on the same DNA sequence, on different reading frames; at some point, the sequence was duplicated and the two copies diverged independently; we assume that the two coding sequences undergo, in their independent evolution, synonymous and non-synonymous point mutations, or full codon insertions and removals.

The insignificant amount of available real data that fits our hypothesis does not allow classical, statistical computation of the foreground and background probabilities. Therefore, instead of doing statistics on real data directly, we will rely on codon frequency tables and codon substitution models.

We assume that codon substitutions in our scenarios can be modeled by a Markov model presented in [16][2] which specifies the relative instantaneous substitution rate from codon i to codon j as:

$$
Q_{ij} = \begin{cases}
0 & \text{if } i \text{ or } j \text{ is a stop codon, or} \\
& \text{if } i \rightarrow j \text{ requires more than 1 nucleotide substitution,} \\
\pi_j & \text{if } i \rightarrow j \text{ is a synonymous transversion,} \\
\pi_j \kappa & \text{if } i \rightarrow j \text{ is a synonymous transition,} \\
\pi_j \omega & \text{if } i \rightarrow j \text{ is a nonsynonymous transversion,} \\
\pi_j \kappa \omega & \text{if } i \rightarrow j \text{ is a nonsynonymous transition.}
\end{cases} \tag{3}
$$

for all $i \neq j$. Here, the parameter ω represents the nonsynonymous-synonymous rate ratio, κ the transition-transversion rate ratio, and π_j the equilibrium frequency of codon j. As in all Markov models of sequence evolution, absolute rates are found by normalizing the relative rates to a mean rate of 1 at equilibrium, that is, by enforcing $\sum_i \sum_{j \neq i} \pi_i Q_{ij} = 1$ and completing the instantaneous rate matrix Q by defining $Q_{ii} = -\sum_{j \neq i} Q_{ij}$ to give a form in which the transition probability matrix is calculated as $P(\theta) = e^{\theta Q}$ [18]. Evolutionary times θ are measured in expected number of nucleotide substitutions per codon.

With this codon substitution model, $f_{t_i t_j}$ can be deduced in several steps. Basically, we first need to identify all pairs of codons with a common subsequence, that have a perfect semi-global alignment (for instance, codons CAT and ATG satisfy this condition, having the common subsequence AT; this example is further explained below). We then assume that the codons from each pair undergo independent evolution, according to the codon substitution model. For the resulting codons, we compute, based on all possible original codon pairs, $p((\alpha_i, p_i, c_i), (\alpha_j, p_j, c_j))$ – the probability that nucleotide α_i, situated on position p_i of codon c_i, and nucleotide α_j, situated on position p_j of codon c_j have a common origin (equation (5)). From these, we can immediately compute, as shown by equation (6), $p((\alpha_i, p_i, a_i), (\alpha_j, p_j, a_j))$, corresponding in fact to the foreground probabilities $f_{t_i t_j}$, where $t_i = (\alpha_i, p_i, a_i)$ and $t_j = (\alpha_j, p_j, a_j)$.

In the following, $\mathbf{p}(\mathbf{c_1} \xrightarrow{\theta} \mathbf{c_2})$ stands for the probability of the event *codon c_1 mutates into codon c_2 in the evolutionary time* θ, and is given by $P_{c_1, c_2}(\theta)$.

$\mathbf{c_1}[\mathbf{interval_1}] \equiv \mathbf{c_2}[\mathbf{interval_2}]$ states that codon c_1 restricted to the positions given by $interval_1$ is a sequence identical to c_2 restricted to $interval_2$. This is equivalent to having a word w obtained by "merging" the two codons. For instance, if $c_1 = CAT$ and $c_2 = ATG$, with their common substring being placed in $interval_1 = [2..3]$ and $interval_2 = [1..2]$ respectively, w is $CATG$.

---

[2] Another, more advanced codon substitution model, targeting sequences with overlapping reading frames, is proposed and discussed in [17]. It does not fit our scenario, because it is designed for overlapping reading frames, where a mutation affects both translated sequences, while in our case the sequences become at one point independent and undergo mutations independently.

Finally, $p(c_1[\mathbf{interval_1}] \equiv c_2[\mathbf{interval_2}])$ is the probability to have c_1 and c_2, in the relation described above, which we compute as the probability of the word w obtained by "merging" the two codons. This function should be symmetric, it should depend on the codon distribution, and the probabilities of all the words w of a given length should sum to 1. However, since we consider the case where the same DNA sequence is translated on two different reading frames, one of the two translated sequences would have an atypical composition. Consequently, the probability of a word w is computed as if the sequence had the known codon composition when translated on the reading frame imposed by the first codon, or on the one imposed by the second. This hypothesis can be formalized as:

$$p(w) = p(w \text{ on } rf_1 \text{ OR } w \text{ on } rf_2) = p^{rf_1}(w) + p^{rf_2}(w) - p^{rf_1}(w) \cdot p^{rf_2}(w) \quad (4)$$

where $p^{rf_1}(w)$ and $p^{rf_2}(w)$ are the probabilities of the word w in the reading frame imposed by the position of the first and second codon, respectively. This is computed as the products of the probabilities of the codons and codon pieces that compose the word w in the established reading frame. In the previous example, the probabilities of $w = CATG$ in the first and second reading frame are:

$$p^{rf_1}(CATG) = p(CAT) \cdot p(G**) = p(CAT) \cdot \sum_{c:c \text{ starts with } G} p(c)$$

$$p^{rf_2}(CATG) = p(**C) \cdot p(ATG) = \sum_{c:c \text{ ends with } C} p(c) \cdot p(ATG)$$

The values of $p((\alpha_i, p_i, c_i), (\alpha_j, p_j, c_j))$ are computed as:

$$\sum_{\substack{c_i', c_j' : c_i'[interval_i] \equiv c_j'[interval_j] \\ p_i \in interval_i, p_j \in interval_j}} p(c_i'[interval_i] \equiv c_j'[interval_j]) \cdot p(c_i' \xrightarrow{\theta} c_i) \cdot p(c_j' \xrightarrow{\theta} c_j) \quad (5)$$

from which obtaining the **foreground probabilities** is straightforward:

$$f_{t_i t_j} = p((\alpha_i, p_i, a_i), (\alpha_j, p_j, a_j)) = \sum_{\substack{c_i \text{ encodes } a_i, \\ c_j \text{ encodes } a_j}} p((\alpha_i, p_i, c_i), (\alpha_j, p_j, c_j)) \quad (6)$$

The **background probabilities** of (t_i, t_j), $b_{t_i t_j}$, can be simply expressed as the probability of the two symbols appearing independently in the sequences:

$$b_{t_i t_j} = b_{(\alpha_i, p_i, a_i), (\alpha_j, p_j, a_j)} = \sum_{\substack{c_i \text{ encodes } a_i, \\ c_j \text{ encodes } a_j}} \pi_{c_i} \pi_{c_j} \quad (7)$$

Substitution matrix for ambiguous symbols. From matrices built as explained above, the versions that use IUPAC ambiguity codes for nucleotides (as proposed in the final paragraph of 3.1) can be computed: the score of pairing two ambiguous symbols is the maximum over all substitution scores for all pairs of nucleotides from the respective sets.

Score evaluation. The score significance is estimated according to the Gumbel distribution, where the parameters λ and K are computed with the method described in [19,20]. Since the forward alignment and the reverse complementary alignment are two independent cases with different score distributions, two parameter pairs, λ_{fw}, K_{fw} and λ_{rc}, K_{rc} are computed and used in practice.

4   Validation

To validate the translation-dependent scoring system we designed in the previous section, we tested it on an artificial data set consisting in 96 pairs of protein sequences of average length 300. Each pair was obtained by translating a randomly generated DNA sequence on two different reading frames. Both sequences in each pair were then mutated independently, according to codon mutation probability matrices corresponding to each of the evolutionary times 0.01, 0.1, 0.3, 0.5, 0.7, 1.0, 1.5, 2.00 (measured in average number of mutations per codon).

To this data set we applied four variants of alignment algorithms: i) classic alignment of DNA sequences using classic base substitution scores and affine gap penalties; ii) classic alignment of DNA sequences using a translation-dependent scoring scheme designed in Section 3.3; iii) alignment of back-translation graphs (Section 3.2) using classic base substitution scores and affine gap penalties; iv) alignment of back-translation graphs using a translation-dependent scoring scheme. For the tests involving translation-dependent scores, we used scoring functions corresponding to evolutionary times from 0.30 to 1.00.

Table 1 briefly shows the e-values of the scores obtained with each setup when aligning sequence pairs with various evolutionary distances. While all variants perform well for highly similar sequences, we can clearly deduce the ability of the

Table 1. Order of the e-values of the scores obtained by aligning artificially diverged pairs of proteins resulted from the translation of the same ancestral sequence on two reading frames. [*] *TDS <evolutionary distance>* = translation-dependent scores; classic substitution scores: match = 3, transversion = -4, transition = -2.

Scores[*]	Input type	Evolutionary distance between the aligned inputs							
		0.01	0.10	0.30	0.50	0.70	1.00	1.50	2.00
TDS 0.30	graphs	10^{-179}	10^{-171}	10^{-149}	10^{-121}	10^{-109}	10^{-83}	10^{-61}	10^{-37}
	known DNAs	10^{-152}	10^{-136}	10^{-110}	10^{-76}	10^{-54}	10^{-21}	10^{-6}	1.00
TDS 0.50	graphs	10^{-166}	10^{-156}	10^{-140}	10^{-118}	10^{-107}	10^{-85}	10^{-55}	10^{-34}
	known DNAs	10^{-140}	10^{-128}	10^{-105}	10^{-75}	10^{-61}	10^{-34}	10^{-6}	10^{-1}
TDS 0.70	graphs	10^{-153}	10^{-145}	10^{-130}	10^{-113}	10^{-102}	10^{-83}	10^{-57}	10^{-51}
	known DNAs	10^{-130}	10^{-120}	10^{-101}	10^{-76}	10^{-64}	10^{-42}	10^{-13}	10^{-7}
TDS 1.00	graphs	10^{-137}	10^{-131}	10^{-118}	10^{-104}	10^{-97}	10^{-80}	10^{-59}	10^{-54}
	known DNAs	10^{-117}	10^{-110}	10^{-93}	10^{-70}	10^{-65}	10^{-46}	10^{-21}	10^{-8}
classic	graphs	10^{-127}	10^{-24}	10^{-12}	10^{-11}	10^{-7}	10^{-5}	10^{-3}	10^{-2}
scores	known DNAs	10^{-86}	10^{-20}	10^{-9}	10^{-7}	10^{-4}	10^{-1}	1.00	1.00

translation-dependent scores to help the algorithm build significant alignments between sequences that underwent important changes.

The resulting alignments reveal that, even after many mutations, the translation-dependent scores manage to recover large parts of the original shared sequence, by correctly aligning most positions. On the other hand, with classic match/mismatch scores, the algorithm usually fails to find these common zones. Moreover, due to the large number of mismatches, the alignment has a low score, comparable to scores that can be obtained for randomly chosen sequences. This makes it difficult to establish whether the alignment is biologically meaningful or it was obtained by chance. This issue is solved by the translation-dependent scores by uneven substitution penalties, according to the codon mutation models.

We conclude that the usage of translation-dependent scores makes the algorithm more robust, able to detect the common origins even after the sequences underwent many modifications, and also able to filter out alignments where the nucleotide pairs match by pure chance and not due to evolutionary relations.

5   Experimental Results

5.1   Tests on Known Overlapping and Frameshifted Genes

We tested the method on pairs of proteins known to be encoded by overlapping genes in viral genomes (phage X174 and Influenza A) and in E.coli plasmids, as well as on the newly identified overlapping genes *yaaW* and *htgA* from E.coli K12 [21]. In all cases, we obtained perfect identification of gene overlaps with simple substitution scores and with translation-dependent scoring matrices corresponding to low evolutionary distances (at most 1 mutation per codon). Translation-dependent scoring matrices of higher evolutionary distances favor, in some (rare) cases, substitutions instead of matches within the alignment. This is a natural consequence of increasing the codon's chance to mutate, and it illustrates the importance of choosing a score matrix corresponding to the real evolutionary distance. Our method was also able to detect, directly on the protein sequences, the frameshifts resulting in functional proteins reported in [1,2,3,4].

5.2   New Divergence Scenarios for Orthologous Proteins

In this section we discuss the application of our method to FMR1NB (Fragile X mental retardation 1 neighbor protein) family. The Ensembl database [22] provides 23 members of this family, from mammalian species, including human, mouse, dog and cow. Their multiple alignment, provided by Ensembl, shows high dissimilarity on the first part (100 amino acids approximately), and good conservation on the rest of the sequence. We apply our alignment algorithm on proteins from several organisms, where the complete sequence is available.

We performed our experiments with translation-dependent scoring matrices corresponding to 0.3, 0.5 and 0.7 mutations per codon. Given that, in our scenario (presented in section 3.3), the divergence is applied on two reading frames,

```
>Q8NOW7|FMR1N_HUMAN[14, 644] / Q8OZA7|FMR1N_MOUSE[0, 623]

...L] [S] [Y] [Y] [L] [C] [S] [G] [S] [S] [Y] [F] [V] [L] [A] [N] [G] [H] [I] [L] [P] [N] [S] [E] [N] [A] [H] [G] [Q] [S] [L] [E] [E] [D] [S] [A] [L] [E] [A] [
...TCTCTTATTACCTTTGTTCTGGATCCTCGTACTTCGTTTTAGCCAACGGGCACATACTGCCAAACTCCGAAAACGCGCACGGGCAGTCGCTGGAAGAGGATTCGGCGTTAGAGGCCT
... +++++++++O+++++++++++O++++++++++---+++++++   --+-O-+++++++++++O+++-O+--O--+++-++++++++  +++--++++++--O+++++++--+
...TCTCTTATTACTTTTGTTCTGGACCCTCGTAATTCTTTTTAGAC---CGGTAAATACTGCCAAACTCCCAATACTCGCAGAGGCAGCCG----AATTGGAACCGCCGTTAGGGAACA
...] [L] [L] [L] [L] [L] [F] [W] [T] [L] [V] [I] [L] [F] [R] [   P] [V] [N] [T] [A] [K] [L] [P] [I] [L] [A] [E] [A] [A] [   E] [L] [E] [P] [P] [L] [G] [N] [

L] [L] [N] [F] [F] [F] [P] [T] [T] [C] [N] [L] [R] [E] [N] [Q] [V] [A] [K] [P] [C] [N] [E] [L] [Q] [D] [L] [S] [E] [S] [E] [C] [L] [R] [H] [K] [C] [C] [F] [...
TGTTAAACTTTTTTTTCCAACAACGTGTAACCTAAGAGAAAATCAAGTAGCGAAGCCGTGTAATGAGCTGCAGGACTTATCAGAATCAGAATGTTTAAGGCACAAATGTTGTTTTTT...
IIIII:IIIIIIIIIIIIIIIIIIII:IIIIII.I.IIIIII.IIIIIIII:I:.I.IIIIIIII:I.II.II.I.IIIIIIIIIIIIIIIIIIII:I::IIIIIIIIIIII:I:.I.
O+++++++++++++++++++++++++000-++++++++++++++++----O++++++-----O--+OO++++++++++++++++++++++++++++---++++++++++++++...
TGTTAGACTTTTTTTTCCAACAGCGTGTATCATAAGAGATAATCAAGTAGTGGTGGCCGTGTAATAACCAGCCGTACTTATCAGAATCAGAATGTTTAAAGAGCAAATGTTGTTCTT...
M] [L] [D] [F] [F] [F] [P] [T] [A] [C] [I] [I] [R] [D] [N] [Q] [V] [V] [V] [A] [C] [N] [N] [Q] [P] [Y] [L] [S] [E] [S] [E] [C] [L] [K] [S] [K] [C] [C] [S] [...
```

Fig. 4. Human and mouse FMR1NB proteins, aligned using a translation-dependent matrix of evolutionary distance 0.7 (the sign of each substitution score appears on the fourth row). The size 4 gap corresponds to a frameshift that corrects the reading frame.

this implies an overall mutation rate of 0.6, 1.0 and 1.4 mutations per codon respectively. Thus, the mutation rate per base reflected by our scores is less than 0.5, which is approximately the nucleotide substitution rate for mouse relative to human [23]. The number of allowed frameshifts was limited to 3. The gap penalties were set in all cases to -20 for codon indels, -20 for size 1 gaps and -5 for the extension of size 1 gaps (size 1 and size 2 gaps correspond to frameshifts). These choices were made so that the penalty for codon indels is higher than the average penalty for 3 substitutions.

Figure 4 presents a fragment of the alignment obtained on the FMR1NB proteins of human (gene ID ENSG00000176988) and mouse (gene ID ENS-MUSG00000062170). The algorithm finds a frameshift near the 100th amino acid, managing to align the initial part of the proteins at the DNA level. Similar frameshifted alignments are obtained for human vs. cow and human vs. dog, while alignments between proteins of primates do not contain frameshifts. The consistency of the frameshift position in these alignments supports the evidence of a frameshift event that might have occurred in the primate lineage.

If confirmed, this frameshift would have modified the first topological domain and the first transmembrane domain of the product protein. Interestingly, the FMR1NB gene occurs nearby the Fragile X mental retardation 1 gene (FMR1), involved in the corresponding genetic disease [24].

6 Conclusions

In this paper, we addressed the problem of finding distant protein homologies, in particular affected by frameshift events, from a codon evolution perspective. We search for protein common origins by implicitly aligning all their putative coding DNA sequences, stored in efficient data structures called back-translation graphs. Our approach relies on a dynamic programming alignment algorithm for these graphs, which involves a non-monotonic gap penalty that handles differently frameshifts and full codon indels. We designed a powerful translation-dependent scoring function for nucleotide pairs, based on codon substitution models, whose purpose is to reflect the expected dynamics of coding DNA sequences.

The method was shown to perform better than classic alignment on artificial data, obtained by mutating independently, according to a codon substitution model coding sequences translated with a frameshift. Moreover, it successfully detected published frameshift mutation cases resulting in functional proteins.

We then described an experiment involving homologous mammalian proteins that showed little conservation at the amino acid level on a large region, and provided possible frameshifted alignments obtained with our method, that may explain the divergence. As illustrated by this example, the proposed method should allow to better explain a high divergence of homologous proteins and to help to establish new homology relations between genes with unknown origins.

An implementation of our method is available at `http://bioinfo.lifl.fr/path/`

References

1. Raes, J., Van de Peer, Y.: Functional divergence of proteins through frameshift mutations. Trends in Genetics 21(8), 428–431 (2005)
2. Okamura, K., et al.: Frequent appearance of novel protein-coding sequences by frameshift translation. Genomics 88(6), 690–697 (2006)
3. Harrison, P., Yu, Z.: Frame disruptions in human mRNA transcripts, and their relationship with splicing and protein structures. BMC Genomics 8, 371 (2007)
4. Hahn, Y., Lee, B.: Identification of nine human-specific frameshift mutations by comparative analysis of the human and the chimpanzee genome sequences. Bioinformatics 21(suppl. 1), i186–i194 (2005)
5. Grantham, R., Gautier, C., Gouy, M., Mercier, R., Pave, A.: Codon catalog usage and the genome hypothesis. Nucleic Acids Research (8), 49–62 (1980)
6. Shepherd, J.C.: Method to determine the reading frame of a protein from the purine/pyrimidine genome sequence and its possible evolutionary justification.. Proceedings National Academy Sciences USA (78), 1596–1600 (1981)
7. Guigo, R.: DNA composition, codon usage and exon prediction. Nucleic Protein Databases, 53–80 (1999)
8. Leluk, J.: A new algorithm for analysis of the homology in protein primary structure. Computers and Chemistry 22(1), 123–131 (1998)
9. Leluk, J.: A non-statistical approach to protein mutational variability. BioSystems 56(2-3), 83–93 (2000)
10. Altschul, S., et al.: Basic local alignment search tool. JMB 215(3), 403–410 (1990)
11. Altschul, S., et al.: Gapped BLAST and PSI-BLAST: a new generation of protein database search programs. Nucleic Acids Res. 25(17), 3389–3402 (1997)
12. Pellegrini, M., Yeates, T.: Searching for Frameshift Evolutionary Relationships Between Protein Sequence Families. Proteins 37, 278–283 (1999)
13. Arvestad, L.: Aligning coding DNA in the presence of frame-shift errors. In: Hein, J., Apostolico, A. (eds.) CPM 1997. LNCS, vol. 1264, pp. 180–190. Springer, Heidelberg (1997)
14. Arvestad, L.: Algorithms for biological sequence alignment. PhD thesis, Royal Institute of Technology, Stocholm, Numerical Analysis and Computer Science (2000)
15. Blake, R., Hess, S., Nicholson-Tuell, J.: The influence of nearest neighbors on the rate and pattern of spontaneous point mutations. JME 34(3), 189–200 (1992)
16. Kosiol, C., Holmes, I., Goldman, N.: An Empirical Codon Model for Protein Sequence Evolution. Molecular Biology and Evolution 24(7), 1464 (2007)

17. Pedersen, A., Jensen, J.: A dependent-rates model and an MCMC-based methodology for the maximum-likelihood analysis of sequences with overlapping reading frames. Molecular Biology and Evolution 18, 763–776 (2001)
18. Lio, P., Goldman, N.: Models of Molecular Evolution and Phylogeny. Genome Research 8(12), 1233–1244 (1998)
19. Altschul, S., et al.: The estimation of statistical parameters for local alignment score distributions. Nucleic Acids Research 29(2), 351–361 (2001)
20. Olsen, R., Bundschuh, R., Hwa, T.: Rapid assessment of extremal statistics for gapped local alignment. In: ISMB, pp. 211–222 (1999)
21. Delaye, L., DeLuna, A., Lazcano, A., Becerra, A.: The origin of a novel gene through overprinting in Escherichia coli. BMC Evolutionary Biology 8, 31 (2008)
22. Hubbard, T., et al.: Ensembl 2007. Nucleic Acids Res. 35 (2007)
23. Clamp, M., et al.: Distinguishing protein-coding and noncoding genes in the human genome. Proc. Natl. Acad. Sci. 104(49), 19428–19433 (2007)
24. Oostra, B., Chiurazzi, P.: The fragile X gene and its function. Clinical genetics 60(6), 399 (2001)

A Markov Classification Model for Metabolic Pathways

Timothy Hancock and Hiroshi Mamitsuka

Bioinformatics Center, Institute for Chemical Research, Kyoto University, Japan
timhancock@kuicr.kyoto-u.ac.jp
mami@kuicr.kyoto-u.ac.jp

Abstract. The size and complexity of metabolic networks has increased past the point where a researcher is able to intuitively understand all interacting components. Confronted with complexity, biologists must now create models of these networks to identify key relationships of specific interest to their experiments. In this paper focus on the problem of identifying pathways through metabolic networks that relate to a specific biological response. Our proposed model, HME3M, first identifies frequently traversed network paths using a Markov mixture model. Then by employing a hierarchical mixture of experts, separate classifiers are built using information specific to each path and combined into an ensemble classifier the response. We compare the performance of HME3M with logistic regression and support vector machines (SVM) in both simulated and realistic environments. These experiments clearly show HME3M is a highly interpretable model that outperforms common classification methods for large realistic networks and high levels of pathway noise.

1 Introduction

The development of highly comprehensive databases detailing the structure of complex biological processes such as metabolism has created a need for accurate and interpretable models for biological networks. Metabolic networks, as described in databases such as KEGG [1], can be represented as directed graphs, with the vertices denoting the compounds and edges labelled by the reactions. The reactions within metabolic networks are catalyzed by specific genes. If a gene is active, then it is possible for the corresponding reaction to occur. If a reaction is active then a pathway is created between two metabolic compounds that is labelled by the gene that catalyzed the reaction. Observations on the activity of genes within metabolic networks can be readily obtained from microarray experiments. The abundance publically available microarray expression data found in databases such as [2] combined with the detailed biological knowledge contained within pathway databases like KEGG [1] has spurred biologists to want to combine these two sources of information and model the metabolic network dynamics under different experimental conditions.

Our proposed model searches for differences in pathways taken between two compounds in a known network structure that are observed within different

S.L. Salzberg and T. Warnow (Eds.): WABI 2009, LNBI 5724, pp. 121–132, 2009.
© Springer-Verlag Berlin Heidelberg 2009

Fig. 1. Example network

experimental conditions. We require the prior specification of the metabolic network with the corresponding gene expression data for each edge and binary response variable that labels the experimental conditions of interest. To construct our model we consider the network to be a directed graph and pathways through the network to be binary strings. For example in the network described in Figure 1 there are 4 possible paths between nodes A and D. In Figure 1 the binary representation of the path that traverses edges $[1, 3, 4]$ is $[1, 0, 1, 1, 0]$. For metabolic networks the set network edges that can be traversed is determined by observing the relative expression of all included genes. The dataset under consideration then consists of N observed paths through the network where each paths is labeled by a binary response $y \in [0, 1]$. The question asked in this paper is *"given a list of observed pathways are there any frequent or dominant pathways that can be used to classify the response?"*

Hidden Markov Models (HMM) are commonly used for identifying structure within sequence information [3]. HMMs assume that the nodes of the network are unknown and the observed sequences are a direct result of transition between these hidden states. However, if the network structure is known, a more direct approach is available through a mixture of Markov chains. Markov mixture models such as 3M [4] directly search for dominant pathways within sequence data by assuming each mixture component is a Markov chain through a known network structure. For metabolic networks, Markov mixture models have been shown to provide an accurate and highly interpretable models of dominant pathways throughout a known network structure [4]. However, both HMM and 3M are unsupervised models and therefore are not able to direct their search to explicity uncover pathways that relate to specific experimental conditions. The creation of a supervised classification technique that exploits the intuitive nature of Markov mixture models would be a powerful interpretable tool for biologists to analyze experimentally significant network pathways.

In this paper we propose a supervised version of the 3M model using the Hierarchical Mixture of Experts (HME) framework [5]. We choose the mixture of experts framework as our supervised model because it provides a complete probabilistic framework for localizing a classification model to specific clusters within a dataset. Our proposed approach, called HME3M employs the 3M model to identify the dominant pathways and logistic regressions within each mixture component to classify the response. In the following section we present the details of our proposed combination of HME and the 3M Markov mixture model which we call HME3M. The results and discussion sections clearly show HME3M significantly outperforms PLR and support vector machines (SVM) with linear, polynomial and radial basis kernels on both simulated and real example datasets.

2   Methods

2.1   Hierarchical Mixture of Experts (HME)

A HME is an ensemble method for predicting the response where each model in the ensemble is weighted by probabilities estimated from a hierarchical framework of mixture models [6]. Our model is the simplest two level HME, where at the top is a mixture model to find clusters within the dataset, and at the bottom are the experts, weighted in the direction of each mixing component, used to classify a response. Given a response variable y and predictor variables x, a 2-layer HME has the following form,

$$p(y|x, \beta_1, \ldots, \beta_m, \theta_1, \ldots, \theta_m) = \sum_{m=1}^{M} p(m|x, \theta_m)p(y|x, \beta_m) . \tag{1}$$

where β_m are the parameters of each expert and θ_m are the parameters of mixture component m. A HME does not restrict the source of the mixture weights $p(m|x, \theta_m)$ and as such can be generated from any model that returns posterior component probabilities for the observations. Taking advantage of this flexibility we propose a HME as a method to supervise the Markov mixture model for metabolic pathways 3M [4].

Combining HME with a Markov mixture model first employs the Markov mixture to find dominant pathways. Posterior probabilities are then assigned to each sequence based on its similarity to the dominant pathway. These are then passed as input weights into the parameter estimation procedure within the supervised technique. Using the posterior probabilities of 3M to weight the parameter estimation of each supervised technique is in effect localizing each expert to summarize the predictive capability of each dominant pathway. Therefore incorperating the 3M Markov mixture model within a HME is creating a method capable of combining network structures with standard data table information. We now formally state the base 3M model and provide the detail of our proposed model, Hierarchical Mixture Experts 3M (HME3M) classifier.

2.2   3M Mixture of Markov Chains

The 3M Markov mixture model assumes that pathway sequences can be represented with a mixture of first order Markov chains [4]. The full model form spanning M components estimating the probabilities of T transitions is,

$$p(x) = \sum_{m=1}^{M} \pi_m p(c_1|\theta_{1m}) \prod_{t=2}^{T} p(c_t, x_t|c_{t-1}; \theta_{tm}) \tag{2}$$

where π_m is the mixture model component probability, $p(c_1|\theta_{1m})$ is the probability of the initial state c_1, and $p(c_t, x_t|c_{t-1}; \theta_{tm})$ is the probability of a path traversing the edge x_t linking states c_{t-1} and c_t. The 3M model is simply a mixture model and as such its parameters are conveniently estimated by an

EM algorithm [4]. The result of 3M is M mixture components, where each component, m, corresponds to a first order Markov model defined by $\theta_m = \{\theta_{1m}, [\theta_{2m}, \ldots, \theta_{tm}, \ldots, \theta_{Tm}]\}$ which are the estimated probabilities for each transition along the mth dominant path.

2.3   HME3M

The HME model combining 3M and a supervised technique for predicting a response vector y can be achieved by using the 3M mixture probabilities $p(m|x, \theta_m)$ (2), for the HME mixture component probabilities in (1). This yields the HME3M likelihood,

$$p(y|x) = \sum_{m=1}^{M} \pi_m p(y|x, \beta_m) p(c_1|\theta_{1m}) \prod_{t=2}^{T} p(c_t, x_t|c_{t-1}; \theta_{tm}) \tag{3}$$

The parameters of (3) can be estimated using the EM algorithm by defining the responsibilities variable h_{im} to be the probability that a sequence i belongs to component m, given x, θ_m, β_m and y. These parameters can be iteratively optimized using an EM algorithm with the following E and M steps:

E-Step: Define the responsibilities h_{im}:

$$h_{im} = \frac{\pi_m p(m|x_i, \theta_m) p(y_i|x_i, \beta_m)}{\sum_{m=1}^{M} \pi_m p(m|x_i, \theta_m) p(y_i|x_i, \beta_m)} \tag{4}$$

M-Step: Estimate the Markov mixture and expert model parameters:

(1) Estimate the mixture parameters

$$\pi_m = \frac{\sum_{i=1}^{N} h_{im}}{\sum_{m=1}^{M} \sum_{i=1}^{N} h_{im}} \quad \text{and} \quad \theta_{tm} = \frac{\sum_{i=1}^{N} \delta(x_{it} = 1) h_{im}}{\sum_{i=1}^{N} h_{im}} \tag{5}$$

where $\delta(x_{it} = 1)$ denotes whether a transition t is active within observation i, or $x_{it} = 1$. This condition enforces the constraint that the probabilities of each set of transitions between any two states must sum to one. Additionally it can be shown that for this model all initial state probabilities $p(c_1|\theta_{1m}) = 1$.

(2) Estimate the expert parameters
The original implementation of HME estimates the expert parameters, β_m, with the Iterative Reweighted Least Squares (IRLS) algorithm, where the HME weights, h_{im} are included multiplicatively by further reweighting the standard IRLS weights [5]. However, in this setting, X is a sparse matrix of binary pathways. The sparseness of X is further exacerbated by the additional reweighting required for inclusion into the HME model. Therefore employing the standard IRLS estimation for the experts is likely to be inaccurate. To overcome this issue we a Penalized Logistic Regression (PLR) for our experts [7].

Penalized Logistic Regression (PLR) uses a penalty [7] to allow for the co-efficients of logistic regression to be run over a sparse or large dataset. PLR maximizes β_{tm} subject to a ridge penalization $|\beta_m|^2$ controlled by $\lambda \in [0, 2]$,

$$l(\beta_m | h_{im}) = \underset{\beta_m}{\arg\max} \left\{ \sum_{i=1}^{N} h_{im} \left(y_i \beta_m^T x_i + log(1 + e^{\beta_m^T x_i}) \right) - \frac{\lambda}{2} |\beta_m|^2 \right\} \quad (6)$$

where X is the dataset of binary paths, y is the binary response variable, h_{im} are HME3M responsibilities for path i in component m (4), and β_{tm} are the PLR coefficients for transition t in component m. In this case we choose the ridge penalty for reasons of computational simplicity. The ridge penalty allows the regularization to be easily included within the estimation by a simple mod-ification to the Netwon-Raphson steps.

The size of λ directly affects the size of the estimates for β_m. As λ approaches 2 the estimates for β_m will become more sparse, and as λ approaches 0 the estimates for β_m approach the IRLS estimates. Additionally our personal ex-perience of IRLS in the HME context indicates the need for additional control over the rate of learning of the experts. This experience suggests that if the PLR iterations converge too quickly the estimates of β_m reach a local optimum. This problem has been noted by [6] and a solution is proposed by the imposition of a learning rate, $\alpha \in [0, 1]$, within the IRLS algorithm. This ensures that at each iteration, a step will be taken to maximize β_{tm}, a sufficient condition for the EM algorithm. Reducing α decreases the step size taken by the IRLS algorithm and allows for smooth convergence. The updates for the learning controlled PLR-IRLS algorithm are now computed by:

$$\beta_{tm}^{new} = \beta_{tm}^{old} + \alpha (X^T W_m X + \Lambda)^{-1} \left(X^T h_{im} (y - \hat{y}) \right) \quad (7)$$

where $\hat{y}$ is the vector of probabilities $p(x; \beta_{tm}^{old})$ and W_m is a diagonal matrix of weights such that $w_{mii} = h_{im} \hat{y}_i (1 - \hat{y}_i)$ and Λ is a diagonal matrix with λ along the diagonal. It is noted by [6] that this method will converge to the same solution as the IRLS method, however the effect of α will increase the number of iterations for convergence.

3   Experiments

Our problem has the following inputs: the network structure, microarray obser-vations and a response variable. A pathway through the network, x_i, is assumed to be a binary vector, where a 1 indicates a traversed edge and 0 represents a non-traversed edge. The decision on which edges can be traversed is made based on the expression of each gene for every microarray observation. Once the set of valid edges have been defined, for each microarray observation all valid pathways are extracted. After extracting all observed pathways and their response class it is possible to set up a supervised classification problem where the response vector y denotes the observed class of each pathway, and the predictor matrix X is an $N \times P$ binary matrix of pathways, where N is the number of pathways and P is the number of edges within the network. The binary predictor matrix, X and its response y can now be directly analyzed by standard classification techniques. We assess the performance of HME3M in four experiments and compare it to PLR and Support Vector Machines (SVM) with three types of kernels, linear, polynomial (degree = 3) and radial basis. The implementation of SVM used for these experiments is sourced from the R package *e1071*, [8].

We point out here that the predictor matrix X is a list of all pathways through the network observed within the original dataset. Therefore X contains all available information on the given network structure contained within the original dataset. Using this information as input into the PLR and SVM models is supplying these methods with the same network information that is provided to the HME3M model. As the supplied information is the same for all models the comparison is fair. The performance differences observed will result from SVM and PLR not considering the Markov nature of the input pathways whereas HME3M explicitly models their order with a first order Markov mixture model.

3.1 Synthetic Data

To construct the simulation experiments we assume that the dataset is comprised of dominant pathways that define the groups and random noise pathways. To ensure that the pathway structure is the major information within the dataset, we specify the network structure and simulate only the binary pathway information. The construction of simulated pathway data requires definition of both the dominant pathways and the percent of noise. A dominant pathway is defined as a frequently observed path within a response class. The level of expression of a dominant pathway is defined to be the number of times it is observed within a group. The percent of noise is defined to be a valid pathway within the network pathway that leads from the start to the end compounds and is not any of the specified dominant pathways. As the percent of noise increases the expression of the dominant paths decreases making correct classification harder.

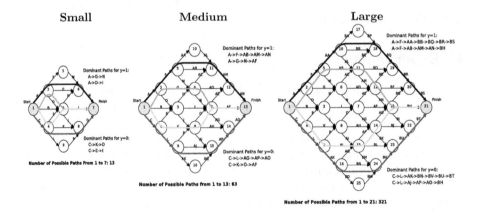

Fig. 2. Simulated network diagrams

We run the simulation experiments on three graphs with the same structure but with increasing complexities as shown in Figure 2. For each network we define two dominant pathways for each response label, $y = 0$ and $y = 1$ and give each dominant pathway equal pathway expression. We simulate a total of 200 pathways per response label which includes observations from the two dominant pathways and noise pathways. Separate simulations are then performed for the specified noise pathway percentages $[0.1, 0.2, 0.3, 0.4, 0.5]$. The performance of each method is evaluated with 10 runs of 10-fold cross-validation. The performance differences between HME3M compared to SVM and PLR are then tested with paired sample t-tests using the test set performances from the cross-validation. We set the HME3M parameters to be $M = [2, 3]$, $\lambda = 1$, $\alpha = 0.5$.

3.2 Real Data

To assess the performance of HME3M in a realistic setting we use the glycolysis metabolic network (Figure 3) extracted from KEGG [1] for the *Arabidopsis thaliana* plant. We deliberately use *Arabidopsis* as it has become a benchmark organism and it is well known that during the developmental stages different metabolic pathways within core metabolic networks such as glycolysis are activated. In Figure 3 we extract the core component of the glycolysis network for *Arabidopsis* between C00668 (*Alpha-D-Glucose*) and C00022 (*Pyruvate*). The extracted network in Figure 3 is a significantly more complex graph than our simulated designs and has 103680 possible pathways between C00668 and C00022. We extract the gene expression observations for all genes on this pathway from the AltGenExpress development series microarray expression data [9] downloaded from the ArrayExpress database [2].

The AltGenExpress database [9] is a microarray expression record of each stage within the growth cycle of *Arabidopsis* and contains expression observations of 22814 genes over 79 replicated conditions. For our purposes we extract observations for *"rosette leaf"* ($n = 21$) and *"flower"* ($n = 15$) and specify "flower" to be target class ($y = 1$) and "rosette leaf" to be the comparison class ($y = 0$). To extract binary instances of the glycolysis pathway within our

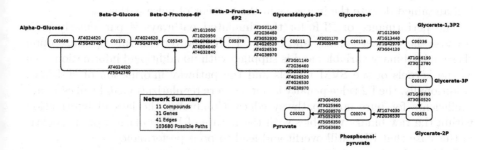

Fig. 3. *Arabidopsis thaliana* glycolysis pathway from Alpha-D-Glucose to Pyruvate. For visual simplicity, we show only a single edge connecting each compound; however in the actual network there is a separate edge for each gene label displayed.

extracted data we scale the observations to have a mean of zero and standard deviation of 1. After scaling the expression values we specify a tolerance of 0 and set any gene that has a positive scaled value to be "1" or overexpressed, otherwise we set its value to "0" or underexpressed. This is a simple discretization as it requires no additional information from the response or external conditions that might limit the number of paths selected. We deliberately choose this simple discretization of the gene expressions as it provides a highly noisy scenario to test the performance of HME3M. Then, based on simple and noisy discretization for each experiment we extract all possible paths within the glycolysis network between C00668 and C00022. This results in 4288 paths observed in the rosette leaf microarrays and 20608 paths observed within the flower microarrays.

To assess the performance of HME3M in classifying glycolysis paths observed in flowers from those observed within rosette leaves we conduct 20-fold inverse cross-validation for model sizes $M = 2$ to $M = 10$ with $\lambda = 1$ and $\alpha = 0.7$. Inverse 20-fold cross-validation firstly divides the observations randomly into 20 groups and then for each group trains using only observations from one group and tests the performance on the observations from the other 19. The performance of HME3M for 20-fold inverse cross-validation is compared to PLR and the SVM models with a matched pairs t-test.

4   Results and Discussion

4.1   Synthetic Data

For the first experiment the correct classification rate (CCR) percentages, ranges and paired sample t-test results for simulated graphs are shown in Table 1. All experiments show HME3M outperforming the trialled SVM kernels and a single PLR model. In fact the only times when the performances of SVM and HME3M are equivalent (P-value < 0.05) is with the small or medium graph with high levels of within group noise. Of particular note is the observation that for the medium and large graphs the median performance for HME3M is always superior to SVM. Furthermore, as the graph complexity increases it is clearly seen that HME3M consistently outperforms SVM and this performance is maintained despite the increases in the percent of noise pathways.

The performance of PLR for the simulated pathways is particularly poor because the dataset is noisy and binary. PLR however can only optimize upon these noisy binary variables and is supplied with no additional information such as the kernels of the SVM models and the pathway information of HME3M. Additionally, the L2 ridge penalty is not a severe regularization and will estimate coefficients for pure noise pathway edges. Combining the lack of information within the raw binary variables with the nature of L2 regularization, it is clear in this case that PLR will overfit and lead to poor performance.

Table 1 also demonstrates that as you increase the number of mixture components in the HME3M model, M, the model's resistance to noise increases. The increased robustness of HME3M is observed in the increase in median performance from $M = 2$ to $M = 3$ when the noise levels are 30% or more (≥ 0.3).

Table 1. The median and range of the 10×10-fold correct classification rates (CCR) for all simulation experiments. The best model performances are emboldened and a * flags models performances that are significantly different from HME3M ($\alpha = 0.05$).

Graph	Model	Percent Within Group Noise									
		$M=2$					$M=3$				
		0.1	0.2	0.3	0.4	0.5	0.1	0.2	0.3	0.4	0.5
Small	HME3M	**96.79**	**90.94**	**85.79**	**79.80**	74.97	96.14	92.64	86.46	79.86	76.13
	CCR Range	4.15	6.99	4.49	7.42	6.73	3.36	7.69	3.98	16.83	7.36
	PLR	50.67*	50.98*	50.68*	50.35*	50.96*	50.83*	50.70*	50.58*	50.62*	50.70*
	CCR Range	1.53	1.54	2.13	1.66	1.55	1.11	1.71	1.35	1.42	0.68
	SVM (linear)	95.14*	90.48*	85.28	78.94	72.53*	94.77*	89.01*	85.16*	78.66	75.21
	CCR Range	0.72	2.31	3.16	2.61	7.02	1.61	4.39	2.72	11.55	14.54
	SVM (polynomial)	95.25*	90.05*	84.19*	79.42	75.03	94.91*	89.71*	84.82*	78.66	75.73
	CCR Range	1.52	1.18	3.74	3.15	6.92	2.33	3.67	4.38	3.09	8.41
	SVM (radial)	95.28*	89.73*	84.19*	79.39	**75.07**	94.37	89.48*	84.82*	78.94	75.35
	CCR Range	1.53	2.81	4.20	2.61	8.02	2.32	4.53	3.28	4.14	7.29
Medium	HME3M	**98.94**	**94.29**	**88.68**	**80.78**	**77.05**	**98.80**	**96.22**	**90.11**	**84.02**	**77.89**
	CCR Range	2.92	6.18	5.73	12.51	5.56	2.40	5.59	10.57	8.98	8.76
	PLR	50.48*	50.62*	50.48*	50.48*	50.55*	50.55*	50.35*	50.62*	50.55*	50.48*
	CCR Range	2.04	0.99	1.30	2.13	1.23	0.97	0.69	1.63	1.31	1.13
	SVM (linear)	94.68*	89.80*	83.91	79.64*	76.02	94.28*	89.52*	84.80*	80.19*	73.41*
	CCR Range	2.74	22.91	3.56	4.17	4.12	1.72	2.60	5.17	2.89	5.40
	SVM (polynomial)	94.92*	90.30*	84.53*	79.32*	75.07*	94.22*	90.35*	84.53*	79.72*	75.21*
	CCR Range	2.73	21.22	4.00	5.80	2.83	1.65	3.69	6.85	2.85	4.91
	SVM (radial)	94.60*	90.30*	84.52*	79.28*	74.70*	94.19*	89.68*	84.31*	80.72*	74.80*
	CCR Range	2.45	4.35	2.82	2.99	4.56	1.90	3.49	4.67	2.93	3.59
Large	HME3M	**99.54**	**97.36**	**93.29**	**83.73**	**77.88**	**99.39**	**97.52**	**94.08**	**84.78**	**83.11**
	CCR Range	2.96	5.00	9.44	10.17	9.97	1.77	6.12	10.23	10.05	11.56
	PLR	50.69*	50.27*	50.49*	50.56*	50.62*	50.41*	50.70*	50.34*	50.83*	50.82*
	CCR Range	1.40	1.27	1.69	1.69	0.82	1.11	1.39	1.51	0.98	1.09
	SVM (linear)	94.91*	89.31*	85.58*	79.55*	73.79*	94.23*	89.34*	85.30*	79.67*	74.34*
	CCR Range	0.76	3.29	4.66	3.72	4.51	2.24	2.18	2.68	2.77	7.30
	SVM (polynomial)	94.78*	89.45*	85.80*	78.67*	74.05*	94.28*	89.86*	84.69*	79.44*	74.33*
	CCR Range	2.22	2.87	4.94	3.46	4.60	2.24	15.39	3.24	4.66	4.86
	SVM (radial)	94.98*	89.74*	85.09*	79.36*	74.21*	94.62*	89.83*	84.47*	79.56*	74.62*
	CCR Range	1.43	4.00	5.29	2.62	2.66	1.69	2.39	2.28	4.99	6.10

A supporting observation of particular note is that when the performances of HME3M with $M = 2$ is compared with the linear kernel SVM on the medium graph and 50% noise there is no significant difference between the model's performances. However, by increasing M to 3, HME3M is observed to significantly outperform linear kernel SVM. Further, in a similar but less significant case, for the small graph with 50% added noise by increasing M from 2 to 3 the median performance of HME3M become greater than that of linear kernel SVM. Although this increase did not prove to be significant the observed increasing trend within the median performance is clearly driving the results of the t-test comparing the model performances.

It is noticeable in Table 1 that the HME3M performance can be less precise than SVM or PLR models. However the larger range of CCR performances is not large enough to affect the significance of the performance gains made by HME3M. The imprecision of HME3M in this case is most likely due to the constant specification of λ, α and M over the course of the simulations. In the glycolysis experiment we show that careful choice of M produces stable model performances with a comparable CCR range than the nearest SVM competitor.

4.2 Real Data

The glycolysis experiment results are displayed in Figure 4. In Figure 4 the left plot corresponds to the mean correct classification rates (CCR) for HME3M and comparison methods. The number of mixture components M is varied from 2 to 10. It is clear from Figure 4 that the mean CCR for HME3M after $M = 2$ is consistently greater than all other methods. This is confirmed by the t-test

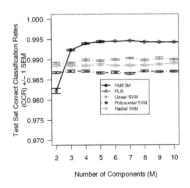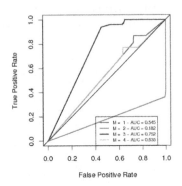

Fig. 4. Right Plot: Inverse cross-validated Correct Classification Rates (CCR) for all models for the Glycolysis pathway for *Arabidopsis*. **Left Plot**: ROC curve of all paths for the optimal model ($M = 4$).

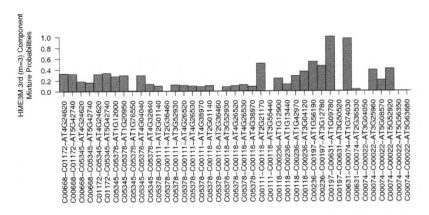

Fig. 5. Transition probabilities for the most expressed glycolysis path that separates flowers from rosette for *Arabidopsis*. Training set CCR = 0.818, AUC = 0.752.

P-values which show HME3M to be significantly outperforming the comparison models after 2 components with the optimal performance being observed at $M = 4$. An interesting feature of Figure 4 is that after the optimal performance has been reached, the addition of more components seems to not affect the overall classification accuracy. This shows HME3M to be resistant to overfitting and complements the results of the noise simulation experiments in Table 1.

The ROC curves for each HME3M component are presented in right plot in Figure 4 and clearly show that the third component is the most important with an AUC of 0.752, whereas the other three components seem to hold limited or no predictive power. A bar plot of the HME3M transition probabilities (θ_m) for the third ($m = 3$) component is presented in Figure 5. Overlaying the transition probabilities from Figure 5 onto the full network in Figure 3 it is found that three compound transitions are reduced to requiring only one possible gene:

- $C00111 \xrightarrow{AT2G21180} C00118$
- $C00197 \xrightarrow{AT1G09780} C00631 \xrightarrow{AT1G74030} C00074$

The second path connecting compounds C00197 through C00631 to C00074 is found by HME3M to have a high probability of being differently expressed when comparing glycolysis in flowers and rosette leaves. The branching of glycolysis at Glycerate-3P ($C00197$) through to Phosphoenol-Pyruvate ($C00074$) corresponds known variants of the glycolysis pathway in *Arabidopis*; the glycolysis I pathway located in the cytosol and the glycolysis II pathway located in the plastids [10]. The key precursors that lead to the branching within cystol variant is defined by two reactions; the first to convert Beta-D-Fructose-6P ($C05378$) to Beta-D-Fructose-1,6P ($C05378$) using diphosphate rather than ATP and the second is a bypass reaction to convert D-Glycerate-1,3P2 ($C00118$) to Glycerate-3P ($C00197$) [10]. Referencing the included pathway gene in Figure 5 within the reference Arabidopsis database TAIR [10] we observe that the genes specific to the percursor reactions for the cystol variant of glycolysis are included within the pathway, i.e. for the first reaction [$AT1G12000, AT1G20950, AT4G0404$] and for the second reaction [$AT1G42970$]. HME3M's identification of the plant cystol variant of the glycolysis pathway confirms this pathway as a flower specific, because the plastids variant is clearly more specific to rosette leaves due to their role in photosynthesis.

A further analysis of the genes identified reveals the interaction between AT1G09780 ($\theta = 1$) and AT1G74030 ($\theta = 0.969$) is of particular importance in stress response of *Arabidopsis*. A literature search on these genes identified both AT1G09780 and AT1G74030 as important in the response of *Arabidopsis* to environmental stresses such as cold exposure, salt and osmotic stress [11,12]. However, AT2G21180, apart from being involved in glycolysis, has not previously been found to be strongly involved in any specific biological function. Interestingly however, a search of TAIR [10] revealed that AT2G21180 is found to be expressed in the same growth and developmental stages as well as in the same plant structure categories as both AT1G09780 and AT1G74030. These findings are indicative of a possible relationship between these three genes in particular in the response to environmental stress.

5   Conclusions

In this paper we have presented a novel approach for the detection of dominant pathways within a network structure for binary classification using the Markov mixture of experts model, HME3M. Simulations clearly show HME3M to outperform both PLR and SVM with linear, polynomial and radial basis kernels. When applied to actual metabolic networks with real microarray data HME3M not only maintained its superior performance but also produced biologically meaningful results.

Acknowledgments

Timothy Hancock was supported by a Japan Society for the Promotion of Science (JSPS) fellowship and BIRD. Hiroshi Mamitsuka was supported in part by BIRD of Japan Science and Technology Agency (JST).

References

1. Kanehisa, M., Goto, S.: KEGG: Kyoto Encyclopedia of Genes and Genomes. Nucleic Acids Res. 28, 27–30 (2000)
2. Brazma, A., Parkinson, H., Sarkans, U., Shojatalab, M., Vilo, J., Abeygunawardena, N., Holloway, E., Kapushesky, M., Kemmeren, P., Lara, G.G., Oezcimen, A., Rocca-Serra, P., Sansone, S.: ArrayExpress–a public repository for microarray gene expression data at the EBI. Nucl. Acids Res. 31(1), 68–71 (2003)
3. Evans, W.J., Grant, G.R.: Statistical methods in bioinformatics: An introduction, 2nd edn. Springer, Heidelberg (2005)
4. Mamitsuka, H., Okuno, Y., Yamaguchi, A.: Mining biologically active patterns in metabolic pathways using microarray expression profiles. SIGKDD Explorations 5(2), 113–121 (2003)
5. Jordan, M., Jacobs, R.: Hierarchical mixtures of experts and the EM algorithm. Neural Computation 6(2), 181–214 (1994)
6. Waterhouse, S.R., Robinson, A.J.: Classification using mixtures of experts. In: IEEE Workshop on Neural Networks for Signal Processing (IV), pp. 177–186 (1994)
7. Park, M.Y., Hastie, T.: Penalized logistic regression for detecting gene interactions. Biostatistics (2007)
8. Dimitdadou, E., Hornik, K., Leisch, F., Meyer, D., Weingessel, A.: e1071 - misc functions of the department of statistics (2002)
9. Schmid, M., Davison, T.S., Henz, S.R., Pape, U.J., Demar, M., Vingron, M., Schölkopf, B., Weigel, D., Lohmann, J.U.: A gene expression map of Arabidopsis thaliana development. Nature Genetics 37(5), 501–506 (2005)
10. Swarbreck, D., Wilks, C., Lamesch, P., Berardini, T.Z., Garcia-Hernandez, M., Foerster, H., Li, D., Meyer, T., Muller, R., Ploetz, L., Radenbaugh, A., Singh, S., Swing, V., Tissier, C., Zhang, P., Huala, E.: The arabidopsis information resource (tair): gene structure and function annotation. Nucl. Acids Res. (2007)
11. Chawade, A., Bräutigam, M., Lindlöf, A., Olsson, O., Olsson, B.: Putative cold acclimation pathways in Arabidopsis thaliana identified by a combined analysis of mRNA co-expression patterns, promoter motifs and transcription factors. BMC Genomics 8(304) (2007)
12. Ndimba, B.K., Chivasa, S., Simon, W.J., Slabas, A.R.: Identification of Arabidopsis salt and osmotic stress responsive proteins using two-dimensional difference gel electrophoresis and mass spectrometry. Proteomics 5(16), 4185–4196 (2005)

Mimosa: Mixture Model of Co-expression to Detect Modulators of Regulatory Interaction

Matthew Hansen, Logan Everett, Larry Singh, and Sridhar Hannenhalli

Penn Center for Bioinformatics
Department of Genetics
University of Pennsylvania
sridharh@pcbi.upenn.edu
http://cagr.pcbi.upenn.edu

Abstract. Functionally related genes tend to be correlated in their expression patterns across multiple conditions and/or tissue-types. Thus co-expression networks are often used to investigate functional groups of genes. In particular, when one of the genes is a transcription factor (TF), the co-expression-based interaction is interpreted, with caution, as a direct regulatory interaction. However, any particular TF, and more importantly, any particular regulatory interaction, is likely to be active only in a subset of experimental conditions. Moreover, the subset of expression samples where the regulatory interaction holds may be marked by presence or absence of a modifier gene, such as an enzyme that post-translationally modifies the TF. Such subtlety of regulatory interactions is overlooked when one computes an overall expression correlation. Here we present a novel mixture modeling approach where a TF-Gene pair is presumed to be significantly correlated (with unknown coefficient) in a (unknown) subset of expression samples. The parameters of the model are estimated using a Maximum Likelihood approach. The estimated mixture of expression samples is then mined to identify genes potentially modulating the TF-Gene interaction. We have validated our approach using synthetic data and on three biological cases in cow and in yeast. While limited in some ways, as discussed, the work represents a novel approach to mine expression data and detect potential modulators of regulatory interactions.

1 Introduction

Eukaryotic gene regulation is carried out, in significant part, at the level of transcription. Many functionally related genes, e.g., members of a pathway, involved in the same biological process, or whose products physically interact, tend to have similar expression patterns [1,2]. Indeed, co-expression has extensively been used to infer functional relatedness [3,4,5,6]. Various metrics have been proposed to quantify the correlated expression, such as Pearson and Spearman correlation [2], and mutual information [5]. However, these metrics are symmetric and they neither provide the causality relationships nor do they discriminate against indirect relations. For instance, two co-expressed genes may be co-regulated, or one may regulate the other, directly or indirectly.

S.L. Salzberg and T. Warnow (Eds.): WABI 2009, LNBI 5724, pp. 133–144, 2009.
© Springer-Verlag Berlin Heidelberg 2009

A critical component of transcription regulation relies on sequence-specific binding of transcription factor (TF) proteins to short DNA sites in the relative vicinity of the target gene [7]. If one of the genes in a pair-wise analysis of co-expression is a TF then the causality is generally assumed to be directed from the TF to the other gene. In the absence of such information, an additional post-processing step [5] can be used to infer directionality between the pair of genes with correlated expression. Moreover, a first order conditional independence metric [4] has been proposed to specifically detect direct interactions.

While TFs are the primary engines of transcription, their activity depends on several other proteins such as modifying enzymes and co-factors, which directly or indirectly interact with the TF to facilitate its activity. For instance, the activity of TF CREB depends on a number of post-translational modifications, most notably, Ser133 phosphorylation by Protein Kinase A [8]. Moreover, for many TFs, the TF activity is likely to be restricted to specific cell types and/or experimental conditions. Thus the common practice of using large compendiums of gene expression data to estimate co-expression and thus functional relatedness has two main limitations: (1) it includes irrelevant expression samples which adds noise to the co-expression signal, and (2) it overlooks the contributions of additional modifier genes and thus fails to detect those modifiers which are critical components of gene regulatory networks.

To infer the dependence of TF activity on histone modification enzymes, Steinfeld et al. analyzed the expression of TF-regulons (putative targets of a TF) in yeast samples where specific histone modification enzymes were knocked out [9]. In a different study, Hudson et al. analyzed two sets of expression data in cow, a less-muscular wild-type and another with mutant TF Myostatin [10]. They found that the co-expression of Myostatin with a differentially expressed gene, MYL2, was significantly different between the mutant and the wild-type sets of expression. This differential co-expression led them to detect Myostatin as the causative TF even though the expression of Myostatin gene itself was not different between the mutant and the wild type. In both of the cited examples [9,10], the two sets of expression were well characterized and known a priori. In fact, Hu et al. have proposed a non-parametric test to detect differentially correlated gene-pairs in two sets of expression samples [11]. However, it is not clear how to detect such differentially co-expressed gene pairs when the appropriate partition of the expression samples is not provided and can not be derived from the description of the experiments. This problem is an important practical challenge for large expression compendiums that cover many diverse experimental conditions. The tremendously growing expression compendium [12], provides a unique opportunity to identify not only co-expressed and functionally related genes, but also to predict putative modifiers of gene regulators.

Here we propose a novel approach, "Mimosa", that seeks to partition expression data into correlated and uncorrelated samples based solely on the co-expression pattern. If found, such a partition suggests the existence of modifier genes, such as TF modifying enzymes, that should be differentially expressed between the correlated and uncorrelated sample partitions. The sample partition is derived from a mixture model of the co-expression data. The free parameters of the mixture model are estimated using a Maximum Likelihood Estimation (MLE) approach. Once the mixture parameters are obtained, we can then compute a weighted partitioning of the samples into the "correlated" and "uncorrelated" sets. In a subsequent step, we detect putative modifier genes that are differentially expressed between "correlated" and "uncorrelated" samples. Using synthetic data we show that Mimosa can partition expression samples and detect modifier genes with high accuracy. We further present three biological applications, one in bovine samples and two in yeast. This work represents a novel approach to mine expression data and detect potential modulators of regulatory interactions.

2   Methods

2.1   Mixture Modeling of Co-expression

Figure 1 illustrates the method. The input data, i.e. the expression profiles, is a matrix $\mathbb{M}$ where the genes, indexed by $i = 1, 2, \ldots, N_g$, are the rows and the expression samples, indexed by $k = 1, 2, \ldots, N_s$, are the columns of the matrix. $M[i, k]$ represents the expression of gene i in expression sample k. All rows are normalized to have mean 0 and variance of 1. For each pair of genes i and j, there are N_s data points of expression value pairs, $(M[i, k], M[j, k])$. For ease of notation, we shall denote the data points as (x_k, y_k). The observed data set for the gene pair, (x_k, y_k), is assumed to be a mixture of two different distributions: the group of uncorrelated samples (group "u") and the group of correlated samples (group "c"), each with its own probability distribution; call these distribution functions $p_u(x, y)$ and $p_c(x, y)$. By definition $p_u(x, y) = p_u(x) \, p_u(y)$, where $p_u(\cdot)$ is the normal distribution.

The observed data is viewed as a random sampling from these two groups with mixing fraction f defined to be the fraction of data points that belong to the uncorrelated group. The total likelihood of a data point (x, y) is $p(x, y) = f \, p_u(x, y) + (1 - f) \, p_c(x, y)$. In the present analysis we assume the uncorrelated distributions to be normal, hence,

$$p_u(x, y) = \frac{1}{2\pi} \exp\left[-\tfrac{1}{2}\left(x^2 + y^2\right)\right] . \tag{1}$$

We derive the distribution of correlated data, $p_c(x, y)$ by assuming that there is some (u, v) coordinate system related to the (x, y) coordinate system by a rotation through an angle θ, such that $p_c(u, v) = \mathcal{N}(u, \sigma_u)\mathcal{N}(v, \sigma_v)$. Here, $\mathcal{N}(x, \sigma)$ is the Gaussian distribution with zero mean and variance σ^2. The coordinate transformations from (x, y) coordinates to (u, v) coordinates are: $u = x \cos\theta - y \sin\theta$ and $v = x \sin\theta + y \cos\theta$. The Jacobian of the transformation is 1, so we have

$$p_c(x, y) = \frac{\exp\left[-\tfrac{1}{2}\left(u^2(x, y)/\sigma_u{}^2 + v^2(x, y)/\sigma_v{}^2\right)\right]}{2\pi\sigma_u\sigma_v} . \tag{2}$$

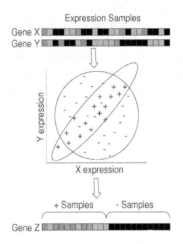

Fig. 1. The figure illustrates the intuition behind Mimosa. Consider a TF gene X and a potential target gene Y. The expression values of X and Y for all expression samples are shown as a heat plot and as a scatter plot. We presume that X and Y expression are correlated only in an unknown subset of samples (depicted by "+") and not in the remaining samples (denoted by "−"). Mimosa computes the maximum likelihood partition of samples. Then given the sample partition, a third gene Z with differential expression between the two partitions may represent a potential modifier. To be precise, we assign a partition probability to each sample as opposed to a binary partition.

There are three unknowns, $\{\theta, \sigma_u, \sigma_v\}$. There are, however, two natural constraints on the form of $p_c(x, y)$; namely, that

$$\int dx \; p_c(x, y) = p_u(y) \tag{3}$$

$$\int dy \; p_c(x, y) = p_u(x) \; . \tag{4}$$

Applying these two constraints to eqn. (2), and assuming that $\sigma_u \neq \sigma_v$, we have

$$p_c(x, y) = \frac{\exp\left[-\frac{1}{2}\left(x^2 + y^2 + 2\alpha xy\right)/(1 - \alpha^2)\right]}{2\pi\sqrt{1 - \alpha^2}}, \tag{5}$$

where $-1 \leq \alpha \leq 1$ is a free parameter of the mixture model that controls the aspect ratio of the correlated distribution. Without loss of generality let $\sigma_v > \sigma_u$; then in terms of α we have $\sigma_u{}^2 = (1 - |\alpha|)$ and $\sigma_v{}^2 = (1 + |\alpha|)$. Note that $\alpha < 0$ corresponds to positively correlated data ($\theta = \pi/4$) and $\alpha > 0$ corresponds to negatively correlated data ($\theta = -\pi/4$). For an aspect ratio defined by $r \equiv \sigma_v/\sigma_u > 1$, we have $|\alpha| = (r^2 - 1)/(r^2 + 1)$. In summary, the mixture model has two free parameters, (f, α), that determine the fraction of uncorrelated points in the observed data and the aspect ratio of the distribution for correlated data.

The log likelihood of the observed data is

$$L(f, \alpha) \equiv \sum_k \ln\left[p(x_k, y_k | f, \alpha)\right] . \tag{6}$$

We maximize L numerically using the quasi Newton-Raphson function optimization routine in the open source Gnu Scientific Library (www.gnu.org/software/gsl). The resulting parameter estimates are $\hat{f}$ and $\hat{\alpha}$. For each selected gene pair, we compute the probability that each sample belongs to the correlated group. For the kth sample, this is given by

$$q_k = \frac{(1 - \hat{f}) \, p_c(x_k, y_k | \hat{\alpha})}{p(x_k, y_k | \hat{f}, \hat{\alpha})} . \tag{7}$$

This vector of probabilities is equivalent to a weighted partitioning of the sample set. Modifier genes are selected based on their correlation with vector q. We compute this correlation with a t-test based on the expected population number, mean, and variance (see below). When computationally feasible, we use non-parametric correlation measures, such as Kendall's Tau.

2.2   Weighted T-Statistic

Given two vectors: (1) the q vector denoting the partition probability for each sample, and (2) expression vector E over all samples for a potential modifier gene, we can, in principle, partition the expression samples into two parts based only on the partition probability, and then compare the expression values in the two parts using a t-statistic or an alternative non-parametric test. However, this approach requires an arbitrary choice of partition probability threshold to partition the sample. We instead used a weighted version of the t-statistic that obviates the need for an arbitrary threshold. The standard t-statistic requires three parameters for each of the two partitions: the two sample-means, the two sample-standard deviations, and the two sample-sizes. We computed all these parameters using a weighted sum. For instance, the sample mean of the "correlated" partition, μ_c, can be estimated as $\mu_c = \frac{1}{n_c} \sum_k q_k E_k$, where $n_c = \sum_k q_k$ is the weighted number of "correlated" samples. Similarly, the standard deviation of the "correlated" partition, σ_c, is given by $\sigma_c^2 = \frac{1}{n_c} \sum_k q_k (E_k - \mu_c)^2$.

2.3   Generating Synthetic Data

To generate a TF-Gene-Modifier triplet for a given f and α we performed the following steps. We first create the modifier and TF expression data independantly by random sampling from a normal distribution. For the given f, we determine the modifier expression threshold m^* such that below this threshold the TF and gene are presumed to be uncorrelated and above this threshold, the TF and the gene are presumed to be correlated. The value of m^* is estimated by $f = \int_{-\infty}^{m^*} dx \mathcal{N}(x, 1)$. We generate the gene expression value as follows. Let m be the modifier expression in the kth sample. If $m < m^*$, then the gene's expression value for that sample, y_k, is drawn from a normal distribution (the uncorrelated distribution). If $m \geq m^*$, then the Gene's expression value is drawn from a Gaussian distribution with mean $-\alpha x_k$ and variance $(1 - \alpha^2)$, where x_k is the expression value of the TF for the kth sample. The latter step follows from the fact that the co-expression distribution for correlated data can be written as $p_c(x, y) = p_u(x)p_c(y|x)$ where $p_c(y|x)$ is a Gaussian with mean $-\alpha x$ and variance $(1 - \alpha^2)$.

3   Results

3.1   Synthetic Data

Given a pair of genes with a mixed set of "correlated" and "uncorrelated" samples, and also a modifier gene whose expression is correlated with the two types of samples, we tested whether our method can detect the correct modifier, which implicitly requires the correct identification of the sample partition. Details of the simulation are provided in §Methods. We generated 1500 non-overlapping TF-Gene-Modifier triplets and for each gene in the triplet we generated the expression data for 300 samples based on an underlying model, parameterized by f and α. We selected a range of parameters and tested the effect of these parameters on the method accuracy. Intuitively, Mimosa will work the best for values of f near $1/2$ and for values of α close to ± 1. Five different values of f were chosen that broadly encompass the value of $f = 0.5$. As the sign of α does not affect Mimosa's ability to partition the data samples, we chose only positive values of α. The three values of α chosen were based on their corresponding aspect ratios (see §Methods); namely aspect ratios of 2, 3, and 5. Not surprisingly, the performance of Mimosa deteriorates for aspect ratios below 2, that is, when the correlation is very poor even for the "correlated" samples (not shown). Each parameter bin contained 100 TF-Gene-Modifier triplets (15 bins X 100 triplets per bin = 1500 triplets, and 3 X 1500 triplets = 4500 total genes). For each of the 1500 TF-Gene pairs, we applied Mimosa to estimate the sample partition and then ranked all 4500 genes based on the weighted t-test p-value of their partitioned expression values (see §Methods). For each 2-dimensional bin (f and α value), we computed the median rank (out of 4500 candidates) of the correct modifier for the 100 TF-Gene pairs in the bin. We also computed the fraction of the 100 TF-Gene pairs for which the correct modifier had the highest rank.

As shown in Table 1, Mimosa detects the correct sample partition and the correct modifier with high accuracy. When the TF-Gene pair is uncorrelated in 90% of the samples (last column) then it is relatively difficult to detect the modifier. Even then, if the correlation is strong (aspect ration of 5) then Mimosa can still detect the modifier with very high accuracy. Note that the highest median rank, 215 for the $\alpha = 0.6$ and $f = 0.9$ bin, when represented as a percentile out of 4500 candidates, is only $215/4500 = 4.8\%$.

Table 1. Performance of Mimosa on synthetic data. Columns represent f ranges and rows represent α ranges (corresponding aspect ratio is shown in parenthesis; see §Methods). Figures in each cell are based on 100 TF-Gene pairs, and shows (1) the median rank of the correct modifier, and (2) the fraction of 100 cases where the correct modifier was top ranked based on the t-test p-value.

α/f	0.1	0.25	0.5	0.75	0.9
0.6 (2)	44, 14%	1, 53%	1, 76%	7, 32%	215, 5%
0.8 (3)	1, 70%	1, 99%	1, 100%	1, 83%	35, 10%
0.923(5)	1, 99%	1, 100%	1, 100%	1, 99%	4, 30%

3.2   Application to Bovine Data

Hudson et al., have compared expression profiles in two different genetic crosses (denoted P and W) of cattle at different developmental time points. The P type has a mutant form of TF Myostatin which results in dysregulation of TGF-β pathway and increased muscle mass [10]. The expression level of Myostatin was not different in these two types. They further identified differentially expressed genes between P and W, and for each such gene, and for each of the 920 putative regulators, they computed the expression correlation between the gene and the regulator, separately in P and in W samples. Based on these pair-wise correlations in the two sets of samples, they identified 424 regulator-gene pairs such that the expression correlation between the two was significantly different when using expression data from P compared with the expression correlation when using expression data from W. This data provides an ideal test bed for our approach.

We tested how well Mimosa partitions the expression samples into P and W without any prior knowledge. We subjected each of the 424 regulator-gene pairs to the mixture modeling, using the 20 expression profiles (10 for P and 10 for W) provided in [10]. This resulted in 424 partition probability vectors q, each of length 20 (see §Methods). If the mixture modeling is effective, we expect $\{q_1, \ldots, q_{10}\}$ (corresponding to P) to be significantly different from $\{q_{11}, \ldots, q_{20}\}$ (corresponding to W), with one being high, and the other being low. We tested this hypothesis using the Wilcoxon test and found that for 109(26%) of the 424 pairs, the p-value ≤ 0.05. Thus the mixture modeling correctly retrieves the hidden sample partition in many cases, even with a small number of expression samples.

3.3   Application to Yeast

We have previously reported a database   PTM-Switchboard [13], which now contains 510 yeast gene triplets, termed "MFG-triplets", where a transcription factor (F) regulates a gene (G) and this regulation is modulated by post-translational modification of F by a modifying enzyme (M). We tested whether, for the given F-G pair, Mimosa can correctly partition a set of expression samples and detect the modifier M. For the expression data, we used 314 *S. cerevisiae*

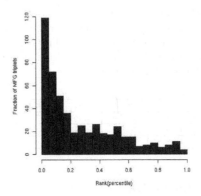

Fig. 2. Distribution of percentile ranks of the correct modifier predicted from among 6000 candidate modifiers, for the 510 experimentally determined TF-Gene-Modifier triplets. Mimosa ranks the correct modifier among the top 5% in 23% of the cases.

expression samples previously compiled in [14] from 18 different studies. These experiments included cell cycle and a variety of stress conditions. We applied Mimosa to each F-G pair and then computed the correlation (using Kendall's Tau) of the sample partition probability vector q (see §Method) and the expression vector of all 6000 yeast genes. We then computed the ranks (in percentile) of the correct modifiers. As shown in Figure 2, we found that Mimosa detects the true modifier among the top 5% in 23% of the cases, a $\sim$ 5-fold enrichment over random expectation.

To test the large-scale applicability of Mimosa, we extracted all yeast TF-Gene pairs detected in a genome-wide ChIP-chip experiment [15]. To reduce the number of gene-pairs to be tested we performed the following filtering steps. For each pair we computed their expression correlation using Kendall's Tau across the 314 expression samples. We retained the pairs for which the Kendall's Tau Bonferroni-corrected p-value ≤ 0.05. After applying Mimosa, we further filtered this set to retain only the cases where the mixing probability parameter f was between 0.45 and 0.55 and the aspect ratio parameter α had an absolute value of at least 0.8 (highly correlated). For each of the 6960 TF-Gene pairs thus obtained we computed the corresponding partition probability vector q. We then calculated the correlation between every gene's expression vector E and each pair's q vector. "Modifiers" for each pair were deemed to be those genes whose correlation qualified a Bonferroni-corrected, weighted t-statistic p-value threshold of 0.05. We used a weighted t-statistic, as opposed to Kendall's Tau, primarily for computational efficiency. We then performed a functional enrichment analysis on the 1356 putative modifier genes thus obtained using the DAVID tool (david.abcc.ncifcrf.gov). Table 2 shows the enriched (FDR < 5%) molecular functions sorted by the fraction of input genes annotated to have that function. The most abundant molecular function category was "catalytic activity", which is consistent with the role of modifying enzymes. This enrichment holds even when we selected the single most significant modifier for each TF-Gene pair.

Table 2. GO molecular functions enriched in the putative modifiers detected for the TF-Gene pairs in *S. cerevisiae* based on ChIP-chip data and 314 expression samples

Molecular function term	% Coverage	p-value	FDR (%)
catalytic activity	43	1.32E-04	0.22
nucleotide binding	14	1.27E-05	0.02
purine nucleotide binding	13	4.74E-05	0.08
purine ribonucleotide binding	12	1.49E-05	0.02
ribonucleotide binding	12	1.49E-05	0.02
RNA binding	11	1.60E-09	2.71E-06
structural molecule activity	10	4.34E-12	7.36E-09
structural constituent of ribosome	9	4.38E-22	7.43E-19
GTP binding	3	7.43E-06	0.01
guanyl nucleotide binding	3	7.43E-06	0.01
guanyl ribonucleotide binding	3	7.43E-06	0.01
oxidoreductase activity, acting on CH-OH group of donors	3	4.66E-05	0.08
translation regulator activity	3	8.71E-07	1.48E-03
oxidoreductase activity, acting on the CH-OH group of donors, NAD or NADP as acceptor	3	1.10E-04	0.19
translation factor activity, nucleic acid	3	2.14E-07	3.62E-04
GTPase activity	2	0.002392	3.98
rRNA binding	2	8.85E-11	1.50E-07
snoRNA binding	2	8.58E-09	1.45E-05
ligase activity, forming aminoacyl-tRNA and related compounds	2	1.33E-06	2.25E-03
ligase activity, forming carbon-oxygen bonds	2	1.33E-06	2.25E-03
aminoacyl-tRNA ligase activity	2	1.33E-06	2.25E-03
RNA helicase activity	2	4.29E-05	0.07
ATP-dependent RNA helicase activity	2	9.12E-07	1.54E-03
RNA-dependent ATPase activity	2	9.12E-07	1.54E-03
translation initiation factor activity	1	4.61E-04	0.78

Further work needs to be done to analyze the biological significance of specific modifiers detected.

4 Discussion

For a pair of co-expressed genes (X and Y), we have presented a mixture modeling approach to partition the expression samples in order to detect the specific subset of samples where X and Y expressions are strongly correlated. In some cases, such a partition may help detect other genes likely to modulate the expression correlation between X and Y. Such a potential modulator is characterized by having differential expression in the two sample partitions. A few previous investigations closely relate to our work. In [10] and in [11], given two sets of expression samples, the authors explicitly search for gene-pairs whose expression correlations are significantly different in the two sets of samples. A different approach, termed Liquid Association, explicitly tries to detect gene triplets (X, Y, Z) where the change in correlation between X and Y varies with the changes in the value of Z [16]. This approach implicitly partitions the expression samples based on the modulator gene expression. In contrast, our approach partitions the expression samples without any knowledge of the modulator gene and proceeds with the search for modulator genes in a subsequent step. In a genome-wide application, such as in the yeast application presented above, in principle, one can apply a Log-Likelihood Ratio (LLR) test, where the likelihood of the mixture model with a free f and α parameters is compared with the likelihood of a model where $f = 0$ and only α is free. The log of the ratio of the two

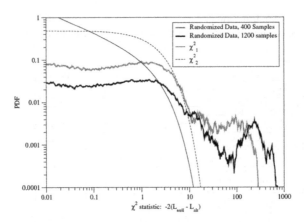

Fig. 3. The figure shows (1) The distribution of Log-Likelihood ratios for randomly generated (normal, i.i.d.) expression data for 400, and 1200 samples, permuted 20,000 times, (2) χ^2 distributions with 1 and 2 degrees of freedom. The "null" is defined by $f = 0$, implying an absence of mixture.

likelihoods can be used to assess significance of the partition based on a χ^2 distribution. While it is appealing to use LLR test to assess the significance of the mixture model, we found that our empirical distribution does not follow a χ^2 distribution (Figure 3). Our next thought was to use an empirically derived p-value for the mixture likelihood by randomly permuting the expression data. However, the empirical distributions of the likelihood itself varied significantly among different gene-pairs and thus we could not use a global distribution. Unfortunately, the number of permutations desired for an adiquately resolved p-value is computationally infeasible if done for each gene-pair separately. Thus, as a practical compromise, in the genome-wide yeast application, we chose to only consider gene-pairs with a Bonferroni-corrected global Kendall's Tau correlation p-value ≤ 0.05.

We face a similar challenge in the second phase of the approach, where, given the mixture model and the sample partition probability vector q, we search for modulator genes based on the correlation of their expression vectors with q. For a large number of trials (number of candidate modulators), a non-parametric test of correlation, such as Kendall's Tau, becomes infeasible. Thus, as another practical compromise, we devised the weighted t-test, which works well for the synthetic data. For the small-scale yeast application on specific (X, Y, Z)-triplets, we used Kendall's Tau but for the large-scale application we used weighted t-statistic. A more detailed study needs to be done to carefully assess the effect of these practical choices on the method's accuracy and efficacy.

Our mixture modeling may be most effective in cases such as the one described in [10], where the sample partition is clearly characterized by a single (unknown) mutant gene. In most practical situations, based on publicly available compendiums of expression data, this may not be the case. Regulatory relationships in eukaryotes have multiple determinants and it is possible that even if the method does detect the "correct" partition, it may be difficult to evaluate the biological significance of the sample partition based on the differential expression of a single modulator gene.

In summary, our work contributes a novel approach to the problem of partitioning expression samples and detecting potential modulators of expression correlation between a pair of genes. While this approach is likely to be effective in specific cases, as discussed above, statistical and computational challenges remain to be resolved and further work needs to be done to harness the approach in a large-scale application.

Acknowledgements

SH is supported by NIH R01-GM-085226, MH is supported by NIH R21-GM-078203, LE is supported by NIH T32-HG-000046 and LS is supported by NIH T32-HG-000046.

References

1. Tornow, S., Mewes, H.W.: Functional modules by relating protein interaction networks and gene expression. Nucleic Acids Res. 31(21), 6283–6289, 1362–4962 (2003); (Electronic) Journal Article Research Support, Non-U.S. Gov't
2. Stuart, J.M., Segal, E., Koller, D., Kim, S.K.: A gene-coexpression network for global discovery of conserved genetic modules. Science 302(5643), 249–255, 1095–9203 (2003); (Electronic) Comparative Study Journal Article Research Support, Non-U.S. Gov't Research Support, U.S. Gov't, Non-P.H.S. Research Support, U.S. Gov't, P.H.S
3. von Mering, C., Jensen, L.J., Kuhn, M., Chaffron, S., Doerks, T., Kruger, B., Snel, B., Bork, P.: String 7–recent developments in the integration and prediction of protein interactions. Nucleic Acids Res. 35(Database issue), D358–D362, 1362–4962 (2007); (Electronic) Journal Article Research Support, Non-U.S. Gov't
4. Magwene, P.M., Kim, J.: Estimating genomic coexpression networks using first-order conditional independence. Genome Biol. 5(12), R100, 1465–6914 (2004); (Electronic) Evaluation Studies Journal Article Research Support, N.I.H., Extramural Research Support, Non-U.S. Gov't Research Support, U.S. Gov't, Non-P.H.S. Research Support, U.S. Gov't, P.H.S
5. Margolin, A.A., Nemenman, I., Basso, K., Wiggins, C., Stolovitzky, G., Dalla Favera, R., Califano, A.: Aracne: an algorithm for the reconstruction of gene regulatory networks in a mammalian cellular context. BMC Bioinformatics 7(suppl. 1), S7, 1471–2105 (2006); (Electronic) Journal Article Research Support, N.I.H., Extramural
6. Basso, K., Margolin, A.A., Stolovitzky, G., Klein, U., Dalla-Favera, R., Califano, A.: Reverse engineering of regulatory networks in human b cells. Nat Genet 37(4), 382–390, 1061–4036 (2005); (Print) Comparative Study Journal Article Research Support, Non-U.S. Gov't Research Support, U.S. Gov't, P.H.S

7. Vaquerizas, J.M., Kummerfeld, S.K., Teichmann, S.A., Luscombe, N.M.: A census of human transcription factors: function, expression and evolution. Nat. Rev. Genet. 10(4), 252–263, 1471–0064 (2009); (Electronic) Journal Article Research Support, Non-U.S. Gov't

8. Khidekel, N., Hsieh-Wilson, L.C.: A 'molecular switchboard'–covalent modifications to proteins and their impact on transcription. Org. Biomol. Chem. 2(1), 1–7, 1477–0520 (2004); (Print) Journal Article Review

9. Steinfeld, I., Shamir, R., Kupiec, M.: A genome-wide analysis in saccharomyces cerevisiae demonstrates the influence of chromatin modifiers on transcription. Nat. Genet. 39(3), 303–309, 1061–4036 (2007); (Print) Journal Article

10. Hudson, N.J., Reverter, A., Dalrymple, B.P.: A differential wiring analysis of expression data correctly identifies the gene containing the causal mutation. PLoS Comput. Biol. 5(5), e1000382, 1553–7358 (2009); (Electronic) Journal Article

11. Hu, R., Qiu, X., Glazko, G., Klebanov, L., Yakovlev, A.: Detecting intergene correlation changes in microarray analysis: a new approach to gene selection. BMC Bioinformatics 10, 20, 1471–2105 (2009); (Electronic) Journal Article Research Support, N.I.H., Extramural Research Support, Non-U.S. Gov't

12. Barrett, T., Troup, D.B., Wilhite, S.E., Ledoux, P., Rudnev, D., Evangelista, C., Kim, I.F., Soboleva, A., Tomashevsky, M., Marshall, K.A., Phillippy, K.H., Sherman, P.M., Muertter, R.N., Edgar, R.: Ncbi geo: archive for high-throughput functional genomic data. Nucleic Acids Res. 37(Database issue), D885–D890, 1362–4962 (2009); (Electronic) Journal Article Research Support, N.I.H., Intramural

13. Everett, L., Vo, A., Hannenhalli, S.: PTM-Switchboard–a database of posttranslational modifications of transcription factors, the mediating enzymes and target genes. Nucleic Acids Res. 37(Database issue), D66–D71, 1362–4962 (2009); (Electronic) Journal Article Research Support, N.I.H., Extramural

14. Chen, G., Jensen, S.T., Stoeckert, C.J.: Clustering of genes into regulons using integrated modeling-cogrim. Genome Biol. 8(1), R4 1465-6914 (2007); (Electronic) Journal Article

15. Harbison, C.T., Gordon, D.B., Lee, T.I., Rinaldi, N.J., Macisaac, K.D., Danford, T.W., Hannett, N.M., Tagne, J.B., Reynolds, D.B., Yoo, J., Jennings, E.G., Zeitlinger, J., Pokholok, D.K., Kellis, M., Rolfe, P.A., Takusagawa, K.T., Lander, E.S., Gifford, D.K., Fraenkel, E., Young, R.A.: Transcriptional regulatory code of a eukaryotic genome. Nature 431(7004), 99–104, 1476–4687 (2004); Journal Article

16. Li, K.C.: Genome-wide coexpression dynamics: theory and application. Proc. Natl. Acad. Sci. USA 99(26), 16875–16880, 0027–8424 (2002); (Print) Journal Article Research Support, U.S. Gov't, Non-P.H.S

Phylogenetic Comparative Assembly

Peter Husemann[1,2] and Jens Stoye[1]

[1] AG Genominformatik, Technische Fakultät, Bielefeld University, Germany
[2] International NRW Graduate School in Bioinformatics and Genome Research
Bielefeld University, Germany
{phuseman,stoye}@techfak.uni-bielefeld.de

Abstract. Recent high throughput sequencing technologies are capable
of generating a huge amount of data for bacterial genome sequencing
projects. Although current sequence assemblers successfully merge the
overlapping reads, often several contigs remain which cannot be assembled any further. It is still costly and time consuming to close all the
gaps in order to acquire the whole genomic sequence. Here we propose
an algorithm that takes several related genomes and their phylogenetic
relationships into account to create a contig adjacency graph. From this
a layout graph can be computed which indicates putative adjacencies of
the contigs in order to aid biologists in finishing the complete genomic
sequence.

1   Introduction

Today the nucleotide sequences of many genomes are known. For the first available genomic sequences, the process of obtaining the sequence was costly and
tedious. The most common approach for de-novo genome sequencing is *whole
genome shotgun sequencing* [1,2]. Here, the genome is fragmented randomly into
small parts. Each of these fragments is sequenced, for example, with recent high
throughput methods. In the next step, overlapping reads are merged with an
assembler software into a contiguous sequence. However, instead of the desired
one sequence of the whole genome, often many *contigs* remain, separated by
gaps. The main reasons for these gaps are lost fragments in the fragmentation
phase and repeating sequences in the genome. In a process called *scaffolding*,
the relative order of the contigs as well as the size of the gaps between them
is estimated. In a subsequent *finishing* phase the gaps between the contigs are
closed with a procedure called *primer walking*. For the ends of two estimated
adjacent contigs, specific primer sequences have to be designed that function as
start points for two polymerase chain reactions (PCRs) for Sanger sequencing.
These PCRs ideally run towards each other until the sequences overlap. To close
a gap completely, new primer pairs have to be generated again and again since
the maximum read length for Sanger sequencing is restricted. This makes the
process expensive and work intensive. It is thus advisable to reduce the pairs of
contigs that have to be considered. If no further information about the order of
the contigs is given there are $\mathcal{O}(n^2)$ possibilities for n contigs to apply primer

S.L. Salzberg and T. Warnow (Eds.): WABI 2009, LNBI 5724, pp. 145–156, 2009.
© Springer-Verlag Berlin Heidelberg 2009

walking. If the order is known, it suffices to do $\mathcal{O}(n)$ primer walks to fill the gaps.

An algorithm that estimates a reasonable order for the contigs is thus a good help for sequencing projects. The estimation is usually based on the sequences of closely related species that are assumed to have a high degree of synteny. A few tools have been developed which use one or several related reference genomes to devise an ordering of the contigs: OSLay [3] for example takes a set of BLAST [4] matches between the contigs and a reference sequence and computes from this a layout for the contigs. The algorithm minimizes the height differences of so-called local diagonal extensions, which are basically matches from the border of a contig to the reference sequence. The program Projector2 [5] is a web service that maps contigs on a reference genome, based on experimentally validated rules, and automatically designs suitable primer sequences. Zhao et al. [6] developed an algorithm to integrate the information of multiple reference genomes. For the ordering, the enhanced ant colony optimization algorithm PGA (pheromone trail-based genetic algorithm) is used.

Our work commenced when analyzing data from inhouse sequencing projects for different species of the *Corynebacteria* genus, where we observed several aspects making it hard to find an ordering of the contigs. Zhao et al. show that poor sequence coverage can be overcome by using multiple reference genomes, but problematic for this approach are major rearrangements in the genomic sequences of more distantly related species. Another challenge are repeating regions in the sequence of the newly sequenced genome. We developed an algorithm that uses the information of all similar regions between a set of contigs and several related reference genomes to estimate an ordering of the contigs. The novel idea is to incorporate the phylogenetic distance of the species in order to alleviate the impact of rearrangements to the ordering. While generating one 'optimal' order of the contigs is the predominant approach to aid the closure of gaps, we propose a more flexible format to describe contig adjacencies that is also capable of dealing with repeating contigs.

The algorithm we present here is based on a simple data structure, the *contig adjacency graph* that is introduced in Sect. 2. There we also give an optimal solution for finding a linear ordering of the contigs using this graph. However, a linear ordering is not sufficient to reflect all relations of real contig data. Therefore we propose a heuristic in Sect. 3 by which the most promising, but not necessarily unique, adjacencies are revealed in a *layout graph*. Sect. 4 shows the results of applying our method to real sequencing data and compares these with a recent approach from the literature.

2   Phylogeny Based Contig Ordering Algorithm

A natural strategy to devise an 'optimal' linear ordering of the contigs, based on one or several related reference genomes, works in three steps: At first, all similar regions between the contigs and each reference genome are determined. Then a graph is created, containing edge weights that reflect the neighborhood of the contigs. In the last step a weight maximizing path through the graph is calculated, which defines the desired order of the contigs. In the following, we describe these three steps in more detail, in particular in Sect. 2.2, we define a novel edge weight function that incorporates the phylogenetic distance of the involved species.

2.1 Matching Contigs against a Reference

Let $\Sigma = \{A, C, G, T\}$ be the alphabet of nucleotides. We denote by Σ^* the set of all finite strings over Σ, by $|s| := \ell$ the *length* of string $s = s_1 \ldots s_\ell$, $s \in \Sigma^\ell$, and by $s[i, j] := s_i \ldots s_j$ with $1 \leq i \leq j \leq \ell$ the *substring* of s that starts at position i and ends at position j. Suppose we are given a set of contigs $\mathcal{C} = \{c_1, \ldots, c_n\}$, $c_i \in \Sigma^*$ and a set of already finished reference genomes $\mathcal{R} = \{g_1, \ldots, g_m\}$, $g_r \in \Sigma^*$. The relation of the reference genomes is given by a phylogenetic tree $\mathcal{T}$ which contains the evolutionary distances of the species. Note that the tree can be generated even if some genomes are not completely assembled yet, for example from 16S-rRNAs.

To infer information about the order and orientation of the contigs, these are mapped onto each reference genome by calculating local alignments. Let $s = c[s_b, s_e]$ be a substring of contig c and $t = g[t_b, t_e]$ be a substring of reference genome g. The tuple $m = ((s_b, s_e), (t_b, t_e))$ is called a *matching region* or simply *match* if s and t share sufficient sequence similarity. The *length* of a match, $|m| := t_e - t_b + 1$, is defined as the length of the covered substring in the reference genome. Sufficient sequence similarity can be defined, for example, by a BLAST hit of significance above a user-defined threshold. We, on the contrary, employ in our implementation the *swift* algorithm [7] for matching. It utilizes a q-gram index and provides for each match m the number of exactly matching q-grams, denoted as $\text{qhits}(m)$, which can be used as a quality estimation for that match. Note that each contig can have several matches on a reference genome. For $s_b > s_e$ we define $c[s_b, s_e]$ to be the reverse complement of $c[s_e, s_b]$ and call m a *reverse match*. Further we assume w.l.o.g. that $t_b < t_e$ for all $g[t_b, t_e]$, otherwise we can replace the involved contig sequence by its reverse complement. For brevity of notation, m_i^r denotes a match between contig $c_i \in \mathcal{C}$ and reference genome $g_r \in \mathcal{R}$, and $\mathcal{M}_i^r = \{m_{i,1}^r, \ldots, m_{i,s}^r\}$ denotes the (possibly empty) set of all such matches.

Each match $m_i^r = ((s_b, s_e), (t_b, t_e))$ implies a projection of the contig c_i onto the reference genome g_r. The *projected contig* $\pi(m_i^r) = ((t_b - s_b), (t_e + |c_i| - s_e))$ refers to the implied pair of index positions on g_r. For reverse complement matches, the projection can be defined similarly. Fig. 1 shows an example of two projected contigs as well as their distance, which will be defined next.

The *distance* of two projected contigs $\pi(m) = (t_b, t_e)$ and $\pi(m') = (t_b', t_e')$ is defined as follows:

$$d\big(\pi(m), \pi(m')\big) = \begin{cases} t_b' - t_e & \text{if } t_b < t_b' \\ t_b - t_e' & \text{if } t_b > t_b' \\ -\min\{|m|, |m'|\} & \text{if } t_b = t_b' \ . \end{cases}$$

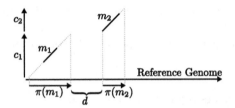

Fig. 1. Illustration of the projected contigs $\pi(m_1)$ and $\pi(m_2)$, which are based on their matches m_1 and m_2

If the matches refer to different reference genomes, the distance of their projections is undefined. Note that the term distance is used here in the sense of displacement, d is *not* a metric in the mathematical sense. For example, d is negative if the projected contigs overlap.

2.2   Contig Adjacency Graph

In the following we define the edge-weighted *contig adjacency graph* $G_{\mathcal{C},\mathcal{R},\mathcal{T}} = (V,E)$ that contains for each contig $c_i \in \mathcal{C}$ two vertices: l_i as the *left connector* and r_i as the *right connector* of contig c_i. The set of vertices V is then defined as $V = L \cup R$, where $L = \{l_1, \ldots, l_n\}$ and $R = \{r_1, \ldots, r_n\}$.

The graph G is fully connected: $E = \binom{V}{2}$. We split these edges into two subsets: the *intra contig edges* $I = \{\{l_1, r_1\}, \ldots, \{l_n, r_n\}\}$ which connect for each contig its left and its right connector; and the set of *adjacency edges* $A = E \setminus I$ that connect the contigs among each other.

Now we define a weight function for the edges. For each intra contig edge $e \in I$ we set the weight $w(e) = 0$. For the remaining edges let $e = \{v_i, v_j\} \in A$ with $v_i \in \{l_i, r_i\}$ and $v_j \in \{l_j, r_j\}$ be an adjacency between contigs c_i and c_j. Then the weight of this adjacency edge is defined as

$$w(e) = \sum_{g_r \in \mathcal{R}} w_r(v_i, v_j)$$

where the function $w_r(v_i, v_j)$ defines a likelihood score for contigs c_i and c_j being adjacent, based on the matches to reference g_r. Moreover, the phylogenetic distance $d_{\mathcal{T}}$ between the contig's species and the reference genome's species is employed as a weight factor. Thus we define

$$w_r(v_i, v_j) = \sum_{m_i^r \in \mathcal{M}_i^r, m_j^r \in \mathcal{M}_j^r} s\Big(d\big(\pi(m_i^r), \pi(m_j^r)\big), d_{\mathcal{T}}\Big) \cdot \mathrm{qhits}(m_i^r) \cdot \mathrm{qhits}(m_j^r)$$

where d is the distance between the projected contigs and $s(d, d_{\mathcal{T}})$ is a suitably defined scoring function.

In order to define s we will give some further biological motivations. The scoring function s models the likelihood that two contigs are adjacent based on

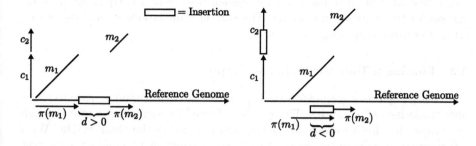

Fig. 2. An insertion in the reference genome leads to a positive distance (left side), whereas an insertion in a contig leads to a negative distance (right side)

the distance d of their projected contigs. Projected contigs which are not adjacent should have a high distance and thus a low score. Adjacent contigs should have a distance close to zero and a high score. But on the one hand, the distance of the two projected contigs reaches positive values due to insertions in the reference's genome. On the other hand, the distances are negative if the projections are overlapping which is the case if there are insertions in the contigs' genome. Both cases can be seen in Fig. 2. Note that an insertion in the one genome corresponds to a deletion in the other.

A second important aspect that is included in our model are rearrangements between the related species, which can lead to misleading adjacencies of projected contigs. Assuming that between closer related species less rearrangements have taken place, we use the phylogenetic tree distance d_T to weight the match information. To model the mentioned two considerations, we use a Gaussian distribution with an expected value of zero:

$$s(d, d_T) := \frac{1}{d_T \cdot \sigma \sqrt{2\pi}} e^{-\frac{1}{2}\left(\frac{d}{d_T \cdot \sigma}\right)^2}$$

where σ is the standard deviation for the size of deletions or insertions. A higher tree distance d_T allows larger insertions and deletions, but scores the reliability of the matches to more distantly related genomes to a lesser degree.

However, this model neglects the fact that in the fragmentation phase, for example in parallel pyrosequencing, often fragments disappear such that there are no reads for this fragment. If a fragment is not sequenced, the same situation arises as if there is an insertion into the reference genome, which causes positive distances. To include this detail we use two superimposed Gaussian distributions for the scoring. The first distribution models insertions into the contigs and into the reference genome, the second models lost fragments during sequence assembly. The influence of each model is determined by a weighting factor φ:

$$s(d, d_T) := \frac{(1 - \varphi)}{d_T \cdot \sigma_1 \sqrt{2\pi}} e^{-\frac{1}{2}\left(\frac{d}{d_T \cdot \sigma_1}\right)^2} + \frac{\varphi}{\sigma_2 \sqrt{2\pi}} e^{-\frac{1}{2}\left(\frac{d - \mu}{\sigma_2}\right)^2} . \qquad (1)$$

The expected value μ of the second Gaussian distribution is equal to the average size of the sequence fragments. The standard deviations σ_1 and σ_2 can be estimated from sequencing projects.

2.3   Finding a Tour through the Graph

The contig adjacency graph with the described edge weights can be used to find a linear ordering of the contigs. This can be achieved by computing a tour through the graph that incorporates all contigs and maximizes the total weight. With minor enhancements of the graph, this becomes equivalent to finding a shortest Hamiltonian cycle.

The modifications are as follows: At first all edge weights have to be converted to distances. This is done by replacing each edge weight w by $m - w$ where m is the maximum weight in the graph. To ensure that each contig is incorporated exactly once, and only in one direction, we add an intermediate node between the left and the right connector of each contig. The modified graph is then defined as $G'_{C,R,T} = (V', E')$ with $V' = V \cup \{v_i \mid 1 \le i \le n\}$ and $E' = A \cup \{\{l_i, v_i\}, \{v_i, r_i\} \mid 1 \le i \le n\}$. The distance of all edges that lead to an intermediate node v_i is set to 0. It is easy to see that a shortest Hamiltonian cycle in the modified graph defines an ordering as well as the orientation of all contigs, and thus any TSP algorithm can be used to find an optimal linear layout of the contigs.

3   Fast Adjacency Discovery Algorithm

As described in the previous section, a linear ordering of the contigs, which is optimal with respect to the adjacency edge weights, can be computed using a suitable optimization algorithm. However, our results on real data in Sect. 4 show that a linear order of the contigs is not necessarily possible, mainly due to arbitrary placement of repeated or rearranged regions. A method that provides a unique layout where possible, but also points out alternative solutions where necessary, may be more useful in practice. We present an approach following this overall strategy in this section.

The basis of our algorithm is a greedy heuristic for the TSP, known as the *multi-fragment heuristic* [8], that proceeds as follows: First the edges of the graph are sorted by increasing distance and then added in this order into an initially empty set of path fragments. Whenever an involved node would exceed the maximal degree of two, or if a path fragment would create a cycle, the edge is skipped. The only exception to the latter is a cycle of length n which constitutes the final Hamiltonian path.

This *best connection first* procedure creates multiple low distance path fragments which are merged sooner or later. We chose this approach because it seems natural to incorporate those adjacencies first into an ordering that are most promising to be investigated for gap closure.

As already indicated, repeating or rearranged regions may prohibit an unambiguous linear ordering of the contigs. Repeating contigs create cycles in a possible path, and rearrangements can lead to conflicting adjacencies of a contig. To model both, we relax the constraints of the multi-fragment heuristic: First, we do not check for cycles, which permits repeating contigs to be incorporated adequately. Secondly, when inserting an edge, we allow one of the incident nodes, but not both, to exceed the degree of two, which allows to also include conflicting information into our layout. The result of this modified heuristic is a subgraph of the contig adjacency graph $L \subset G$ that we call the *layout graph*. The procedure to generate the layout graph is formally described in Algorithm 1. Note that the resulting layout graph is not necessarily connected.

Algorithm 1. Contig adjacency discovery algorithm

Input: set of contigs C, set of related genomes R, phylogenetic tree T
Output: layout graph L of the contigs
1 create contig adjacency graph $G_{C,R,T} = (V, I \cup A)$;
2 create empty layout graph $L = (V_L, E_L)$ with $V_L = \emptyset$ and $E_L = \emptyset$;
3 **foreach** *adjacency edge* $e = \{v_i, v_j\} \in A$, *sorted by decreasing weight* $w(e)$ **do**
4   **if** $|V_L \cap \{v_i, v_j\}| < 2$ **then**
5        $V_L = V_L \cup \{v_i, v_j\}$;
6        $E_L = E_L \cup e$;
7   **end**
8 **end**
9 $E_L = E_L \cup I$;

The layout graph can be analyzed to make assumptions about repeating contigs and rearrangements. Conflicting edges can give hints about these two problems. The information about unambiguously incorporated contigs can be used to generate primer pairs for gap closure. Displaying also the ambiguities allows to investigate the conflicting connections further. Instead of pinning the result down to a single, possibly wrong, order of the contigs we prefer to output the best possibilities. Nonetheless it should be kept in mind that rearrangements can cause seemingly good adjacencies that do not belong to a correct layout.

4   Results

We tested our algorithm on real contig data of the *Corynebacteria* genus. For the genomic sequences of *C. aurimucosum* (unpublished), *C. urealyticum* [9], and *C. kroppenstedtii* [10] we obtained the finished genomic sequences as well as the underlying contig data from sequencing projects that were conducted at Bielefeld University. Additionally we chose four more publicly available *Corynebacteria* genomes, *C. diphtheriae*, *C. efficiens*, *C. glutamicum ATCC 13032 Bielefeld*, and *C. jeikeium* that we downloaded from NCBI [11,12]. Figure 3 shows the evolutionary tree of all employed genomes. The tree was generated with the EDGAR framework [13] applying Neighbor Joining [14] to a set of *core* genes. As an illustration of the major rearrangements that happened between the employed

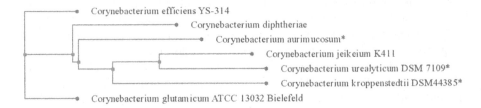

Fig. 3. Phylogenetic tree of the employed Corynebacteria. Contig data was available for the species marked with an asterisk (*). Tree calculated with EDGAR [13], image generated with PHY.FI [15].

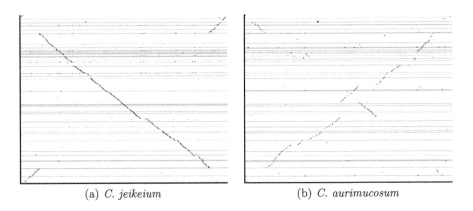

(a) *C. jeikeium*   (b) *C. aurimucosum*

Fig. 4. Synteny plots of the above species and the contigs of *C. urealyticum*. The latter are stacked on the vertical axis in reference order, separated by horizontal lines.

species, Fig. 4 shows two example synteny plots. It is clearly observable that due to rearrangements a mapping of the contigs on the displayed related genomes would provide incorrect adjacencies of some contigs.

All finished sequences, except the genome of the contigs to be layouted, served as references to find a layout for the contig sets. As a standard of truth we computed for each set of contigs a *reference order* by mapping them onto the corresponding finished genome. We would like to note here that this reference order is not unique since many contigs contain or even consist of repeating regions which map non-uniquely to different locations.

We implemented our proposed algorithm in Java. The software *tree-cat* (tree based contig arrangement tool) contains a re-implementation of the fast local alignment algorithm *swift* [7], the contig adjacency graph creation, a branch and bound exact TSP algorithm, and the fast layout graph heuristic described in Sect. 3. The software is open source (GPL) and available upon request.

Input to *tree-cat* are the FASTA sequences of the contigs and of the related references as well as a phylogenetic tree in NEWICK format. Each reference can consist of several sequences, for example several chromosomes. When the algorithm is run, first all matches from the contigs to each reference are computed. For the following results, matches were considered to have a minimal length of 64 bases and a maximum error rate of 8%. The matches are cached which allows a visualization like in Fig. 4 and avoids a new computation if subsequent steps are re-run with different parameters. As the second step, after the matching, the contig adjacency graph is constructed as defined in Sect. 2.2. The following (empirically estimated) parameters were used for the scoring function (1) to compute the results: The standard deviation of the insertion/deletion size was set to $\sigma_1 = 10\,000$ bases and the expected lost fragment size to $\mu = 2\,000$ bases with a standard deviation of $\sigma_2 = 1\,000$ bases. The lost fragment weighting factor φ was set to 0.1. In the last step, the computed adjacency graph is used to devise the contig layout graph which can then be visualized with the open source software package GraphViz [16].

We compared the output of our method with the results of applying PGA [6] to the same data sets. OSLay [3] and Projector2 [5] were not included in the comparison since they use only one single genome as reference. Due to space restrictions we give in Fig. 5 only the results for the set of *C. urealyticum* contigs, but the other results are comparable. The data set consists of 69 contigs with a total size of 2.3 Mb and six reference genomes which have a total size of 16.6 Mb. To compute the results, a sparcv9 processor operating at 750 MHz under Solaris was used. Figures 5(a) and 5(b) show two PGA results and Fig. 5(c) contains our resulting layout graph. The node labels of all graphs are the rank of the corresponding contig with respect to the reference order of the contigs. The correct path should therefore be $0, 1, \ldots, 68$.

PGA uses BLAST to match the contigs on each genome. After that it computes five paths for the contigs that optimize a *fitness matrix* which is comparable to our contig adjacency graph. For that purpose a genetic algorithm is used, possibly giving different connections with each run. The connections of all five paths are included into the result and the edge weights give the percentage how often a connection occurred. Some nodes are missing in the first two graphs since PGA filters all contigs of length less than 3\,500 bases.

In Fig. 5(a) PGA was applied with only one reference, the already finished genome of *C. urealyticum*, which ought to provide the 'perfect' information. The resulting graph shows that it is impossible to find a unique linear ordering of the contigs, even if the best possible reference sequence is available.

A comparison between PGA and *tree-cat* is given by the graphs in Figs. 5(b) and 5(c). Both are created with all other related genomes of our data set as reference genomes. For the graph in Fig. 5(b) the matching using BLAST needed 457 seconds and after that PGA took 161 seconds to compute its result, using the standard parameters given in [6]. Our algorithm, within *tree-cat*, required 95 seconds for the matching and about two seconds for the creation of the contig adjacency graph as well as the calculation of the layout graph, shown in Fig. 5(c). Although the reference order is not completely reliable, we used it as a rough estimation for the quality of the results. For each edge in the graphs we tested whether it is present in the reference order. PGAs graph contains 19 true positive

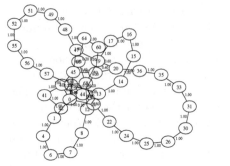

(a) PGA with perfect information.

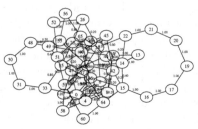

(b) PGA using multiple references.

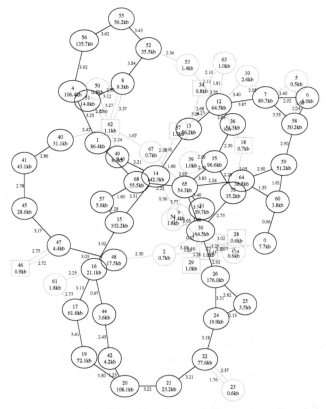

(c) *tree-cat* using multiple references. Contigs smaller than
3.5 kb have gray nodes, repeating contigs for which at least
95% of the sequence occurs more than once on a reference
genome have rectangular nodes. Edge weights are given in
logarithmic scale.

Fig. 5. *C. urealyticum* contig connections generated by PGA and *tree-cat*. The contig
nodes are numbered in reference order.

connections and 75 false positive. *tree-cat* computed 27 correct and 70 false connections. This shows that our method achieved a better sensitivity (0.39 to 0.28) as well as a better precision (0.28 to 0.2), while being much faster. Additionally, PGA's result contains connections that are obviously incorrect, like, for example, placing contig 26 next to contig 52. Our approach does not show this connection. Manual inspection shows that this is due to the evolutionary distances that we incorporate for the edge scoring since phylogenetically closer genomes do not contain this adjacency. This is further supported by the fact that the connection is also not present when PGA uses the 'perfect' reference, see Fig. 5(a).

5   Conclusion

The contribution of our paper is twofold. On the one hand our results demonstrate that the common approach of searching one linear optimal ordering of the contigs is not feasible for real world data, due to repeating regions and rearrangements between the species. Therefore, we propose a more flexible output for the ordering of contigs and give an algorithm which generates such results.

On the other hand we developed a novel scoring function for the contig adjacency estimation based on multiple reference genomes that is biologically motivated in two ways: Firstly, it contains a sophisticated weighting scheme for the distances of projected contigs, and secondly it integrates the phylogenetic distances of the species to alleviate the effects caused by rearrangements.

A first evaluation of our algorithm shows that its implementation *tree-cat* is considerably faster than a recent approach from the literature while it is at the same time generating better results. We believe that with our approach of including phylogenetic information into the problem of contig layouting, we have gone one step further in using all available information for this important task within genome finishing.

Nevertheless, in sequencing projects, often additional information emerges which is not yet included in our approach. For example, information derived from mate pairs, fosmid libraries or radiation hybrid maps might give valuable hints on the orientation and the distance of contigs while not being biased by evolutionary events. Concerning the phylogenetic tree, rearrangements between the genomes were not predicted by the methods presented in this paper. This leads to ambiguous information for the ordering of contigs and thus to weak or misleading adjacency scores which need to be curated manually. A strategy for the discovery of rearrangements is thus desired in future work. Furthermore, due to horizontal gene transfer some regions of a genome can have different evolutionary histories than others. Detecting such regions and treating them in a special way might be advisable in an even more sophisticated approach.

Acknowledgments

The authors wish to thank Christian Rückert, Susanne Schneiker-Bekel, and Andreas Tauch for the sequence data, Jochen Blom and Burkhard Linke for the phylogenetic tree, and Travis Gagie and Roland Wittler for helpful discussions.

References

1. Staden, R.: A strategy of DNA sequencing employing computer programs. Nucleic Acids Res. 6(7), 2601–2610 (1979)
2. Anderson, S.: Shotgun DNA sequencing using cloned DNase I-generated fragments. Nucleic Acids Res. 9(13), 3015–3027 (1981)
3. Richter, D.C., Schuster, S.C., Huson, D.H.: OSLay: optimal syntenic layout of unfinished assemblies. Bioinformatics 23(13), 1573–1579 (2007)
4. Altschul, S., Gish, W., Miller, W., Myers, E., Lipman, D.: Basic local alignment search tool. J. Mol. Biol. 215, 403–410 (1990)
5. van Hijum, S.A.F.T., Zomer, A.L., Kuipers, O.P., Kok, J.: Projector 2: contig mapping for efficient gap-closure of prokaryotic genome sequence assemblies. Nucleic Acids Res. 566, W560–W566 (2005)
6. Zhao, F., Zhao, F., Li, T., Bryant, D.A.: A new pheromone trail-based genetic algorithm for comparative genome assembly. Nucleic Acids Res. 36(10), 3455–3462 (2008)
7. Rasmussen, K.R., Stoye, J., Myers, E.W.: Efficient q-gram filters for finding all epsilon-matches over a given length. J. Comp. Biol. 13(2), 296–308 (2006)
8. Bentley, J.J.: Fast algorithms for Geometric Traveling Salesman Problems. Informs. J. Comp. 4(4), 387–411 (1992)
9. Tauch, A., et al.: The lifestyle of *Corynebacterium urealyticum* derived from its complete genome sequence established by pyrosequencing. J. Biotechnol. 136(1-2), 11–21 (2008)
10. Tauch, A., et al.: Ultrafast pyrosequencing of *Corynebacterium kroppenstedtii DSM44385* revealed insights into the physiology of a lipophilic corynebacterium that lacks mycolic acids. J. Biotechnol. 136(1-2), 22–30 (2008)
11. Wheeler, D.L., Chappey, C., Lash, A.E., Leipe, D.D., Madden, T.L., Schuler, G., Tatusova, T.A., Rapp, B.A.: Database resources of the national center for biotechnology information. Nucleic Acids Res. 28(1), 10–14 (2000)
12. Benson, D.A., Karsch-Mizrachi, I., Lipman, D.J., Ostell, J., Rapp, B.A., Wheeler, D.L.: Genbank. Nucleic Acids Res. 28(1), 15–18 (2000)
13. Blom, J., Albaum, S.P., Doppmeier, D., Pühler, A., Vorhölter, F.J., Goesmann, A.: EDGAR: A software framework for the comparative analysis of microbial genomes. BMC Bioinformatics (to appear, 2009)
14. Saitou, N., Nei, M.: The neighbor-joining method: a new method for reconstructing phylogenetic trees. Mol. Biol. Evol. 4(4), 406–425 (1987)
15. Fredslund, J.: PHY.FI: fast and easy online creation and manipulation of phylogeny color figures. BMC Bioinformatics 7, 315 (2006)
16. Gansner, E.R., North, S.C.: An open graph visualization system and its applications to software engineering. SPE 30, 1203–1233 (1999)

K-Partite RNA Secondary Structures*

Minghui Jiang**, Pedro J. Tejada, Ramoni O. Lasisi, Shanhong Cheng,
and D. Scott Fechser

Department of Computer Science, Utah State University, Logan, UT 84322-4205, USA
mjiang@cc.usu.edu, p.tejada@aggiemail.usu.edu,
ramoni.lasisi@aggiemail.usu.edu, s.cheng@aggiemail.usu.edu,
s.fechser@aggiemail.usu.edu

Abstract. RNA secondary structure prediction is a fundamental problem in
structural bioinformatics. The prediction problem is difficult because RNA secondary structures may contain pseudoknots formed by crossing base pairs. We
introduce k-partite secondary structures as a simple classification of RNA secondary structures with pseudoknots. An RNA secondary structure is k-partite if
it is the union of k pseudoknot-free sub-structures. Most known RNA secondary
structures are either bipartite or tripartite. We show that there exists a constant
number k such that any secondary structure can be modified into a k-partite secondary structure with approximately the same free energy. This offers a partial
explanation of the prevalence of k-partite secondary structures with small k. We
give a complete characterization of the computational complexities of recognizing k-partite secondary structures for all $k \geq 2$, and show that this recognition
problem is essentially the same as the k-colorability problem on circle graphs. We
present two simple heuristics, iterated peeling and first-fit packing, for finding k-
partite RNA secondary structures. For maximizing the number of base pair stackings, our iterated peeling heuristic achieves a constant approximation ratio of at
most k for $2 \leq k \leq 5$, and at most $\frac{6}{1-(1-6/k)^k} \leq \frac{6}{1-e^{-6}} < 6.01491$ for $k \geq$
6. Experiment on sequences from PseudoBase shows that our first-fit packing
heuristic outperforms the leading method HotKnots in predicting RNA secondary
structures with pseudoknots. Source code, data set, and experimental results are
available at http://www.cs.usu.edu/~mjiang/rna/kpartite/.

1 Introduction

RNA secondary structure prediction is a fundamental problem in structural bioinformatics. The main theme of classical research on this problem has been the design of
dynamic programming algorithms that, given an RNA sequence, compute the "optimal"
secondary structure of the sequence with the lowest free energy. The first dynamic programming algorithm for predicting pseudoknot-free RNA secondary structures dates
back to the 1970s [20], and has been implemented by well-known software packages
such as Mfold [31] and Vienna [12]. During the last decade, several dynamic programming algorithms have also been designed for the much more difficult problem of predicting RNA secondary structures with pseudoknots [24,27,19,2,8,22,15]. These algorithms

---

* Supported in part by NSF grant DBI-0743670.
** Corresponding author.

S.L. Salzberg and T. Warnow (Eds.): WABI 2009, LNBI 5724, pp. 157–168, 2009.

CAGUGUUUUGAAGUCCACUUAAAUAGAACUUCU
:((((:::::[[[[[))))::::::::]:]]]]:

Fig. 1. A bipartite secondary structure

are often ad-hoc, and can handle only pseudoknots of certain restricted types implicit in their dynamic programming formulations [6,21]. If arbitrary pseudoknots are allowed in the secondary structures, then the prediction problem becomes computationally intractable even in very simple models [14,18], and we have to resort to approximation algorithms [16] and heuristics [26,23].

In this paper, we study k-partite RNA secondary structures. Intuitively, an RNA secondary structure is k-*partite* if it is the union of k pseudoknot-free sub-structures. For the $k = 2$ case, bipartite secondary structures have been previously studied by Stadler et al. as the bi-secondary structures [11,30] and by Ieong et al. as the planar secondary structures [14]. A recent result by Rødland on the enumeration of RNA pseudoknots [25] implies that most known RNA secondary structures are either bipartite or tripartite, i.e., k-partite for $k = 2$ or 3. Indeed, among the hundreds of known RNA secondary structures with pseudoknots in PseudoBase [5], only one structure[1] is tripartite, and the others are all bipartite (see Fig. 1 for an example[2]). The prevalence of bipartite and tripartite secondary structures has motivated us to consider k-partite RNA secondary structures. The main results of this paper are the following:

1. We show that there exists a constant number k such that any secondary structure of an RNA sequence can be modified into a k-partite secondary structure with approximately the same free energy. This offers a partial explanation of the prevalence of k-partite secondary structures with small k.
2. We give a complete characterization of the computational complexities of recognizing k-partite secondary structures for all $k \geq 2$.
3. We present two simple heuristics, iterated peeling and first-fit packing, for RNA secondary structure prediction. The iterated peeling heuristic achieves the first nontrivial constant approximation ratios for computing k-partite secondary structures with the maximum numbers of base pair stackings. The first-fit packing heuristic outperforms the leading method HotKnots in predicting RNA secondary structures with pseudoknots.

Preliminaries. The *primary structure* of an RNA molecule is the sequence of nucleotides (that is, the four different bases *A*, *C*, *G*, and *U*) in its single-stranded polymer. In its natural state, an RNA molecule folds into a three-dimensional structure by forming hydrogen bonds between non-consecutive bases that are complementary. The

---

[1] PKB00071: pseudoknot of regulatory region of alpha ribosomal protein operon of E.coli.
[2] PKB00116: pseudoknot PKb of upstream pseudoknot domain of 3'-UTR of beet soil-borne virus RNA 1.

three-dimensional arrangement of the atoms in the folded RNA molecule is the *tertiary structure*; the collection of base pairs in the tertiary structure is the *secondary structure*. Two bases are *complementary* if they are either a Watson-Crick pair $\{A, U\}$ or $\{G, C\}$, or a wobble pair $\{G, U\}$. Two complementary bases with sequence indices i and j form a *base pair* (i, j) if they satisfy the *hairpin constraint* that $j - i > 3$, that is, they are separated by at least three other bases in the sequence. Two base pairs (i_1, j_1) and (i_2, j_2) are *crossing* if either $i_1 < i_2 < j_1 < j_2$ or $i_2 < i_1 < j_2 < j_1$, are *nesting* if either $i_1 < i_2 < j_2 < j_1$ or $i_2 < i_1 < j_1 < j_2$, and are *sequential* if either $i_1 < j_1 < i_2 < j_2$ or $i_2 < j_2 < i_1 < j_1$. A *secondary structure* is a set of disjoint base pairs, where each base participates in at most one base pair. Two crossing base pairs form a *pseudoknot*. A secondary structure is *pseudoknot-free* if all base pairs in the structure are either nesting or sequential. A *base pair stacking* is formed by two consecutive base pairs (i, j) and $(i + 1, j - 1)$. A *helix* (i, j, l), $l \geq 2$, is a set of l consecutive base pairs $(i, j), (i + 1, j - 1), \ldots, (i + l - 1, j - l + 1)$, forming $l - 1$ base pair stackings.

2   Prevalence of k-Partite Secondary Structures with Small k

The following proposition partially explains the interesting fact that most real RNA secondary structures are k-partite for a small k:

Proposition 1. *There exists a constant number k such that any secondary structure of an RNA sequence can be modified into a k-partite secondary structure with approximately the same free energy.*

This proposition is deliberately stated in an imprecise way, considering the diversity and complexity of the many energy models for RNA folding. To make more precise statements with rigorous proofs, we have to restrict our discussion to simple models. In the following theorem, we focus on the simplest energy model in which the free energy of an RNA secondary structure is proportional to its number of Watson-Crick base pairs $\{A, U\}$ and $\{G, C\}$.

Theorem 1. *If an RNA sequence has a secondary structure with n Watson-Crick base pairs, then it must also have a 4-partite secondary structure with at least $n - 8$ Watson-Crick base pairs. Moreover, each part of the 4-partite secondary structure consists of all nesting pairs.*

Proof. Let $\mathcal{P}$ be a set of n Watson-Crick base pairs in a secondary structure. Partition $\mathcal{P}$ into four subsets $\mathcal{P}_{AU}$, $\mathcal{P}_{UA}$, $\mathcal{P}_{GC}$, and $\mathcal{P}_{CG}$ of the four different types of base pairs AU, UA, GC, and CG, respectively. Let x be the number of base pairs in $\mathcal{P}_{AU} \cup \mathcal{P}_{UA}$. We next show that $\mathcal{P}_{AU} \cup \mathcal{P}_{UA}$ can be transformed into a bipartite secondary structure with at least $x - 4$ base pairs. Then a similar transformation for $\mathcal{P}_{GC} \cup \mathcal{P}_{CG}$ completes the proof.

   We first consider the subset $\mathcal{P}_{AU}$. Let (i_1, j_1) and (i_2, j_2) be the two farthest non-nesting pairs in $\mathcal{P}_{AU}$ such that the index i_1 is the smallest and the index j_2 is the largest. Then the base pairs (i, j) with $i < i_1$ or $j > j_2$ must be nesting, and the other base

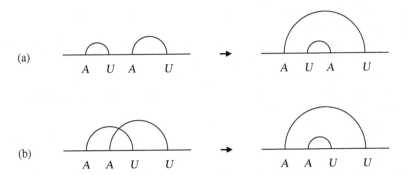

Fig. 2. (a) Swap two sequential base pairs to two nesting pairs. (b) Swap two crossing base pairs to two nesting pairs.

pairs including (i_1, j_1) and (i_2, j_2) are nested inside them. If $j_2 - i_1 > 3$, we *swap* the two non-nesting pairs into two nesting pairs as in Fig. 2: (a) if (i_1, j_1) and (i_2, j_2) are sequential, swap them to two nesting pairs (i_1, j_2) and (j_1, i_2); (b) if (i_1, j_1) and (i_2, j_2) are crossing, swap them to two nesting pairs (i_1, j_2) and (i_2, j_1). In either case, the outer one of the two nesting pairs, (i_1, j_2), is a valid base pair that satisfies the hairpin constraint. In case (a), the inner one of the two nesting pairs is a UA pair, so we move it from $\mathcal{P}_{AU}$ to $\mathcal{P}_{UA}$. Define this swap operation similarly for $\mathcal{P}_{UA}$.

Each base pair (i, j) has a *pair distance* $j - i$. The x pairs in $\mathcal{P}_{AU} \cup \mathcal{P}_{UA}$ correspond to x pair distances, which can be sorted in non-ascending order to a *distance vector* of x fields. In either case (a) or case (b) of the swap operation, the larger one of the two pair distances always increases after the swap, thus the distance vector of the x pairs in $\mathcal{P}_{AU} \cup \mathcal{P}_{UA}$ always increases lexicographically. Since each pair distance is at least 1 and at most $n - 1$, the total number of different distance vectors is at most $(n-1)^x$, which is a finite number. Repeat the swap operation on the two subsets $\mathcal{P}_{AU}$ and $\mathcal{P}_{UA}$ whenever possible. Eventually, after at most $(n - 1)^x$ swap operations, each subset consists of all nesting base pairs except for at most two pairs in the middle that may not satisfy the hairpin constraint and are discarded. We have thus transformed $\mathcal{P}_{AU} \cup \mathcal{P}_{UA}$ into a bipartite secondary structure with at least $x - 4$ base pairs. A similar transformation for $\mathcal{P}_{GC} \cup \mathcal{P}_{CG}$ completes the proof.

We note that the number $n - 8$ in Theorem 1 is not optimized and can be improved to at least $n - 4$ by a tighter analysis. On the other hand, if we want to keep the same number n of base pairs in the k-partite secondary structure, then the number $k = 4$ in Theorem 1 is best possible, as we can see from the sequence $AUGCUACG$ with 4 crossing base pairs $(1, 5)$, $(2, 6)$, $(3, 7)$, and $(4, 8)$.

We next consider a more realistic model in which the free energy of an RNA secondary structure is proportional to the total length of its helices, i.e., the total number of stacking base pairs. Assume again for simplicity that only Watson-Crick pairs are allowed in the helices. Recall that a helix must have length at least 2. Then any helix of length $l \geq 4$ can be decomposed into $\lfloor l/2 \rfloor$ short helices of lengths 2 and 3. Since there are 4^2 different doubles and 4^3 different triples of RNA bases, any secondary structure can be partitioned into at most $4^2 + 4^3 = 80$ subsets, and every two subsets of complementary bases can be transformed into a bipartite secondary structure as we did for $\mathcal{P}_{AU}$ and $\mathcal{P}_{UA}$ in the proof of Theorem 1. Thus Proposition 1 remains valid for this model.

3 Recognizing k-Partite Secondary Structures

Any RNA sequence of bases can be represented by a circular sequence of points on a circle such that two bases consecutive in the RNA sequence correspond to two points consecutive on the circle. Then each base pair of a secondary structure can be represented by a chord in the circle connecting the two corresponding points. The intersection graph of n chords in a circle is a *circle graph* of n vertices, one vertex for each chord. Two vertices are adjacent in the circle graph if and only if the two corresponding chords intersect, that is, if and only if the two corresponding base pairs cross. Thus the problem of recognizing k-partite secondary structures is equivalent to the problem of deciding whether a circle graph (given the explicit chord representation) has a k-coloring. Indeed the two problems can be reduced to each other in linear time.

The equivalence of an RNA secondary structure and a circle graph implies the following charaterization of k-partite secondary structures: A secondary structure is k-*partite* if and only if its corresponding circle graph has chromatic number at most k. We note here a related concept called k-noncrossing secondary structures[3], which are defined as secondary structures containing no more than k mutually crossing base pairs [13]. In terms of circle graphs, we have the following charaterization: A secondary structure is k-*noncrossing* if and only if its corresponding circle graph has clique number at most k. Recall that a circle graph is not a perfect graph, thus in general the chromatic number and the clique number of a circle graph are not equal.

We now proceed to the recognition of k-partite secondary structures using the charaterization based on circle graphs. It is known that coloring circle graphs is NP-complete [10]. Moreover, for any constant $k \geq 4$, deciding whether a circle graph has a k-coloring is also NP-complete [28]. For $k = 3$, the 3-coloring problem admits an $O(n \log n)$ time algorithm given the explicit representation of n chords [29]. For $k = 2$, the 2-coloring problem is the same as the problem of testing bipartiteness[4] of circle graphs. It is well known that bipartiteness testing, even for general graphs, admits a linear-time algorithm (linear in the total number of vertices and edges); see [17, Section 3.4]. But since a circle graph of n vertices can have as many as $\binom{n}{2}$ edges, a straight-forward application of the generic linear-time algorithm for bipartiteness testing gives only an $O(n^2)$ time algorithm for the 2-coloring of circle graphs.

An alternative way to obtain an $O(n^2)$ time algorithm for the 2-coloring problem is to encode the circle graph as a 2-SAT formula: the color of each chord corresponds to the true/false value of a boolean variable; two crossing chords of variables x_1 and x_2

---

[3] We thank an anonymous referee for bringing this very recent paper to our attention.

[4] Haslinger and Stadler [11, Theorem 2(vi)] incorrectly characterized this problem as testing the triangle-free property. Note that the five base pairs $(1, 5)$, $(3, 7)$, $(6, 10)$, $(8, 12)$, and $(2, 11)$ induce a 5-cycle in the circle graph, which is triangle-free but not bipartite. Indeed, Ageev [1] even constructed a triangle-free circle graph with chromatic number 5.

correspond to two clauses $x_1 \vee x_2$ and $\neg x_1 \vee \neg x_2$. Since 2-SAT also admits a linear-time algorithm [3], we again have an $O(n^2)$ time algorithm for the 2-coloring of circle graphs.

A circle graph is the intersection graph of a set of chords in a circle. In general, we can have intersection graphs of arbitrary line segments in the plane. The 2-coloring problem for circle graphs is thus a special case of the problem of testing bipartiteness of segment intersection graphs. Recently, Eppstein [9] designed a rather sophisticated $O(n \log n)$ time algorithm for this general problem, given the explicit representation of n segments. This immediately implies an $O(n \log n)$ time algorithm for the 2-coloring of circle graphs. We are not aware of any simple algorithm for the 2-coloring of circle graphs that runs in $o(n^2)$ time and is easy to implement.

We summarize the results in the following theorem:

Theorem 2. *Recognizing k-partite secondary structures is NP-complete for any constant $k \geq 4$, and admits $O(n \log n)$ time algorithms for $k = 2$ and 3, where n is the number of base pairs in the input secondary structure.*

4   Iterated Peeling

A k-partite secondary structure consists of k disjoint pseudoknot-free sub-structures. We now present an *iterated peeling* heuristic that, given a sequence S of RNA bases, extracts k disjoint sub-structures one by one in a greedy fashion as follows. First find a pseudoknot-free structure $\mathcal{P}_1$ with the lowest energy of the given sequence. Next find a pseudoknot-free structure $\mathcal{P}_2$ with the lowest energy of the subsequence of bases that are not paired up in $\mathcal{P}_1$. In general, at step t, find a pseudoknot-free structure $\mathcal{P}_t$ with the lowest energy of the subsequence of bases that are not paired up in any previous structure $\mathcal{P}_s$, $1 \leq s \leq t - 1$. After k steps, return the k-partite secondary structure $\mathcal{P}_1 \cup \cdots \cup \mathcal{P}_k$.

Since the stackings of base pairs contribute the most to the overall free energy of a secondary structure, we use dynamic programming at each step of the iterated peeling heuristic to find a pseudoknot-free sub-structure with the minimum *stacking* energy. Our dynamic programming algorithm is similar in spirit to an algorithm by Ieong et al. [14], but the latter does not distinguish the bases yet unpaired from the bases already paired up at the previous steps, and has the optimization goal of maximizing the (positive) number of base pair stackings instead of minimizing the (negative) stacking energy.

We now describe the dynamic programming algorithm. Denote by $W(i, j)$ the lowest stacking energy of a pseudoknot-free secondary structure of the subsequence $S[i, j]$. Denote by $V(i, j)$ the lowest stacking energy of a pseudoknot-free secondary structure of the subsequence $S[i, j]$ with the additional constraint that the two bases i and j must either form a base pair or do not participate in any base pairs at all. Let $E_2(i, j, i + 1, j - 1)$ be the stacking energy of the two pairs (i, j) and $(i + 1, j - 1)$ if none of the four bases $i, j, i + 1, j - 1$ are paired at the previous steps, and let it be $+\infty$ otherwise. Recall that any base pair (i, j) must satisfy the hairpin constraint $j - i \geq 4$. The base case and the recurrence of the dynamic programming algorithm are as follows:

Base case: For all (i, j) such that $i \leq j \leq i + 4$,

$$W(i, j) = V(i, j) = 0.$$

Recurrence: For all (i, j) such that $j \geq i + 5$,

$$V(i, j) = \min \left\{ W(i + 1, j - 1), V(i + 1, j - 1) + E_2(i, j, i + 1, j - 1) \right\},$$
$$W(i, j) = \min \left\{ V(i, j), \min_{i \leq x < j} \left\{ W(i, x) + W(x + 1, j) \right\} \right\}.$$

The dynamic programming algorithm clearly runs in $O(\ell^3)$ time, where ℓ is the sequence length. Thus the overall running time of the iterated peeling heuristic is $O(k\ell^3)$. We next analyze the approximation ratio of the iterated peeling heuristic. For simplicity, we consider the unweighted version of the problem that maximizes the number of base pair stackings (that is, with an energy of -1 for each base pair stacking, minimizes the stacking energy). Ieong et al. [14] have shown that the problem of finding a bipartite secondary structure with the maximum number of base pair stackings is NP-complete and admits a 2-approximation.

Let $\mathcal{S}^*$ be an optimal k-partite secondary structure with the maximum number N^* of base pair stackings. By the Pigeon-hole principle, at least one of the k pseudoknot-free sub-structures of $\mathcal{S}^*$ has at most N^*/k base pair stackings. Let n_s be the number of base pair stackings in the sub-structure $\mathcal{P}_s$ obtained at step s of the heuristic. Then $\sum_{s=1}^{k} n_s \geq n_1 \geq N^*/k$, that is, even the first step of the heuristic alone already achieves a k-approximation. By Proposition 1, the optimal k-partite secondary structure $\mathcal{S}^*$, even for a potentially unbounded number k, can be modified into a k'-partite secondary structure $\mathcal{S}'$ with approximately the same free energy, such that k' is a constant. Thus the heuristic indeed achieves a constant approximation even for arbitrary k. We next show, without using the implication of Proposition 1, that the iterated peeling heuristic still achieves a small constant approximation for arbitrary k. For this we need to consider the contributions n_s by the subsequent steps $s = 2, \ldots, k$. The idea of our analysis comes from the analysis of a similar heuristic for multi-layer routing in integrated circuit layout design [7]. In the following, we consider only the case that $k \geq 6$ since the approximation ratio of k achieved by the first step of the heuristic is already quite small for $2 \leq k \leq 5$.

Consider an arbitrary base pair stacking found by the iterated peeling heuristic. If this stacking is not in the optimal structure $\mathcal{S}^*$, then it intersects at most 6 stackings in $\mathcal{S}^*$ (see Figure 3 for an example). Thus, by the Pigeon-hole principle, we have the following inequalities:

Fig. 3. One base pair stacking intersects six other base pair stackings

$$n_1 \geq \frac{N^*}{k}, \quad n_2 \geq \frac{N^* - 6n_1}{k}, \quad \cdots, \quad n_k \geq \frac{N^* - 6(n_1 + \cdots + n_{k-1})}{k}.$$

Let

$$x_1 = \frac{N^*}{k}, \quad x_2 = \frac{N^* - 6x_1}{k}, \quad \cdots, \quad x_k = \frac{N^* - 6(x_1 + \cdots + x_{k-1})}{k}.$$

We now prove that $\sum_{s=1}^{t} n_s \geq \sum_{s=1}^{t} x_s$ for $1 \leq t \leq k$ by induction. The base case $n_1 \geq x_1$ for $t = 1$ is obvious. For the inductive step from t to $t+1$, we have (for $k \geq 6$)

$$\sum_{s=1}^{t+1} n_s \geq \sum_{s=1}^{t} n_s + \frac{N^* - 6\sum_{s=1}^{t} n_s}{k} = \frac{N^* + (k-6)\sum_{s=1}^{t} n_s}{k}$$

$$\geq \frac{N^* + (k-6)\sum_{s=1}^{t} x_s}{k} = \sum_{s=1}^{t} x_s + \frac{N^* - 6\sum_{s=1}^{t} x_s}{k} = \sum_{s=1}^{t+1} x_s.$$

It remains to bound $\sum_{s=1}^{k} x_s$ in terms of N^*. Note that $k(x_s - x_{s-1}) = -6x_{s-1}$ for $2 \leq s \leq k$. Thus

$$x_s = \left(1 - \frac{6}{k}\right) x_{s-1} = \cdots = \left(1 - \frac{6}{k}\right)^{s-1} x_1.$$

It follows that

$$\sum_{s=1}^{k} x_s = x_1 \cdot \sum_{s=1}^{k} \left(1 - \frac{6}{k}\right)^{s-1} = \frac{N^*}{k} \cdot \frac{1 - (1 - 6/k)^k}{1 - (1 - 6/k)} = \frac{1 - (1 - 6/k)^k}{6} \cdot N^*.$$

Note that

$$(1 - 6/k)^k = \left((1 - 6/k)^{k/6}\right)^6 \leq \left(\lim_{x \to \infty} (1 - 1/x)^x\right)^6 = (1/e)^6.$$

Thus the approximation ratio of the iterated peeling heuristic is

$$\frac{N^*}{\sum_{s=1}^{k} n_s} \leq \frac{N^*}{\sum_{s=1}^{k} x_s} = \frac{6}{1 - (1 - 6/k)^k} \leq \frac{6}{1 - e^{-6}} = 6.014909\ldots.$$

We have proved the following theorem:

Theorem 3. *Given an RNA sequence of length ℓ, the iterated peeling heuristic finds a k-partite secondary structure in $O(k\ell^3)$ time. The approximation ratio of the heuristic for maximizing the number of base pair stackings is at most k for $2 \leq k \leq 5$, and is at most $\frac{6}{1-(1-6/k)^k} \leq \frac{6}{1-e^{-6}} < 6.01491$ for $k \geq 6$.*

5   First-Fit Packing

The iterated peeling heuristic discussed in the previous section uses the greedy approach to find k sub-structures that compose a k-partite secondary structure. We now present another heuristic that applies the greedy approach in an entirely different way. Our *first-fit packing* heuristic is based on the first-fit strategy for the classical combinatorial optimization problem of multiple knapsack packing. Intuitively, the k sub-structures are the k knapsacks, and each helix of at least two stacking base pairs is an item to be packed into one of the knapsacks; the helices (items) in each sub-structure (knapsack) must be disjoint and non-crossing.

The first-fit packing heuristic runs in four steps: first extract base pairs from the sequence, next group the stacking base pairs into helices, then extend each helix until maximal, and finally select a disjoint set of base pairs from the helices.

To extract base pairs from the sequence, we use a simple criterion: If two stacking base pairs (i, j) and $(i + 1, j - 1)$ are both Watson-Crick pairs, then both base pairs are extracted. If a base pair (i, j) is a wobble pair and is stacked with two other base pairs $(i - 1, j + 1)$ and $(i + 1, j - 1)$ that are both Watson-Crick pairs, then all three base pairs are extracted.

Consecutive base pairs are grouped into helices. We do not allow bulges or internal loops in the helices. Thus each helix consists exclusively of stacking base pairs, and can be specified by three numbers (i, j, l), where (i, j) is the outer-most base pair in the helix, and l is the helix length, i.e., the number of base pairs in the helix. For example, the helix $(4, 12, 3)$ consists of three stacking base pairs $(4, 12)$, $(5, 11)$, and $(6, 10)$.

We then extend each helix into a maximal helix. First extend (i, j, l) inward to $(i, j, l + c)$ incrementally, for $c = 1, 2, \ldots$, while the inner-most base pair satisfies the hairpin constraint $(j - l - c + 1) - (i + l + c - 1) > 3$ and the helix energy becomes lower. Next extend the helix outward, from (i, j, l') to $(i - c, j + c, l' + c)$, in a similar manner.

Finally, we select a disjoint set of base pairs from these maximal helices. This is done by repeating the following selection round until every candidate helix is either discarded, or added to a sub-structure $\mathcal{P}_s$, $1 \leq s \leq k$:

1. Find a helix H with the lowest stacking energy among the candidate helices not yet added to any sub-structure.
2. Find the sub-structure $\mathcal{P}_s$ with the smallest index s such that H does not cross any helices already in $\mathcal{P}_s$.
3. If no such sub-structure $\mathcal{P}_s$ exists, discard the helix H. Otherwise, add H to $\mathcal{P}_s$, then trim all remaining candidate helices such that their base pairs that overlap with H are removed. If a candidate helix is shortened to less than three base pairs after the trimming, discard it.

Unlike the iterated peeling heuristic, the first-fit packing heuristic does not seem to achieve any constant approximation ratio for maximizing the number of base pair stackings. When $k = 2$, for example, the sequence $A^4 G^4 (A^3 G^3)^r U^4 C^4 (U^3 C^3)^r$ has an optimal bipartite secondary structure of $4r + 4$ stackings, but the first-fit packing heuristic finds only 6 stackings (see Fig. 4 for an illustration where $r = 2$). Nevertheless, as we

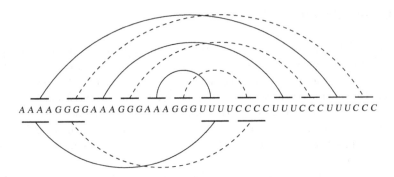

Fig. 4. An example for finding bipartite secondary structures. Upper: The optimal structure. Lower: the structure found by the first-fit packing heuristic. The two sub-structures of each bipartite secondary structure are depicted by solid and dashed arcs respectively.

will show in the next section, the first-fit packing heuristic beats the iterated peeling heuristic in predicting RNA secondary structures.

6   Experiment on RNA Secondary Structure Prediction

We compare our two greedy heuristics, code-named IPeel for iterated peeling and FPack for first-fit packing, with the HotKnots algorithm of Ren, Rastegari, Condon, and Hoos [23] on RNA secondary structure prediction. HotKnots is a leading method for RNA secondary structure prediction that has been shown [23] to outperform many other methods [24,8,26,4,22].

For the test data, we use the RNA sequences with known secondary structures from PseudoBase [5]. As of May 25, 2008, the PseudoBase contained 279 partial RNA sequences, of which 252 are consecutive sequences without gaps. We collected these 252 sequences and their known secondary structures as the data set. The sequences in the data set have maximum length 131 and minimum length 20; the average length is about 43 with a standard deviation of 24.

For each sequence in the data set, we use each of the three methods, HotKnots, IPeel, and FPack, to predict a secondary structure, then compare the predicted secondary structures with the known secondary structure from PseudoBase. The parameter k is set to 2 for our methods IPeel and FPack. The three methods are all very efficient; the entire experiment finishes in less than four minutes on an Apple iMac with 2 GHz PowerPC G5 Processor and 2 GB RAM running Mac OS X 10.4.11.

To compare a predicted secondary structure with a known secondary structure, we count the *true positives* (tp) as the base pairs that are known and are predicted, the *false negatives* (fn) as the base pairs that are known but are not predicted, and the *false positives* (fp) as the base pairs that are not known but are predicted. Then we define three quality measures:

$$\text{Sensitivity} = \frac{\text{tp}}{\text{tp} + \text{fn}}, \quad \text{Selectivity} = \frac{\text{tp}}{\text{tp} + \text{fp}}, \quad \text{Accuracy} = \frac{\text{tp}}{\text{tp} + \text{fn} + \text{fp}}.$$

To compare the overall performances of the three methods (HotKnots, IPeel, and FPack), we calculate the three quality measures (sensitivity, selectivity, and accuracy) of each method using the equations above with the three numbers (true positives, false negatives, and false positives) summed over all 252 sequences in the data set. We obtained the following results:

	Sensitivity	Selectivity	Accuracy
HotKnots	71.69%	78.47%	59.90%
IPeel	51.09%	42.33%	30.12%
FPack	80.50%	75.39%	63.75%

The results are a little surprising. The iterated peeling heuristic is the most theoretically interesting among the three methods, but it has the worst performance in this experiment. On the other hand, although not achieving any constant approximation ratio, the first-fit packing heuristic performs the best. Note that FPack is slightly worse than Hot-Knots in selectivity but outperforms HotKnots in both sensitivity and accuracy. It is impressive that a rather simple (indeed almost naive) greedy method can have such a good performance. Our explanation for this unexpected success is the following:

1. For RNA folding, the kinetic aspect of the folding process is as important as the thermodynamic stability of the folded structure.
2. For the simulation of RNA folding kinetics, the greedy choice of selecting helices one by one (first-fit packing) is more realistic than the greedy choice of selecting sub-structures one by one (iterated peeling).

References

1. Ageev, A.A.: A triangle-free circle graph with chromatic number 5. Discrete Mathematics 152, 295–298 (1996)
2. Akutsu, T.: Dynamic programming algorithms for RNA secondary structure prediction with pseudoknots. Discrete Applied Mathematics 104, 45–62 (2000)
3. Aspvall, B., Plass, M.F., Tarjan, R.E.: A linear-time algorithm for testing the truth of certain quantified boolean formulas. Information Processing Letters 8, 121–123 (1979)
4. van Batenburg, F.H.D., Gultyaev, A.P., Pleij, C.W.A.: An APL-programmed genetic algorithm for the prediction of RNA secondary structure. Journal of Theoretical Biology 174, 269–280 (1995)
5. van Batenburg, F.H.D., Gultyaev, A.P., Pleij, C.W.A., Ng, J., Oliehoek, J.: Pseudobase: a database with RNA pseudoknots. Nucleic Acids Research 28, 201–204 (2000)
6. Condon, A., Davy, B., Rastegari, B., Zhao, S., Tarrant, F.: Classifying RNA pseudoknotted structures. Theoretical Computer Science 320, 35–50 (2004)
7. Cong, J., Hossain, M., Sherwani, N.A.: A provably good multilayer topological planar routing algorithms in IC layout designs. IEEE Transactions on Computer Aided Design of Integrated Circuits and Systems 12, 70–78 (1993)
8. Dirks, R.M., Pierce, N.A.: A partition function algorithm for nucleic acid secondary structure including pseudoknots. Journal of Computational Chemistry 24, 1664–1677 (2003)
9. Eppstein, D.: Testing bipartiteness of geometric intersection graphs. In: Proceedings of the 15th Annual ACM-SIAM Symposium on Discrete Algorithms (SODA 2004), pp. 860–868 (2004)

10. Garey, M.R., Johnson, D.S., Miller, G.L., Papadimitriou, C.H.: The complexity of coloring circular arcs and chords. SIAM Journal on Algebraic & Discrete Methods 1, 216–227 (1980)
11. Haslinger, C., Stadler, P.F.: RNA structures with pseudo-knots: graph theoretical, combinatorial, and statistical properties. Bulletin of Mathematical Biology 61, 437–467 (1999)
12. Hofacker, I.L.: Vienna RNA secondary structure server. Nucleic Acids Research 31, 3429–3431 (2003)
13. Huang, F.W.D., Peng, W.W.J., Reidys, C.M.: Folding 3-noncrossing RNA pseudoknot structures (2008), http://arxiv.org/abs/0809.4840v1
14. Ieong, S., Kao, M.-Y., Lam, T.-W., Sung, W.-K., Yiu, S.-M.: Predicting RNA secondary structure with arbitrary pseudoknots by maximizing the number of stacking pairs. Journal of Computational Biology 10, 981–995 (2003)
15. Jabbari, H., Condon, A., Zhao, S.: Novel and efficient RNA secondary structure prediction using hierarchical folding. Journal of Computational Biology 15, 139–163 (2008)
16. Jiang, M.: Approximation algorithms for predicting RNA secondary structures with arbitrary pseudoknots. ACM/IEEE Transactions on Computational Biology and Bioinformatics (to appear), http://dx.doi.org/10.1109/TCBB.2008.109
17. Kleinberg, J., Tardos, E.: Algorithm Design. Addison-Wesley, Reading (2005)
18. Lyngsø, R.B.: Complexity of pseudoknot prediction in simple models. In: Díaz, J., Karhumäki, J., Lepistö, A., Sannella, D. (eds.) ICALP 2004. LNCS, vol. 3142, pp. 919–931. Springer, Heidelberg (2004)
19. Lyngsø, R.B., Pedersen, C.N.S.: RNA pseudoknot prediction in energy-based models. Journal of Computational Biology 7, 409–427 (2000)
20. Nussinov, R., Pieczenik, G., Griggs, J.R., Kleitman, D.J.: Algorithms for loop matching. SIAM Journal on Applied Mathematics 35, 68–82 (1978)
21. Rastegari, B., Condon, A.: Parsing nucleic acid pseudoknotted secondary structure: algorithm and applications. Journal of Computational Biology 14, 16–32 (2007)
22. Reeder, J., Giegerich, R.: Design, implementation and evaluation of a practical pseudoknot folding algorithm based on thermodynamics. BMC Bioinformatics 5, 104 (2004)
23. Ren, J., Rastegari, B., Condon, A., Hoos, H.H.: HotKnots: Heuristic prediction of RNA secondary structures including pseudoknots. RNA 11, 1494–1504 (2005)
24. Rivas, E., Eddy, S.R.: A dynamic programming algorithm for RNA structure prediction including pseudoknots. Journal of Molecular Biology 285, 2053–2068 (1999)
25. Rødland, E.A.: Pseudoknots in RNA secondary structures: representation, enumeration, and prevalence. Journal of Computational Biology 13, 1197–1213 (2006)
26. Ruan, J., Stormo, G.D., Zhang, W.: An iterated loop matching approach to the prediction of RNA secondary structure with pseudoknots. Bioinformatics 20, 58–66 (2004)
27. Uemura, Y., Hasegawa, A., Kobayashi, S., Yokomori, T.: Tree adjoining grammars for RNA structure prediction. Theoretical Computer Science 210, 277–303 (1999)
28. Unger, W.: On the k-colouring of circle-graphs. In: Cori, R., Wirsing, M. (eds.) STACS 1988. LNCS, vol. 294, pp. 61–72. Springer, Heidelberg (1988)
29. Unger, W.: The complexity of colouring circle graphs. In: Finkel, A., Jantzen, M. (eds.) STACS 1992. LNCS, vol. 577, pp. 389–400. Springer, Heidelberg (1992)
30. Witwer, C., Hofacker, I.L., Stadler, P.F.: Prediction of consensus RNA secondary structures including pseudoknots. IEEE/ACM Transactions on Computational Biology and Bioinformatics 1, 66–77 (2004)
31. Zuker, M.: Mfold web server for nucleic acid folding and hybridization prediction. Nucleic Acids Research 31, 3406–3415 (2003)

Efficient Algorithms for Analyzing Segmental Duplications, Deletions, and Inversions in Genomes

Crystal L. Kahn[1,*], Shay Mozes[1,**], and Benjamin J. Raphael[1,2,*]

[1] Department of Computer Science, Brown University, and
[2] Center for Computational Molecular Biology, Brown University,
Providence, RI 02912, USA
{clkahn,shay,braphael}@cs.brown.edu

Abstract. Segmental duplications, or low-copy repeats, are common in mammalian genomes. In the human genome, most segmental duplications are mosaics consisting of pieces of multiple other segmental duplications. This complex genomic organization complicates analysis of the evolutionary history of these sequences. Earlier, we introduced a genomic distance, called duplication distance, that computes the most parsimonious way to build a target string by repeatedly copying substrings of a source string. We also showed how to use this distance to describe the formation of segmental duplications according to a two-step model that has been proposed to explain human segmental duplications. Here we describe polynomial-time exact algorithms for several extensions of duplication distance including models that allow certain types of substring deletions and inversions. These extensions will permit more biologically realistic analyses of segmental duplications in genomes.

1 Introduction

Genomes evolve via many types of mutations ranging in scale from single nucleotide mutations to large genome rearrangements. Computational models of the mutational process allow researchers to derive similarity measures between genome sequences and to reconstruct evolutionary relationships between genomes. For example, considering chromosomal inversions as the only type of mutation leads to the so-called reversal distance problem of finding the minimum number of inversions/reversals that transform one genome into another [1]. Several elegant polynomial-time algorithms have been found to solve this problem (cf. [2] and references therein). Developing genome rearrangement models that are both biologically realistic *and* computationally tractable remains an active area of research.

---
* Work supported by funding from the ADVANCE Program at Brown University, under NSF Grant No. 0548311, and by a Career Award at the Scientific Interface from the Burroughs Wellcome Fund to BJR.
** Work supported by NSF Grant CCF-0635089.

S.L. Salzberg and T. Warnow (Eds.): WABI 2009, LNBI 5724, pp. 169–180, 2009.
© Springer-Verlag Berlin Heidelberg 2009

Duplicated sequences in genomes present a particular challenge for genome rearrangement analysis and often make the underlying computational problems more difficult. For instance, computing reversal distance in genomes with duplicated segments is NP-hard [3]. Moreover, models that include multiple types of mutations–such as inversions and duplications–often result in similarity measures that cannot be computed efficiently. Current approaches for duplication analysis rely on heuristics, approximation algorithms, or restricted models of duplication [3,4,5,6,7]. For example, there are efficient algorithms for computing tandem duplication [8,9,10,11] and whole-genome duplication [12,13] histories.

Here we consider another class of duplications: large segmental duplications (also known as low-copy repeats) that are common in many mammalian genomes [14]. These segmental duplications can be quite large (up to hundreds of kilobases), but their evolutionary history remains poorly understood. The mystery surrounding them is due in part to their complex organization; many segmental duplications are mosaic patterns of smaller repeated segments, or *duplicons*. One hypothesis proposed to explain these mosaic patterns is a two-step model of duplication [14]. In this model, a first phase of duplications copies duplicons from the ancestral genome and aggregates these copies into an array of contiguous duplication blocks. Then in a second phase, portions of these duplication blocks are copied and reinserted into the genome at disparate loci forming secondary duplication blocks.

In [15], we introduced a measure called duplication distance, which we used in [16] to find the most parsimonious duplication scenario consistent with the two-step model of segmental duplication. The duplication distance from a source string $\mathbf{x}$ to a target string $\mathbf{y}$ is the minimum number of substrings of $\mathbf{x}$ that can be sequentially copied from $\mathbf{x}$ and pasted into an initially empty string in order to construct $\mathbf{y}$. Note that the string $\mathbf{x}$ does not change during the sequence of duplication events. We derived an efficient exact algorithm for computing the duplication distance between a pair of strings. Here, we extend the duplication distance measure to include certain types of deletions and inversions. These extensions make our model less restrictive and permit the construction of more rich, and perhaps more biologically plausible, duplication scenarios. In particular, our contributions are the following.

Summary of Contributions. Let $\mu(\mathbf{x})$ denote the maximal number of times a character appears in the string $\mathbf{x}$. Let $|\mathbf{x}|$ denote the length of $\mathbf{x}$.

1. We provide an $O(|\mathbf{y}|^2|\mathbf{x}|\mu(\mathbf{x})\mu(\mathbf{y}))$-time algorithm to compute the distance between (signed) strings $\mathbf{x}$ and $\mathbf{y}$ when duplication and certain types of deletion operations are permitted.
2. We provide an $O(|\mathbf{y}|^2|\mathbf{x}|\mu(\mathbf{x})\mu(\mathbf{y}))$-time algorithm to compute the distance between signed strings $\mathbf{x}$ and $\mathbf{y}$ when duplicated strings may be inverted before being inserted into the target string. Deletion operations are also permitted.
3. We provide an $O(|\mathbf{y}|^2|\mathbf{x}|^3\mu(\mathbf{x})\mu(\mathbf{y}))$-time algorithm to compute the distance between signed strings $\mathbf{x}$ and $\mathbf{y}$ when any substring of the duplicated string may be inverted before being inserted into the target string. Deletion operations are also permitted.

4. We provide a formal proof of correctness of the duplication distance recurrence presented in [16]. No proof of correctness was previously given.

2   Preliminaries

We begin by reviewing some definitions and notation that were introduced in [15] and [16]. Let $\emptyset$ denote the empty string. For a string $\mathbf{x} = x_1 \ldots x_n$, let $\mathbf{x}_{i,j}$ denote the substring $x_i x_{i+1} \ldots x_j$. We define a *subsequence* S of $\mathbf{x}$ to be a string $x_{i_1} x_{i_2} \ldots x_{i_k}$ with $i_1 < i_2 < \cdots < i_k$. We represent S by listing the indices at which the characters of S occur in $\mathbf{x}$. For example, if $\mathbf{x} = abcdef$, then the subsequence $S = (1, 3, 5)$ is the string ace. Note that every substring is a subsequence, but a subsequence need not be a substring since the characters comprising a subsequence need not be contiguous. For a pair of subsequences S_1, S_2, denote by $S_1 \cap S_2$ the maximal subsequence common to both S_1 and S_2.

Definition 1. *Subsequences* $S = (s_1, s_2)$ *and* $T = (t_1, t_2)$ *of a string* $\mathbf{x}$ *are* ***alternating*** *in* $\mathbf{x}$ *if either* $s_1 < t_1 < s_2 < t_2$ *or* $t_1 < s_1 < t_2 < s_2$.

Definition 2. *Subsequences* $S = (s_1, \ldots, s_k)$ *and* $T = (t_1, \ldots, t_l)$ *of a string* $\mathbf{x}$ *are* ***overlapping*** *in* $\mathbf{x}$ *if there exist indices* i, i' *and* j, j' *such that* $1 \le i < i' \le k$, $1 \le j < j' \le l$, *and* $(s_i, s_{i'})$ *and* $(t_j, t_{j'})$ *are alternating in* $\mathbf{x}$. *See Figure 1.*

Definition 3. *Given subsequences* $S = (s_1, \ldots, s_k)$ *and* $T = (t_1, \ldots, t_l)$ *of a string* $\mathbf{x}$, S *is* ***inside*** *of* T *if there exists an index* i *such that* $1 \le i < l$ *and* $t_i < s_1 < s_k < t_{i+1}$. *That is, the entire subsequence* S *occurs in between successive characters of* T. *See Figure 2.*

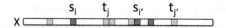

Fig. 1. Two blue (light) and red (dark) subsequences are overlapping in $\mathbf{x}$. The indices $(s_i, s_{i'})$ and $(t_j, t_{j'})$ are alternating in $\mathbf{x}$.

Fig. 2. The red (dark) subsequence is inside the blue (light) subsequence T. All the characters of the red subsequence occur between the indices t_i and t_{i+1} of T.

Definition 4. *A* ***duplicate operation*** *from* $\mathbf{x}$, $\delta_{\mathbf{x}}(s, t, p)$, *copies a substring* $x_s \ldots x_t$ *of the source string* $\mathbf{x}$ *and pastes it into a target string at position* p. *Specifically, if* $\mathbf{x} = x_1 \ldots x_m$ *and* $\mathbf{z} = z_1 \ldots z_n$, *then* $\mathbf{z} \circ \delta_{\mathbf{x}}(s, t, p) = z_1 \ldots z_{p-1} x_s \ldots x_t z_p \ldots z_n$. *See Figure 3.*

The cost associated with a duplicate operation is 1.[1]

---

[1] For simplicity of exposition we fix the cost of each duplicate operation to be one (i.e. the duplication distance just counts the number of duplicate operations). It is not difficult to extend the cost to an affine function of the length of the duplicated substring. The technique is similar to that of handling the affine cost of inversion in section 5. We omit the details.

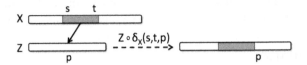

Fig. 3. A duplicate operation, $\delta_{\mathbf{x}}(s,t,p)$. A substring of the source string $\mathbf{x}$ ranging between indices s and t is copied and inserted into the target string $\mathbf{z}$ at index p.

Definition 5. *The **duplication distance** from a source string x to a target string y is the minimum cost of a sequence of duplicate operations from x, $\delta_x(s_1,t_1,p_1)$, $\delta_x(s_2,t_2,p_2)$, $\delta_x(s_3,t_3,p_3)$, ..., that generates y from an initially empty target string[2] (that is, $y = \emptyset \circ \delta_x(s_1,t_1,p_1) \circ \delta_x(s_2,t_2,p_2) \circ \dots$).*

Lemma 1 (Non-overlapping Property). *Consider a source string x, an initial target string z, and a sequence of duplicate operations of the form $\delta_x(s_i,t_i,p_i)$ that transforms z into the final target string y. The substrings of x that are duplicated during the construction of y appear as mutually non-overlapping subsequences of y.*

Proof. (Sketch) First note that a substring S that is duplicated from $\mathbf{x}$ and inserted into a target string $\mathbf{z}$ is initially a substring of the newly augmented target, but subsequent duplicate operations may insert other strings in between the characters of S transforming it into a subsequence of the final target string $\mathbf{y}$. Consider now two strings S_1 and S_2 that are copied from $\mathbf{x}$ and inserted into an intermediate target string in succession yielding an augmented target string $\mathbf{z}'$. There are two cases: (1) $S_1 \cap S_2 = \emptyset$ or (2) S_2 is inside of S_1 in $\mathbf{z}'$ (Definition 3). In either case, S_1 and S_2 are not overlapping in $\mathbf{z}'$. This argument can be extended for any sequence of duplicate operations; as each string is inserted into an intermediate target string $\mathbf{z}'$, it does not overlap any of the subsequences of $\mathbf{z}'$ that correspond to previously-inserted strings; we maintain the invariant that the subsequences of $\mathbf{z}'$ corresponding to characters inserted in individual duplicate operations are mutually non-overlapping in $\mathbf{z}'$, and therefore also in the final string $\mathbf{y}$.   □

3   Duplication Distance

In this section we review the basic recurrence for computing duplication distance that was introduced in [16]. The recurrence is based on the observation that y_1 must be the first (i.e. leftmost) character to be copied from $\mathbf{x}$ in some duplicate operation. The recurrence defines two quantities: $d(\mathbf{x},\mathbf{y})$ and $d_i(\mathbf{x},\mathbf{y})$. We shall show, by induction, that for a pair of strings, $\mathbf{x}$ and $\mathbf{y}$, the value $d(\mathbf{x},\mathbf{y})$ is equal to the duplication distance from $\mathbf{x}$ to $\mathbf{y}$ and that $d_i(\mathbf{x},\mathbf{y})$ is equal to the duplication distance from $\mathbf{x}$ to $\mathbf{y}$ under the restriction that the character y_1 is copied

---

[2] We assume that every character in $\mathbf{y}$ appears at least once in $\mathbf{x}$.

from index i in $\mathbf{x}$, i.e. x_i *generates* y_1. $d(\mathbf{x}, \mathbf{y})$ is found by considering the minimum among all characters x_i of $\mathbf{x}$ that can generate y_1, see Eq. 1. To compute $d_i(\mathbf{x}, \mathbf{y})$, there are two possible cases to consider: either (1) y_1 was duplicated as a substring of $\mathbf{x}$ of length one, namely just x_i, or (2) x_{i+1} is also copied in the same duplicate operation. In the former case the minimum cost is 1 (for this duplicate operation) plus the minimum cost to generate the remaining suffix of $\mathbf{y}$, namely $1 + d(\mathbf{x}, \mathbf{y}_{2,|\mathbf{y}|})$. In the latter case, the set of candidate characters in $\mathbf{y}$ to be generated by x_{i+1} is $\{j : y_j = x_{i+1}, j > 1\}$. For each such j, the minimum cost of generating $\mathbf{y}$ provided that y_1 is generated by x_i and y_j is generated by x_{i+1}, is the minimum cost of generating $\mathbf{y}_{2,j-1}$ (i.e., $d(\mathbf{x}, \mathbf{y}_{2,j-1})$) plus the minimum cost of generating the string $y_1 \mathbf{y}_{j,|\mathbf{y}|}$ using x_i and x_{i+1} to generate y_1 and y_j. Since x_i and x_{i+1} are copied in the same duplicate operation, the cost of generating $y_1 \mathbf{y}_{j,|\mathbf{y}|}$ using x_i and x_{i+1} is equal to the cost of generating just $\mathbf{y}_{j,|\mathbf{y}|}$ using x_{i+1}, namely $d_{i+1}(\mathbf{x}, \mathbf{y}_{j,|\mathbf{y}|})$, see Eq. 2. The recurrence is, therefore:

$$d(\mathbf{x}, \emptyset) = 0$$

$$d(\mathbf{x}, \mathbf{y}) = \min_{\{i : x_i = y_1\}} d_i(\mathbf{x}, \mathbf{y}) \tag{1}$$

$$d_i(\mathbf{x}, \emptyset) = 0$$

$$d_i(\mathbf{x}, \mathbf{y}) = \min \begin{cases} 1 + d(\mathbf{x}, \mathbf{y}_{2,|\mathbf{y}|}) \\ \min_{\{j : y_j = x_{i+1}, j > 1\}} \{d(\mathbf{x}, \mathbf{y}_{2,j-1}) + d_{i+1}(\mathbf{x}, \mathbf{y}_{j,|\mathbf{y}|})\} \end{cases} \tag{2}$$

Theorem 1. $d(\boldsymbol{x}, \boldsymbol{y})$ *is the minimum cost of a sequence of duplicate operations that generate $\boldsymbol{y}$ from $\boldsymbol{x}$. For $\{i : x_i = y_1\}$, $d_i(\boldsymbol{x}, \boldsymbol{y})$ is the minimum cost of a sequence of operations that generate $\boldsymbol{y}$ from $\boldsymbol{x}$ such that y_1 is generated by x_i.*

Proof. Let $OPT(\mathbf{x}, \mathbf{y})$ denote minimum cost of a sequence of duplicate operations that generate $\mathbf{y}$ from $\mathbf{x}$. Let $OPT_i(\mathbf{x}, \mathbf{y})$ denote the minimum cost of a sequence of operations that generate $\mathbf{y}$ from $\mathbf{x}$ such that y_1 is generated by x_i. We prove by induction on $|\mathbf{y}|$ that $d(\mathbf{x}, \mathbf{y}) = OPT(\mathbf{x}, \mathbf{y})$ and $d_i(\mathbf{x}, \mathbf{y}) = OPT_i(\mathbf{x}, \mathbf{y})$.

For $|\mathbf{y}| = 1$, since we assume there is at least one i for which $x_i = y_1$, $OPT(\mathbf{x}, \mathbf{y}) = OPT_i(\mathbf{x}, \mathbf{y}) = 1$. By definition, the recurrence also evaluates to 1. For the inductive step, assume that $OPT(\mathbf{x}, \mathbf{y}') = d(\mathbf{x}, \mathbf{y}')$ and $OPT_i(\mathbf{x}, \mathbf{y}') = d_i(\mathbf{x}, \mathbf{y}')$ for any string $\mathbf{y}'$ shorter than $\mathbf{y}$. We first show that $OPT_i(\mathbf{x}, \mathbf{y}) \leq d_i(\mathbf{x}, \mathbf{y})$. Since $OPT(\mathbf{x}, \mathbf{y}) = \min_i OPT_i(\mathbf{x}, \mathbf{y})$, this also implies $OPT(\mathbf{x}, \mathbf{y}) \leq d(\mathbf{x}, \mathbf{y})$. We describe different sequences of duplicate operations that generate $\mathbf{y}$ from $\mathbf{x}$, using x_i to generate y_1:

- Consider a minimum-cost sequence of duplicates that generates $\mathbf{y}_{2,|\mathbf{y}|}$. By the inductive hypothesis its cost is $d(\mathbf{x}, \mathbf{y}_{2,|\mathbf{y}|})$. By duplicating y_1 separately using x_i we obtain a sequence of duplicates that generates $\mathbf{y}$ whose cost is $1 + d(\mathbf{x}, \mathbf{y}_{2,|\mathbf{y}|})$.
- For every $\{j : y_j = x_{i+1}, j > 1\}$ consider a minimum-cost sequence of duplicates that generates $\mathbf{y}_{j,|\mathbf{y}|}$ using x_{i+1} to produce y_j, and a minimum-cost sequence of duplicates that generates $\mathbf{y}_{2,j-1}$. By the inductive hypothesis their costs are $d_{i+1}(\mathbf{x}, \mathbf{y}_{j,|\mathbf{y}|})$ and $d(\mathbf{x}, \mathbf{y}_{2,j-1})$ respectively. By extending the

start index s of the duplicate operation that starts with x_{i+1} to produce y_j to start with x_i and produce y_1 as well, we can produce $\mathbf{y}$ with the same number of duplicate operations, and hence the same cost.

Since the cost of $OPT_i(\mathbf{x}, \mathbf{y})$ is at most the cost of any of these options, it is also at most their minimum. Hence,

$$OPT_i(\mathbf{x}, \mathbf{y}) \leq \min \begin{cases} 1 + d(\mathbf{x}, \mathbf{y}_{2,|\mathbf{y}|}) \\ \min_{\{j:y_j=x_{i+1},j>1\}} \{d(\mathbf{x}, \mathbf{y}_{2,j-1}) + d_{i+1}(\mathbf{x}, \mathbf{y}_{j,|\mathbf{y}|})\} \end{cases}$$
$$= d_i(\mathbf{x}, \mathbf{y}).$$

To show the other direction (i.e. that $d(x, y) \leq OPT(x, y)$ and $d_i(x, y) \leq OPT_i(x, y)$), consider a minimum-cost sequence of duplicate operations that generate $\mathbf{y}$ from $\mathbf{x}$, using x_i to generate y_1. There are a few cases:

- If y_1 is generated by a duplicate operation that only duplicates x_i, then $OPT_i(\mathbf{x}, \mathbf{y}) = 1 + OPT(\mathbf{x}, \mathbf{y}_{2,|\mathbf{y}|})$. By the inductive hypothesis this equals $1 + d(\mathbf{x}, \mathbf{y}_{2,|\mathbf{y}|})$ which is at least $d_i(\mathbf{x}, \mathbf{y})$.
- Otherwise, y_1 is generated by a duplicate operation that copies x_i and also duplicates x_{i+1} to generate some character y_j. In this case the sequence Δ of duplicates that generates $\mathbf{y}_{2,j-1}$ must appear after the duplicate operation that generates y_1 and y_j because $\mathbf{y}_{2,j-1}$ is inside (Definition 3) of (y_1, y_j). Without loss of generality, suppose Δ is ordered after all the other duplicates so that first $y_1 y_j \ldots y_{|\mathbf{y}|}$ is generated, and then Δ generates $y_2 \ldots y_{j-1}$ between y_1 and y_j. Hence, $OPT_i(\mathbf{x}, \mathbf{y}) = OPT_i(\mathbf{x}, y_1 \mathbf{y}_{j,|\mathbf{y}|}) + OPT(\mathbf{x}, \mathbf{y}_{2,j-1})$. Since in the optimal sequence x_i generates y_1 in the same duplicate operation that generates y_j from x_{i+1}, we have $OPT_i(\mathbf{x}, y_1 \mathbf{y}_{j,|\mathbf{y}|}) = OPT_{i+1}(\mathbf{x}, \mathbf{y}_{j,|\mathbf{y}|})$. By the inductive hypothesis, $OPT(\mathbf{x}, \mathbf{y}_{2,j-1}) + OPT_{i+1}(\mathbf{x}, \mathbf{y}_{j,|\mathbf{y}|}) = d(\mathbf{x}, \mathbf{y}_{2,j-1}) + d_{i+1}(\mathbf{x}, \mathbf{y}_{j,|\mathbf{y}|})$ which is at least $d_i(\mathbf{x}, \mathbf{y})$. □

This recurrence naturally translates into a dynamic programing algorithm that computes the values of $d(\mathbf{x}, \cdot)$ and $d_i(\mathbf{x}, \cdot)$ for various target strings. To analyze the running time of this algorithm, note that both $\mathbf{y}_{2,j}$ and $\mathbf{y}_{j,|\mathbf{y}|}$ are substrings of $\mathbf{y}$. Since the set of substrings of $\mathbf{y}$ is closed under taking substrings, we only encounter substrings of $\mathbf{y}$. Also note that since i is chosen from the set $\{i : x_i = y_1\}$, there are $O(\mu(\mathbf{x}))$ choices for i, where $\mu(\mathbf{x})$ is the maximal multiplicity of a character in $\mathbf{x}$. Thus, there are $O(\mu(\mathbf{x})|\mathbf{y}|^2)$ different values to compute. Each value is computed by considering the minimization over at most $\mu(\mathbf{y})$ previously computed values, so the total running time is bounded by $O(|\mathbf{y}|^2 \mu(\mathbf{x}) \mu(\mathbf{y}))$, which is $O(|\mathbf{y}|^3 |\mathbf{x}|)$ in the worst case.

4   Duplication-Deletion Distance

In this section we generalize the model to include deletions. Consider the intermediate string $\mathbf{z}$ generated after some number of duplicate operations. A deletion operation removes a contiguous substring $z_i, \ldots, z_j$ of $\mathbf{z}$, and subsequent duplicate and deletion operations are applied to the resulting string.

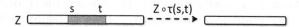

Fig. 4. A delete operation, $\tau(s,t)$. The substring of $\mathbf{z}$ that ranges between indices s and t is deleted.

Definition 6. *A **delete operation**, $\tau(s,t)$, deletes a substring $z_s \ldots z_t$ of the target string $\mathbf{z}$, thus making $\mathbf{z}$ shorter. Specifically, if $\mathbf{z} = z_1 \ldots z_s \ldots z_t \ldots z_m$, then $\mathbf{z} \circ \tau(s,t) = z_1 \ldots z_{s-1} z_{t+1} \ldots z_m$. See Figure 4.*

The cost associated with $\tau(s,t)$ depends on the number $t - s + 1$ of characters deleted and is denoted $\Phi(t - s + 1)$.

Definition 7. *The **duplication-deletion distance** from a source string $\mathbf{x}$ to a target string $\mathbf{y}$ is the cost of a minimum sequence of duplicate operations from $\mathbf{x}$ and deletion operations, in any order, that generates $\mathbf{y}$.*

We now show that although we allow arbitrary deletions from the intermediate string, it suffices to consider just deletions from the duplicated strings before they are pasted into the intermediate string, provided that the cost function for deletion, $\Phi(\cdot)$ is non-decreasing and obeys the triangle inequality.

Definition 8. *A **duplicate-delete** operation from $\mathbf{x}$, $\eta_x(i_1, j_1, i_2, j_2, \ldots, i_k, j_k, p)$, for $i_1 \leq j_1 < i_2 \leq j_2 < \cdots < i_k \leq j_k$ copies the subsequence $x_{i_1} \ldots x_{j_1} x_{i_2} \ldots x_{j_2} \ldots x_{i_k} \ldots x_{j_k}$ of the source string $\mathbf{x}$ and pastes it into a target string at position p. Specifically, if $\mathbf{x} = x_1 \ldots x_m$ and $\mathbf{z} = z_1 \ldots z_n$, then $\mathbf{z} \circ \eta_x(i_1, j_1, \ldots, i_k, j_k, p) = z_1 \ldots z_{p-1} x_{i_1} \ldots x_{j_1} x_{i_2} \ldots x_{j_2} \ldots x_{i_k} \ldots x_{j_k} z_p \ldots z_n$.*

The cost associated with such a duplication-deletion is $1 + \sum_{\ell=1}^{k-1} \Phi(i_{\ell+1} - j_\ell - 1)$.

Lemma 2. *If $\Phi(\cdot)$ is non-decreasing and obeys the triangle inequality then the cost of a minimum sequence of duplicate and delete operations that generates a target string $\mathbf{y}$ from a source string $\mathbf{x}$ is equal to the cost of a minimum sequence of duplicate-delete operations that generates $\mathbf{y}$ from $\mathbf{x}$.*

Proof. Since duplicate operations are a special case of duplicate-delete operations, the cost of a minimal sequence of duplicate-delete operations and delete operations that generates $\mathbf{y}$ cannot be more than that of a sequence of just duplicate operations and delete operations. We show the (stronger) claim that an arbitrary sequence of duplicate-delete and delete operations that produces a string $\mathbf{y}$ with cost c can be transformed into a sequence of just duplicate-delete operations that generates $\mathbf{y}$ with cost at most c by induction on the number of delete operations. The base case, where the number of deletions is zero, is trivial. Consider the first delete operation, τ. Let k denote the number of duplicate-delete operations that precede τ, and let $\mathbf{z}$ be the intermediate string produced by these k operations. For $i = 1, \ldots, k$, let S_i be the subsequence of $\mathbf{x}$ that was used in the ith duplicate-delete operation. $S_1, \ldots, S_k$ form a partition of $\mathbf{z}$ into disjoint, non-overlapping subsequences of $\mathbf{z}$. Let d denote the substring of $\mathbf{z}$ to be deleted. Since d is a contiguous substring, $S_i \cap d$ is a (possibly empty) substring of S_i for each i. There are several cases:

1. $S_i \cap d = \emptyset$. In this case we do not change any operation.
2. $S_i \cap d = S_i$. In this case all characters produced by the ith duplicate-delete operation are deleted, so we may omit the ith operation altogether and decrease the number of characters deleted by τ. Since $\Phi(\cdot)$ is non-decreasing, this generates $\mathbf{z}$ (and hence $\mathbf{y}$) with a lower cost.
3. $S_i \cap d$ is a prefix (or suffix) of S_i. Assume it is a prefix. The case of suffix is similar. Instead of deleting the characters $S_i \cap d$ we can avoid generating them in the first place. Let r be the smallest index in $S_i \setminus d$ (that is, the first character in S_i that is not deleted by τ). We change the ith duplicate-delete operation to start at r and decrease the number of characters deleted by τ. Since $\Phi(\cdot)$ is non-decreasing, the cost of generating $\mathbf{z}$ may only decrease.
4. $S_i \cap d$ is a non-empty substring of S_i that is neither a prefix nor a suffix of S_i. We claim that this case applies to at most one value of i. This implies that after taking care of all the other cases τ only deletes characters in S_i. We can then change the ith duplicate-delete operation to also delete the characters deleted by τ, and omit τ. Since $\Phi(\cdot)$ obeys the triangle inequality, this will not increase the total cost of deletion. By the inductive hypothesis, the rest of $\mathbf{y}$ can be generated by just duplicate-delete operations with at most the same cost. It remains to prove the claim. Recall that the set $\{S_i\}$ is comprised of mutually non-overlapping subsequences of $\mathbf{z}$. Suppose that there exist indices $i \neq j$ such that $S_i \cap d$ is a non-prefix/suffix substring of S_i and $S_j \cap d$ is a non-prefix/suffix substring of S_j. There must exist indices of both S_i and S_j in $\mathbf{z}$ that precede d, are contained in d, and succeed d. Let $i_p < i_c < i_s$ be three such indices of S_i and let $j_p < j_c < j_s$ be similar for S_j. It must be the case also that $j_p < i_c < j_s$ and $i_p < j_c < i_s$. Without loss of generality, suppose $i_p < j_p$. It follows that (i_p, i_c) and (j_p, j_s) are alternating in $\mathbf{z}$. So, S_i and S_j are overlapping which contradicts Lemma 1.   □

To extend the recurrence from the previous section to duplication-deletion distance, we must observe that because we allow deletions in the string that is duplicated from $\mathbf{x}$, if we assume character x_i is copied to produce y_1, it may not be the case that the character x_{i+1} also appears in $\mathbf{y}$; the character x_{i+1} may have been deleted. Therefore, we minimize over all possible locations $k > i$ for the next character in the duplicated string that is not deleted. The extension of the recurrence from the previous section to duplication-deletion distance is:

$$\hat{d}(\mathbf{x}, \emptyset) = 0 \quad , \quad \hat{d}(\mathbf{x}, \mathbf{y}) = \min_{\{i : x_i = y_1\}} \hat{d}_i(\mathbf{x}, \mathbf{y}), \tag{3}$$

$$\hat{d}_i(\mathbf{x}, \emptyset) = 0 \quad ,$$

$$\hat{d}_i(\mathbf{x}, \mathbf{y}) = \min \begin{cases} 1 + \hat{d}(\mathbf{x}, \mathbf{y}_{2, |\mathbf{y}|}), \\ \min_{k > i} \min_{\{j : y_j = x_k, j > 1\}} \left\{ \begin{array}{l} \hat{d}(\mathbf{x}, \mathbf{y}_{2, j-1}) + \hat{d}_k(\mathbf{x}, \mathbf{y}_{j, |\mathbf{y}|}) \\ + \Phi(k - i - 1) \end{array} \right\} \end{cases}. \tag{4}$$

Theorem 2. $\hat{d}(x, y)$ *is the duplication-deletion distance from* x *to* y. *For* $\{i :$ $x_i = y_1\}$, $\hat{d}_i(x, y)$ *is the duplication-deletion distance from* x *to* y *under the additional restriction that* y_1 *is generated by* x_i.

The proof of Theorem 2 is almost identical to that of Theorem 1 in the previous section and is omitted. However, the running time increases; while the number of entries in the dynamic programming table does not change, the time to compute each entry is multiplied by the possible values of k in the recurrence, which is $O(|\mathbf{x}|)$. Therefore, the running time is $O(|\mathbf{y}|^2|\mathbf{x}|\mu(\mathbf{x})\mu(\mathbf{y}))$, which is $O(|\mathbf{y}|^3|\mathbf{x}|^2)$ in the worst case. Note that if the cost function is the identity $\Phi(\cdot) = 1$, then the duplication-deletion distance is equal to duplication distance without deletions. We omit the proof.

5 Duplication-Inversion-Deletion Distance

In this section we extend the duplication-deletion distance recurrence to allow inversions. We now explicitly define characters and strings as having two orientations: forward (+) and inverse (-). An inversion of the signed string $(+1 + 2 + 3)$ yields $(-3 - 2 - 1)$. In a duplicate-invert operation a substring is copied from $\mathbf{x}$ and *inverted* before being inserted into the target string $\mathbf{y}$. We allow the cost of inversion to be an affine function in the length ℓ of the duplicated inverted string, which we denote $\Theta_1 + \ell\Theta_2$, where $\Theta_1, \Theta_2 \geq 0$. We still allow for normal duplicate operations and delete operations from the target string at any time during the generation process. Note that we only handle deletions after inversions of the same substring. The order of operations might be important, at least in terms of costs. The cost of inverting $(+1 + 2 + 3)$ and then deleting -2 may be different than the cost of first deleting $+2$ from $(+1 + 2 + 3)$ and then inverting $(+1 + 3)$.

Definition 9. *A* **duplicate-invert operation** *from* x, $\bar{\delta}_x(s, t, p)$, *copies an inverted substring* $-x_t, \ldots, -x_s$ *of the source string* x *and pastes it into a target string at position* p. *Specifically, if* $x = x_1 \ldots x_m$ *and* $z = z_1 \ldots z_n$, *then* $z \circ \bar{\delta}_x(s, t, p) = z_1 \ldots z_{p-1}\bar{x}_t\bar{x}_{t-1} \ldots \bar{x}_s z_p \ldots z_n$.

The cost associated with each duplicate-invert operation is $1 + \Theta_1 + (t - s + 1)\Theta_2$.

Definition 10. *The* **duplication-inversion-deletion distance** *from a source string* x *to a target string* y *is the cost of a minimum sequence of duplicate and duplicate-invert operations from* x *and deletion operations, in any order, that generates* y.

Definition 11. *A* **duplicate-invert-delete** *operation from* x, $\bar{\eta}_x(i_1, j_1, i_2, j_2, \ldots, i_k, j_k, p)$, *for* $i_1 \leq j_1 < i_2 \leq j_2 < \cdots < i_k \leq j_k$ *pastes the string* $-x_{j_k} - x_{j_k - 1} \cdots -x_{i_k} - x_{j_{k-1}} - x_{j_{k-1}-1} \cdots - x_{i_{k-1}} \cdots \cdots - x_{j_1} - x_{j_1-1} \cdots - x_{i_1}$ *into a target string at position* p. *Specifically, if* $x = x_1 \ldots x_m$ *and* $z = z_1 \ldots z_n$, *then* $z \circ \bar{\eta}_x(i_1, j_1, i_2, j_2, \ldots, i_k, j_k, p) = z_1 \ldots z_{p-1} - x_{j_k} - x_{j_k - 1} \ldots$ $- x_{i_k} - x_{j_{k-1}} - x_{j_{k-1}-1} \cdots - x_{i_{k-1}} \cdots \cdots - x_{j_1} - x_{j_1-1} \cdots - x_{i_1} z_p \ldots z_n$.

The cost of such an operation is $1 + \Theta_1 + (j_k - i_1 + 1)\Theta_2 + \sum_{\ell=1}^{k-1} \Phi(i_{\ell+1} - j_\ell - 1)$. Similar to the previous section, it suffices to consider just duplicate-invert-delete operations, rather than duplicate-invert operations and delete operations.

Lemma 3. *If $\Phi(\cdot)$ is non-decreasing and obeys the triangle inequality and if the cost of inversion is an affine non-decreasing function as defined above, then the cost of a minimum sequence of duplicate, duplicate-invert and delete operations that generates a target string y from a source string x is equal to the cost of a minimum sequence of duplicate and duplicate-invert-delete operations that generates y from x.*

The proof of the lemma is essentially the same as that of Lemma 2. Note that in that proof we did not require all duplicate operations to be from the same string x. Therefore, the arguments in that proof apply to our case, where we can regard some of the duplicates from x and some from the inverse of x.

There are a few differences between the recurrence for duplication-inversion-deletion distance and the one for duplication-deletion distance (Eqs. 3, 4). First, when considering the possible characters to generate y_1, we consider characters in x that match either y_1 or its inverse, $-y_1$. In the former case, then, we use $\bar{d}_i^+(x,y)$ to denote the duplication-inversion-deletion distance with the additional restriction that y_1 is generated by x_i without an inversion. The recurrence for $\bar{d}_i^+$ is the same as for $\hat{d}_i$ in Eq. 4. In the latter case, we consider an inverted duplicate in which y_1 is generated by $-x_i$. This is denoted by $\bar{d}_i^-$, which follows a similar recurrence. In this recurrence, since an inversion occurs, x_i is the *last* character of the duplicated string, rather than the first one. Therefore, when considering k, the next character in x to be used in this operation, we look for an index smaller than i rather than larger than i. The recurrence for $\bar{d}_i^-$ also differs in the cost term. If x_k is the next character used by this operation, then all the characters between x_k and x_i are part of the inversion. We therefore add $(i-k)\Theta_2$ to the cost. If x_i is the last character used by this operation, we add $\Theta_2 + \Theta_1$ to the cost. The extension of the recurrence from the previous section to duplication-inversion-deletion distance is therefore:

$$\bar{d}(x,\emptyset) = 0 \quad, \quad \bar{d}(x,y) = \min\left\{ \min_{\{i:x_i=y_1\}} \bar{d}_i^+(x,y), \min_{\{i:x_i=-y_1\}} \bar{d}_i^-(x,y)\right\},$$

$$\bar{d}_i^+(x,\emptyset) = 0 \quad, \quad \bar{d}_i^-(x,\emptyset) = 0,$$

$$\bar{d}_i^+(x,y) = \min\begin{cases} 1 + \bar{d}(x,y_{2,|y|}), \\ \min_{k>i}\min_{\{j:y_j=x_k,j>1\}}\left\{\begin{matrix}\bar{d}(x,y_{2,j-1}) + \bar{d}_k^+(x,y_{j,|y|}) \\ +\Phi(k-i-1)\end{matrix}\right\} \end{cases},$$

$$\bar{d}_i^-(x,y) = \min\begin{cases} 1 + \Theta_1 + \Theta_2 + \bar{d}(x,y_{2,|y|}), \\ \min_{k<i}\min_{\{j:y_j=-x_k,j>1\}}\left\{\begin{matrix}\bar{d}(x,y_{2,j-1}) + \bar{d}_k^-(x,y_{j,|y|}) \\ +(i-k)\Theta_2 + \Phi(i-k-1)\end{matrix}\right\} \end{cases}.$$

Theorem 3. *$\bar{d}(x,y)$ is the duplication-inversion-deletion distance from x to y. For $\{i:x_i=y_1\}$, $\bar{d}_i^+(x,y)$ is the duplication-inversion-deletion distance from x to y under the additional restriction that y_1 is generated by x_i. For $\{i:x_i=$*

$-y_1\}$, $\bar{d}_i^-(\boldsymbol{x}, \boldsymbol{y})$ *is the duplication-inversion-deletion distance from $\boldsymbol{x}$ to $\boldsymbol{y}$ under the additional restriction that y_1 is generated by $-x_i$.*

We omit the proof. The running time of the corresponding dynamic programming algorithm remains $O(|\mathbf{y}|^2|\mathbf{x}|\mu(\mathbf{y})\mu(\mathbf{x}))$, where the multiplicity $\mu(S)$ is maximal number of times a character appears in the string S, regardless of its sign.

We note here without further explanation that if a sequence of duplicate and duplicate-invert operations is desired (without any deletions), it suffices to modify the recurrence that computes duplication distance in section 3, increasing its running time by only a constant factor. Restricting the model of rearrangement to allow only duplicate and duplicate-invert operations (instead of duplicate-invert-delete operations) may be desirable from a biological perspective because each duplicate and duplicate-invert requires only three breakpoints in the genome, whereas a duplicate-invert-delete operation can be significantly more complicated, requiring more breakpoints.

Other Variants of Duplication with Inversions. Above we handled the model where the duplicated substring of **x** may be inverted in its entirety before being inserted into the target. We now generalize to models in which only a substring of the duplicated string is inverted before being inserted into **y**. If the length of the inverted substring is ℓ, then the cost of this operation is $1 + \Theta(\ell)$. For example, we allow the substring $(+1 + 2 + 3 + 4 + 5 + 6)$ to become $(+1 + 2 - 5 - 4 - 3 + 6)$ before being inserted into **y**.

It is possible to extend the duplication-inversion-deletion recurrence to handle an inversion of any prefix or suffix of the duplicated substring, without changing the asymptotic running time of the recurrence. We omit the details.

Another variant allows for inversion of an arbitrary substring of the duplicated string at the price of asymptotically longer running time. For $1 \leq s \leq t \leq |\mathbf{x}|$, let $\tilde{\mathbf{x}}^{s,t}$ be the string $x_1 \ldots x_{s-1} - x_t \cdots - x_s x_{t+1} \ldots x_{|\mathbf{x}|}$. That is, the string that is obtained from **x** by inverting (in-place) $\mathbf{x}_{s,t}$. For convenience, define also $\tilde{\mathbf{x}}^{0,0} = \mathbf{x}$. The duplication-deletion with arbitrary-substring-duplicate-inversions distance satisfies the following recurrence. The running time is $O(|\mathbf{y}|^2|\mathbf{x}|^3\mu(\mathbf{x})\mu(\mathbf{y}))$.

$$\tilde{d}(\mathbf{x}, \emptyset) = 0 \quad , \quad \tilde{d}(\mathbf{x}, \mathbf{y}) = \min_{\{s,t:s=0,t=0 \text{ or } 1 \leq s \leq t \leq |\mathbf{x}|\}} \min_{\{i:\tilde{\mathbf{x}}_i^{s,t}=y_1\}} \tilde{d}_i^{st}(\mathbf{x}, \mathbf{y}),$$

$$\tilde{d}_i(\mathbf{x}, \emptyset) = 0 \quad ,$$

$$\tilde{d}_i^{st}(\mathbf{x}, \mathbf{y}) = \min \begin{cases} 1 + \Theta(t - s + 1) + \tilde{d}(\mathbf{x}, \mathbf{y}_{2,|\mathbf{y}|}), \\ \min_{k>i} \min_{\{j:y_j=\tilde{\mathbf{x}}_k^{s,t}, j>1\}} \begin{Bmatrix} \tilde{d}(\mathbf{x}, \mathbf{y}_{2,j-1}) + \tilde{d}_k^{st}(\mathbf{x}, \mathbf{y}_{j,|\mathbf{y}|}) \\ + \varPhi(k - i - 1) \end{Bmatrix} \end{cases}.$$

6   Conclusion

We have shown how to generalize duplication distance to include certain types of deletions and inversions and how to compute these new distances efficiently via dynamic programming. In earlier work [15,16], we used duplication distance to derive phylogenetic relationships between human segmental duplications. We plan to apply the generalized distances introduced here to the same data to determine if these richer computational models yield new biological insights.

References

1. Sankoff, D., Leduc, G., Antoine, N., Paquin, B., Lang, B., Cedergren, R.: Gene Order Comparisons for Phylogenetic Inference: Evolution of the Mitochondrial Genome. Proc. Natl. Acad. Sci. U.S.A. 89(14), 6575–6579 (1992)
2. Pevzner, P.: Computational molecular biology: an algorithmic approach. MIT Press, Cambridge (2000)
3. Chen, X., Zheng, J., Fu, Z., Nan, P., Zhong, Y., Lonardi, S., Jiang, T.: Assignment of Orthologous Genes via Genome Rearrangement. IEEE/ACM Trans. Comp. Biol. Bioinformatics 2(4), 302–315 (2005)
4. Marron, M., Swenson, K.M., Moret, B.M.E.: Genomic Distances Under Deletions and Insertions. TCS 325(3), 347–360 (2004)
5. El-Mabrouk, N.: Genome Rearrangement by Reversals and Insertions/Deletions of Contiguous Segments. In: Giancarlo, R., Sankoff, D. (eds.) CPM 2000. LNCS, vol. 1848, pp. 222–234. Springer, Heidelberg (2000)
6. Zhang, Y., Song, G., Vinar, T., Green, E.D., Siepel, A.C., Miller, W.: Reconstructing the Evolutionary History of Complex Human Gene Clusters. In: Vingron, M., Wong, L. (eds.) RECOMB 2008. LNCS (LNBI), vol. 4955, pp. 29–49. Springer, Heidelberg (2008)
7. Ma, J., Ratan, A., Raney, B.J., Suh, B.B., Zhang, L., Miller, W., Haussler, D.: Dupcar: Reconstructing contiguous ancestral regions with duplications. Journal of Computational Biology 15(8), 1007–1027 (2008)
8. Bertrand, D., Lajoie, M., El-Mabrouk, N.: Inferring Ancestral Gene Orders for a Family of Tandemly Arrayed Genes. J. Comp. Biol. 15(8), 1063–1077 (2008)
9. Chaudhuri, K., Chen, K., Mihaescu, R., Rao, S.: On the Tandem Duplication-Random Loss Model of Genome Rearrangement. In: Proceedings of the Seventeenth Annual ACM-SIAM Symposium on Discrete Algorithms (SODA), pp. 564–570. ACM, New York (2006)
10. Elemento, O., Gascuel, O., Lefranc, M.P.: Reconstructing the Duplication History of Tandemly Repeated Genes. Mol. Biol. Evol. 19(3), 278–288 (2002)
11. Lajoie, M., Bertrand, D., El-Mabrouk, N., Gascuel, O.: Duplication and Inversion History of a Tandemly Repeated Genes Family. J. Comp. Bio. 14(4), 462–478 (2007)
12. El-Mabrouk, N., Sankoff, D.: The reconstruction of doubled genomes. SIAM J. Comput. 32(3), 754–792 (2003)
13. Alekseyev, M.A., Pevzner, P.A.: Whole Genome Duplications and Contracted Breakpoint Graphs. SICOMP 36(6), 1748–1763 (2007)
14. Bailey, J., Eichler, E.: Primate Segmental Duplications: Crucibles of Evolution, Diversity and Disease. Nat. Rev. Genet. 7, 552–564 (2006)
15. Kahn, C.L., Raphael, B.J.: Analysis of Segmental Duplications via Duplication Distance. Bioinformatics 24, i133–i138 (2008)
16. Kahn, C.L., Raphael, B.J.: A Parsimony Approach to Analysis of Human Segmental Duplications. In: Pacific Symposium on Biocomputing, pp. 126–137 (2009)

Predicting Gene Structures from Multiple RT-PCR Tests
(Extended Abstract)

Jakub Kováč[1], Tomáš Vinař[2], and Broňa Brejová[1]

[1] Department of Computer Science, Comenius University, Mlynská Dolina,
842 48 Bratislava, Slovakia
kuko@ksp.sk, brejova@dcs.fmph.uniba.sk
[2] Department of Applied Informatics, Comenius University, Mlynská Dolina,
842 48 Bratislava, Slovakia
vinar@ii.fmph.uniba.sk

Abstract. It has been demonstrated that the use of additional information such as ESTs and protein homology can significantly improve accuracy of gene prediction. However, many sources of external information are still being omitted from consideration. Here, we investigate the use of product lengths from RT-PCR experiments in gene finding. We present hardness results and practical algorithms for several variants of the problem and apply our methods to a real RT-PCR data set in the *Drosophila* genome. We conclude that the use of RT-PCR data can improve the sensitivity of gene prediction and locate novel splicing variants.

Keywords: gene finding, RT-PCR, NP-completeness, dynamic programming, splicing graph.

1 Introduction

In spite of recent progress, gene finding remains a difficult problem, particularly in the presence of ubiquitous alternative splicing (Guigo et al., 2006). Nonetheless, prediction accuracy of gene finders can be significantly improved by incorporating various sources of experimental evidence. Reverse transcription-polymerase chain reaction (RT-PCR) is an experimental method often used to confirm or reject predicted gene structures (Siepel et al., 2007). However, the results of RT-PCR tests could also be used as additional evidence in gene finding to propose new transcripts. In this paper, we study the problem of using the estimated lengths of RT-PCR products for this purpose. We also present a proof-of-concept study on a recently acquired data set of RT-PCR products in the *Drosophila* genome (Brent et al., 2007) designed to verify previously unknown transcripts predicted by CONTRAST (Gross et al., 2007) and NSCAN-EST (Wei and Brent, 2006) gene finders.

In an RT-PCR experiment, we select two short sequences, called primers, from two predicted exons of the same gene. If both primers are indeed present in the

S.L. Salzberg and T. Warnow (Eds.): WABI 2009, LNBI 5724, pp. 181–193, 2009.

same transcript of the gene, RT-PCR will amplify the region of the transcript between the primers, and we can estimate its length (with the introns spliced out) on an electrophoresis gel. In case of alternative splicing, we will observe multiple lengths. RT-PCR products can be further cloned and sequenced; however, this process incurs further costs.

We propose a new computational problem in which we want to select a set of transcripts that best agrees with the observed results of several RT-PCR tests. In particular, we represent an RT-PCR test by the positions p_1, p_2 of the two primers in the sequence, and a (potentially empty) list of product lengths. Each length is an interval $[m, M]$, since it is impossible to estimate the length of the product exactly.

As an input to our algorithm, we use a set of potential transcripts of one gene represented as a *splicing graph* (Heber et al., 2002). Vertices of this graph correspond to non-overlapping segments of the DNA sequence. An edge (u, v) in the splicing graph indicates that segment v immediately follows segment u in some transcript (examples of splicing graphs can be seen in Figures 3 and 4). Moreover, two special vertices s and t mark the beginning and the end of a transcript. Thus, every transcript is represented by an (s, t) path in the splicing graph. Note that the vertices do not necessarily correspond to whole exons: a single exon can be split into several vertices, for example due to alternative splice sites. Vertices and edges of the splicing graph are assigned scores, which correspond to the confidence we have that a particular predicted intron or exon segment is correct.

We say that a path π through the splicing graph *explains* test $T = (p_1, p_2, [m_1, M_1], [m_2, M_2], \ldots, [m_k, M_k])$ if the path contains both primers p_1 and p_2, and the distance between the two primers in the transcript defined by π belongs to one of the intervals $[m_1, M_1], \ldots, [m_k, M_k]$. If the path contains both primers, but the distance is not covered by any of the intervals associated with T, we say that the path π is *inconsistent with* test T. We can now define a *score of a path π with respect to a set of tests S* as a sum of the scores of all of its vertices and edges, plus a bonus B for each explained test from S, and minus a penalty P for each inconsistent test.

Definition 1 (Gene finding with RT-PCR tests). *For a given splicing graph G and a set of RT-PCR tests S, find the path π with the highest score.*

There are several practical ways to obtain a splicing graph at a given locus in the genome using *ab initio* gene finders based on hidden Markov models (HMMs). Besides the actual highest probability gene structure in the HMM, Genscan (Burge and Karlin, 1997) can output additional exons that have reasonably high posterior probability. These exons can be joined to potential transcript based on compatibility of their reading frames. Gene finder Augustus (Stanke et al., 2006) uses an HMM path sampling algorithm to output multiple predictions that may correspond to alternative transcripts of the same gene. These methods can also be generalized to take into account additional information, such as ESTs or protein similarity (Stanke et al., 2008). Regardless of the method of generating candidate exons or transcripts, our goal is to set the parameters of these algorithms to achieve high sensitivity even at the cost of low specificity.

A similar problem has been previously investigated by Agrawal and Stormo (2006) who designed a heuristic to find a transcript that explains a single RT-PCR test. Our approach improves on their work by using an exact algorithm rather than a heuristic and by integrating information from multiple overlapping tests.

We have also investigated a simpler problem, where we do not consider path scores and product lengths. Instead, the tests are simply pairs of vertices in the splicing graph, and we either seek a path that contains as many pairs as possible (paths passing useful pairs, PPUP), or try to avoid these pairs altogether (paths avoiding forbidden pairs, PAFP). While even these simpler versions of the problem are strongly NP-hard in general, we show polynomial-time algorithms for several special cases which also extend to the original scenario with lengths and scores and result in pseudo-polynomial algorithms. The PAFP problem has also been studied in connection with automated software testing (Krause et al., 1973; Srimani and Sinha, 1982; Gabow et al., 1976) and tandem mass spectrometry (Chen et al., 2001). We explore several new variants and improve the recent results of Kolman and Pankrác (2009) in this area.

2   General Algorithm

In general, the problem of finding a path through the splicing graph consistent with even a single RT-PCR test is NP-hard. This is easy to see, since one can create an instance with a set of disjoint exons of various lengths, and an edge between any two exons. The (s, t) path in such a graph will correspond to a solution of the NP-hard subset-sum problem (Garey and Johnson, 1979), where the single RT-PCR test specifies the target sum.

Fortunately, we can easily design a simple practical dynamic programming algorithm. First, assume a single RT-PCR test $(s, t, [m, M])$. Our task is to find the highest scoring (s, t) path of length between m and M. Let $H[i, \ell]$ be the highest score we can achieve by an (s, v_i) path of length ℓ, or $-\infty$ if there is no such path. Let $\ell(i)$ be the length of vertex v_i and $S(i, j)$ be the score of edge (v_i, v_j) plus the score of vertex v_i. The values of $H[i, \ell]$ can be computed by a simple recurrence: $H[i, \ell] = \max_{j:(v_j, v_i) \in E} H[j, \ell - \ell(i)] + S(j, i)$, with base cases $H[0, 0] = 0$ and $H[0, \ell] = -\infty$ for $\ell > 0$. The highest value of $H[n + 1, \ell]$ for $m \leq \ell \leq M$ represents the desired path. The running time is $O(ME)$, where E is the number of edges in the splicing graph. We can further improve this algorithm by considering only achievable lengths ℓ for each vertex i. We can also eliminate lengths that cannot achieve the target length. In particular for each vertex, we calculate the minimum and maximum length of a (v, t) path $\text{mind}(v)$ and $\text{maxd}(v)$ and ignore all lengths ℓ for which $\ell + \text{mind}(v) > M$ or $\ell + \text{maxd}(v) < m$. Since H will often be sparse, and values of M are small due to limitations of RT-PCR, this algorithm is a practical solution to the problem.

The algorithm is easily extended to multiple tests and to the general problem with bonuses and penalties as defined above. Even though the running time

$O(M^k E)$ grows exponentially with the maximum number of overlapping RT-PCR tests k, this number is typically small in real data sets (such as the data set that we consider below, where $k = 4$).

3 Primer Positioning and Hardness

In the previous section, we have demonstrated that the general problem is NP-hard. We have shown an algorithm that is pseudo-polynomial for a single test, and exponential in the number of overlapping tests in the general case. Here, we demonstrate several complexity results for various special cases of interest characterized by a particular placement of primer pairs.

First, we examine two cases illustrated in Fig. 1. We say that two tests *halve* each other if the corresponding intervals of the DNA sequence overlap each other, but neither is completely included in the other. In the *ordered case*, all the left primers are arranged in the same order as the corresponding right primers (i.e., every two tests are either disjoint, or they halve each other). In the *ordered halving case*, the tests are ordered and moreover every two tests halve each other.

Even though the distinction between these two cases seems rather subtle, we will show that in the ordered case, the problem is strongly NP-hard, while in the ordered halving case there exist a pseudo-polynomial solution. Moreover, this is true even if we do not consider lengths of the RT-PCR products and scores associated with the vertices and edges of the splicing graph.

In this simplified scenario, every pair of primers is either *useful* or *forbidden*. We consider two versions of the problem. The *PAFP problem* seeks paths avoiding all forbidden pairs, i.e. from each pair we are allowed to include at most one end vertex. The *PPUP problem* seeks paths passing as many useful pairs as possible. The PAFP problem corresponds to RT-PCR tests without any products, while the PPUP corresponds to successful RT-PCR tests, but without considering product lengths.

Theorem 1. *The PAFP and PPUP problems with an ordered set of primer pairs are strongly NP-complete.*

Proof. We will prove the claim by reduction from 3-SAT. Let φ be a conjunction of n clauses $\varphi_1 \wedge \cdots \wedge \varphi_n$ over m variables $x_1, \ldots, x_m$, where $\varphi_i = \ell_{i,1} \vee \ell_{i,2} \vee \ell_{i,3}$ and each literal $\ell_{i,j}$ is either x_k or $\overline{x_k}$. We will construct graph G and a set of pairs S such that the solution of the corresponding PPUP problem gives the satisfying assignment of φ.

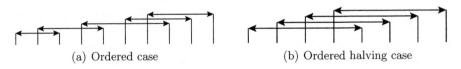

(a) Ordered case   (b) Ordered halving case

Fig. 1. Two special cases of primer pair positions

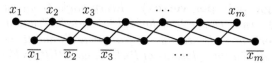

(a) Block B – vertices of this graph correspond
to positive and negative literals; path through this
graph corresponds to a valuation of variables

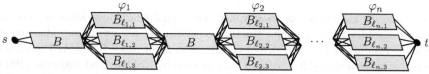

(b) Construction of G from the blocks B and $B_{\ell_{i,j}}$. Blocks $B_{\ell_{i,j}}$ are interleaved so
that vertices for each literal from $B_{\ell_{i,1}}$, $B_{\ell_{i,2}}$, and $B_{\ell_{i,3}}$ form a continuous interval.

Fig. 2. Construction of the graph G for a 3-SAT formula φ

Graph G consists of several copies of a block B of $2m$ vertices as shown in
Fig. 2(a). Any left-to-right path through the block B naturally corresponds to an
assignment of the variables. For each literal $\ell_{i,j}$, we also construct a block $B_{\ell_{i,j}}$
which is identical to B, except that the vertex corresponding to $\overline{\ell_{i,j}}$ is missing.
The blocks are joined together as outlined in Fig. 2(b). The path passing through
the construct corresponding to a clause φ_i must pass through one of the three
blocks, and thus choose an assignment that satisfies the clause.

The set of pairs S will enforce that the assignment of the variables is the
same in all blocks. This is done by adding a useful pair between corresponding
literals in block $B_{\ell_{i,j}}$ and the preceding B-block and the following B-block. A
path corresponding to the solution of PPUP in (G, S) will thus give a unique
satisfying assignment of the variables $x_1, \ldots, x_m$ if the path contains at least
$(2n - 1)m$ pairs (otherwise there is no satisfying assignment).

The resulting set S is not ordered, since three nested intervals start in each
node of B. The issue can be easily fixed by splitting each vertex of B into a
path of length three and using a different vertex of the path for each of the three
intervals.

The reduction is analogous for PAFP. In this case, we reverse the order of
vertices x_i and $\overline{x_i}$ in B-blocks (but we keep the order in $B_{\ell_{i,j}}$ blocks the same).
The set of ordered pairs S will now be composed from forbidden pairs between
atoms x_i in $B_{\ell_{i,j}}$ blocks with their counterparts $\overline{x_i}$ in the previous and the
following B-block.   □

We have demonstrated that PAFP and PPUP are strongly NP-hard on ordered
pairs. Consequently, the general problem explored in the previous section is also
strongly NP-hard, since PPUP is a special case of that problem. However, both
PAFP and PPUP can be solved in polynomial time for the ordered halving pairs.
Before we demonstrate the algorithm, we note a simplifying transformation on
graph G.

Lemma 1 (Single pair per vertex). *Every graph G and a set S of either forbidden or useful pairs can be transformed to a graph G' and a set S' such that in each vertex starts or ends exactly one pair from S', and there is a one-to-one correspondence between the solutions of PAFP and PPUP problems for G, S and for G', S'. Moreover, the halving and ordering structure of the graph is preserved, and the transformation can be done efficiently in $O(PE)$ time, where P is the number of pairs in S.*

The above lemma allows us to simplify the algorithms below, since we do not need to consider multiple pairs starting or ending at a particular node, and at the same time we reduce the size of the graph in most cases. The PAFP problem has recently been solved for ordered halving case by Kolman and Pankrác (2009) in $O(PE + P^5)$ time. Here, we show an algorithm for the PPUP problem with ordered halving pairs.

Theorem 2. *The PPUP problem with ordered halving useful pairs can be solved in $O(PE + P^3)$ time.*

Proof. First, we can remove all useful pairs (v_i, v_j) for which there is no (s, v_i) path or no (v_j, t) path because these pairs will never be used on any (s, t) path. Furthermore, due to Lemma 1, we can assume without loss of generality that the graph contains exactly one pair starting or ending at each vertex of the graph G. Such a graph G will have $2P$ vertices $v_1, \ldots, v_{2P}$, and useful pairs $(v_1, v_{P+1}), (v_2, v_{P+2}), \ldots, (v_P, v_{2P})$.

To search for the (s, t) path containing the largest number of useful pairs, we construct a new graph H with vertices $w_1, \ldots, w_P$, each vertex corresponding to a single useful pair. We will say that w_i and w_j are connected by a *blue edge* if there are left-to-right (v_i, v_j) and (v_{P+i}, v_{P+j}) paths in graph G. Moreover, vertices w_i and w_j are connected by a *red edge* if there is a left-to-right (v_j, v_{P+i}) path in G. Graph H can be constructed in $O(PE + P^2)$ time.

Searching for the PPUP in G now translates into searching for the longest (w_i, w_j) left-to-right blue path in H, where w_i and w_j are also connected by a red edge. The longest such paths for all pairs of w_i and w_j can be easily found by dynamic programming in $O(P^3)$ time, and thus the total running time of the algorithm is $O(PE + P^3)$ (including the preprocessing time required by the transformation in Lemma 1). □

We have also investigated the complexity of PAFP and PPUP problem for other special conformations of pairs: *disjoint pairs* (no two intervals defined by the pairs overlap), *well-parenthesized pairs* (no two intervals halve each other), *halving structure* (every two intervals halve each other or they are nested), and *nested pairs* (for any two intervals, the smaller interval is nested in the larger interval). The results are summarized in Table 1 and the proofs are omitted due to the space restrictions. Note that the NP-hardness result on general PAFP problem is due to Gabow et al. (1976), and several of the other forms of the PAFP problem have been recently investigated by Kolman and Pankrác (2009); all the other results are new.

Table 1. Complexity of the PAFP and PPUP problem, where E is the number of edges in the input graph and P is the number of forbidden/useful pairs. NP-hardness of the general problem (*) was proved by Gabow et al. (1976); results marked by [†] were first proved by Kolman and Pankrác (2009).

PROBLEM	FORBIDDEN (PAFP)	USEFUL (PPUP)	EXAMPLE
general problem	NP-hard [*]	NP-hard	
halving structure	NP-hard [†]	NP-hard	
ordered	NP-hard	NP-hard	
well-parenthesized	$O(PE + P^3)$ [†]	$O(PE + P^3)$	
ordered halving	$O(PE + P^5)$ [†]	$O(PE + P^3)$	
nested	$O(PE + P^3)$	$O(PE)$	
disjoint	$O(E)$	$O(E)$	

Table 2. Complexity of the general problem of finding an (s, t) path in a splicing graph with zero penalty, where E is the number of edges, P is the number RT-PCR tests, M is the maximum measured length and Δ is the maximum number of lengths in an RT-PCR test that get any bonus (e.g. $\Delta = M - m + 1$ for one measured length $[m, M]$)

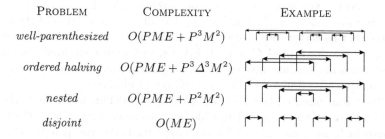

PROBLEM	COMPLEXITY	EXAMPLE
well-parenthesized	$O(PME + P^3 M^2)$	
ordered halving	$O(PME + P^3 \Delta^3 M^2)$	
nested	$O(PME + P^2 M^2)$	
disjoint	$O(ME)$	

The hardness proofs obviously carry over to the more general problem of finding the best (highest score) (s, t) path in a splicing graph, where each vertex has a length, each edge has a score and we get bonus B for explaining a length of an RT-PCR test. Thus we cannot hope for even a pseudopolynomial algorithm for the halving or ordered case (unless P=NP). On the other hand, positive results for the PPUP problem can be extended to pseudopolynomial algorithms for the more general setting. Our results are summarized in Table 2; we omit the proofs due to the space restrictions. In the well-parenthesized case we were able to further generalize the algorithm by considering penalties for inconsistent RT-PCR tests, achieving the same running time.

4   Finding Genes in *D. melanogaster*

To test the improvement in gene finding accuracy achieved by incorporating RT-PCR product lengths, we have used a set of 2784 actual RT-PCRs from the Brent laboratory on the *Drosophila melanogaster* genome (Brent et al., 2007). These experiments were designed to test novel transcripts predicted by CON-TRAST (Gross et al., 2007) and NSCAN-EST (Wei and Brent, 2006). Each RT-PCR product was sequenced and aligned to the genome. We have estimated the length of each product by locating likely positions of the two primers in the sequenced product. We have added ±15% margin or at least ±50 nucleotides to simulate the uncertainty in length estimates from electrophoresis gels. After discarding primers spanning exon boundaries or located in the predicted un-translated regions, and merging primer pairs with identical primer coordinates, we were left with 1955 RT-PCR tests in 1159 loci, each locus corresponding to a set of overlapping predicted or known genes. Overall, 942 tests have produced at least one product and fewer than ten tests have produced two products with significantly different lengths. Note that approximately 2% of the products mapped to other than the intended genomic locus; thus the estimated lengths in our test do not always correspond to real transcripts at the locus of interest. We expect that this type of error would also occur in practice, since without sequencing we cannot determine whether the product indeed maps to the expected location.

To obtain the splicing graph, we have used Augustus gene finder (Stanke et al., 2006) capable of sampling random gene structures from the posterior probability distribution defined by the underlying generalized HMM. For each locus, we have created the splicing graph based on 1000 samples. For each vertex of the graph, we have estimated the posterior probability p as a fraction of the samples that contain this vertex. Score of the vertex was then set to $p - C(1-p)$ for some constant C (we have used $C = 0.5$). Edge scores were computed analogously. A similar scoring scheme was used as an optimization criterion in the recent discriminative gene finder CONTRAST (Gross et al., 2007).

We have implemented the general $O(M^k E)$ algorithm, which finds the (s, t) path with the maximum score that aside from scores of vertices and edges on the path also includes bonus $B = 5$ for each explained test and penalty $P = 1$ for each inconsistent test. This path may explain several of the observed RT-PCR product lengths. For each length not explained by the path, we run the algorithm again, this time finding the highest scoring path that explains this length, if there is any. In this way, we may obtain multiple alternative transcripts.

Table 3 shows the results of our algorithm on 1022 loci for which Augustus sampling produced at least two different transcripts. We compare our algorithm with the most probable transcript produced by Augustus run in the Viterbi mode and also with the highest scoring path in our splicing graph (without considering any RT-PCR bonuses or penalties). All three versions have almost identical accuracy compared to the known RefSeq genes. In particular, although our version is capable of producing multiple transcripts per gene, this does not lead to significant decrease in specificity.

Table 3. Gene prediction accuracy on 1022 loci. Sensitivity is the fraction of annotated coding nucleotides, exons, or splice sites that were correctly predicted and specificity is the fraction of predictions that are correct.

Compared to RefSeq:	with PCR	w/o PCR	Augustus	
Exon Sensitivity	65%	64%	63%	
Exon Specificity	58%	58%	59%	
Nucleotide Sensitivity	84%	84%	83%	
Nucleotide Specificity	75%	75%	76%	
Compared to RT-PCR products:	with PCR	w/o PCR	Augustus	RefSeq
Acceptor Sensitivity	75%	73%	72%	73%
Donor Sensitivity	76%	74%	73%	74%

Although RefSeq annotation is often used for evaluating gene finding accuracy, it is not perfect and may miss some real transcripts. Indeed, the RT-PCR tests used here were designed specifically to discover new transcripts not contained in RefSeq. Thus we also compare the predictions to the exons defined by aligning the sequenced RT-PCR products to the *Drosophila* genome. Since the products often do not span whole exons, we compare individual splice sites (donors and acceptors). In this test, we see some improvement of sensitivity: for example Augustus predicts 72% of the acceptor sites while our program 75%. This is even more than RefSeq, which includes only 73% of these sites (Table 3).

The accuracy of our program is limited by two factors. First, we can only predict transcripts that have an (s, t)-path in the splicing graph. In this test, Augustus splicing graphs contained 85% of all donors and acceptors supported by aligned RT-PCR products, so even under the ideal conditions our approach could not exceed this level of sensitivity. Moreover, we rely on the Augustus scores together with bonuses and penalties to choose among possible transcripts, and therefore improved quality of these scores would also improve our prediction. Second limitation stems out of the density and quality of the RT-PCR tests. In our data set, 62% of the loci have only one RT-PCR test, and only 16% have three or more tests. Also, we add minimum of 100 bp tolerance to the observed lengths which means that we are unable to detect presence or absence of smaller exons unless they contain a primer. Perhaps this problem can be alleviated by a careful study of errors in length estimates obtained from electrophoresis gels.

Figures 3 and 4 illustrate both advantages and problems of our approach. In Fig.3, our algorithm successfully predicts RefSeq exons omitted by Augustus. However, one of the downstream predicted exons is shorter on its 3' end because the splicing graph does not contain the correct form of the exon. Moreover, even the mispredicted shorter form satisfies the length tolerance of the test and gets a bonus. In Fig.4, the lengths from two tests allowed inference of a gene structure quite different from both RefSeq and Augustus transcripts. The prediction agrees quite well with the sequenced RT-PCR products.

190   J. Kováč, T. Vinař, and B. Brejová

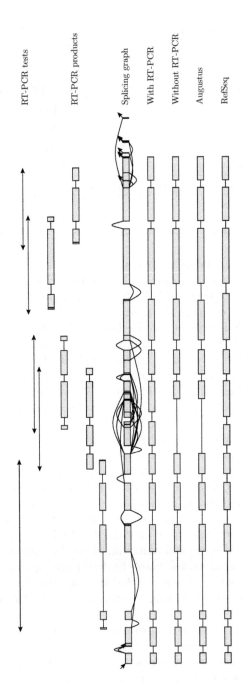

Fig. 3. Predicted and annotated transcripts and RT-PCR products in the locus of the GluRIIB gene (glutamate receptor). The locus has five RT-PCR tests, each with a sequenced product mapping to this region.

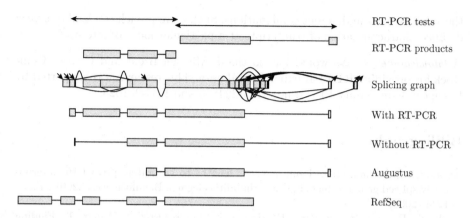

Fig. 4. Locus of the Q7KTW2 gene belonging to the amiloride-sensitive sodium channel family. Our algorithm predicts a transcript different from RefSeq structure yet agreeing well with one of the sequenced RT-PCR products.

5   Discussion and Future Work

In this paper, we introduced a new computational problem inspired by integrating RT-PCR information into gene finding. We have shown a practical algorithm and explored the boundary between NP-hard and polynomially solvable special cases of the problem. Finally, we have demonstrated that this method is indeed able to locate new splicing variants.

One problem we have not explored in this paper is the design of RT-PCR experiments. Currently one typically first uses a gene finder to predict individual transcripts, and then concentrates on predictions that are novel or different from established annotations (Siepel et al., 2007). In contrast, our approach suggests the possibility of designing primers based on a splicing graph representing exponential number of potential transcripts. In fact, we have previously investigated a theoretical problem related to this question (Biedl et al., 2004). The results presented here suggest that while the general problem of gene finding with RT-PCR product lengths is NP-hard, it is possible to position the queries in such a way that they can be analyzed efficiently.

It has been suggested (Agrawal and Stormo, 2006) that RT-PCR experiments followed by estimation of the product lengths on an electrophoresis gel can be considered a high-throughput method, especially if the gels could be analyzed computationally. In principle, the products of RT-PCR can be sequenced, and the cost of this is no longer prohibitive. In addition, new sequencing technologies suggest the possibility to exhaustively sequence large cDNA libraries in the near future. Nonetheless, we believe that the approach described in this paper will remain relevant for some time for smaller laboratories wishing to concentrate on non-model organisms or particular genomic loci. In the RT-PCR dataset explored in this paper, 84% of the loci have fewer than three RT-PCR tests, but instead one could cover a single locus in a more exhaustive way, possibly reusing the same primers in different combinations. Such an approach can lead to a very detailed characterization of transcripts at a particular locus of interest.

Acknowledgments. We would like to thank Michael Brent and Charles Comstock for providing RT-PCR experimental data. This research was supported by European Community FP7 grants IRG-224885 and IRG-231025.

Bibliography

Agrawal, R., Stormo, G.D.: Using mRNAs lengths to accurately predict the alternatively spliced gene products in Caenorhabditis elegans. Bioinformatics 22(10), 1239–1244 (2006)

Biedl, T., Brejova, B., Demaine, E., Hamel, A., Lopez-Ortiz, A., Vinar, T.: Finding hidden independent sets in interval graphs. Theoretical Computer Science 310(1-3), 287–307 (2004)

Brent, M., Langton, L., Comstock, C.L., van Baren, J.: Exhaustive RT-PCR and sequencing of all novel NSCAN predictions in Drosophila melanogaster. Personal communication (2007)

Burge, C., Karlin, S.: Prediction of complete gene structures in human genomic DNA. Journal of Molecular Biology 268(1), 78–94 (1997)

Chen, T., Kao, M.Y., Tepel, M., Rush, J., Church, G.M.: A dynamic programming approach to de novo peptide sequencing via tandem mass spectrometry. Journal of Computational Biology 8(3), 325–327 (2001)

Gabow, H.N., Maheswari, S.N., Osterweil, L.J.: On two problems in the generation of program test paths. IEEE Trans. Soft. Eng. 2(3), 227–231 (1976)

Garey, M.R., Johnson, D.S.: Computers and Intractability: A Guide to the Theory of NP-Completeness. W.H. Freeman, New York (1979)

Gross, S.S., Do, C.B., Sirota, M., Batzoglou, S.: CONTRAST: a discriminative, phylogeny-free approach to multiple informant de novo gene prediction. Genome Biology 8(12), R269 (2007)

Guigo, R., et al.: EGASP: the human ENCODE genome annotation assessment project. Genome Biology 7(suppl. 1), S2 (2006)

Heber, S., Alekseyev, M., Sze, S.H., Tang, H., Pevzner, P.A.: Splicing graphs and EST assembly problem. Bioinformatics 18(suppl. 1), S181–S188 (2002)

Kolman, P., Pankrác, O.: On the complexity of paths avoiding forbidden pairs. Discrete Applied Mathematics 157(13), 2871–2876 (2009)

Krause, K.W., Smith, R.W., Goodwin, M.A.: Optional software test planning through automated network analysis. In: Proceedings 1973 IEEE Symposium on Computer Software Reliability, pp. 18–22 (1973)

Siepel, A., Diekhans, M., Brejova, B., Langton, L., Stevens, M., Comstock, C.L., Davis, C., Ewing, B., Oommen, S., Lau, C., Yu, H.C., Li, J., Roe, B.A., Green, P., Gerhard, D.S., Temple, G., Haussler, D., Brent, M.R.: Targeted discovery of novel human exons by comparative genomics. Genome Research 17(12), 1763–1763 (2007)

Srimani, P.K., Sinha, B.P.: Impossible pair constrained test path generation in a program. Information Sciences 28(2), 87–103 (1982)

Stanke, M., Diekhans, M., Baertsch, R., Haussler, D.: Using native and syntenically mapped cDNA alignments to improve de novo gene finding. Bioinformatics 24(5), 637–644 (2008)

Stanke, M., Keller, O., Gunduz, I., Hayes, A., Waack, S., Morgenstern, B.: AUGUS-TUS: ab initio prediction of alternative transcripts. Nucleic Acids Research 34(Web Server issue), W435–W439 (2006)

Wei, C., Brent, M.R.: Using ESTs to improve the accuracy of de novo gene prediction. BMC Bioinformatics 7, 327 (2006)

A Tree Based Method for the Rapid Screening of Chemical Fingerprints

Thomas G. Kristensen, Jesper Nielsen, and Christian N.S. Pedersen

Bioinformatics Research Center (BiRC)
Aarhus University, C.F. Møllers Allé 8,
DK-8000 Århus C, Denmark
{tgk,jn,cstorm}@birc.au.dk

Abstract. The fingerprint of a molecule is a bitstring based on its structure, constructed such that structurally similar molecules will have similar fingerprints. Molecular fingerprints can be used in an initial phase for identifying novel drug candidates by screening large databases for molecules with fingerprints similar to a query fingerprint. In this paper, we present a method which efficiently finds all fingerprints in a database with Tanimoto coefficient to the query fingerprint above a user defined threshold. The method is based on two novel data structures for rapid screening of large databases: the kD grid and the Multibit tree. The kD grid is based on splitting the fingerprints into k shorter bitstrings and utilising these to compute bounds on the similarity of the complete bitstrings. The Multibit tree uses hierarchical clustering and similarity within each cluster to compute similar bounds. We have implemented our method and tested it on a large data set from the industry. Our experiments show that our method yields a three-fold speed-up over previous methods.

1 Introduction

When developing novel drugs, researchers are faced with the task of selecting a subset of all commercially available molecules for further experiments. There are more than 8 million available molecules [1], and it is therefore not possible to perform computationally expensive calculations on each one. Therefore, the need arise for fast screening methods for identifying the molecules that are most likely to have an effect on a disease or illness. It is often the case that a molecule with some effect is already known, e.g. from an already existing drug. An obvious initial screening method presents itself, namely to identify the molecules which are similar to this known molecule. To implement this screening method one must decide on a representation of the molecules and a similarity measure between representations of molecules. Several representations and similarity measures have been proposed [2,3,4]. We focus on *molecular fingerprints*. A fingerprint for a given molecule is a bitstring of size N which summarises structural information about the molecule [3]. Fingerprints should be constructed such that if two fingerprints are very similar, so are the molecules which they represent.

S.L. Salzberg and T. Warnow (Eds.): WABI 2009, LNBI 5724, pp. 194–205, 2009.

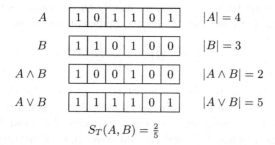

Fig. 1. Example of calculation of the Tanimoto coefficient $S_T(A, B)$, where $A = 101101$ and $B = 110100$

There are several ways of measuring the similarity between fingerprints [4]. We focus on the *Tanimoto coefficient*, which is a normalised measure of how many bits two fingerprints share. It is 1.0 when the fingerprints are the same, and strictly smaller than 1.0 when they are not. Molecular fingerprints in combination with the Tanimoto coefficient have been used successfully in previous studies [5].

We focus on the screening problem of finding all fingerprints in a database with Tanimoto coefficient to a query fingerprint above a given threshold, e.g. 0.9. Previous attempts have been made to improve the query time. One approach is to reduce the number of fingerprints in the database for which the Tanimoto coefficient to the query fingerprint has to be computed explicitly. This includes storing the fingerprints in the database in a vector of bins [6], or in a trie like structure [7], such that searching certain bins, or parts of the trie, can be avoided based on an upper-bound on the Tanimoto coefficient between the query fingerprint and all fingerprints in individual bins or sub tries. Another approach is to store an XOR summary, i.e. a shorter bitstring, of each fingerprint in the database, and use these as rough upper bounds on the maximal Tanimoto coefficients achievable, before calculating the exact coefficients [8].

In this paper, we present an efficient method for the screening problem, which is based on an extension of an upper bound given in [6] and two novel tree based data structures for storing and retrieving fingerprints. To further reduce the query time we also utilise the XOR summary strategy [8]. We have implemented our method and tested it on a realistic data set. Our experiments clearly demonstrate that it is superior to previous strategies, as it yields a three-fold speed-up over the previous best method.

2   Methods

A fingerprint is a bitstring of length N. Let A and B be bitstrings, and let $|A|$ denote the number of 1-bits in A. Let $A \wedge B$ denote the *logical and* of A and B, that is, $A \wedge B$ is the bitstring that has 1-bits in exactly those positions where both A and B do. Likewise, let $A \vee B$ denote the *logical or* of A and B, that is,

$A \vee B$ is the bitstring that has 1-bits in exactly those positions where either A or B do. With this notation the Tanimoto coefficient becomes:

$$S_T(A, B) = \frac{|A \wedge B|}{|A \vee B|}.$$

Figure 1 shows an example the usage of this notation. In the following, we present a method for finding all fingerprints B in a database of fingerprints with a Tanimoto coefficient above some query-specific threshold $S_{\min}$ to a query fingerprint A. The method is based on two novel data structures, the kD grid and the Multibit tree, for storing the database of fingerprints.

2.1 kD Grid

Swamidass *et al.* showed in [6] that if $|A|$ and $|B|$ are known, $S_T(A, B)$ can be upper-bounded by

$$S_{\max} = \frac{\min(|A|, |B|)}{\max(|A|, |B|)}.$$

This bound can be used to speed up the search, by storing the database of fingerprints in $N + 1$ buckets such that bitstring B is stored in the $|B|$th bucket. When searching for bitstrings similar to a query bitstring A it is sufficient to examine the buckets where $S_{\max} \geq S_{\min}$.

We have generalised this strategy. Select a number of dimensions k and split the bitstrings into k equally sized fragments such that

$$A = A_1 \cdot A_2 \cdot \ldots \cdot A_k$$
$$B = B_1 \cdot B_2 \cdot \ldots \cdot B_k,$$

where $X \cdot Y$ is the concatenation of bitstrings X and Y.

The values $|A_1|, |A_2|, ..., |A_k|$ and $|B_1|, |B_2|, ..., |B_k|$ can be used to obtain a tighter bound than $S_{\max}$. Let N_i be the length of A_i and B_i. The kD grid is an k-dimensional cube of size $(N_1 + 1) \times (N_2 + 1) \times \ldots \times (N_k + 1)$. Each grid point is a bucket and the fingerprint B is stored in the bucket at coordinates $(n_1, n_2, ..., n_k)$, where $n_i = |B_i|$. By comparing the partial coordinates $(n_1, n_2, ..., n_i)$ of a given bucket to $|A_1|, |A_2|, ..., |A_i|$, where $i \leq k$, it is possible to upper-bound the Tanimoto coefficient between A and every B in that bucket. By looking at the partial coordinates $(n_1, n_2, ..., n_{i-1})$, we can use this to quickly identify those partial coordinates $(n_1, n_2, ..., n_i)$ that may contain fingerprints B with a Tanimoto coefficient above $S_{\min}$.

Assume the algorithm is visiting a partial coordinate at level i in the data structure. The indices $n_1, n_2, ..., n_{i-1}$ are known, but we need to compute which n_i to visit at this level. The entries to be visited further down the data structure $n_{i+1}, ..., n_k$ are, of course, unknown at this point. A bound can be calculated in the following manner.

$$S_T(A, B) = \frac{|A \wedge B|}{|A \vee B|}$$

$$= \frac{\sum_{j=1}^{k} |A_j \wedge B_j|}{\sum_{j=1}^{k} |A_j \vee B_j|}$$

$$\leq \frac{\sum_{j=1}^{k} \min(|A_j|, n_j)}{\sum_{j=1}^{k} \max(|A_j|, n_j)}$$

$$= \frac{\sum_{j=1}^{i-1} \min(|A_j|, n_j) + \min(|A_i|, n_i) + \sum_{j=i+1}^{k} \min(|A_j|, n_j)}{\sum_{j=1}^{i-1} \max(|A_j|, n_j) + \max(|A_i|, n_i) + \sum_{j=i+1}^{k} \max(|A_j|, n_j)}$$

$$\leq \frac{\sum_{j=1}^{i-1} \min(|A_j|, n_j) + \min(|A_i|, n_i) + \sum_{j=i+1}^{k} |A_j|}{\sum_{j=1}^{i-1} \max(|A_j|, n_j) + \max(|A_i|, n_i) + \sum_{j=i+1}^{k} |A_j|}$$

$$= S_{\max}^{\text{grid}}$$

The n_is to visit lie in an interval and it is thus sufficient to compute the upper and lower indices n_u and n_l respectively. Setting $S_{\min} = S_{\max}^{\text{grid}}$, isolating n_i and ensuring that the result is an integer in the range $0...N_i$ gives:

$$n_l = \max\left(\left\lceil S_{\min}(A_i^{\max} + |A_i| + A_i^{|\cdot|}) - (A_i^{\min} + A_i^{|\cdot|})\right\rceil, 0\right)$$

and

$$n_u = \min\left(\left\lfloor \frac{A_i^{\min} + |A_i| + A_i^{|\cdot|} - S_{\min}(A_i^{\max} + A_i^{|\cdot|})}{S_{\min}} \right\rfloor, N_i\right)$$

where $A_i^{\min} = \sum_{j=1}^{i-1} \min(|A_j|, n_j)$ is a bound on the number of 1-bits in the *logical and* in the first part of the bitstrings. $A_i^{\max} = \sum_{j=1}^{i-1} \max(|A_j|, n_j)$ is a bound for the *logical or* in the first part of the bitstrings. Similarly, $A_i^{|\cdot|} = \sum_{j=i+1}^{k} |A_j|$ is a bound on the last part.

We have implemented the kD grid as a list of lists, where any list containing no fingerprints is omitted. See Fig. 2 for an example of a 4D grid containing four bitstrings. The fingerprints stored in a single bucket in the kD grid can be organised in a number of ways. The most naive approach is to store them in a simple list which has to be search linearly. We propose to store them in tree structures as explained below.

2.2 Singlebit Tree

The *Singlebit tree* is a binary tree which stores the fingerprints of a single bucket from a kD grid. At each node in the tree a position in the bitstring is chosen. All fingerprints with a zero at that position are stored in the left subtree while all those with a one are stored in the right subtree. This division is continued recursively until all the fingerprints in a given node are the same. When searching for a query bitstring A in the tree it now becomes possible, by comparing A to the path from the root of the tree to a given node, to compute an upper bound $S_{\max}^{\text{single}}$ on $S_T(A, B)$ for every fingerprint B in the subtree of that given node.

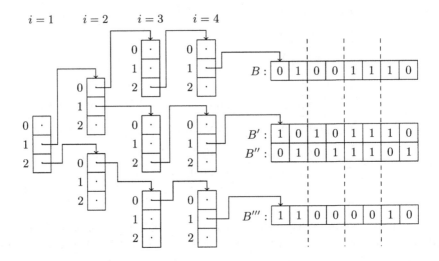

Fig. 2. Example of a 4D grid containing four bitstrings, stored as in our implementation. The dotted lines indicate the splits between B_i and B_{i+1}.

Given two bitstring A and B let M_{ij} be the number of positions where A has an i and B has a j. There are four possible combinations of i and j, namely M_{00}, M_{01}, M_{10} and M_{11}.

The path from the root of a tree to a node defines lower limits m_{ij} on M_{ij} for every fingerprint in the subtree of that node. Let u_{ij} denote the unknown difference between M_{ij} and m_{ij}, that is $u_{ij} = M_{ij} - m_{ij}$. Remember that, although B is unknown, $|B| = \sum_{i=1}^{k} n_k$ is known when processing a given bucket.

By using

$$u_{10} + u_{11} = |A| - m_{11} - m_{10}$$
$$u_{01} + u_{11} = |B| - m_{11} - m_{01}$$
$$u_{11} \leq \min(u_{11} + u_{01}, u_{11} + u_{10})$$
$$u_{11} + u_{10} + u_{01} \geq \max(u_{11} + u_{01}, u_{11} + u_{10})$$

an upper bound on the Tanimoto coefficient of any fingerprint B in the subtree can then be calculated as

$$
\begin{aligned}
S_T(A, B) &= \frac{M_{11}}{M_{11} + M_{01} + M_{10}} \\
&= \frac{m_{11} + u_{11}}{m_{11} + u_{11} + m_{01} + u_{01} + m_{10} + u_{10}} \\
&\leq \frac{m_{11} + \min(u_{11} + u_{01}, u_{11} + u_{10})}{m_{11} + m_{01} + m_{10} + \max(u_{11} + u_{01}, u_{11} + u_{10})} \\
&= S_{\max}^{\text{single}}.
\end{aligned}
$$

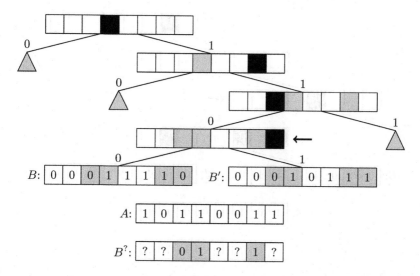

Fig. 3. Example of a Singlebit tree. The black squares mark the bits chosen for the given node, while the grey mark bits chosen at an ancestor. The grey triangles represent subtrees omitted to keep this example simple. Assume we are searching for the bitstring A in the example. When examining the node marked by the arrow we have the knowledge shown in $B^?$ about all children of that node. Comparing A against $B^?$ gives us $m_{00} = 0$, $m_{01} = 0$, $m_{10} = 1$ and $m_{11} = 2$. Thus $S_{max}^{single} = \frac{2}{3}$. Indeed we find that $S_T(A, B) = \frac{2}{7}$ and $S_T(A, B') = \frac{1}{2}$.

When building the tree data structure it is not immediately obvious how best to choose which bit positions to split the data on, at a given node. The implemented approach is to go through all the children of the node and choose the bit which best splits them into two parts of equal size, in the hope that this creates a well-balanced tree. However, it should be noted that the tree structure that gives the fastest searches is not necessarily a well-balanced tree. Figure 3 shows an example of a Singlebit tree.

The Singlebit tree can also be used to store all the fingerprints in the database without a kD grid. In this case, however, $|B|$ is no longer available and thus the S_{max}^{single} bound cannot be used. A less tight bound can be formulated, but preliminary experiments (not included in this paper) indicate that this is a poor strategy.

2.3   Multibit Tree

The experiments illustrated in Fig. 5 in Sec. 3 unfortunately show that using the kD grid combined with Singlebit trees decrease performance compared to using the kD grid and simple lists. The fingerprints used in our experiments have a length of 1024 bits. In our experiments no Singlebit tree was observed to contain more the 40,000 fingerprints. This implies that the expected height of the Singlebit trees is no more than 15 (as we aim for balanced trees cf. above).

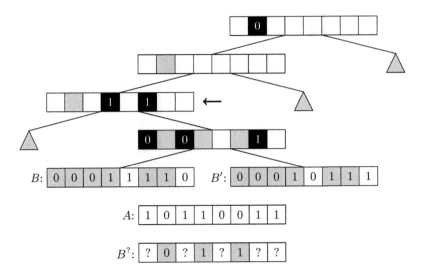

Fig. 4. An example of a Multibit tree. The black squares marks the match-bits and their annotation. Grey squares show bits that were match-bits at an ancestor. Grey triangles are subtrees omitted to keep this example simple. When visiting the node marked by the arrow we get $m_{00} = 1$, $m_{01} = 1$, $m_{10} = 0$ and $m_{11} = 1$, thus $S_{\max}^{\text{multi}} = \frac{3}{4}$. Still $S_T(A, B) = \frac{2}{7}$ and $S_T(A, B') = \frac{1}{2}$.

Consequently, the algorithm will only obtain information about 15 out of 1024 bits before reaching the fingerprints. A strategy for obtaining more information is to store a list of bit positions, along with an annotation of whether each bit is zero or one, in each node. The bits in this list are called the *match-bits*.

The *Multibit tree* is an extension of the Singlebit tree, where we no longer demand that all children of a given node are split according to the value of a single bit. In fact we only demand that the data is arranged in *some* binary tree. The match-bits of a given node are computed as all bits that are not a match-bit in any ancestor and for which all children of the node have the same value. Note that a node could easily have no match-bits. When searching through the Multibit tree, the query bitstring A is compared to the match-bits of each visited node and m_{00}, m_{01}, m_{10} and m_{11} are updated accordingly. $S_{\max}^{\text{multi}}$ is compute the same way as $S_{\max}^{\text{single}}$ and again only branches for which $S_{\max}^{\text{multi}} \geq S_{\min}$ are visited.

Again, the best way to build the tree is not obvious. Currently, the same method as Singlebit trees, where the data is split on the bit which best splits it in half, is used. Figure 4 shows an example of a Multibit tree. To reduce the memory consumption of the inner nodes, the splitting is stopped and leaves created, for any node that has less than some limit l children. Based on empirical data (not included in this paper) l is chosen as 6, which reduces memory consumption by more than a factor of two and has no significant impact on speed. An obvious alternative way to build the tree would be to base it on some hierarchical clustering method, such as Neighbour joining.

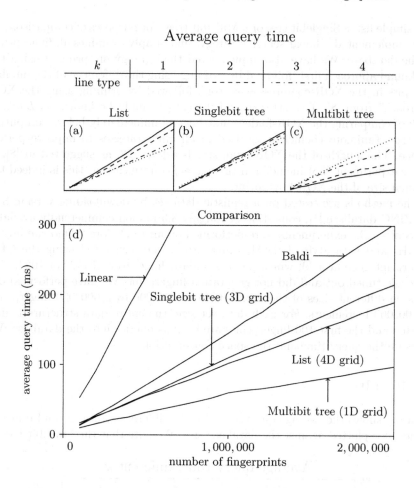

Fig. 5. Different strategies tested with $k = 1, \ldots, 4$. Each experiment is performed 100 times, and the average query time is presented. All experiments are performed with a $S_{\min}$ of 0.9. The three graphs (a) – (c) show the performance of the three bucket types for the different values of k. The best k for each method is presented in graph (d) along with the simple linear search results and Baldi.

3   Experiments

We have implemented the kD grid and the Single- and Multibit tree in Java. The implementation along with all test data is available at

`http://cs.au.dk/~tgk/TanimotoQuery/`

Using these implementations, we have constructed several search methods corresponding to the different combinations of the data structures. We have examined the kD grid for $k = 1, 2, 3$ and 4, where the fingerprints in the buckets are stored in a simple list, a Singlebit tree or a Multibit tree. For purposes of comparison, we have implemented a linear search strategy, that simply examines all fingerprints in the database. We have also implemented the strategy of "pruning using the bit-bound approach first, followed by pruning using the difference of the number of 1-bits in the XOR-compressed vectors, followed by pruning using the XOR approach" from [8]. This strategy will hereafter simply be known as *Baldi*. A trick of comparing the XOR-folded bitstrings [8] immediately before computing the true Tanimoto coefficient, is used in all our strategies to improve performance. The length of the XOR summary is set to 128, as suggested in [8]. A minor experiment (not included in this paper) confirmed that this is indeed the optimal size of the XOR fingerprint.

The methods are tested on a realistic data set by downloading version 8 of the ZINC database [1], consisting of roughly 8.5 million commercially available molecules. The experiments were performed on an Intel Core 2 Duo running at 2.5GHz and with 2GB of RAM. Fingerprints were generated using the CDK fingerprint generator [9] which has a standard fingerprint size N of 1024. One molecule timed out and did not generate a fingerprint. We have performed our tests on different sizes of the data set, from $100,000$ to $2,000,000$ fingerprints in $100,000$ increments. For each data set size, the entire data structure is first created and the first 100 fingerprints are used as queries into the database. We measure the query time and the space consumption.

4   Results

Figure 5 shows the average query time for the different strategies and different values of k plotted against the database size. We note that the Multibit tree in

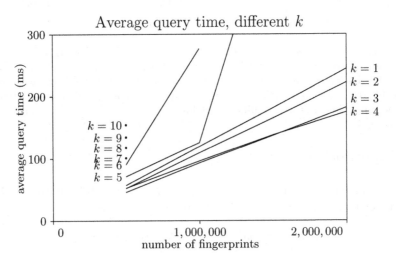

Fig. 6. Experiments with simple lists for $k = 1, \ldots, 10$ Each test is performed 100 times, and the average query time is presented. All experiments are performed with a $S_{\min}$ of 0.9. Missing data points are from runs with insufficient memory.

a 1D grid is best for all sizes. Surprisingly the simple list, for an appropriately high value of k, is faster than the Singlebit tree, yet slower than the Multibit tree. This is probably due to the fact that the Singlebit trees are too small to contain sufficient information for an efficient pruning: the entire tree is traversed, which is slower than traversing the corresponding list implementation. All three approaches (List, Singlebit- and Multibit trees) are clearly superior to the Baldi approach, which in turn is better than a simple linear search (with the XOR folding trick).

From Fig. 5a we notice that the List strategy seems to become faster for increasing k. This trend is further investigated in Fig. 6, which indicate that a k of three or four seems optimal. As k grows the grid becomes larger and more time consuming to traverse while the lists in the buckets become shorter. For

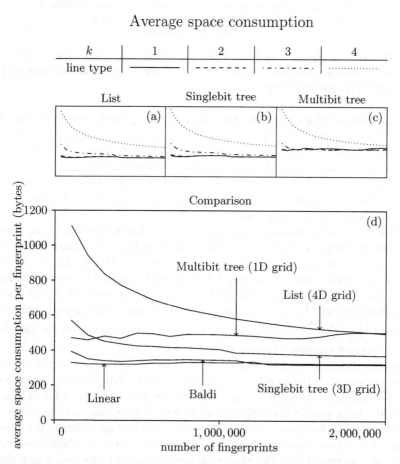

Fig. 7. The memory consumption of the data structure for different strategies tested with $k = 1, \dots, 4$. The three graphs (a) – (c) show the performance of the three bucket types for the different values of k. The k yielding the fastest query time for each method is presented in graph (d) along with the simple linear search results and Baldi.

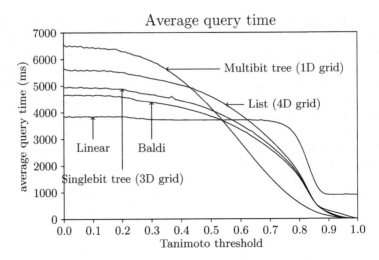

Fig. 8. The best strategies from Figure 5 tested for different values of S_{min}. All experiments are performed 100 times, with $2,000,000$ fingerprints in the database, and the average query time is presented.

sufficiently large values of k, the time spent pruning buckets exceeds the time visiting buckets containing superfluous fingerprints. The Singlebit tree data in Fig. 5b indicates that the optimal value of k is three. It seems the trees become too small to contain enough information for an efficient pruning, when k reaches four. In Fig. 5c we see the Multibit tree. With arguments similar to those for the Singlebit tree, a too large k will actually slow down the data structure. It seems a k as low as one is optimal.

Figure 7 shows the memory usage per fingerprint as a function of the number of loaded fingerprints. The first thing we note is that the Multibit tree uses significantly more memory than the other strategies. This is due to the need to store a variable number of match-bits in each node. The second thing to note is the space usage for different k's. In the worst case, where all buckets contain fingerprints, the memory consumption per fingerprint, for the grid alone, becomes $\mathcal{O}\left(\frac{1}{n}\left(\frac{N}{k}\right)^{k}\right)$, where n is the number of fingerprints in the database. Thus we are not surprised by our actual results.

Figure 8 shows the search time as a function of the Tanimoto threshold. In general we note that the simpler and more naive data structures performs better for a low Tanimoto threshold. This is due to the fact that, for a low Tanimoto threshold a large part of the entire database will be returned. In these cases very little pruning can be done, and it is faster to run through a simple list than to traverse a tree and compare bits at each node. Of course we should remember that we are interested in performing searches for similar molecules, which means large Tanimoto thresholds.

The reason why linear search is not constant time for a constant data set is that, while it will always visit all fingerprints, the time for visiting a given fingerprint is not constant due to the XOR folding trick.

5   Conclusion

In this paper we have presented a method for finding all fingerprints in a database with Tanimoto coefficient to a query fingerprint above a user defined threshold. Our method is based on a generalisation of the bounds developed in [6] to multiple dimensions. Experiments indicate that this generalisation results in a performance increase. Furthermore, we have examined the possibility of utilising trees as secondary data structures in the buckets. Again, our experiments clearly demonstrate that this leads to a significant performance increase.

References

1. Irwin, J.J., Shoichet, B.K.: Zinc: A free database of commercially available compounds for virtual screening. Journal of Chemical Information and Modeling 45(1), 177–182 (2005)
2. Gillet, V.J., Willett, P., Bradshaw, J.: Similarity searching using reduced graphs. Journal of Chemical Information and Computer Sciences 43(2), 338–345 (2003)
3. Leach, A.R., Gillet, V.J.: An Introduction to Chemoinformatics. Rev. edn. Kluwer Academic Publishers, Dordrecht (2007)
4. Willett, P.: Similarity-based approaches to virtual screening. Biochem. Soc. Trans. 31(Pt 3), 603–606 (2003)
5. Willett, P., Barnard, J.M., Downs, G.M.: Chemical similarity searching. Journal of Chemical Information and Computer Sciences 38(6), 983–996 (1998)
6. Swamidass, S.J., Baldi, P.: Bounds and algorithms for fast exact searches of chemical fingerprints in linear and sublinear time. Journal of Chemical Information and Modeling 47(2), 302–317 (2007)
7. Smellie, A.: Compressed binary bit trees: A new data structure for accelerating database searching. Journal of Chemical Information and Modeling 49(2), 257–262 (2009)
8. Baldi, P., Hirschberg, D.S., Nasr, R.J.: Speeding up chemical database searches using a proximity filter based on the logical exclusive or. Journal of Chemical Information and Modeling 48(7), 1367–1378 (2008)
9. Steinbeck, C., Han, Y., Kuhn, S., Horlacher, O., Luttmann, E., Willighagen, E.: The chemistry development kit (cdk): An open-source java library for chemo- and bioinformatics. Journal of Chemical Information and Computer Sciences 43(2), 493–500 (2003)

Generalizing the Four Gamete Condition and Splits Equivalence Theorem: Perfect Phylogeny on Three State Characters

Fumei Lam[1], Dan Gusfield[1], and Srinath Sridhar[2]

[1] Department of Computer Science, University of California, Davis
[2] Department of Computer Science, Carnegie Mellon University

Abstract. We study the three state perfect phylogeny problem and show that there is a three state perfect phylogeny for a set of input sequences if and only if there is a perfect phylogeny for every subset of three characters. In establishing these results, we prove fundamental structural features of the perfect phylogeny problem on three state characters and completely characterize the obstruction sets that must occur in input sequences that do not have a perfect phylogeny. We also give a proof for a stated lower bound involved in the conjectured generalization of our main result to any number of states.

1   Introduction

One of the fundamental problems in biology is the construction of phylogenies, or evolutionary trees, to describe ancestral relationships between a set of observed taxa. Each taxon is represented by a sequence and the evolutionary tree provides an explanation of branching patterns of mutation events transforming one sequence into another. There have been many elegant theoretical and algorithmic results on the problem of reconstructing a plausible history of mutations that generate a given set of observed sequences and to determine the minimum number of such events needed to explain the sequences.

A widely used model in phylogeny construction and population genetics is the *infinite sites* model, in which the mutation of any character can occur at most once in the phylogeny. This implies that the data must be binary (a character can take on at most two states), and that without recombination the phylogeny must be a tree, called a (binary) Perfect Phylogeny. The problem of determining if a set of binary sequences fits the infinite sites model without recombination, corresponds to determining if the data can be derived on a binary Perfect Phylogeny. A generalization of the infinite sites model is the *infinite alleles* model, where any character can mutate many times but each mutation of the character must lead to a different allele (state). Again, without recombination, the phylogeny is tree, called a *multi-state* Perfect Phylogeny. Correspondingly, the problem of determining if multi-state data fits the infinite-alleles model without recombination corresponds to determining if the data can be derived on a multi-state perfect phylogeny.

S.L. Salzberg and T. Warnow (Eds.): WABI 2009, LNBI 5724, pp. 206–219, 2009.

In the case of binary sequences, the well-known Splits Equivalence Theorem (also known as the *four gamete condition*) gives a necessary and sucient condition for the existence of a (binary) perfect phylogeny.

Theorem 1 (Splits Equivalence Theorem, Four Gamete Condition [14,18,33]). *A perfect phylogeny exists for binary input sequences if and only if no pair of characters contains all four possible binary pairs 00, 01, 10, 11.*

It follows from this theorem that for binary input, it is possible to either construct a perfect phylogeny, or output a pair of characters containing all four gametes as an obstruction set witnessing the nonexistence of a perfect phylogeny. This test is the building block for many theoretical results and practical algorithms. Among the many applications of this theorem, Gusfield et al. [21,22] and Huson et al. [28] apply the theorem to achieve decomposition theorems for phylogenies, Gusfield, Hickerson, and Eddhu [24] Bafna and Bansal [2,3], and Hudson and Kaplan [27] use it to obtain lower bounds for recombination events, Gusfield et al. [20,23] use it to obtain algorithms for constructing networks with constrained recombination, Sridhar et al. [6,38,39] and Satya et al. [35] use it to achieve a faster near-perfect phylogeny reconstruction algorithm, Gusfield [19] uses it to infer phase inference (with subsequent papers by Gusfield et al. [4,5,11,21], Eskin, Halperin, and Karp [13,26], Satya and Mukherjee [34] and Bonizzoni [9]), and Sridhar [37] et al. use it to obtain phylogenies from genotypes.

This work focuses on extending results for the binary perfect phylogeny problem to the multiple state character case, addressing the following natural questions arising from the Splits Equivalence Theorem. Given a set of sequences on r states ($r \geq 3$), is there a necessary and sufficient condition for the existence of a perfect phylogeny analogous to the Splits Equivalence Theorem? If no perfect phylogeny exists, what is the size of the smallest witnessing obstruction set?

In 1975, Fitch gave an example of input S over three states such that every *pair* of characters in S allows a perfect phylogeny while the entire set of characters S does not [15,16,17,36]. In 1983, Meacham generalized these results to characters over r states ($r \geq 3$)[33], constructing a class of sequences called *Fitch-Meacham examples*, which we examine in detail in Section 6. Meacham writes:

"The Fitch examples show that any algorithm to determine whether a set of characters is compatible must consider the set as a whole and cannot take the shortcut of only checking pairs of characters." [33]

However, while the Fitch-Meacham construction does show that checking pairs of characters is not sufficient for the existence of a perfect phylogeny, our main result will show that for three state input, there is a sufficient condition which does *not* need to consider the entire set of characters simultaneously. In particular, we give a complete answer to the questions posed above for the three state case, by

1. showing the existence of a necessary and sufficient condition analogous to the Splits Equivalence Theorem (Sections 3, 4),

2. in the case no perfect phylogeny exists, proving the existence of a small obstruction set as a witness (Section 4),
3. giving a complete characterization of all minimal obstruction sets (Section 5), and
4. giving the proof for a stated lower bound involved in the conjectured generalization of our main result to any number of states (Section 6).

In establishing these results, we prove fundamental structural features of the perfect phylogeny problem on three state characters.

2   Perfect Phylogenies and Partition Intersection Graphs

The input to our problem is a set of n sequences (representing taxa), where each sequence is a string of length m over r states. Throughout this paper, the states under consideration will be the set $\{0, 1, 2, \ldots r - 1\}$ (in particular, in the case $r = 2$, the input are sequences over $\{0, 1\}$). The input can be considered as a matrix of size $n \times m$, where each row corresponds to a sequence and each column corresponds to a character (or site). We denote characters by $\mathcal{C} = \{\chi^1, \chi^2, \chi^3, \ldots \chi^m\}$ and the states of character χ^i by χ^i_j for $0 \leq j \leq r - 1$. A *species* is a sequence $s_1, s_2, \ldots s_m \in \chi^1_{j_1} \times \chi^2_{j_2} \times \cdots \chi^m_{j_m}$, where s_i is the *state* of character χ^i for s.

The *perfect phylogeny problem* is to determine whether an input set S can be displayed on a tree such that

1. each sequence in input set S labels exactly one leaf in T
2. each vertex of T is labeled by a species
3. for every character χ^i and for every state χ^i_j of character χ^i, the set of all vertices in T such that the state of character χ^i is χ^i_j forms a connected subtree of T.

The general perfect phylogeny problem (with no constraints on r, n, and m) is NP-complete [7,40]. However, the perfect phylogeny problem becomes polynomially solvable (in n and m) when r is fixed. For $r = 2$, this follows from the Splits Equivalence Theorem 1. For larger values of r, this was shown by Dress and Steel for $r = 3$ [12], by Kannan and Warnow for $r = 3$ or 4 [31], and by Agarwala and Fernández-Baca for all fixed r [1] (with an improved algorithm by Kannan and Warnow [32]).

Definition 1 ([10,36]). *For a set of input sequences S, the* partition intersection graph $G(S)$ *is obtained by associating a vertex for each character state and an edge between two vertices χ^i_j and χ^k_l if there exists a sequence s with state j in character $\chi^i \in \mathcal{C}$ and state l in character $\chi^k \in \mathcal{C}$. We say s is a row that* witnesses *edge (χ^i_j, χ^k_l). For a subset of characters $\Phi = \{\chi^{i_1}, \chi^{i_2}, \ldots \chi^{i_k}\}$, let $G(\Phi)$ denote the partition intersection graph $G(S)$ restricted to the characters in Φ.*

Note that by definition, there are no edges in the partition intersection graph between states of the same character.

Definition 2. *A graph H is* chordal, *or* triangulated, *if there are no induced chordless cycles of length four or greater in H.*

Consider coloring the vertices of the partition intersection graph $G(S)$ in the following way. For each character χ^i, assign a single color to the vertices $\chi_0^i, \chi_1^i, \ldots \chi_{r-1}^i$. A *proper triangulation* of the partition intersection graph $G(S)$ is a chordal supergraph of $G(S)$ such that every edge has endpoints with different colors. In [10], Buneman established the following fundamental connection between the perfect phylogeny problem and triangulations of the partition intersection graph.

Theorem 2. *[10,36] A set of taxa S admits a perfect phylogeny if and only if the corresponding partition intersection graph $G(S)$ has a proper triangulation.*

We will use Theorem 2 to extend the Splits Equivalence Theorem to a test for the existence of a perfect phylogeny on trinary state characters. In a different direction, Theorem 2 and triangulation were also recently used to obtain an algorithm to handle perfect phylogeny problems with missing data [25].

To outline our approach, suppose a perfect phylogeny exists for S and consider every subset of three characters. Then each of these $\binom{m}{3}$ subsets also has a perfect phylogeny. We show that this necessary condition is also sufficient and moreover, we can systematically piece together the proper triangulations for each triple of characters to obtain a triangulation for the entire set of characters. On the other hand, if no perfect phylogeny exists, then we show there exists a witness set of three characters for which no perfect phylogeny exists. This extends the Splits Equivalence Theorem to show that for binary and trinary state input, the number of characters needed for a witness obstruction set is equal to the number of character states. The following is the main theorem of the paper.

Theorem 3. *Given an input set S on m characters with at most three states per character ($r \leq 3$), S admits a perfect phylogeny if and only if every subset of three characters of S admits a perfect phylogeny.*

This theorem demonstrates that to verify that a trinary state input matrix S has a perfect phylogeny, it suffices to verify that partition intersection graphs $G[\chi^i, \chi^j, \chi^k]$ have proper triangulations for all triples $\chi^i, \chi^j, \chi^k \in \mathcal{C}$. In Section 6, we will show that the Fitch-Meacham examples [16,33] demonstrate that the size of the witness set in Theorem 3 is best possible.

3 Structure of Partition Intersection Graphs for Three Characters

We begin by studying the structure of partition intersection graphs on three characters with at most three states per character ($m \leq 3$, $r \leq 3$). For convenience, we will denote the three characters by the letters a, b, c (interchangeably referring to them as characters and colors) and denote the states of these characters by a_i, b_i, c_i ($i \in \{0, 1, 2\}$).

210   F. Lam, D. Gusfield, and S. Sridhar

The problem of finding proper triangulations for graphs on at most three colors and arbitrary number of states ($m = 3$, r arbitrary) has been studied in a series of papers [8,29,30]. However, it will be unnecessary in our problem to employ these triangulation algorithms, as our instances will be restricted to those arising from character data on at most three states ($m = 3, r \leq 3$). In such instances, we will show that if a proper triangulation exists, then the structure of the triangulation is very simple. The following lemmas characterize the possible cycles contained in the partition intersection graph. Due to space considerations, we omit the proofs of most of the lemmas in this extended abstract; the full version with all proofs can be found on the arXiv (http://arxiv.org/abs/0905.1417).

Lemma 1. *Let S be a set of input species on three characters a, b, and c with at most three states per character. Suppose every pair of characters induces a properly triangulatable character partition intersection graph (i.e., $G[a, b]$, $G[b, c]$ and $G[a, c]$ are properly triangulatable) and let C be a chordless cycle in $G[a, b, c]$. Then C cannot contain all three states of any character.*

Lemma 2. *Let S be a set of input species on three characters a, b, and c with at most three states per character. If the partition intersection graph $G[a, b, c]$ is properly triangulatable, then for every chordless cycle C in $G[a, b, c]$, there exists a color (a, b, or c) that appears exactly once in C.*

Lemmas 1 and 2 show that if C is a chordless cycle in a properly triangulatable graph $G[a, b, c]$, then no color can appear in all three states and one color appears uniquely. This leaves two possibilities for chordless cycles in $G[a, b, c]$ (see Figure 1):

- a chordless four cycle, with two colors appearing uniquely and the remaining color appearing twice
- a chordless five cycle, with one color appearing uniquely and the other two colors each appearing twice

In the next lemma, we show that if $G[a, b, c]$ is properly triangulatable, the second case cannot occur, i.e., $G[a, b, c]$ cannot contain a chordless five cycle.

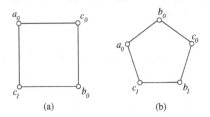

(a)   (b)

Fig. 1. The only possible chordless cycles in $G[a, b, c]$: (a) characters a and b appear uniquely while character c appears twice; (b) character a appears uniquely while characters b and c each appear twice

Lemma 3. *Let S be a set of input species on three characters a, b, and c with at most three states per character. If the partition intersection graph $G[a, b, c]$ is properly triangulatable, then $G[a, b, c]$ cannot contain chordless cycles of length five or greater.*

Lemma 4. *Let S be a set of input species on three characters a, b, and c with at most three states per character. If the partition intersection graph $G[a, b, c]$ is properly triangulatable, then every chordless cycle in $G[a, b, c]$ is uniquely triangulatable.*

Proof. By Lemma 3, if C is a chordless cycle in $G[a, b, c]$, then C must be a four cycle with the color pattern shown in Figure 2 (up to relabeling of the colors). Then C is uniquely triangulatable by adding the edge between the two colors appearing uniquely (in Figure 2, these are colors a and b).

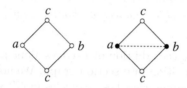

Fig. 2. Color pattern for chordless cycle C

□

For any three colors a, b, c, Lemma 4 gives a simple algorithm to properly triangulate $G[a, b, c]$: for each chordless cycle C in $G[a, b, c]$, check that C is a four cycle with two nonadjacent vertices having colors that appear exactly once in C and add an edge between these two vertices.

4   The 3-SNP Test

We now consider the case of trinary input sequences S on m characters (for m greater or equal to 4). Our goal is to prove that the existence of proper triangulations for all subsets of three characters at a time is a sufficient condition to guarantee existence of a proper triangulation for *all* m characters.

By Lemma 2, if a set of three characters χ^i, χ^j, χ^k is properly triangulatable, then there is a unique set of edges $F(\chi^i, \chi^j, \chi^k)$ that must be added to triangulate the chordless cycles in $G[\chi^i, \chi^j, \chi^k]$. Construct a new graph $G'(S)$ on the same vertices as $G(S)$ with edge set $E(G(S)) \cup \{\cup_{1 \leq i < j < k \leq m} F(\chi^i, \chi^j, \chi^k)\}$. $G'(S)$ is the partition intersection graph $G(S)$ together with all of the additional edges used to properly triangulate chordless cycles in $G[\chi^i, \chi^j, \chi^k]$ ($1 \leq i < j < k \leq m$). In $G'(S)$, edges from the partition intersection graph $G(S)$ are called E-edges and edges that have been added as triangulation edges for some triple of columns are called F-edges. We call a cycle consisting only of E-edges an E-cycle.

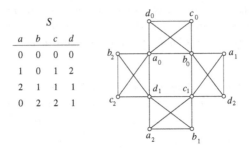

S			
a	b	c	d
0	0	0	0
1	0	1	2
2	1	1	1
0	2	2	1

Fig. 3. Example 1. Partition intersection graph $G'(S)$ contains a chordless four cycle.

Example 1. Input set S and partition intersection graph $G(S)$ are shown in Figure 3. Each triple of characters in S induces a chordal graph while the entire partition intersection graph $G(S)$ contains a chordless cycle of length four. Since each triple of characters induces a chordal graph, no F-edges are added and $G(S) = G'(S)$.

As Example 1 illustrates, the addition of F-edges alone may not be sufficient to triangulate the entire partition intersection graph. We now turn to the problem of triangulating the remaining chordless E-cycles in $G'(S)$.

Consider any E-cycle C that is chordless in $G'(S)$ satisfying the properties

1. C has length equal to four
2. all colors of C are distinct

For every such chordless cycle, add the chords between the two pairs of non-adjacent vertices in C (note that these are legal edges). Call this set of edges F'-edges and let $G''(S)$ denote the graph $G'(S)$ with the addition of F'-edges. Note that the sets of E-edges, F-edges, and F'-edges are pairwise disjoint; we call the set of F and F'-edges *non-E edges*.

We begin by investigating structural properties of cycles in $G'(S)$ and $G''(S)$ containing at least one F-edge or F'-edge. These structural properties will imply a sequence of lemmas eliminating the possibilities for chordless cycles in graph $G'(S)$.

Lemma 5. $G'(S)$ *cannot contain a chordless cycle with exactly one F-edge.*

Lemma 6. $G''(S)$ *cannot contain a chordless cycle with exactly one non-E edge.*

Lemma 7. $G''(S)$ *cannot contain a chordless cycle with two or more non-E edges.*

Lemmas 5, 6, and 7 eliminate the possibility of chordless cycles in $G''(S)$ containing non-E edges. To show that $G''(S)$ is properly triangulated, we proceed to show that $G''(S)$ does not contain chordless E-cycles.

Lemma 8. *If C is an E-cycle that is chordless in $G'(S)$, then C has length exactly four with four distinct colors.*

Lemma 8 implies all chordless E-cycles in $G'(S)$ have length four containing four distinct colors. We have triangulated all such cycles by F'-edges in $G''(S)$, implying the following corollary.

Corollary 1. $G''(S)$ *cannot contain a chordless E-cycle.*

Lemmas 5, 6, 7, and Corollary 1 together imply that $G''(S)$ is properly triangulated, proving the main theorem.

Theorem 3 *Given an input set S on m characters with at most three states per character ($r = 3$), S admits a perfect phylogeny if and only if every subset of three characters of S admits a perfect phylogeny.*

5 Enumerating Obstruction Sets for Three State Characters

We now turn to the problem of enumerating all minimal obstruction sets to perfect phylogenies on three-state character input. By Theorem 3, it follows that the minimal obstruction sets are input sequences on at most three characters; we enumerate all instances S on three characters a, b, and c satisfying the following conditions:

(i) each character a, b and c has at most three states
(ii) every pair of characters allows a perfect phylogeny
(iii) the three characters a, b, and c together do not allow a perfect phylogeny.

Note that Condition (ii) implies the partition intersection graph $G(S)$ does not contain a cycle on exactly two colors and Condition (iii) implies $G(S)$ contains at least one chordless cycle. Let C be the largest chordless cycle in $G(S)$. Condition (ii) and Lemma 1 together imply C cannot contain all three states of any character. Therefore, C has length at most six. If $G(S)$ contains a chordless six-cycle C, then each color appears exactly twice in C and C must have one of the color patterns shown in Figure 4.

In Figures 4(a) and 4(b), there is one state in each character that does not appear in C (states a_2, b_2, and c_2). Since C is chordless, the witness for each

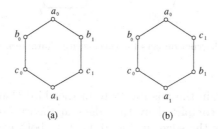

(a)   (b)

Fig. 4. Color patterns for chordless cycle of length six

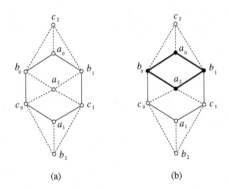

Fig. 5. Forced patterns for row witnesses of Figure 4(a)

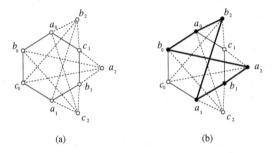

Fig. 6. Forced patterns for row witnesses of Figure 4(b)

a	b	c
0	0	2
0	2	1
1	0	0
2	1	0
2	1	1

a	b	c
0	0	2
0	2	1
1	0	0
2	1	0
1	1	1

a	b	c
0	0	2
0	2	1
1	0	0
1	1	0
1	1	1

a	b	c
0	1	0
1	0	0
2	0	1
0	2	1
1	2	2

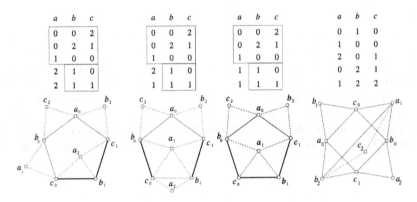

Fig. 7. Minimal obstruction sets for three-state characters up to relabeling

edge is forced to contain the missing state in the third character. This implies Figure 4(a) must be completed by the edges in Figure 5(a) and Figure 4(b) must be completed by the edges in Figure 6(a). In both cases, there is a cycle on two characters a and b (see Figures 5(b) and 6(b)). This implies the pair of characters a and b is not properly triangulatable, a contradiction to condition (ii). Therefore, $G(S)$ cannot contain chordless cycles of length six.

If C has length four or five, a case analysis on the row witnesses for the edges in the cycle results in the exhaustive list of minimal obstruction sets shown in Figure 7 (up to relabeling of the character states). We emphasize the similarity between the first three examples by highlighting the input entries with the same values in all three examples.

6   Construction of Fitch-Meacham Examples

In this section, we examine in detail the class of Fitch-Meacham examples, which were first introduced by Fitch [16,17] and later generalized by Meacham [33]. The goal of these examples is to demonstrate a lower bound on the number of characters that must be simultaneously examined in any test for perfect phylogeny. The natural conjecture generalizing our main result is that for any r, there is a perfect phylogeny on r-state characters if and only if there is one for every subset of r characters. We show here that such a result would be the best possible, for any r. While the general construction of these examples and the resulting lower bounds were stated by Meacham [33], to the best of our knowledge, the proof of correctness for these lower bounds has not been established. We fill this gap by explicitly describing the complete construction for the entire class of Fitch-Meacham examples and providing a proof for the lower bound claimed in [33].

For each integer r ($r \geq 2$), the Fitch-Meacham construction F_r is a set of $r+2$ sequences over r characters, where each character takes r states. We describe the construction of the partition intersection graph $G(F_r)$; the set of sequences F_r can be obtained from $G(F_r)$ in a straightforward manner, with each taxon corresponding to an r-clique in $G(F_r)$.

Label the r characters in F_r by $0, 1, \ldots r - 1$; each vertex labeled by i will correspond to a state in character i. The construction starts with two cliques EC_1 and EC_2 of size r, called end-cliques, with the vertices of each clique labeled by $0, 1, \ldots r-1$. The vertex labeled i in EC_1 is adjacent to the vertex labeled $(i+1)$ mod r in EC_2. For each such edge $(i, (i+1) \mod r)$ between the two end-cliques, we create a clique of size $r - 2$ with vertices labeled by $\{0, 1, \ldots r-1\} \setminus \{i, (i+1) \mod r\}$. Every vertex in this $(r - 2)$-clique is then attached to both i (in end-clique 1) and $(i+1)$ mod r (in end-clique 2), creating an r-clique whose vertices are labeled with integers $0, 1, \ldots r - 1$. There are a total of r such cliques, called tower-cliques, and denoted by TC_1, TC_2, $\ldots TC_r$. Note that for each i ($0 \leq i \leq r - 1$), there are exactly r vertices labeled by i; we give each such vertex a distinct state, resulting in r states for each character.

Note that the graph corresponding to the four gamete obstruction set is an instance of the Fitch-Meacham construction with $r = 2$. In this case, the four binary sequences $00, 01, 10, 11$ have two states, two colors and four taxa and the partition intersection graph for these sequences is precisely the graph $G(F_2)$. Note that in this case, every subset of $r - 1 = 1$ characters has a perfect phylogeny, while the entire set of characters does not. Similarly, the fourth graph shown in Figure 7 illustrating the obstruction set for 3-state input is the graph $G(F_3)$ corresponding to the Fitch-Meacham construction for $r = 3$ (in the figure, $EC_1 = \{a_0, b_2, c_1\}$ and $EC_2 = \{a_1, b_0, c_0\}$). As shown in Section 5, every $r-1 = 2$ set of characters in the corresponding input set allows a perfect phylogeny while the entire set of characters does not. The following theorem generalizes this property to the entire class of Fitch-Meacham examples. Because the theorem was stated without proof in [33], we provide a proof of the result here.

Theorem 4. *[33] For every $r \geq 2$, F_r is a set of input sequences over r state characters such that every $r - 1$ subset of characters allows a perfect phylogeny while the entire set F_r does not allow a perfect phylogeny.*

Proof. We first show that $G(F_r)$ does not allow a proper triangulation for any r. As observed above, $G(F_2)$ is a four cycle on two characters and therefore, does not allow a proper triangulation (since any proper triangulation for a graph containing cycles must have at least three colors). Suppose $G(F_r)$ is properly triangulatable for some $r \geq 3$, let s be the smallest integer such that $G(F_s)$ has a proper triangulation, and let $G'(F_s)$ be a minimal proper triangulation of $G(F_s)$.

For each tower-clique TC_i in $G(F_s)$, consider the set of vertices in TC_i that are not contained in either end-clique; call these vertices *internal tower-clique vertices* and the remaining two tower vertices *end tower-clique vertices*. Note that the removal of the two end tower-clique vertices disconnects the internal tower-clique vertices from the rest of the graph. This implies that the internal tower-clique vertices cannot be part of any chordless cycle: otherwise, such a chordless cycle C must contain *both* end tower-clique vertices i and $(i+1)$ mod s. However, the two end tower-clique vertices are connected by an edge and therefore induce a chord in C, a contradiction since C is a chordless cycle.

In the graph $G(F_s)$, consider the following cycle of length four: $s - 2$ (in EC_1) $\rightarrow s - 1$ (in EC_1) $\rightarrow 0$ (in EC_2) $\rightarrow s - 1$ (in EC_2) $\rightarrow s - 2$ (in EC_1). This four-cycle has a unique proper triangulation, which forces the edge e between vertex $s - 2$ in EC_1 and vertex 0 in EC_2 to be included in $G'(F_s)$. Consider removing all vertices labeled $s - 1$ from $G'(F_s)$, and for the two vertices labeled $s - 1$ in end-cliques EC_1 and EC_2, remove all interior tower-clique vertices (but not end tower-clique vertices) adjacent to $s - 1$. Then edge e between vertices $s - 2$ and 0 is still present and we can expand e into a tower-clique of size $s - 1$ (by forming a clique with new vertices $1, 2, \ldots s - 3$ adjacent to both $s - 2$ and 0 of the two end-cliques).

In the resulting graph, the vertices are exactly those of $G(F_{s-1})$ and all edges in $G(F_{s-1})$ are present. Furthermore, if there is a chordless cycle in this graph, then it would create a chordless cycle in $G'(F_s)$ since no internal tower-clique vertex can be part of any chordless cycle (and in particular, the new vertices $1, 2, \ldots s - 3$ cannot be part of any chordless cycle). Therefore, the resulting graph is a proper triangulation for $G(F_{s-1})$, a contradiction since s was chosen to be the smallest integer such that $G(F_s)$ allows a proper triangulation.

To prove the second part of the theorem, we show that in F_r, any subset of $r - 1$ characters does allow a perfect phylogeny by proving that the partition intersection graph on any subset of $r - 1$ characters has a proper triangulation. By the symmetry of the construction of F_r, we can assume without loss of generality that the $r - 1$ characters under consideration are $\{0, 1, \ldots r - 2\}$. Consider the graph obtained by connecting every vertex i $(0 \le i \le r - 3)$ in EC_1 to every vertex j satisfying $j > i$ in EC_2. Note the asymmetry between the first and second end-cliques in this construction and observe that none of the added edges are between characters with the same label.

Suppose the resulting graph contains a chordless cycle C. Then C cannot contain three or more vertices in either end-clique and cannot contain any internal tower-clique vertices (as noted earlier), so must have length exactly four with two vertices in each end-clique. It cannot be the case that two nonadjacent vertices of C are in the same end-clique, since these vertices would be adjacent and C would not be chordless. Therefore, cycle C must be formed as follows: i (in EC_1) $\rightarrow j$ (in EC_2) $\rightarrow j'$ (in EC_2) $\rightarrow i'$ (in EC_1). Since i and j are adjacent, we have $i < j$ and since i' and j' are adjacent, we have $i' < j'$. If $i < j'$, then i and j' are adjacent and the cycle C is not chordless, a contradiction. Therefore, $i' < j' \le i < j$, which implies i' and j are adjacent and the cycle C is not chordless, again a contradiction. It follows that there are no chordless cycles and the added edges form a proper triangulation for the partition intersection graph on the subset of $r - 1$ characters $\{0, 1, \ldots r - 2\}$. □

Acknowledgments. The authors gratefully acknowledge I. Coskun, G. Blelloch, R. Ravi, R. Schwartz, and T. Warnow for stimulating discussions and suggestions. This research was partially supported by NSF grants SEI-BIO 0513910, CCF-0515378, and IIS-0803564.

References

1. Agarwala, R., Fernandez-Baca, D.: A polynomial-time algorithm for the perfect phylogeny problem when the number of character states is fixed. SIAM Journal on Computing 23, 1216–1224 (1994)
2. Bafna, V., Bansal, V.: Improved recombination lower bounds for haplotype data. In: Miyano, S., Mesirov, J., Kasif, S., Istrail, S., Pevzner, P.A., Waterman, M. (eds.) RECOMB 2005. LNCS (LNBI), vol. 3500, pp. 569–584. Springer, Heidelberg (2005)
3. Bafna, V., Bansal, V.: The number of recombination events in a sample history: Conflict graph and lower bounds. IEEE/ACM Transactions on Computational Biology and Bioinformatics 1, 78–90 (2004)
4. Bafna, V., Gusfield, D., Hannenhalli, G., Yooseph, S.: A note on efficient computation of haplotypes via perfect phylogeny. Journal of Computational Biology 11, 858–866 (2004)
5. Bafna, V., Gusfield, D., Lancia, G., Yooseph, S.: Haplotyping as perfect phylogeny: A direct approach. Journal of Computational Biology 10, 323–340 (2003)
6. Blelloch, G.E., Dhamdhere, K., Halperin, E., Ravi, R., Schwartz, R., Sridhar, S.: Fixed parameter tractability of binary near-perfect phylogenetic tree reconstruction. In: Bugliesi, M., Preneel, B., Sassone, V., Wegener, I. (eds.) ICALP 2006. LNCS, vol. 4051, pp. 667–678. Springer, Heidelberg (2006)

7. Bodlaender, H., Fellows, M., Warnow, T.: Two strikes against perfect phylogeny. In: International Colloquium on Automata, Languages and Programming, pp. 273–283 (1992)
8. Bodlaender, H., Kloks, T.: A simple linear time algorithm for triangulating three-colored graphs. J. Algorithms 15(1), 160–172 (1993)
9. Bonizzoni, P.: A linear-time algorithm for the perfect phylogeny haplotype problem. Algorithmica 48, 267–285 (2007)
10. Buneman, P.: A characterization of rigid circuit graphs. Discrete Math. 9, 205–212 (1974)
11. Ding, Z., Filkov, V., Gusfield, D.: A linear-time algorithm for the perfect phylogeny haplotyping problem. J. of Computational Biology 13, 522–553 (2006)
12. Dress, A., Steel, M.: Convex tree realizations of partitions. Applied Math. Letters 5, 36 (1993)
13. Eskin, E., Halperin, E., Karp, R.M.: Efficient reconstruction of haplotype structure via perfect phylogeny. Journal of Bioinformatics and Computational Biology, 1–20 (2003)
14. Estabrook, G., Johnson, C., McMorris, F.: A mathematical formulation for the analysis of cladistic character compatibility. Math. Bioscience 29 (1976)
15. Felsenstein, J.: Inferring Phylogenies. Sinauer Associates, Sunderland (2004)
16. Fitch, W.M.: Toward finding the tree of maximum parsimony. In: Estabrook, G.F. (ed.) The Eighth International Conference on Numerical Taxonomy, pp. 189–220. W. H. Freeman and Company, San Francisco (1975)
17. Fitch, W.M.: On the problem of discovering the most parsimonious tree. American Naturalist 11, 223–257 (1977)
18. Gusfield, D.: Efficient algorithms for inferring evolutionary trees. Networks 21, 19–28 (1991)
19. Gusfield, D.: Haplotyping as a perfect phylogeny: Conceptual framework and efficient solutions. In: Research in Computational Molecular Biology (2002)
20. Gusfield, D.: Optimal, efficient reconstruction of Root-Unknown phylogenetic networks with constrained and structured recombination. JCSS 70, 381–398 (2005)
21. Gusfield, D., Bansal, V.: A fundamental decomposition theory for phylogenetic networks and incompatible characters. In: Miyano, S., Mesirov, J., Kasif, S., Istrail, S., Pevzner, P.A., Waterman, M. (eds.) RECOMB 2005. LNCS (LNBI), vol. 3500, pp. 217–232. Springer, Heidelberg (2005)
22. Gusfield, D., Bansal, V., Bafna, V., Song, Y.: A decomposition theory for phylogenetic networks and incompatible characters. Journal of Computational Biology 14(10), 1247–1272 (2007)
23. Gusfield, D., Eddhu, S., Langley, C.: Optimal, efficient reconstruction of phylogenetic networks with constrained recombination. J. Bioinformatics and Computational Biology 2(1), 173–213 (2004)
24. Gusfield, D., Hickerson, D., Eddhu, S.: An efficiently-computed lower bound on the number of recombinations in phylogenetic networks: Theory and empirical study. Discrete Applied Math. 155, 806–830 (2007); Special issue on Computational Biology
25. Gusfield, D.: The multi-state perfect phylogeny problem with missing and removable data. In: Research in Computational Molecular Biology, RECOMB (2009)
26. Halperin, E., Eskin, E.: Haplotype reconstruction from genotype data using imperfect phylogeny. Bioinformatics (2004)
27. Hudson, R., Kaplan, N.: Statistical properties of the number of recombination events in the history of a sample of DNA sequences. Genetics 111, 147–164 (1985)

28. Huson, D., Klopper, T., Lockhart, P.J., Steel, M.A.: Reconstruction of reticulate networks from gene trees. In: Miyano, S., Mesirov, J., Kasif, S., Istrail, S., Pevzner, P.A., Waterman, M. (eds.) RECOMB 2005. LNCS (LNBI), vol. 3500, pp. 233–249. Springer, Heidelberg (2005)
29. Idury, R.M., Schäffer, A.A.: Triangulating three-colored graphs in linear time and linear space. SIAM J. Discret. Math. 6(2), 289–293 (1993)
30. Kannan, S., Warnow, T.: Triangulating three-colored graphs. In: SODA 1991: Proc. ACM-SIAM Symposium on Discrete algorithms (SODA), pp. 337–343 (1991)
31. Kannan, S., Warnow, T.: Inferring evolutionary history from DNA sequences. SIAM J. on Computing 23, 713–737 (1994)
32. Kannan, S., Warnow, T.: A fast algorithm for the computation and enumeration of perfect phylogenies. SIAM Journal on Computing 26, 1749–1763 (1997)
33. Meacham, C.: Theoretical and computational considerations of the compatibility of qualitative taxonomic characters. Nato ASI series, vol. G1 on Numerical Taxonomy. Springer, Heidelberg (1983)
34. Satya, R.V., Mukherjee, A.: An optimal algorithm for perfect phylogeny haplotyping. Journal of Computational Biology 13, 897–928 (2006)
35. Satya, R.V., Mukherjee, A., Alexe, G., Parida, L., Bhanot, G.: Constructing near-perfect phylogenies with multiple homoplasy events. Bioinformatics 22, e514–i522 (2006); Bioinformatics Suppl., Proceedings of ISMB 2006 (2006)
36. Semple, C., Steel, M.: Phylogenetics. Oxford University Press, Oxford (2003)
37. Sridhar, S., Blelloch, G.E., Ravi, R., Schwartz, R.: Optimal imperfect phylogeny reconstruction and haplotyping. In: Proceedings of Computational Systems Bioinformatics, CSB (2006)
38. Sridhar, S., Dhamdhere, K., Blelloch, G.E., Halperin, E., Ravi, R., Schwartz, R.: Simple reconstruction of binary near-perfect phylogenetic trees. In: International Workshop on Bioinformatics Research and Applications (2006)
39. Sridhar, S., Dhamdhere, K., Blelloch, G.E., Halperin, E., Ravi, R., Schwartz, R.: Algorithms for efficient near-perfect phylogenetic tree reconstruction in theory and practice. ACM/IEEE Transactions on Computational Biology and Bioinformatics (2007)
40. Steel, M.A.: The complexity of reconstructing trees from qualitative characters and subtrees. Journal of Classification 9, 91–116 (1992)

Decoding Synteny Blocks and Large-Scale Duplications in Mammalian and Plant Genomes

Qian Peng[1,*], Max A. Alekseyev[3], Glenn Tesler[2], and Pavel A. Pevzner[1]

[1] Department of Computer Science and Engineering
qpeng@cs.ucsd.edu
[2] Department of Mathematics
University of California, San Diego, 9500 Gilman Drive, La Jolla, CA 92093
[3] Department of Computer Science and Engineering
University of South Carolina, 315 Main St., Columbia, SC 29208

Abstract. The existing synteny block reconstruction algorithms use *anchors* (e.g., orthologous genes) shared over *all* genomes to construct the synteny blocks for multiple genomes. This approach, while efficient for a few genomes, cannot be scaled to address the need to construct synteny blocks in many mammalian genomes that are currently being sequenced. The problem is that the number of anchors shared among *all* genomes quickly decreases with the increase in the number of genomes. Another problem is that many genomes (plant genomes in particular) had extensive duplications, which makes decoding of genomic architecture and rearrangement analysis in plants difficult. The existing synteny block generation algorithms in plants do not address the issue of generating *non-overlapping* synteny blocks suitable for analyzing rearrangements and evolution history of duplications. We present a new algorithm based on the *A-Bruijn* graph framework that overcomes these difficulties and provides a unified approach to synteny block reconstruction for multiple genomes, and for genomes with large duplications.

Supplementary material: http://grimm.ucsd.edu/ABS

1   Introduction

Plant genomes exhibit an unusually large proportion of duplicated regions [1]. The large number of duplications makes decoding of genomic architecture and rearrangement analysis in plants difficult. In particular, segmental duplications represent a major obstacle to reconstruction of *synteny blocks* (i.e., conserved regions across the genomes), resulting in relatively few published results on synteny blocks in plant genomes as compared to vertebrate genomes (and especially to mammalian genomes) where segmental duplications are less prevalent and can therefore be largely ignored while constructing synteny blocks.[1] It is estimated that segmental duplications account only for less than 10% of the human

---

* Corresponding author.
[1] Segmental duplications in the human genome are usually represented as a set of pairwise alignments and are masked out by synteny block generation algorithms.

S.L. Salzberg and T. Warnow (Eds.): WABI 2009, LNBI 5724, pp. 220–232, 2009.

genome [2] and 2.9% of the mouse genome [3]. By contrast, in plant genomes, duplications are prevalent (they account for more than 70% of the *Arabidopsis thaliana* genome [4]), and ignoring duplicated regions would render a synteny block analysis meaningless. This represents an intrinsic difficulty in constructing syntenies in plant genomes.

From an algorithmic perspective, the problems of finding synteny blocks between two genomes and duplicated blocks are very similar. In fact, finding synteny blocks of multiple genomes can be converted into the problem of finding duplicated (or multi-copy) blocks within a single genome by concatenating the multiple genomes in an arbitrary order. In the past, this problem of reconstructing synteny blocks in k mammalian genomes was addressed by constructing k-way anchors shared between all genomes [5]. However, this approach is limited to small k since with the growing number of genomes, the number of k-way anchors sharply decreases. The disappearing k-way anchors may lead to disappearing synteny blocks. Short synteny blocks (which are important in studies of chromosome evolution [5,6]) are particularly vulnerable to this effect. In this paper, we propose a unified approach to synteny block reconstruction for two or multiple genomes, synteny block reconstruction for genomes with large duplications, and duplicated block reconstruction within a genome.

A typical synteny block generation algorithm takes as an input a set of *alignment anchors* (i.e., local alignments or pairs of similar genes) between two genomes (or two copies of the same genome) and outputs a set of synteny blocks (or duplicated blocks) that cover (without overlaps) most of each participating genome. As a result, each genome is represented as a shuffled sequence of the constructed synteny blocks that enables further rearrangement analysis of the genomes (e.g., computing the rearrangement distance between them). For two genomes, most existing synteny blocks generation algorithms employ a 2-dimensional *genomic dot-plot* where two genomes (or two copies of the same genome) are placed along the axes on the plane and their alignment anchors are represented as dots (Fig. 1(a)). These algorithms further decompose the dot-plot into a collection of "long" diagonal-like segments constituting *2-D synteny blocks* (Fig. 1(b)). The conventional (1-D) synteny blocks for each genome can be obtained as projections of the 2-D synteny blocks onto a corresponding axis (Fig. 1(b)). The notions of 2-dimensional dot-plots and synteny blocks generalize to k-dimensions when there are k genomes. This simple description hides a number of computational details that make the problem of synteny block generation non-trivial [7,8].

Nadeau *et al.* [10] introduced the notion of *conserved segments*. Waterston *et al.* [11] and Pevzner *et al.* [7] described two approaches to synteny block generation, that produce similar results. There are many other studies describing different methods of "synteny block" generation [12,13,14,15,16,17,18]. While these approaches proved to be adequate for small sets of mammalian genomes,[2] and in some cases prokaryotic genomes, they do not particularly address issues that stem

---

[2] Since the number of duplications in mammalian genomes is small, the 2-D synteny blocks usually do not overlap in 1-D.

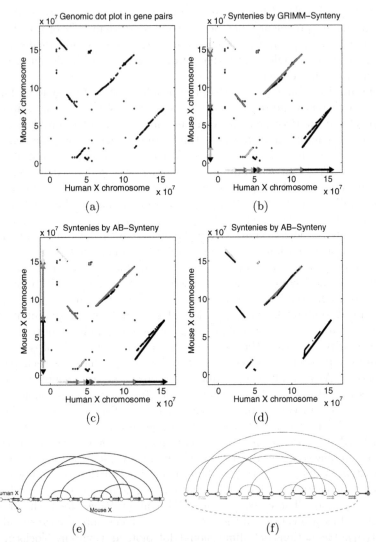

Fig. 1. (a) Genomic dot-plot between human (x axis) and mouse (y axis) X chromosomes for all 714 orthologous gene pairs in Ensembl. Each dot represents a pair of "similar" genes between the two species. (b,c) Synteny blocks in 2-D (diagonals) and 1-D (bar with arrow) produced by (b) GRIMM-Synteny [9]. (c) AB-Synteny. Each color represents a synteny (duplication on the concatenated genome). (d) Gene pairs removed. Black diagonals are the human-mouse part of the human-mouse-rat-dog-chicken 5-way syntenies. (e) A-Bruijn graph after simplifications. The bold red and blue edges correspond to the syntenies in (c). In the A-Bruijn graph, each pair of red/blue edges is a single edge of multiplicity 2 (illustrated as parallel edges). The black edge concatenates the human and mouse chromosomes. (f) Redraw of (e) with syntenies represented in different colors. An end node (shaded) is added. The A-Bruijn graph is equivalent to the breakpoint graph if edges representing syntenies (colored edges) and the edge transitioning from mouse to human (dashed arc) are removed.

from extensive duplications in plant genomes. In the presence of duplications, the 2-D synteny blocks may overlap in 1-D, i.e., along one of the genomes. There are a few previous efforts to generate synteny blocks for genomes with large duplications [4,19,20,21,22,23,24,25]. They do not directly address the issue of generating *non-overlapping synteny blocks in 1-D* either, which are more suitable for analyzing rearrangements and evolution history of duplications.

Pevzner *et al.* [26] introduced the *A-Bruijn graph* approach to repeat classification, and the approach was later found useful in other problems [27,28,29]. In this paper we demonstrate that the A-Bruijn graph framework can be also applied to the problem of synteny block generation for genomes with large duplications. Our algorithm produces non-overlapping synteny blocks in both 2-D and 1-D representations. By simply concatenating multiple genomes, we can generalize this approach to synteny block generation for multiple genomes. Previous efforts to generate synteny blocks for k genomes often required k-way alignment anchors, e.g., orthologous genes present in all k genomes [30]. While recent approaches [16,31] allow missing anchors, they post other constrains. Our approach is not subject to constraints such as all-vs-all alignment as it uses pairwise anchors as an input, nor does it require a reference genome.

We benchmarked our algorithm on five vertebrate genomes to reconstruct 5-way syntenies, and on the plant genome *A. thaliana* to find duplicated blocks. We compared the results to the published syntenies or duplicated blocks. While there is no gold standard to what constitutes "correct" synteny blocks, all synteny block generation algorithms are parameter-dependent and may produce different synteny blocks on the same input data. To evaluate the performance of synteny block generation algorithms, we simulated genomes with large duplications and known synteny blocks and analyzed how well our algorithm reconstructs the underlined synteny blocks (see supplemental material).

2   Approach

Fig. 2(a) shows a hypothetical sequence of genes, resulting from multiple segmental duplications. In reality, we are given only the resulting genomic sequence and know nothing about the structure of its segments (marked by the colors in Fig. 2(a)). It is natural to ask what evolutionary events (including rearrangements and duplications) created the given genomic sequence. Before answering this question, we need to understand the duplication structure of the given genome, i.e., to represent it as a sequence of non-overlapping blocks, each of which may appear one or more times.

The diagonals in Fig. 2(b) are what conventional synteny block construction methods would produce as synteny blocks from the genomic dot-plot of a genome against itself. Since these blocks overlap along the sequence, the duplication structure is unclear. Ideally, we would like to see diagonals that do not overlap along the sequence (Fig. 2(c)). One natural approach is for every pair of partially overlapped blocks along each axis to cut the overlapping region off these blocks into two new entirely overlapping blocks. As newly created blocks

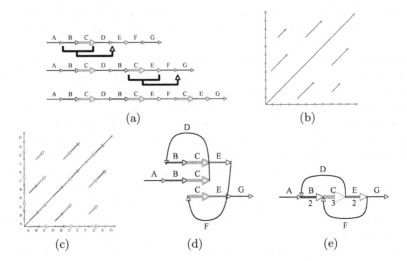

Fig. 2. (a) Hypothetical sequence with multiple duplications. (b,c) Genomic dot-plot and the resulting synteny blocks of the sequence. The 2-D representations overlap in 1-D. (d) Generate A-Bruijn graph of the sequence. (e) A-Bruijn graph of the sequence. Edges with multiplicity greater than 1 are synteny blocks. Blocks B, E each have two copies, and C has three copies. The algorithm outputs B,C,E as separate paths/blocks.

may partially overlap with other blocks, to eliminate all such partial overlaps a number of subsequent cuts may be required. The problem with such an approach, however, is that in some cases the initial synteny blocks might result in the iterative fragmenting and shrinking of synteny blocks. While this phenomenon is well known in repeat classification (e.g., RECON algorithm [32] follows a similar scheme), it has not been addressed yet in synteny block reconstruction. This simple and seemingly sensible approach does not work well in complex cases [26]. For example, early attempts to use a similar approach for constructing "duplication subunits" (analogs of synteny blocks for segmental duplications) failed and elaborate techniques were used to resolve this challenge [33]. While the complexity of synteny reconstruction so far is nowhere close to the complexity of the repeat analysis, the addition of every new species will soon make the synteny reconstruction more difficult, thus calling for techniques to overcome the limits of tools based on iterative splitting. In addition, synteny blocks (different from repeats) are subject to microrearrangements, further complicating the problem.

Although repeats and duplicated synteny blocks result from different biological events, they both represent sub-sequences appearing multiple times in the genomes. Repeats and duplicated synteny blocks differ mostly in length and in the number of occurrences in the genomes. Therefore, the problem of constructing non-overlapping synteny blocks for genomes with duplications is similar to the problem of *de novo* repeat classification and can be solved accordingly.

The same approach can be used for generation of synteny blocks across multiple genomes by simply concatenating them into a single genome. If there are no duplications in the original genomes, a k-copy synteny block in the concatenated genome corresponds to a k-way synteny block of the original genomes.

AB-Synteny

Without loss of generality we will generate synteny blocks for a single genome. Suppose that the given genome is represented as a sequence of elements (base pairs or genes) $v_1, v_2, \ldots, v_n$. These elements form the vertices of a *path-graph* P where every pair of consecutive vertices v_i and v_{i+1} (for $i = 1, \ldots, n - 1$) are connected with a directed edge. To obtain an A-Bruijn graph [26] A from the graph P, one needs to "glue" all vertices of P belonging to the same anchor into a single vertex (Fig. 2(d)). The resulting A-Bruijn graph A inherits all edges from the path-graph P, counting multiplicity of each edge as its weight (hence, the edges in A are weighted and there are no parallel edges (Fig. 2(e)).

The A-Bruijn graph A has one source and one sink[3] such that the original genome can be read along some path from the source to the sink. Every edge with weight greater than one corresponds to a *syntenic region* (i.e., a region that may belong to at most one synteny block), and its weight gives the number of copies of this syntenic region in the genome.

Unfortunately, such an interpretation of the A-Bruijn graph meets a number of obstacles. Inconsistencies in alignments and tandem duplications may create *whirls*[4] in A, while gaps in alignments may create *bulges* in A [26]. As a result, the constructed A-Bruijn graph can be exceedingly complicated. For example, the A-Bruijn graph constructed from *Arabidopsis* gene pairs has 6394 vertices and 12761 edges. To overcome these difficulties, [26] suggested a heuristic routine simplifying an A-Bruijn graph, which we partially apply to the graph A. In the process, we simplify the A-Bruijn graph by substituting every simple path in the graph by a single edge with its length equal to the length of the path.

Overall, our synteny block generation algorithm AB-Synteny(G, C, g, B, L) has the following parameters: G and C are gap and block size thresholds in pre-processing to eliminate noisy anchors; the *girth* g specifies a distance threshold for removing whirls; B specifies the threshold for bulge removal; L is the block size threshold (minimum number of elements in the blocks):

1. For two or more genomes, concatenate all genomes forming a single genome.
2. Pre-processing: run GRIMM-Synteny(G, C) [9] to produce non-overlapping syntenic blocks in 2-D (blocks may overlap in 1-D). GRIMM-Synteny removes all anchors within "small" blocks (smaller than C).
3. Construct an A-Bruijn graph: run A-Bruijn(g, B)—a simpler version of the graph clean up routines detailed in [26]—on the remaining anchors, removing whirls shorter than g and simple bulges shorter than B.
4. Output non-overlapping paths whose multiplicities are greater than one and whose lengths are equal or greater than L. These are syntenic regions.

---

[3] In practice, it is convenient to have also an inverted sequence of the same genome (for reverse DNA strand), in which case there are two sources and two sinks.

[4] Whirls refer to short directed cycles and bulges short undirected cycles in A.

5. Post-processing: merge neighboring syntenic regions of same orientation interrupted by short gaps into syntenies. Assign each block a unique ID.

We remark that since the constructed paths (syntenic regions) do not overlap in the A-Bruijn graph, they also do not overlap in 1-D (both before and after the post-processing step). As a result, AB-Synteny produces a number of synteny blocks non-overlapping in 1-D and a representation of the given genome as a mosaic of these blocks (each block may appear in multiple copies). In other words, an entire genome is represented as a *word* over the alphabet of synteny blocks, which facilitates further duplication and rearrangement studies.

3   Syntenic Analysis of Vertebrate Genomes

As an illustration and a validation of the AB-Synteny algorithm, we extracted and analyzed 714 gene pairs between human and mouse X chromosomes from Ensembl database version 39. The gene pairs are described as "orthologs" by Ensembl. After highly repetitive gene pairs (present in ≥ 10 copies) are removed (as they do not normally contribute to synteny blocks but would instead increase noises), 606 gene pairs remain. The human X chromosome has a total of 1360 genes and mouse 1267 genes. We further constructed 5-way synteny blocks for human, mouse, rat, dog and chicken genomes using all available pairwise orthologous (one-to-one) genes from Ensembl 44 (supplementary Table S1).

We concatenated 1360 genes from the human X chromosome and 1267 genes from the mouse X chromosome, forming a genome of 2627 genes. During concatenation, a number of elements (larger than the gap threshold) were inserted between two chromosomes to prevent synteny blocks from forming across boundaries of chromosomes or genomes. An A-Bruijn graph was constructed on the concatenated genomes using the 606 gene pairs between human and mouse X chromosomes as gluing instructions. The A-Bruijn graph has 906 vertices and 1636 edges (not shown) The graph was further simplified with the parameters $(g, B, L) = (10, 20, 4)$. The remaining graph has 46 vertices and 64 edges, from which synteny blocks were extracted as shown in Fig. 1(e).[5] Fig. 1(f) illustrates that the A-Bruijn graph is actually equivalent to the breakpoint graph for analyzing rearrangement scenarios [34]. After joining the neighboring syntenies of the same orientation, a total of 8 strips of syntenies emerged (Fig. 1(c)), covering 85.64% of human and 89.72% of mouse X chromosomes. The syntenies are similar to the published results [7] with small differences mainly caused by correcting fragment assembly errors in the latest versions of the human and mouse genomic sequences. The GRIMM-Synteny results on this dataset are shown in 1(b). The blocks from the two algorithms largely coincide.

For human-mouse-rat-dog-chicken, we did two sets of synteny block generations. In the first set, we concatenated all genes (31101 in human, 28157 in

---

[5] The numbers of vertices and edges reflect the final A-Bruijn graph including both the forward and inverted sequences. The final A-Bruijn graph shown however only includes forward sequence for illustration purpose.

mouse, 27264 in rat, 22602 in dog, and 15936 in chicken) into a single genome, and applied AB-Synteny $(G, C, g, B, L = 30, 3, 10, 20, 4)$ to the resulting genome using total of 125347 available gene pairs (1-to-1 orthologs) between any two genomes as gluing instructions. In the second set, we removed all genes that do not belong to any gene pair, and concatenated the remaining genes (16196 in human, 17196 in mouse, 16464 in rat, 15792 in dog, and 10908 in chicken) into a single genome, and applied AB-Synteny with the same parameters. The results are very similar and we report the results from the second set. After the vertices with 1-in and 1-out edges are merged, the A-Bruijn graph has 23228 vertices and 41833 edges. After simplification, 3564 vertices and 6814 edges remain, resulting in 666 5-way synteny blocks. We also extracted 8735 5-way orthologous genes from the gene pairs and applied GRIMM-Synteny$(G, C = 100, 4)$, where G is the total gap threshold. The results from the two algorithms are compared in Table 1. The shared coverage refers to regions of a genome that belong to synteny blocks reconstructed by both algorithms. Fig. 1(d) compares the human-mouse X chromosome portion of the 5-way synteny blocks to those synteny blocks derived from human-mouse data alone as shown in Fig. 1(c). As expected, the blocks from 5-way syntenies are shorter and more fragmented.

GRIMM-Synteny requires k-way anchors when there are k species. Some of the synteny blocks recovered by AB-Synteny but missed by GRIMM-Synteny

Table 1. 5-way synteny blocks constructed by AB-Synteny and GRIMM-Synteny

	Genome length (Mb)	AB-Synteny (666) length (Mb)	%	GRIMM-Synteny (466) length (Mb)	%	Shared coverage length (Mb)	%
human	3080	2017	65.47	2091	67.90	1837	59.63
mouse	2644	1792	67.79	1810	68.45	1604	60.66
rat	2719	1884	69.30	1915	70.43	1692	62.24
dog	2445	1669	68.26	1776	72.62	1527	62.44
chicken	1032	790	76.52	790	76.57	702	68.03

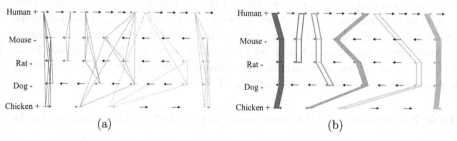

(a)   (b)

Fig. 3. A region of chromosome 1 of five species that is recovered by AB-Synteny but missed by GRIMM-Synteny due to the small number of 5-way anchors. (a) one-to-one orthologous gene pairs from Ensembl database. (b) 5-way anchors input to GRIMM-Synteny (3 filled): block missed; and all available anchors used by AB-Synteny (+3 unfilled): block recovered.

are due to a reduced number of k-way (5-way) anchors as the number of species increases. Fig. 3 illustrates such an example. The region is on chromosome 1 of the five species and consists of 13 genes from human, 9 from mouse, 9 from rat, 10 from dog and 5 from chicken. There are three 5-way anchors, two 4-way anchors and one 3-way anchor. Only the 5-way anchors can be used as inputs to GRIMM-Synteny or any other algorithms that require k-way anchors for k genomes. Since the number of such anchors is below the block threshold, the syntenic region is missed by GRIMM-Synteny. On the other hand, AB-Synteny requires only pairwise anchors between any two genomes. All six anchors therefore can be used as inputs. With equivalent parameter settings, AB-Synteny is able to recover the block as a result of more supporting anchors. This feature of AB-Synteny allows the algorithm to scale more easily to a large number of genomes.

4   Duplication Analysis of Plant Genome

We analyzed 5700 paralogous gene pairs in *A. thaliana* from [20], selected from about 30503 *A. thaliana* genes, and compared the results to published duplicated blocks generally accepted by the plant research community.

We applied AB-Synteny $(G, C, g, B, L = 20, 6, 10, 100, 4)$ to the *A. thaliana* genome with 30503 genes and 5700 anchors (gene pairs). It generated 223 non-overlapping segments in 1-D, making up 103 synteny blocks. Tables 2 and 3 compare AB-Synteny results with the synteny blocks from [20] and [4]. Almost all our synteny blocks are inside blocks of Bowers *et al.*, and about 2.66% of our blocks (covering 812 genes) are outside blocks of Blanc *et al.*

Fig. 4 shows several synteny blocks generated by AB-Synteny for the *A. thaliana* genome. Notice that the blocks appearing in more than two copies (blue colored blocks) are delineated from the 2-copy blocks (magenta blocks). The single red block is one of the synteny blocks (referred to as *chromosomal segment pairs*) reported in [20]. Careful inspection of the genomic segment shown in Fig. 4 reveals a large gap of 499 genes in chromosome 3 (x axis) and a corresponding gap of 12 genes in chromosome 4 (y axis). We argue that the AB-Synteny representation provides a more accurate view of the *A. thaliana* genomic architecture.

The last step in the synteny block generation algorithm in [20] combines adjacent syntenic regions with opposite orientation and order that may be explained by local inversions, although it is not clear which inversions are considered local. The separation of segments in such cases can partially explain the comparatively low coverage of AB-Synteny as shown in Table 2.

Table 2. Comparison of AB-Synteny results to published *A. thaliana* synteny blocks

Methods	# of Synteny blocks	Coverage # genes	Coverage %	Overlap in 1-D # genes	Overlap in 1-D %
1. [20]	34	26034	85.35	5069	16.62
2. [4]	91	24370	79.89	7118	23.34
3. AB-Synteny	103	21862	71.67	0	0

Table 3. Synteny block coverage shared between methods in Table 2

Methods	# genes	%
1 & 2	24089	78.97
1 & 3	21847	71.62
2 & 3	21050	69.01

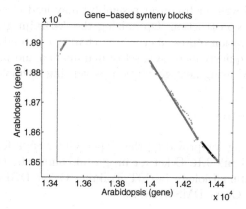

Fig. 4. A local view of synteny blocks of *A. thaliana* generated by AB-Synteny. the synteny block from [20] is a red box; and other colored diagonals are synteny blocks generated by AB-Synteny: magenta: 2 copies, blue: 3 copies (extra copy not shown).

There is partial agreement of our synteny blocks with those generated by LineUp [21] (data not shown). While the syntenic regions reported by LineUp do in general overlap with the regions generated by AB-Synteny, LineUp reports all statistically significant syntenic regions without trying to define the boundaries of the regions. As these regions overlap significantly, they cannot be used for reconstruction of rearrangement and duplication scenarios.

5   Discussion

The uniqueness of our new synteny block generation algorithm AB-Synteny stems from the fact that it produces synteny blocks that do not overlap in 1-D representations, an essential property for further analysis of the synteny blocks to study rearrangement and duplication history. AB-Synteny can also be used for generation of synteny blocks across multiple genomes. Given k genomes, one simply concatenates them into a single genome. If there are no duplications in the original genomes, then the edges with multiplicity k in the A-Bruijn graph correspond to synteny blocks shared by all k genomes.

Our AB-Synteny algorithm constructs A-Bruijn graphs using the the RepeatGluer code initially developed for repeat classification and DNA fragment assembly. Extending RepeatGluer to new research domains typically requires new application-specific algorithmic developments (e.g., constructing A-Bruijn graphs in mass spectrometry applications [29]). Similarly, the synteny block reconstruction may benefit from the modifications of the A-Bruijn graph approach that take into account the specific challenges of analyzing large highly duplicated genomes. We found that while most RepeatGluer steps (e.g., bulge removal) work well for synteny block generation, some steps need to be further optimized for the new application domain. In particular, we found that the *threading* heuristic from [26] (which worked well for fragment assembly) may lead to suboptimal results in synteny block reconstruction. Optimizing the A-Bruijn graph approach for synteny block generation represents the next challenge in analyzing the genomic architecture of the quickly increasing set of mammalian and plant genomes that are being sequenced using next generation sequencing technologies.

Acknowledgment

We thank H.X. Tang for providing the RepeatGluer code for A-Bruijn graph construction. We thank J.R. Ecker for insightful comments. QP was supported by NSF Plant System Biology IGERT Training Grant DGE0504645. GT was supported by NSF Grant DMS-0718810.

References

1. Vision, T.J., Brown, D.G., Tanksley, S.D.: The Origins of Genomic Duplications in Arabidopsis. Science 290(5499), 2114–2117 (2000)
2. Lander, E., Linton, L., Birren, B., Nusbaum, C., et al.: Initial sequencing and analysis of the human genome. Nature 409, 860–921 (2001)
3. Bailey, J., Baertsch, R., Kent, W., Haussler, D., Eichler, E.: Hotspots of mammalian chromosomal evolution. Genome Biol. 5(4), R23 (2004)
4. Blanc, G., Hokamp, K., Wolfe, K.H.: A recent polyploidy superimposed on older large-scale duplications in the arabidopsis genome. Genome Res. 13(2), 137–144 (2003)
5. Bourque, G., Pevzner, P.A., Tesler, G.: Reconstructing the Genomic Architecture of Ancestral Mammals: Lessons From Human, Mouse, and Rat Genomes. Genome Res. 14(4), 507–516 (2004)
6. Pevzner, P., Tesler, G.: Human and mouse genomic sequences reveal extensive breakpoint reuse in mammalian evolution. PNAS 100(13), 7672–7677 (2003)
7. Pevzner, P., Tesler, G.: Genome rearrangements in mammalian evolution: Lessons from human and mouse genomes. Genome Res. 13, 37–45 (2002)
8. Peng, Q., Pevzner, P., Tesler, G.: The fragile breakage versus random breakage models of chromosome evolution. PLoS Comput. Biol. 2(2), e14 (2006)
9. Tesler, G.: Grimm: genome rearrangements web server. Bioinf. 18(3), 492–493 (2002)
10. Nadeau, J., Taylor, B.: Lengths of chromosomal segments conserved since divergence of man and mouse. PNAS 81, 814–818 (1984)
11. Waterston, R., Lindblad-Toh, K., Birney, E., Rogers, J., Abril, J., Agarwal, P., Agarwala, R., Ainscough, R., Alexanderson, M., An, P., et al.: Initial sequencing and comparative analysis of the mouse genome. Nature 420, 520–562 (2002)

12. Kent, W.J., Baertsch, R., Hinrichs, A., Miller, W., Haussler, D.: Evolution's cauldron: Duplication, deletion, and rearrangement in the mouse and human genomes. PNAS 100(20), 11484–11489 (2003)
13. Brudno, M., Malde, S., Poliakov, A., Do, C., Couronne, O., et al.: Glocal alignment: Finding rearrangements during alignment. Bioinf. 19, i54–i62 (2003)
14. Darling, A., Mau, B., Blattner, F., Perna, N.T.: Mauve: Multiple alignment of conserved genomic sequence with rearrangements. Genome Res. 14, 1394–1403 (2004)
15. Bourque, G., Yacef, Y., El-Mabrouk, N.: Maximizing synteny blocks to identify ancestral homologs. In: McLysaght, A., Huson, D.H. (eds.) RECOMB 2005. LNCS (LNBI), vol. 3678, pp. 21–34. Springer, Heidelberg (2005)
16. Ma, J., Zhang, L., Suh, B.B., Raney, B.J., Burhans, R.C., Kent, W.J., Blanchette, M.: Reconstructing contiguous regions of an ancestral genome. Genome Res. 16, 1557–1565 (2006)
17. Sinha, A., Meller, J.: Cinteny: flexible analysis and visualization of synteny and genome rearrangements in multiple organisms. BMC Bioinf. 8(1), 82 (2007)
18. Hachiya, T., Osana, Y., Popendorf, K., Sakakibara, Y.: Accurate identification of orthologous segments among multiple genomes. Bioinf. 25(7), 853–860 (2009)
19. Kellis, M., Birren, B.W., Lander, E.S.: Proof and evolutionary analysis of ancient genome duplication in the yeast saccharomyces cerevisiae. Nature 428(6983), 617–624 (2004)
20. Bowers, J.E., Chapman, B.A., Rong, J., Paterson, A.H.: Unravelling angiosperm genome evolution by phylogenetic analysis of chromosomal duplication events. Nature 422, 433–438 (2003)
21. Hampson, S., McLysaght, A., Gaut, B., Baldi, P.: LineUp: Statistical Detection of Chromosomal Homology With Application to Plant Comparative Genomics. Genome Res. 13(5), 999–1010 (2003)
22. Haas, B.J., Delcher, A.L., Wortman, J.R., Salzberg, S.L.: DAGchainer: a tool for mining segmental genome duplications and synteny. Bioinf. 20(18), 3643–3646 (2004)
23. Vandepoele, K., Saeys, Y., Simillion, C., Raes, J., Van de Peer, Y.: The Automatic Detection of Homologous Regions (ADHoRe) and Its Application to Microcolinearity between Arabidopsis and Rice. Genome Res. 12(11), 1792–1801 (2002)
24. Simillion, C., Janssens, K., Sterck, L., Van de Peer, Y.: i-ADHoRe 2.0: an improved tool to detect degenerated genomic homology using genomic profiles. Bioinf. 24(1), 127–138 (2008)
25. Soderlund, C., Nelson, W., Shoemaker, A., Paterson, A.: SyMAP: A system for discovering and viewing syntenic regions of FPC maps. Genome Res. 16(9), 1159–1168 (2006)
26. Pevzner, P.A., Tang, H., Tesler, G.: De Novo Repeat Classification and Fragment Assembly. Genome Res. 14(9), 1786–1796 (2004)
27. Raphael, B., Zhi, D., Tang, H., Pevzner, P.: A novel method for multiple alignment of sequences with repeated and shuffled elements. Genome Res. 14(11), 2336–2346 (2004)
28. Zhi, D., Raphael, B., Price, A., Tang, H., Pevzner, P.: Identifying repeat domains in large genomes. Genome Biol. 7(1), R7 (2006)
29. Bandeira, N., Clauser, K.R., Pevzner, P.A.: Shotgun Protein Sequencing: Assembly of Peptide Tandem Mass Spectra from Mixtures of Modified Proteins. Mol. Cell Proteomics 6(7), 1123–1134 (2007)
30. Bourque, G., Zdobnov, E.M., Bork, P., Pevzner, P.A., Tesler, G.: Comparative architectures of mammalian and chicken genomes reveal highly variable rates of genomic rearrangements across different lineages. Genome Res. 15(1), 98–110 (2005)

31. Dewey, C.N., Pachter, L.: Mercator: Multiple whole-genome-orthology map construction (2006), http://bio.math.berkeley.edu/mercator
32. Bao, Z., Eddy, S.R.: Automated De Novo Identification of Repeat Sequence Families in Sequenced Genomes. Genome Res. 12(8), 1269–1276 (2002)
33. Jiang, Z., Tang, H., Ventura, M., Cardone, M.F., Marques-Bonet, T., She, X., Pevzner, P.A., Eichler, E.E.: Ancestral reconstruction of segmental duplications reveals punctuated cores of human genome evolution. Nat. Genet. 11, 1361–1368 (2007)
34. Hannenhalli, S., Pevzner, P.: Transforming cabbage into turnip: polynomial algorithm for sorting signed permutations by reversals. J ACM 46, 1–27 (1999)

A General Framework for Local Pairwise Alignment Statistics with Gaps

Pasi Rastas

Department of Computer Science & HIIT Basic Research Unit
P.O. Box 68 (Gustaf Hällströmin katu 2b)
FIN-00014 University of Helsinki, Finland
`firstname.lastname@cs.helsinki.fi`

Abstract. We present a novel dynamic programming framework that allows one to compute tight upper bounds for the p-values of gapped local alignments in pseudo–polynomial time. Our algorithms are fast and simple and unlike most earlier solutions, require no curve fitting by sampling. Moreover, our new methods do not suffer from the so–called edge effects, a by–product of the common practice used to compute p-values. These new methods also provide a way to get into very small p-values, that are needed when comparing sequences against large databases. Based on our experiments, accurate estimates of small p-values are difficult to get by curve fitting.

1 Introduction

The most basic and fundamental sequence analysis task is to find out if two sequences are related [1]. This can be accomplished by aligning sequences, to get the alignment score as a sequence similarity measure. Based on the score, it is decided whether the gained alignment occurs because of the sequences are related or just by chance. The most important similarity measure of two sequences is their best local alignment score [2], used for example in the BLAST software [3,4]. The score of a local alignment is the sum of the scores of aligned characters subtracted by the gap penalties. Local alignment is used routinely to compare new DNA and protein sequences against large sequence databases.

In this paper, a dynamic programming framework is sketched that allows one to compute tight upper bounds for the p-values of gapped local alignment scores. The gap models include linear, affine, and constant penalty models. Our algorithms use dynamic programming and unlike most earlier solutions, require no curve fitting. Moreover, our new methods do not suffer from the so–called edge effects, a by–product of the common practice to reduce the two dimensional alignment problem to a one–dimensional case [5]. Our methods provide a way to get into any meaningful p-value, including very small ones needed when comparing sequences against large databases. Based on our experiments, accurate estimates of small p-values are difficult to get by curve fitting. We present the framework for the simple Bernoulli distribution, but so other distributions would be possibe as well. For example, one of the strings could be fixed and only the

S.L. Salzberg and T. Warnow (Eds.): WABI 2009, LNBI 5724, pp. 233–245, 2009.

other could be drawn from the null distribution. Also a fixed order Markov chain could be used as the null model.

The organization of this paper is following. Section 2 gives some basic definitions of local alignments and Section 3 recalls the problem of statistical significance of alignment scores. Section 3 also reviews of some previous methods to compute the significance of local alignment scores. In Section 4 we present our framework for the computation of gapped alignment statistics. The computational hardness is considered in Section 5 followed by experimental results in Section 6. Last, the discussion and future work part concludes this paper in Section 7.

2   Definition

Let $x = x_1, \ldots, x_n$ and $y = y_1, \ldots, y_m$ be two strings of length n and m from a finite alphabet Σ. Let $s(a, b)$ be the score of aligning, or matching residues or letters a and b from the alphabet. Furthermore, let d be the linear gap penalty, i.e. the penalty of deleting or inserting a single character.

A local alignment of x and y can be described as a path in the alignment grid (i, j), $i = 0, \ldots, n$, $j = 0, \ldots, m$, taking vertical, diagonal or horizontal steps. A diagonal step from (i, j) to $(i + 1, j + 1)$ corresponds to aligning residues x_{i+1} and y_{j+1} and has score $s(x_{i+1}, y_{j+1})$, a horizontal step from (i, j) to $(i + 1, j)$ deletes character x_{i+1} and has score $-d$, and a vertical step from (i, j) to $(i, j+1)$ inserts character y_{j+1} and has score $-d$. The total score of the alignment given by the path is the sum of the scores of individual steps.

As an example, consider the alignment of $x = ACGT$ and $y = ACCT$:

```
x: AC-G
y: ACCT
```

The score of this alignment is $s(A, A) + s(C, C) - d + s(G, T)$. The path taken in this example is $(0, 0) \mapsto (1, 1) \mapsto (2, 2) \mapsto (2, 3) \mapsto (3, 4)$.

Let z be a path in the alignment grid from (i', j') to (i, j) ($0 \le i' \le i \le n$ and $0 \le j' \le j \le m$). Denote by $\text{score}(x, y, z)$ the score of x and y locally aligned along path z. The best local alignment score of x and y, $\text{score}(x, y)$, is defined as $\max_z \text{score}(x, y, z)$. The value of $\text{score}(x, y)$ can be computed by Smith–Waterman algorithm in time $O(mn)$ [2,6,1].

3   Statistical Significance of a Local Alignment

Having computed that the best (local) alignment score of some x and y is t, a natural question to ask is how significant the alignment is. To answer this, we use the classical statistical framework and compute a p-value of getting a alignment score as good or better than t, given the null hypothesis or the null model.

The null model used for sequence x (and for y) from the alphabet Σ is the random (Bernoulli) model, i.e. the probability of x in the null model is

$P(x) = \prod_{i=1}^{n} q_{x_i}$, where q_a is the (arbitrary) probability of an alphabet symbol a. The model fixes probabilities only for fixed length sequences, i.e. the length distribution of x (nor y) is not modeled.

Let D_t consist of fixed length string pairs (x, y) with best score t or better, i.e. $D_t = \{(x, y) \mid \text{score}(x, y) \geq t\}$. The significance measure, i.e. the p-value of alignment score t, is $p_t = \sum_{(x,y) \in D_t} P(x)P(y)$.

3.1   Significance of Alignments without Gaps

The significance of alignments without gaps is typically solved approximately as a one–dimensional problem. A single string of length mn is created from the two dimensional alignment problem. Strings x and y are aligned (globally) in all $m + n - 1$ different ways and resulting string pairs are put together to form a string $Y = Y_1, \ldots, Y_{mn}$ of mn letter pairs ($Y_i \in \Sigma \times \Sigma$). Then a probability $q_a q_b$ is assigned for each letter pair (a, b). This dimension reduction can also be thought as concatenating diagonals of the alignment grid (but skipping first row and column) to get a single linear chain.

The best one–dimensional local alignment score is the maximal segment score

$$M(Y) = \max_{1 \leq i \leq j \leq mn} \sum_{k=i}^{j} s(Y_k). \tag{1}$$

This dimension reduction does not give an exact solution because of the so–called edge effects. Some alignments can overlap the concatenation points, i.e. these alignments do not match to any real alignments. Moreover, the mn letter pairs of Y are not independent, as they depend on the two underlying strings x and y. In fact, if a string of mn letter pairs is positionally independent, there are Σ^{2mn} different strings of independent mn pairs of letters, while there are only Σ^{m+n} different Y as the number of Y's equals the number of different string pairs (x, y).

Karlin-Altschul Statistics. The most common way to approximate the distribution of $M(Y)$ for Bernoulli distributed x and y, used in BLAST as well [3], is to use the Karlin-Altschul statistics [5,7]. Assuming that expected score $\sum_{a,b} q_a q_b s(a, b)$ is negative and there is some letter pair (a, b) with $s(a, b) > 0$, and m and n are about equal size, the p-value for the alignment can be approximated as

$$P(M(Y) > t) \approx 1 - \exp(-Kmne^{-\lambda t}), \tag{2}$$

for some K and λ [8]. In the gapless case, parameters K and λ can be solved analytically [5,8]. The term $Kmne^{-\lambda t}$ is the E-value or E-score of BLAST [3]. Equation (2) can also be used for alignments with gaps [4].

Exact One–Dimensional Solution. In [9], a pseudo–polynomial time algorithm is given for evaluating the probability of $P(M(Y) \geq t)$ exactly [9]. The method is quite simple and uses a Markov chain illustrated in Figure 1. In the Markov chain, there is a state for each score value $0, 1, \ldots, t$ (possible scores must be integers). The transition probability from state i to state j is $P(s(a, b) = j - i)$,

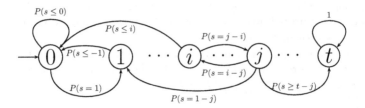

Fig. 1. A Markov chain used in [9] to compute $P(M(Y) \geq t)$. The number on each state denotes the alignment score. For clarity, all transitions have not been drawn.

if $i, j = 1, \ldots, t-1$. Transitions from and to states 0 and t are handled as shown in Figure 1. After taking nm steps in the Markov chain, the probability of being in the state t is $P(M(Y) \geq t)$.

A straightforward implementation of this Markov chain approach runs in time $O(mnt) = O(n^3)$ (assuming the absolute values of $s(\cdot, \cdot)$ are bounded by a constant). The algorithm of [9] runs in time $O(t^{2.376} \log(mn))$.

3.2   Gapped Significance

As already mentioned, Formula (2) is also used for alignments with gaps. Many experimental studies show that the distribution of gapped alignment scores follow equation (2) reasonable accurately, e.g. [10]. However, to our best knowledge, there is no analytic, general way to solve K and λ with gaps. Some steps have been made towards analytic solution for these parameters [11].

4   Dynamic Programming Framework for Gapped Significance

In this section, we present our general framework, which allows us to compute expected number of alignments with various types of distributions for the sequences. This expectation is an upper bound for the p-value when the seqences are distributed according to the null model.

Let X_t be a random variable counting the number of alignments of x and y with score $\geq t$, when x and y are distributed according to the null model. The p-value p_t is $P(X_t \geq 1)$. The expectation of X_t, $E(X_t)$, can be computed by summing over all possible alignment paths z and adding up the probability that the alignment score of x and y is $\geq t$ along each z. To justify this, consider a set A_t^z of string pairs (x, y) with best score t or better on a given local alignment z, i.e. $A_t^z = \{(x, y) \mid \text{score}(x, y, z) \geq t\}$. Now the random variable X_t can be written as a sum of indicator functions of A_t^z, i.e. $X_t = \sum_z 1_{A_t^z}$. As the expectation is linear, $E(X_t) = E(\sum_z 1_{A_t^z}) = \sum_z E(1_{A_t^z}) = \sum_z P(A_t^z)$.

We can compute $E(X_t)$ by dynamic programming. Let $L'(i, j, s)$ be the expected number of paths with score s and ending at (i, j) for a sequence pair with lengths i and j. Consider the linear gap model and assume that we somehow know the values of $L'(i-1,j,\cdot)$, $L'(i,j-1,\cdot)$, and $L'(i-1,j-1,\cdot)$. How could we then compute $L'(i,j,s)$? It is not hard to see that

$$L'(i,j,s) = L'(i,j-1,s+d) + L'(i-1,j,s+d) + \sum_{a,b} q_a q_b L'(i-1,j-1,s-s(a,b)).$$

This gives our framework to work on. We can utilize this idea to compute value $L(i,j,s) = L'(i,j,s) - L'(i,j,s+1)$, by dynamic programming as follows (the unassigned values are zeros):

$$L(0,0,0) = L(0,j,0) = L(i,0,0) = L(i,j,0) = 1$$
$$L(i,j,s) = L(i-1,j,s+d) + L(i,j-1,s+d)$$
$$+ \sum_{a,b\in\Sigma} q_a q_b L(i-1,j-1,s-s(a,b)), \tag{3}$$

where $i = 1,\ldots,n$, $j = 1,\ldots,m$, and $s = 1,\ldots,(n\max_{a,b} s(a,b))$. Then the expectation $E(X_t)$ can be computed as $E(X_t) = \sum_{i,j}\sum_{s\geq t} L(i,j,s)$. We have obtained an upper bound for the p-value, as by using Markov inequality $p_t = P(X_t \geq 1) \leq E(X_t)$.

This framework is a generalization of the one in [12] used to count and enumerate (sub)optimal alignments, explained also in [13] (chapter 13). The presented framework is pseudo–polynomial on the score values; if the possible score values are not bounded, the algorithms become very slow. All values of $s(\cdot,\cdot)$ and the gap penalties (d and e) are assumed to be integers, whose absolute values are bounded by B. Assuming that the (possible exponential) large numbers used in the computation fit to a fixed size floating point representation with sufficient accuracy, the time complexity of directly evaluating (3), assuming $n \leq m$, is $O(mn^2 B)$ and space $O(n^2 B)$ suffices. Later on, when time and space complexities are considered, B is assumed to be a small constant. This assumption seems reasonable in the bioinformatics domain. Moreover, the values of m and n are assumed to be of about equal size, i.e. $n = \Theta(m)$. Thus, evaluating (3) needs $O(n^3)$ time and $O(n^2)$ space.

For $E(X_t)$ to be meaningful, the expected score, $\sum_{a,b\in\Sigma} q_a q_b s(a,b)$, must be negative and there must be some characters $a,b \in \Sigma$ for which $s(a,b) > 0$. Moreover, useful thresholds t are ≥ 1.

Equation (3) generalizes to affine gaps as well. Let d and e be the penalties for gap opening and gap extension, respectively. The dynamic programming becomes as follows:

$$M(0,0,0) = M(0,j,0) = M(i,0,0) = M(i,j,0) = 1$$
$$M(i,j,s) =$$
$$\sum_{a,b\in\Sigma} q_a q_b \Big(I(i-1,j-1,s-s(a,b)) + M(i-1,j-1,s-s(a,b)) \Big) \tag{4}$$
$$I(i,j,s) = M(i,j-1,s+d) + M(i-1,j,s+d)$$
$$+ I(i,j-1,s+e) + I(i-1,j,s+e),$$

where $i = 1, \ldots, n$, $j = 1, \ldots, m$ and $s = 1, \ldots, (n \max_{a,b} s(a,b))$. The value of $E(X_t)$ is now $\sum_{i,j} \sum_{s \geq t} (M(i,j,s) + I(i,j,s))$.

This is the simplest formulation for affine gaps from [1]. However, with these equations insertion of k symbols followed by k' deletions gets penalty $d + (k + k' - 1)e$, not $2d - (k + k' - 2)e$ as it should. This can be fixed by splitting table I into two tables I_x, and I_y, where I_x (I_y) accounts insertions to x (y). The time and space complexities of evaluating (4) are $O(n^3)$ and $O(n^2)$, respectively.

4.1 Faster in Linear Space

Both dynamic programming algorithms (3) and (4) compute $O(n^3)$ values of $E(X_t)$, as this expectation is computed for each t and for each sequence lengths up to m and n. In this sense they have an optimal time complexity of $O(n^3)$. However, in practice one is usually interested in only one t for some fixed m and n. Moreover, the quadratic space consumption of the above methods is quite restrictive. Next we show how to compute a single $E(X_t)$ faster using only linear space.

First, consider a global gapless alignment of strings x and y of equal length n. The path of this alignment travels along the diagonal of the alignment graph or grid. The partial score from 1 to i of this alignment is simply $\sum_{j=1}^{i} s(x_j, y_j)$. The probability that partial score stays positive and is s at position i, denoted by $H(i,s)$, can be computed by dynamic programming:

$$H(0,0) = 1\,, H(0,s) = 0$$
$$H(i,s) = \sum_{a,b \in \Sigma} q_a q_b H(i-1, s - s(a,b))\,, \tag{5}$$

where $i = 1, \ldots, n$ and $s = 1, \ldots, n \max_{a,b} s(a,b)$. All values of this table H can be computed in $O(n^2)$ time. The space consumption is linear, as we can compute $H(i, \cdot)$ by keeping only $H(i-1, \cdot)$ in the memory.

If the scoring function $s(a,b)$ is s_1 when $a = b$ and $s(a,b) = s_2$ otherwise, the table H can be computed more efficiently, even when the absolute values s_1 and s_2 are not bounded by a constant. This is because for each i, at most n different values for each $H(i, \cdot)$ need to be stored and evaluated.

4.2 Linear Gaps

With table H, it is possible to compute $E(X_t)$ with linear gaps in $O(n^2)$ time and in $O(n)$ space. The trick is to use the linearity of expectation, and compute the expectation for each $m' \times n'$ submatrix only once, and multiply each expectation by $(n - n' + 1)(m - m' + 1)C$, where C is the number of alignments on each submatrix.

The number of alignments C for a submatrix size $m' \times n'$ is $\binom{n'}{k}\binom{m'}{k}$, where k is the number of matched characters, i.e. this the number of ways to choose k positions from both n' and m' possibilities independently. Term $(n - n' + 1)(m - m' + 1)$ is the number of ways to choose $m' \times n'$ sized submatrix (box) inside a matrix (box) of size $m \times n$.

By denoting $H'(i,t) = \sum_{s' \geq t} H(i,s')$, we obtain the following formula

$$E(X_t) = \sum_{k=0}^{\min\{m,n\}} \sum_{n'=k}^{n} \sum_{m'=k}^{m} (m-m'+1)(n-n'+1)\binom{n'}{k}\binom{m'}{k} H'(k,t+d(n'+m'-2k)).$$

By noticing that the value of $H'(k,t+d(n'+m'-2k))$ stays the same for all n', m' for which $n'+m' = l$, we get

$$E(X_t) = \sum_k \sum_l D(k,l) H'(k,t+d(l-2k)),$$

where $D(k,l) = \sum_{m',n':m'+n'=l}(m-m'+1)(n-n'+1)\binom{n'}{k}\binom{m'}{k}$.

Lemma 1. [14] $\sum_{0 \leq k \leq l}\binom{l-k}{m}\binom{q+k}{n} = \binom{l+q+1}{m+n+1}$

Lemma 2. $(m+1)\binom{m}{k} = (k+1)\binom{m+1}{k+1}$

Now we can write $D(k,l)$ as $\sum_{m'}\binom{l-m'}{k}\binom{m'}{k}\Big(-(m'+2)(m'+1) + (l-n+m+3)(m'+1) + (m+2)(n-l)\Big)$. By using Lemma 2 three times and then using Lemma 1 we get form $D(k,l) =$

$$(k+1)(-(k+2)\binom{l+3}{2k+3} + (l-n+m+3)\binom{l+2}{2k+2}) + (m+2)(n-l)\binom{l+1}{2k+1},$$

which can be evaluated incrementally on l in $O(1)$ time. Thus, we have an algorithm to compute $E(X_t)$ in $O(n^2)$ time and in linear space.

Note that the actual value computed in this way is slightly bigger than one would get from Equation (3). This is due to the fact that gaps are handled as they would occur at the end of the alignment. Depending on how the gaps are put, some alignments with a total score $\geq t$, have a non–positive partial score at some point. This difference can be made smaller by taking into account the fact that the first gap cannot occur at the first position. To take this into account, the term $\binom{n'}{k}\binom{m'}{k}$ is replaced by $\binom{n'-1}{k-1}\binom{m'-1}{k-1}$. Actually, depending on the scoring model and gap penalties, some k' first (and without loss of generality k' last) positions must be matched. This could also be taken into account.

4.3   Affine Gaps

With linear gaps the number of alignments was easy to compute as $\binom{n}{k}$ for length n (sub)sequence with k matches, i.e. positions selected to be aligned. With affine gaps, one has to compute the number of alignments with certain number of consecutive gaps. Next we count the number of these alignments.

Consider a string x to be aligned with k positions selected for matches. Then there are at most $k+1$ gap openings, and exactly $k+1$ gap openings if no two matches are adjacent and no match occurs at the beginning or at the end of x. Consider that the length of x is n and k positions have been selected from x to

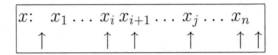

Fig. 2. A string x with four (4) matches, three (3) gap openings and $n-7$ gap extensions. Positions 0 and $n+1$ are marked as matches with arrows.

be matched. Further, consider that there are two additional matches, one just before position 1, and the other just after position n (at positions 0 and $n+1$). Now there are $k+1-l$ gap openings if there are l adjacent matches. Figure 2 illustrates these gap openings.

Let us now focus on a related problem of computing the number of binary strings of length n with k ones and l adjacent ones, denoted as $\mathcal{N}(n,k,l)$. Ones in this string correponds to matched positions and zeros are gaps. Further on, $\mathcal{N}(n,k,l)$ can be composed as a sum $\mathcal{Z}(n,k,l)+\mathcal{O}(n,k,l)$, where $\mathcal{O}$ counts strings that end to one and $\mathcal{Z}$ those that end to zero. Formulas for $\mathcal{Z}$ and $\mathcal{O}$ are

$$\mathcal{Z}(n,k,l) = \binom{n-k}{k-l}\binom{k-1}{l}, \text{ and } \mathcal{O}(n,k,l) = \binom{n-k}{k-l-1}\binom{k-1}{l},$$

defining $\binom{n}{k}$ to be non–zero only when $0 \le k \le n$. These formulas are derived (and can be proved) from the fact that $\mathcal{O}(n,k,l) = \mathcal{Z}(n-1,k-1,l) + \mathcal{O}(n-1,k-1,l-1)$ and $\mathcal{Z}(n,k,l) = \mathcal{Z}(n-1,k,l) + \mathcal{O}(n-1,k,l)$.

With $\mathcal{O}$, one can compute the number ways to choose k positions from x with l adjacent positions. This number is $C'(n,k,l) = \mathcal{O}(n+2,k+2,l)-\mathcal{O}(n+1,k+2,l)$, as it is the number of strings of length $n+2$, with $k+2$ ones, from which l are adjacent, and with ones at both ends. When $n-k-1 \ge 0$, C' can be simplified to $\binom{n-k-1}{k-l}\binom{k+1}{l}$, and thus

$$C'(n,k,l) = \begin{cases} \binom{n-k-1}{k-l}\binom{k+1}{l} & \text{, if } n > k \\ 1 & \text{, if } n = k \text{ and } l = k+1 \\ 0 & \text{, otherwise.} \end{cases}$$

Constant Gap Penalty. The constant gap penalty model is a special case of affine gaps, in which the gap extension penalty is zero. Next we show, that $E(X_t)$ can be computed in $O(n^2 \log n)$ time with constant gap penalties.

As earlier, value $E(X_t)$ can be computed as $\sum_k \sum_{l_x=0}^{k+1} \sum_{l_y=0}^{k+1} \sum_{n'} \sum_{m'} (m - m' + 1)(n - n' + 1)C'(n',k,l_x)C'(m',k,l_y)H'(k, t + (2k + 2 - l_x - l_y)d)$ using variables l_x and l_y to account the number of adjacent matches in x and y. Regrouping gives us

$$\sum_k \sum_{l_x} \sum_{l_y} C(m,k,l_y)C(n,k,l_x)H'(k, t + (2k + 2 - l_x - l_y)d),$$

where $C(m, k, l) = \sum_{m'=k}^{m}(m - m' + 1)C'(m', k, l)$. Using Lemmas 2 and 3, the value of C can be written as

$$C(m, k, l) = \binom{k+1}{l}\left((m - k + 1)\binom{m-k}{k-l+1} - (k - l + 1)\binom{m-k+1}{k-l+2}\right).$$

Thus, this formulation gives an $O(n^3)$ time algorithm.

Lemma 3. $\sum_{i=0}^{m}\binom{i}{k} = \binom{m+1}{k+1}$

This algorithm can be improved by noticing that $H'(k, t + (2k + 2 - l_x - l_y)d)$ is the same for all l_x and l_y that sum to l. This can be utilized by writing $E(X_t)$ as

$$\sum_{k}\sum_{l} C(m, n, k, l)H'(k, t + (2k + 2 - l)d),$$

where $C(m, n, k, l) = \sum_{l_x, l_t : l_x + l_t = l} C(m, k, l_y)C(n, k, l_x)$. The value of $C(m, n, k, l)$ for all l can be computed as a convolution $f * g$ of two vectors f and g, $f[l'] = C(m, k, l')$ and $g[l'] = C(n, k, l')$ (As $(f * g)[l] = \sum_{l'} f[l']g[l - l']$). The convolution can be computed in $O(n \log n)$ time using fast Fourier transformation (FFT) techniques [15]. With these techniques, one achieves an $O(n^2 \log n)$ time algorithm with linear space to compute $E(X_t)$ with constant penalty gaps. In our experiments, we observed that for n and m less than 2000, it is faster to use integer multiplication algorithms [16], like the Karatsuba's algorithm ($O(n^{1.585})$), instead of FFT ($O(n \log n)$).

General Affine Gaps. In this part, we consider general gap penality model with penalties d for opening and e for extending a gap. By combining all tricks encountered so far allows us to write $E(X_t)$ as

$$\sum_{k}\sum_{n'}\sum_{m'}\sum_{l_x}\sum_{l_y}(m - m' + 1)(n - n' + 1)C'(n', k, l_x)C'(m', k, l_y)$$
$$H'\big(k, t + (2k + 2 - l_x - l_y)(d - e) + e(m' + n' - 2k)\big).$$

Then, using FFT based convolution, we get $O(n^4 \log n)$ time with linear space. This algorithm is too slow, so with general affine gaps one should use (4). It is left as an open problem, whether it is possible to improve the above algorithm's time complexity, by using only linear space.

4.4   Relation between $E(X_t)$ and E-Score

The E-score of BLAST [3] is similar to expectation $E(X_t)$. However, the E-score is the expected number of high-scoring segment pairs, HSPs. A HSP is a maximal alignment that cannot be improved by extension or trimming [3].

A necessary condition for an alignment path z to be maximal, is that the score decreases if matches at either ends of z are added. This can be utilized in Formula (3) to get an upper bound for p_t that is tighter than $E(X_t)$. The idea is that a match with negative score is added to both ends of each path

z, if possible. However, value $c^2 E(X_t)$ is also a good approximation of this new bound, where $c = P(s(a,b) < 0)$. Moreover, $c^2 E(X_t)$ is also approximately equal to the E-score of BLAST.

The E-score E of BLAST can be changed into $1 - e^{-E} \approx p$-value. Similarly, $e^{-E(X_t)} = e^{-\sum_z P(A_t^z)} = \prod_z e^{-P(A_t^z)} \approx \prod_z (1 - P(A_t^z))$. If events $(x,y) \in A_t^z$ would be independent (they are not), then $e^{-E(X_t)} \approx P(X_t = 0) = 1 - p_t$.

5   Computational Hardness

It is not a coincidence that all presented algorithms are pseudo–polynomial on the score values. It turns out that the problem of computing $E(X_t)$ is NP-hard.

Without loss of generality, consider the expected number of alignments with the score being exactly t, i.e. the values $E(X_t') = E(X_t) - E(X_{t+1})$. Now, it is NP-hard to decided whether $E(X_t') > 0$.

The polynomial–time reductions are from the NP-complete Subset Sum [17].

Subset Sum

Instance: Positive integers $a_1, \ldots, a_n$ and a positive integer A.

Question: Is there a subset $I \subset \{1, \ldots, n\}$, such that $\sum_{i \in I} a_i = A$?

The alignment instance with string lengths n: First, define alignment scores as

$$s(i,j) = \begin{cases} 2^n a_i + 2^{i-1} & \text{if } i = j \\ 2^{i-1} & \text{if } i \equiv j+1 \ (\mathrm{mod}\ n) \\ 0 & \text{otherwise} . \end{cases}$$

The background probability is uniform over the alphabet $\Sigma = \{1, \ldots, n\}$, i.e $q_i = \frac{1}{n}$. Further, define the gap penalty $d = 2^n \sum_{i=0}^n a_i + 1$. We do not have to consider gaps as even a single gap would introduce higher penalty than any match of n alphabet pairs.

Now, there is a subset I with $\sum_{i \in I} a_i = A$, if and only if $E(X'_{2^n(A+1)-1}) > 0$. Due to the space limitations, the proof is omitted.

$\square$

Similar result can be found for $p_t' = p_t - p_{t+1}$, if $E(X'_{2^n(A+1)-1}) > 0$ is replaced by $p'_{2^n(A+1)-1} > 0$.

6   Experimental Results

We have sampled 10^7 random DNA sequence pairs of equal lengths of 125, 250, 500, 1000, and 2000. The distribution for each A, C, G and T was uniform (we have succesfully tested some slightly non–uniform distributions as well). For each sampled pair, we have computed the maximum local alignment score using Smith–Waterman algorithm [2,6]. The scoring schema is the same that is used in BLASTn, i.e. the score of match is +1 and mismatch -3, gap opening penalty

Table 1. Comparison of sampled p-values ($\hat{p}_t$) and $E(X_t)$ for affine gaps for each sequence length from 125 to 2000. Sampled values are accurate when $p_t \geq 10^{-5}$. In the equivalent table with linear gaps, the numbers were almost the same (not shown).

t	$m = n =$ 125		250		500		1000		2000	
	$\hat{p}_t$	$E(X_t)$	$\hat{p}_t$	$E(X_t)$	$\hat{p}_t$	$E(X_t)$	$\hat{p}_t$	$E(X_t)$	$\hat{p}_t$	$E(X_t)$
6	0.9332	5.083	1	21.32	1	87.31	1	353.3	1	1422
7	0.4894	1.267	0.9405	5.363	1	22.06	1	89.46	1	360.3
8	0.1542	0.3156	0.5079	1.349	0.9458	5.572	1	22.65	1	91.31
9	0.04092	0.07862	0.1631	0.3392	0.5208	1.408	0.9498	5.734	1	23.14
10	0.0103	0.01958	0.04379	0.0853	0.1693	0.3556	0.5306	1.452	0.9528	5.866
11	0.00257	0.004877	0.01121	0.02145	0.0458	0.08983	0.174	0.3675	0.5385	1.487
12	6.34e-4	0.001214	0.002811	0.005394	0.01172	0.02269	0.04731	0.09305	0.1777	0.3768
13	1.576e-4	3.023e-4	7.077e-4	0.001356	0.002991	0.005732	0.01219	0.02356	0.04831	0.09549
14	3.74e-05	7.524e-5	1.764e-4	3.41e-4	7.482e-4	0.001448	0.003116	0.005964	0.01245	0.0242
15	9.8e-6	1.872e-5	4.16e-5	8.574e-5	1.889e-4	3.658e-4	7.938e-4	0.00151	0.003183	0.006134
16	2.3e-6	4.659e-6	1.08e-5	2.156e-5	4.78e-5	9.24e-5	1.972e-4	3.822e-4	8.145e-4	0.001555
17	4e-7	1.159e-6	2.3e-6	5.419e-6	1.4e-5	2.334e-5	4.73e-5	9.677e-5	2.142e-4	3.940e-4
18	1e-7	2.883e-7	5e-7	1.362e-6	2.5e-6	5.895e-6	1.12e-5	2.45e-5	5.37e-5	9.986e-5
19	0	7.168e-8	1e-7	3.425e-7	8e-7	1.489e-6	2.9e-6	6.203e-6	1.34e-5	2.531e-5
20	0	1.782e-8	0	8.609e-8	1e-7	3.761e-7	5e-7	1.57e-6	3.4e-6	6.415e-6
21	0	4.43e-9	0	2.164e-8	0	9.501e-8	1e-7	3.975e-7	1.1e-6	1.626e-6
22	0	1.101e-9	0	5.44e-9	0	2.4e-8	0	1.006e-7	2e-7	4.12e-7

Table 2. Comparison of p-values computed by the Karlin-Altschul statistics (KA-N), the method of Mercier et.al.[9] (without gaps), and by our expectation $E(X_t)$ for score $\geq t = 25$ and with varying sequence lengths. Value $c^2 E(X_t)$ is included as the closest estimate of p_t (implied by the Table 1). In KA-N, parameters K and λ are fitted to a sample of N random alignment scores (as the learning is done for each sequence length separately, learned K and λ should reduce the effect of edges). Affine gaps were used.

Method	$m = n =$ 125	250	500	1000	2000
KA-100	1.086e-12	1.454e-10	9.196e-12	1.348e-8	2.615e-9
KA-1000	4.906e-12	2.241e-11	2.919e-10	1.687-9	4.963e-9
KA-10000	3.009e-11	2.936e-11	2.168e-10	8.975e-10	3.760e-9
Mercier et.al.	1.337e-11	5.353e-11	2.142e-10	8.569e-10	3.428e-9
$E(X_t)$	1.687e-11	8.636e-11	3.868e-10	1.633-9	6.708-9
$c^2 E(X_t)$	9.491e-12	4.858e-11	2.176e-10	9.186e-10	3.773e-9

d is 5 and gap extension has penalty $e = 2$. For the linear gap model we have used gap penalty d of 5.

The experimental results in Table 1 show than when $E(X_t) < \frac{1}{2}$, the ratio of $E(X_t)$ and p_t is about 2. This is very close to value $c^{-2} = \frac{16}{9}$ suggested in Subsection 4.4. In Table 2, we have compared small p-values computed by the Karlin-Altschul statistics, by the method of [9], and by our methods. The p-values computed by the Karlin-Altschul statistics seem to vary quite much, depending on the sample size used to fit parameters K and λ. Also the p-values computed with Karlin-Altschul are probably sometimes larger, and sometimes smaller than the actual p_t (suggested by the fact $p_t \leq E(X_t)$). For the experiments of Table 2, the true p_t was not known, as the sampling would require about 10^{13} sequence pairs, but based on Table 1, it seems plausible that the value $c^2 E(X_t)$ is very close to the actual p_t.

7   Discussion and Future Work

The results with our framework seem promising and might have implications we do not yet know. The framework is best suited for linear gap penalties, as then the algorithm needs quadratic time and only linear space. Thus, the algorithm is as fast as the Smith–Waterman algorithm [2], that computes the best local alignment score between sequences of interest. For our java implementation, it took about 2.8 seconds to compute the upper bound for linear gapped p-value of two sequences of length 10000 on a standard desktop computer.

It was left as an open problem, how to improve the computation time of $E(X_t)$ with affine gaps. Moreover, the null model used in the framework could be order or position dependent instead of position independent Bernoulli. The problem of having one string fixed and the other being distributed according to the null model is another interesting problem to study in the future.

We have also tried our framework for global alignment score statistics, but it turned out that $E(X_t)$ does not give a good approximation of p_t in this case.

Acknowledgements. The author would like to thank Professor Esko Ukkonen and the anonymous referees for useful comments on this paper.

References

1. Durbin, R., Eddy, S., Krogh, A., Mitchison, G.: Biological Sequence Analysis: Probabilistic Models of Proteins and Nucleic Acids, 1st edn. Cambridge University Press, Cambridge (1998)
2. Smith, T., Waterman, M.: Identification of common molecular subsequences. J. Mol. Biol. 147, 195–197 (1981)
3. Altschul, S., Gish, W., Miller, W., Myers, E., Lipman, D.: Basic local alignment search tool. J. Mol. Biol. 215, 403–410 (1990)
4. Altschul, S., Madden, T., Schaffer, A., Zhang, J., Zhang, Z., Miller, W., Lipman, D.: Gapped BLAST and PSI-BLAST: a new generation of protein database search programs. Nucl. Acids Res. 25, 3389–3402 (1997)
5. Karlin, S., Altschul, S.F.: Methods for assessing the statistical significance of molecular sequence features by using general scoring schemes. PNAS 87, 2264–2268 (1990)
6. Gotoh, O.: An improved algorithm for matching biological sequences. J. Mol. Biol. 162, 705–708 (1982)
7. Karlin, S., Dembo, A., Kawabata, T.: Statistical composition of high–scoring segments from molecular sequences. The Annals of Statistics 18, 571–581 (1990)
8. Karlin, S.: Statistical signals in bioinformatics. Proc. Natl. Acad. Sci. USA 102, 13355–13362 (2005)
9. Mercier, S., Cellier, D., Charlot, F., Daudin, J.J.: Exact and asymptotic distribution of the local score of one i.i.d. Random sequence. In: Gascuel, O., Sagot, M.-F. (eds.) JOBIM 2000. LNCS, vol. 2066, pp. 74–83. Springer, Heidelberg (2001)
10. Pearson, W.: Empirical statistical estimates for sequence similarity searches. J. Mol. Biol. 276, 71–84 (1998)
11. Mitrophanov, A., Borodovsky, M.: Statistical significance in biological sequence analysis. Briefings in Bioinformatics 7, 2–24 (2006)

12. Naor, D., Brutlag, D.: On suboptimal alignments of biological sequences. In: Apostolico, A., Crochemore, M., Galil, Z., Manber, U. (eds.) CPM 1993. LNCS, vol. 684, pp. 179–196. Springer, Heidelberg (1993)
13. Gusfield, D.: Algorithms on Strings, Trees, and Sequences: Computer Science and Computational Biology, 1st edn. Cambridge University Press, Cambridge (1997)
14. Graham, R., Knuth, D., Patashnik, O.: Concrete mathematics: A Foundation for Computer Science, 2nd edn. Addison-Wesley, Reading (1994)
15. Cooley, J., Tukey, J.: An algorithm for the machine calculation of complex Fourier series. Mathematics of Computation 19, 297–301 (1965)
16. Bernstein, D.: Multidigit multiplication for mathematicians (2001)
17. Garey, M., Johnson, D.: Computers and Intractability: A Guide to the Theory on NP-Completeness. W. H. Freeman and Company, New York (1979)

MPSCAN: Fast Localisation of Multiple Reads in Genomes

Eric Rivals, Leena Salmela, Petteri Kiiskinen, Petri Kalsi, and Jorma Tarhio

LIRMM, CNRS and Université de Montpellier 2, Montpellier, France
rivals@lirmm.fr
Helsinki University of Technology, P.O. Box 5400, FI-02015 TKK, Finland
{lsalmela,iridian,tarhio}@cs.hut.fi

Abstract. With Next Generation Sequencers, sequence based transcriptomic or epigenomic assays yield millions of short sequence reads that
need to be mapped back on a reference genome. The upcoming versions of these sequencers promise even higher sequencing capacities; this
may turn the *read mapping* task into a bottleneck for which alternative
pattern matching approaches must be experimented. We present an algorithm and its implementation, called MPSCAN, which uses a sophisticated
filtration scheme to match a set of patterns/reads exactly on a sequence.
MPSCAN can search for millions of reads in a single pass through the
genome without indexing its sequence. Moreover, we show that MPSCAN
offers an optimal average time complexity, which is sublinear in the text
length, meaning that it does not need to examine all sequence positions.
Comparisons with BLAT-like tools and with six specialised read mapping programs (like BOWTIE or ZOOM) demonstrate that MPSCAN also
is the fastest algorithm in practice for exact matching. Our accuracy and
scalability comparisons reveal that some tools are inappropriate for read
mapping. Moreover, we provide evidence suggesting that exact matching
may be a valuable solution in some read mapping applications. As most
read mapping programs somehow rely on exact matching procedures to
perform approximate pattern mapping, the filtration scheme we experimented may reveal useful in the design of future algorithms. The absence
of genome index gives MPSCAN its low memory requirement and flexibility that let it run on a desktop computer and avoids a time-consuming
genome preprocessing.

1 Introduction

Next-generation sequencers (NGS), able to yield millions of sequences in a single
run, are presently being applied in a variety of innovative ways to assess crucial
biological questions: to interrogate the transcriptome with high sensitivity [1], to
assay protein-DNA interactions at a genome wide scale [2], or to investigate the
open chromatine structure of human cells [3,4]. Due to their wide applicability,
cost effectiveness, and small demand in biological material, these techniques
become widespread and generate massive data sets [5]. These experiments yield
small sequence reads, also called *tags*, which need to be positioned on the genome.
For instance, one transcriptomics experiment delivered $\simeq 8$ million different 27

S.L. Salzberg and T. Warnow (Eds.): WABI 2009, LNBI 5724, pp. 246–260, 2009.
© Springer-Verlag Berlin Heidelberg 2009

bp tags, which were then mapped back to the genome. Only the tags mapping to a unique genomic location served to predict novel transcribed regions and alternative transcripts [6]. Generally, further analyses concentrate on those tags mapped to a unique genomic location [7].

The goal of tag mapping is to find for each tag the best matching genomic position. The ELAND program, which belongs to the bioinformatic pipeline delivered with the Solexa/Illumina® 1G sequencer, reports first an exact matching location if one is found, and otherwise seeks for locations that differ by 1 or 2 mismatches.

In the vast pattern matching literature, numerous guaranteed algorithms have been described to match exactly or approximately a pattern in a text (*i.e.* a read in a sequence), but only a few have been implemented to process efficiently tens of thousands of patterns [8]. In the context of read mapping, tools must be able to process millions of reads and thus, programs that exploit a precomputed genome index often prove more efficient [9,10,11,12]. Read mapping tools offer possibilities of approximate matching up to a limited number of differences (generally a few mismatches). However, they usually trade off a guaranteed accuracy for efficiency [13,10,11,12].

Another specificity of read mapping applications is that further processing considers only reads mapping to a unique position in the genome [7]. From a statistical viewpoint, exact matching of a 20 bp read is sufficient to identify a unique position in the human genome [14,15]. This implies that, instead of approximately matching full length reads, it may be as adequate to match, *i.e.* read prefixes, exactly. This would allow to keep the 100% accuracy, while still being efficient. Thus, it is desirable to further investigate whether exact set pattern matching algorithms can be adapted to meet the requirements of read mapping. For instance, it remains open whether an efficient pattern matching algorithm able to process huge read sets without indexing the genome exists.

To perform the mapping task, the user chooses either fast BLAST-like similarity search programs (BLAT [16], MEGABLAST [17], or SSAHA [18]), or specialised mapping tools (ELAND, TAGGER [19], RMAP [11], SEQMAP [13], SOAP [10], MAQ [9], BOWTIE [12], and ZOOM [20]). ELAND is probably the most used one [6,3,2]. While mappers were designed to process the huge tag sets output by NGS and allow only a few of substitutions and/or indels, similarity search tools were intended to find local alignments for longer query sequences, but can be twisted to map tags [3,21]. To speed up the search, both categories of tools follow a filtration strategy that eliminates quickly non-matching regions. The filtration usually requires to match exactly or approximately a short piece of the sequence (*e.g.*, BLAT or SEQMAP). All mappers but one [20] use variants of the *PEX* filter (as called in [8]), which consists in splitting the tag in $k + 1$ adjacent pieces, knowing that at least one will match exactly when a maximum of k errors are allowed. Logically to accelerate the filtration step, several of these tools exploit an index of the genome's words of length q (or q-mers) [16,18,19,12][1], which is stored on disk, loaded in memory once before all searches, and requires a computer intensive preprocessing of the genome [12]. The construction of a human

---

[1] As well as the version 2.0 of SOAP.

genome index lasts several hours even on powerful servers [12]. Among mapping tools, ZOOM distinguishes itself with a filtration relying on spaced seeds, *i.e.* matching subsequences instead of pieces [20].

Here we present a computer program MPSCAN[2], short for Multi-Pattern Scan, that is able to locate multiple reads in a single pass through the searched sequence and study its average time complexity (Section 2). In Section 3, we compare MPSCAN with the fastest of BLAST-like tools and mapping programs in terms of speed and scalability on large tag sets, and also evaluate the accuracy of similarity search tools for this task. We conclude by discussing the practical and algorithmical implications of our findings.

2   MPSCAN Algorithm

MPSCAN, short for Multi-Pattern Scan, is a program for set pattern matching: it searches simultaneously in a text for a set of words (*i.e.* tags) on a single computer (no parallelisation, no special hardware). To enable fast matching of large tag sets, we combine a *filtration/verification* approach with a search procedure based on bitwise comparisons, and a compact representation of the tag set. The tags are loaded in memory at the start and indexed in a trie-like structure, while the text is scanned on-the-fly by pieces.

The filtration strategy, which was explored for sets of up to $100,000$ patterns in [22], is the clue of MPSCAN efficiency. Filtration aims at eliminating most positions that cannot match any tag with an easy criterion. Then, verification checks whether the remaining positions truly match a tag. MPSCAN can handle a tag set in which tags differ in length. However, the filtration strategy works with tags of identical length; thus, it creates internally a set in which all tags are cut to the size of the smallest one (call this size l). The verification tests whether a complete tag matches. Filtration has been extensively applied to speed up similarity search algorithms, as in BLAST or BLAT [16]. MPSCAN's criterion relies on the fact that a matching window must share subwords of length q with the tags. Subwords of length q are called q-*mers*.

For verification purposes we index the tag set with a trie [8]. To save space, we prune the trie at nodes where the remaining suffixes can be stored in approximately 512 bytes. The suffixes are sorted for easier access during the verification phase. The pruned trie allows for efficient lookup speed and memory usage with patterns sharing common prefixes, and the remaining suffixes are efficiently packed, without compromising efficiency. Without pruning, the trie alone would result in unacceptably huge memory usage, as a single trie node takes up dozens of bytes in the form of pointers alone.

2.1   Filtration Strategy

Let us explain the filtration scheme with an example. Assume a set of 3 tags of length $l = 8$: $\{P_1, P_2, P_3\} = \{accttggc, gtcttggc, accttcca\}$, and set q to 5. The

---

[2] MPSCAN is freely available for academic users and can be downloaded at http://www.atgc-montpellier.fr/mpscan

overlapping 5-mers of each pattern are given in Figure 2. For a text window W of length 8 to match P_1, we need that the subword starting at position i in W matches the ith q-mer of P_1 for all possible i, and conversely. Now, we want to filter out windows that do not match any tag. If the subword starting at position i in W does not match the ith q-mer of neither P_1, P_2, nor P_3, then we are sure W cannot match any of the tags. Thus, our filtration criterion to surely eliminate any non-matching window W is to find if there exists a position i such that the previous condition is true.

From a set of tags, MPSCAN builds a single q-mer generalised pattern (Fig. 2). A generalised pattern allows several symbols to match at a position (like in a PROSITE pattern where a position e.g. $[DENQ]$ matches symbols D, E, N, or Q). However, here each q-mer is processed as a single symbol. Then, MPSCAN searches for this generalised pattern in the text with the *Backward Nondeterministic DAWG Matching (BNDM)* algorithm [8], which efficiently uses bit-parallelism. The basic idea of the algorithm is to recognize reversed factors (or substrings) of the pattern when scanning a window backward. When the scanned suffix of a window matches a prefix of the pattern, we store this position as a potential start of the next window. When we reach a point in the backward scanning where the suffix of the window is not a factor of the pattern, we shift the window forward based on the longest recognized prefix of the pattern except for the whole pattern. If no prefix was recognized, the length of the shift is $l - q + 1$. To achieve this efficiently, we initialize during preprocessing a bit vector $B[s]$ for each q-mer s, where the ith bit in the bit vector is one if the q-mer appears in the reversed pattern in position i. During searching the algorithm maintains a state vector E, where the ith bit is one if the scanned q-mers match the pattern starting at position i. When we read a new q-mer s, the state vector is updated as follows:

$$E = (E \ll 1) \ \& \ B[s] \ ,$$

where $\ll$ shifts the bits to the left and $\&$ performs a bitwise and of the two bit vectors. If the first bit in E is one, we have read a prefix of the pattern, and if all the bits in E are zero, the scanned suffix of the window does not match any factor of the pattern. Figure 1 gives the pseudo code for the filtration phase.

2.2 Optimal Average Complexity of MPSCAN

For a single pattern, BNDM has a sublinear average complexity with respect to the text length n; in other words, it does not examine all characters of the text. The combination of the BNDM algorithm with q-mers was first studied in [22], where it was shown sublinear. Here we prove that, if one sets the value of q relatively to the total number of tags r, MPSCAN average time complexity is not only sublinear with respect to n, but optimal. Indeed, the average complexity of the set pattern matching problem is $\Omega(n \log_c(rl)/l)$ (cf. [23]) and we prove:

Theorem 1. *The average time complexity of* MPSCAN *for searching r patterns of size l in a text of length n over an alphabet of size c is* $\mathcal{O}(n \log_c(rl)/l)$ *if* $q = \Theta(\log_c(rl))$.

```
1: i ← l − q + 1
2: while i ≤ n − q + 1 do
3:     j = 1; last ← l − q + 1
4:     E = B[s_i]  {s_i is the i^th q-mer of the scanned sequence}
5:     while true do
6:         if first bit in E is one then
7:             {the scanned window is a prefix of the pattern}
8:             if j = l − q + 1 then
9:                 verify an occurrence; break
10:            end if
11:            last ← l − q + 1 − j
12:        end if
13:        if E = 0 then
14:            break {the scanned window is not a factor of the pattern}
15:        end if
16:        E ← (E ≪ 1) & B[s_{i−j}]  {s_{i−j} is the (i − j)^th q-mer of the scanned sequence}
17:        j ← j + 1
18:    end while
19:    i ← i + last
20: end while
```

Fig. 1. Pseudo code for the filtration phase of MPSCAN

Proof. We want to prove that the average time complexity of MPSCAN for search-ing r patterns of size l in a text of length n over an alphabet of size c is $\mathcal{O}(n \log_c(rl)/l)$ if $q = \Theta(\log_c(rl))$. Practically, c equals 4 for DNA sequences.

Remember that MPSCAN processes the text in windows and it always reads the windows from right to left. We will call a window *good* if the last q-mer of the window does not match any pattern in any position. All other windows are called *bad*. In a good window, MPSCAN reads only the last q-mer and then shifts the window by $l − q + 1$ characters. In a bad window MPSCAN reads up to l characters and then shifts the window by at least one position (but often more than that).

For the purposes of the proof, the filtering phase of MPSCAN is divided into subphases that we define as follows. Let W_i, $i = 1, 2, \ldots$ be the windows scanned by MPSCAN. The first subphase starts with W_1. Let W_s be the first window of a subphase. Only a good window can end a subphase, but not all of them do. Indeed, the first good window in the series of windows indexed with $i := s + qk$, i.e. W_{s+qk}, with $k = 0, 1, \ldots$ is the last window of that subphase. The next window starts a new subphase. It follows that each subphase consists of X groups of q windows and one good window, with $X \geq 0$ being a random variable. Each of the X groups of q windows starts with a bad window and the rest $q − 1$ windows may be of any type. Figure 3 shows an example of dividing the windows into subphases.

The type of a window following a group of q windows is independent of the first window of the group, because the pattern has been shifted by at least q

$$\{P_1, P_2, P_3\} = \{accttggc, gtcttggc, accttcca\}$$

(a)

```
 1 2 3 4 5 6 7 8          1 2 3 4 5 6 7 8          1 2 3 4 5 6 7 8
P₁ a c c t t g g c       P₂ g t c t t g g c       P₃ a c c t t c c a
   a c c t t                 g t c t t                 a c c t t
     c c t t g                 t c t t g                 c c t t c
       c t t g g                 c t t g g                 c t t c c
         t t g g c                 t t g g c                 t t c c a
```

(b)

$$[acctt, gtctt][ccttg, tcttg, ccttc][cttgg, cttcc][ttggc, ttcca]$$

(c)

Fig. 2. Filtration scheme of MPSCAN. (a) A set of 3 tags of length $l = 8$. (b) The overlapping 5-mers starting at position 1 to 4 (resp. in light, dark, normal, very dark gray) of each tag. (c) The generalised 5-mers pattern for the set of tags.

bad good **good** **good** **bad** bad **bad** good **good** **bad** bad **good** ...

Fig. 3. Dividing the search phase into subphases when $q = 2$. The windows, whose type influences the division, are shown in boldface.

positions between them and the type of a window is determined solely by the last q-mer of the window. If $q \leq l - q + 1$, the type of a window after a good window is also independent of the good window, *i.e.* the q-mer determining the type of the next window contains only characters that have not been previously read. Because each subphase contains at least one good window, the text of length n will surely be covered after $\mathcal{O}(n/(l-q+1))$ subphases.

The probability that a random q-mer matches any of the patterns in any position is at most rl/c^q, because there are c^q different q-mers and at most rl of these can occur in the patterns (r patterns each of length l). This is also the probability that a window is bad. In a bad window MPSCAN reads the q-mers from right to left. It surely stops when it encounters a q-mer that does not match any q-mer in any of the patterns. In the worst case, MPSCAN reads the whole window and compares it against all the patterns taking $\mathcal{O}(rl)$ time. Note that this is a very pessimistic estimate. In practise, verification is not triggered in all bad windows and even then MPSCAN compares the window against only a few patterns.

In a good window, MPSCAN reads q characters. Therefore in one subphase of filtering, the number of characters read by MPSCAN is less than

$$\mathcal{O}(q) \cdot P(X = 0) + \sum_{i=1}^{\infty} (\mathcal{O}(q) + i \cdot q \cdot \mathcal{O}(rl)) \cdot P(X = i)$$

$$= \mathcal{O}(q) + \sum_{i=1}^{\infty} i \cdot q \cdot \mathcal{O}\left(rl\right) \cdot P(X=i) \le \mathcal{O}(q) + q \cdot \mathcal{O}\left(rl\right) \sum_{i=1}^{\infty} i \left(\frac{rl}{c^q}\right)^i.$$

This sum will converge if $rl/c^q < 1$ or equally if $q > \log_c(rl)$ and then

$$\mathcal{O}(q) + q \cdot \mathcal{O}\left(rl\right) \sum_{i=1}^{\infty} i \left(\frac{rl}{c^q}\right)^i = \mathcal{O}(q) + q \cdot \mathcal{O}\left(rl\right) \frac{\frac{rl}{c^q}}{\left(1 - \frac{rl}{c^q}\right)^2} = \mathcal{O}(q) + q \cdot \mathcal{O}\left(rl\right) \frac{rl \cdot c^q}{(c^q - rl)^2}.$$

If we choose $q \ge a \log_c(rl)$, where $a > 1$ is a constant, then $c^q \ge r^a l^a$. Because $a > 1$, $c^q - rl = \Omega(c^q)$ and therefore

$$\frac{1}{c^q - rl} = \mathcal{O}\left(\frac{1}{c^q}\right).$$

Now, the work done by the algorithm in one subphase takes less than

$$\mathcal{O}(q) + q \cdot \mathcal{O}\left(rl\right) \frac{rl \cdot c^q}{(c^q - rl)^2} = \mathcal{O}(q)\left(1 + \mathcal{O}\left(\frac{r^2 l^2 c^q}{c^{2q}}\right)\right) = \mathcal{O}\left(q \cdot \frac{r^2 l^2}{c^q}\right) = \mathcal{O}(q)$$

if $a \ge 2$. There are $\mathcal{O}(n/(l - q + 1)) = \mathcal{O}(n/l)$ subphases and the average complexity of one subphase is $\mathcal{O}(q)$. Overall the average case complexity of filtering in MPSCAN is thus

$$\mathcal{O}\left(\frac{n}{l} \cdot q\right) = \mathcal{O}(n \log_c(rl)/m)$$

if $q = a \log_c(rl) \le l - q + 1$ for a constant $a \ge 2$. The condition $q \le l - q + 1$ is equivalent to $q \le (l+1)/2$. Such a q can be found if $2 \log_c(rl) < (l+1)/2$ or equally if $r < c^{\frac{1}{4}(l+1)}/l$. □

The above proof predicts that a good choice for q would be $2 \log_c(rl)$, but in practice a good choice for q seems to be roughly $\log_c(rl)$. If we analysed the complexity of bad windows and verification more carefully, we could bring the theoretical result closer to the practical one.

3   Comparison

The MPSCAN algorithm offers a good theoretical average complexity, but how does it behave in practice and compare to other solutions? We performed search tests to investigate MPSCAN behavior and to compare it to either ultra-fast similarity search tools used for this task (BLAT, MEGABLAST, and SSAHA) or to mapping tools. For each tool, we set its parameters to let it search only for exact matches (which is for instance impossible with MAQ). ELAND could not be included in the comparison for we do not have a copy of the program.

Let us first recall some distinguishing features of those similarity search programs. They were designed to search for highly similar sequences faster than BLAST and exploit this high level of similarity to speed up the search. All are heuristic, but are by design better and faster than BLAST for searching exact occurrences of reads in a genome. MEGABLAST proceeds like BLAST: it scans the genome with the query read, which takes time proportional to the genome size. BLAT and SSAHA both use an index of the genome that records the occurrence positions of q-mers in the genome for some length q. Then, they search for all q-mers of the query in the index to determine which regions of the genome likely contain occurrences. This requires a time proportional to the read size. Note that q is the key parameter to balance between sensitivity and speed. Hence, BLAT and SSAHA avoid scanning repeatedly the whole genome, but require to precompute an index.

3.1   Speed and Memory with Respect to Text Length

First, we compared the running times of MPSCAN and of all similarity search programs with a set of 200 K-tags and texts of increasing length (Fig. 6, time in log scale). For all programs except BLAT, the running time increases less than linearly with the text length (but BLAT follows the same trend above 50 Mbps). For instance, MPSCAN takes 1.1 sec to search in 10 Mbps of Human chromosome 1, but only 5.6 sec in 247 Mbps: a 5-fold increase of the running time for a 25-fold increase in length. This illustrates well the sublinear time complexity of MPSCAN (Th. 1), which proves to be faster than the reference methods. The behavior is logical: MEGABLAST and MPSCAN first build their search engine, and then scan the text by pieces. The time spent for initialisation of the engine is better amortised with longer texts. This also explains why the memory used by MPSCAN is independent of the text length.

Second, we measured the time and memory footprint needed by MPSCAN and mapping tools to search the complete human genome with one million 27 bp tags. ZOOM requires 17 minutes and 0.9 Gigabytes, RMAP takes 30 min and 0.6 Gb, SEQMAP performs the task in 14 min with 9 Gb, BOWTIE runs in > 6 min with 1.4 Gb and MPSCAN needs < 5 min using 0.3 Gb. MPSCAN runs faster than BOWTIE although the latter uses a precomputed index, and it is three times faster than SEQMAP, the third most efficient tool.

3.2   Scalability with Respect to Number of Patterns

The central issue is the scalability in terms of number of tags. To investigate this issue, we plot their running times when searching for increasing tag sets (Fig 4). The comparison with similarity search tools is shown in Figure 4a. BLAT is by far the slowest tool, while MEGABLAST's time increases sharply due an internal limitation on the maximal number of tags searched at once, which forces it to perform several scans. SSAHA takes full advantage of its index with large pattern sets, and becomes 10 times faster than MEGABLAST. However, MPSCAN runs always faster than BLAT, MEGABLAST, and SSAHA. Especially for more than 400 K-tags, it outperforms other programs by almost an order of magnitude (9.8 s for 700 K-tags instead of 78 for SSAHA, 670 for MEGABLAST and 4, 234 s for BLAT). Importantly, the times needed by other programs increase more sharply with the number of tags than that of MPSCAN, especially after 100K, auguring ill for their scalability beyond a million tags.

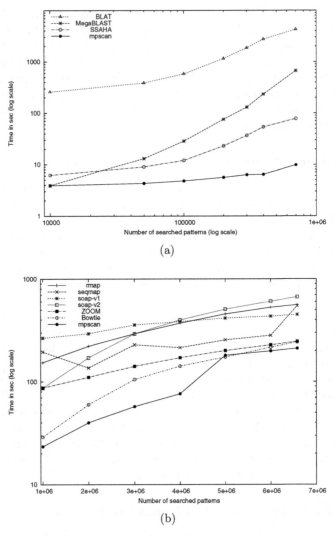

Fig. 4. Evaluation of scalability. Search times on chromosome 1 (247 Mbp) for increasing tag sets. (a) Comparison with similarity search tools. Search times of BLAT, MEGABLAST, SSAHA, MPSCAN in seconds for 21 bp LongSAGE tags, for sets of 10, 50, 100, 200, 300, 400, and up to 700 Kilo-tags (K-tags). Both axes have logarithmic scales. The curve of MPSCAN running time is almost flat: for instance doubling the tag set from 200 to 400 K-tags yields a small increase from 5.6 to 6.4 s. Its time increases in a sublinear fashion with the number of tags. For all other tools, the increase of the tag set gives rise to a proportional growth of the running time. *E.g.*, SSAHA needs 23 s for 200 K-tags and 54 s for 400 K-tags. (b) Comparison with mapping tools: Search times of RMAP, SEQMAP, SOAP (v1 & v2), ZOOM, BOWTIE and MPSCAN in seconds (log scale) for increasing subsets of 27 bp ChIP-seq tags. All tools behave similarly and offer acceptable scalability. MPSCAN remains the most efficient of all and can be 10 times faster than tools like SEQMAP or RMAP. The times do not include the index construction time.

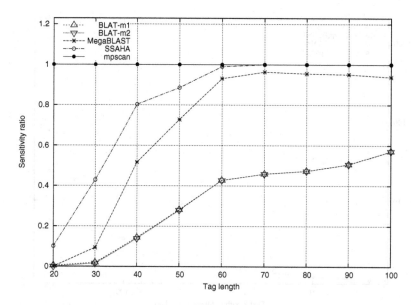

Fig. 5. Sensitivity of MEGABLAST, BLAT, and SSAHA compared to MPSCAN as the percentage of found matches after filtering. Influence of tag length on sensitivity (with $r = 10,000$ tags). BLAT-m1 and BLAT-m2 give BLAT's sensitivity when the filtration criterion asks for one or two seed matches, respectively; BLAT-m1 found necessarily more matches than BLAT-m2. However, here their curves are superimposed. The sensitivity of similarity search tools is low (< 0.5) for tags ≤ 30 bp and reaches a maximum for MEGABLAST and SSAHA at ≥ 60 bp.

Beyond that, we consider specialised mapping tools whose behavior is illustrated in Figure 4b. For this, we used 6.5M 27 bp RNA Polymerase II ChIP-seq tags sequenced in an erythroleukemia cell line (HK652, GEO GSM325934) and took increasing subsets every million tags. All tools exhibit a running time that increases linearly with the number of tags: a much better scalability than similarity search tools. Compared to similarity search tools, all mappers behave similarly, probably due to the resemblance of their filtration algorithm.

Both BOWTIE and SOAP-v2 use a Burrows-Wheeler-Transform index with a similar exact matching algorithm, but it benefits more BOWTIE than SOAP-v2, making BOWTIE the faster of mapping tools. This emphasises how much implementation issues influence efficiency. Among non-index based programs, ZOOM exhibits a behavior close to that of BOWTIE above 3M tags, showing that ultrafast running times are not bound to an index. For moderate tag sets (< 4M tags) MPSCAN is two to four times faster than ZOOM, its fastest competitor in this category. Even if MPSCAN's running time increases from 4 to 5M tags due to a multiplication by 5 of the number of matches, it remains the fastest of all tools for exact matching. This shows that exact set pattern matching can be highly efficient even without a genome index and answers the question asked in the introduction. MPSCAN's filtration strategy is logically sensitive to the

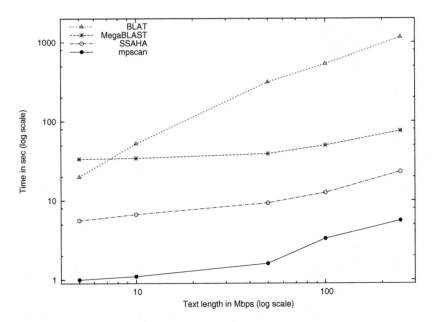

Fig. 6. Search times of BLAT, MegaBlast, SSAHA, mpscan in seconds for 200 Kilo-tags (LongSAGE tags of 21 bp) on increasing pieces of length 5, 10, 50, 100, and 247 Mbp of Human chromosome 1. Both axes have logarithmic scales. These curves illustrate the sublinear increase of time with respect to text length for all tools except BLAT, and the superiority of mpscan in running time.

ratio $\#reads/4^q$, which suggests that using longer computer-words (on 64-bit processors) will improve its efficiency and scalability.

3.3   Accuracy

The mpscan algorithm is guaranteed 100% accurate (and extensive tests have shown that the mpscan program also is): it reports all patterns' occurrences (100% sensitive) and only these (100% selective) [22,8].

Despite the availability of specialised mapping tools, popular heuristic similarity search program like BLAT are still used for read mapping [21], for they can find distant alignments. However to our knowledge, their accuracy has never been assessed in this context. We performed a thorough comparison of their exact matching capacity, since it should be the easiest part of the task. Our results show it is a complex matter: especially their sensitivity is influenced by the numbers of occurrences, the relative length of seeds compared to matches, the parameters set for both building the index and searching.

While all tools (SSAHA, BLAT, and MegaBlast) achieve their best accuracy for long patterns (for ≥ 60 bp, *i.e.* when the seed is guaranteed to fall in each occurrence), all encounter problems finding short patterns (≤ 30 bp). Index and parameters must be adapted to gain sensitivity at the expense of time and flexibility (one cannot exploit the same index for different tag lengths), which is an issue for digital transcriptomic and ChIP-seq data (≤ 25 bp in [1,3,2]). For instance, with 30 bp patterns, all are less than 50% sensitive with pattern sets $\geq 10,000$ (Fig. 5). For both parameter sets used with BLAT, its sensitivity remains below 0.6 whatever the tag length. The number of tags also has a negative influence on the sensitivity of similarity search tools (data not shown). However, it is logical that similarity search tools have limited accuracy, since they were not designed for exact pattern matching.

The accuracy of mapping tools that allow both exact and approximate matching should be evaluated globally and their dependence to several parameters (tag length, error types, tag number, genome length) should be investigated. Indeed, the underlying definitions of a read best match, the strategies for finding it, as well as the notion of approximation differ among tools, hampering this comparison. This is beyond the scope of this paper. Nevertheless, we have analysed the accuracy of ELAND. Although, we do not have access to the program, some of ELAND's raw results can be downloaded from public repositories like GEO. ELAND searches for exact and approximate matches with ≤ 2 mismatches. We analysed the subset of mapped tags in the ELAND output of the NRSF ChIP-seq data set [2]. ELAND finds only approximate matches for $442,766$ tags, while MPSCAN locates an exact match for $59,571$ of these tags (13% of them).

Such an inaccuracy may impact the final positioning of protein binding or DNA modification sites. This comparison illustrates the difficulty of searching for large tag sets in sequences and the benefit of using a guaranteed pattern matching algorithm for this task.

3.4   Relevance of Exact vs. Approximate Mapping

Currently, new sequencers yield short tags (*i.e.* < 30 bp) in Digital Gene Expression, RNA-Seq, and ChIP-seq experiments (14, 20, and 27 bp in [1,3,6] respectively). Technological developments aim at increasing the tag length to improve the probability of a read to match a unique genomic location. However, the error probability also increases with tag length [21,15]. Altogether, the tag length has an opposite influence on the probabilities of a tag to be mapped and to be mapped at a unique location.

To evaluate the relevance of exact versus approximate matching, we did the following experiment with a Pol-II ChIP-seq set of 34 bp tags (GEO GSM325934). If one maps with MPSCAN the full length tags, 86% remain unmapped and 11% are uniquely mapped. With at most two mismatches, ELAND finds 14% of additional uniquely mapped tags (categories U1 and U2), while mapping the 20 bp prefix of each tag with MPSCAN allows to map 25% of all tags at unique positions (14% more sites than with full length tags).

This result suggests that optimising the final output of a sequence census assay in terms of number of uniquely mapped locations is a complex issue. Approximate mapping is seen as a solution to detect more genomic sites, but it often maps tags at multiple locations [24]. In fine, exact matching may turn out to be a relevant alternative strategy compared to approximate matching. Thus, the proposed filtration algorithm may be useful in read mapping applications, especially if one considered mapping a substring of the original reads. A more in-depth investigation of this issue is exposed in [15].

4   Discussion

Key biological questions can be investigated at genome scale with new sequencing technologies. Whether in genomic, transcriptomic or epigenomic assays, millions of short sequence reads need first to be mapped on a reference genome. This is a compulsory step in the bioinformatic analysis. We presented an efficient program, MPSCAN, for mapping tags exactly on a genome, evaluated its relevance for read mapping, and compared it to two classes of alternative solutions: i) ultrafast similarity search tools and ii) specifically designed mapping tools. We summarise below some valuable evidence and take-home messages brought by this study.

Similarity search tools are inappropriate for mapping exactly short patterns ≤ 40 bp, since their sensitivity remains too low (< .5 for 30 bp long tags). Whatever the number of seeds required to examine a hit, BLAT is the least sensitive among the tested similarity search tools. Its sensitivity never reaches 0.6, even with patterns up to 100 bp. In other words, similarity search tools miss many exact matching locations, which are considered to be the most secure locations in many applications [3,2]. In general, the scalability of similarity search tools is not satisfactory for tag mapping: both the speed of processing and the sensitivity suffer when the number of tags becomes large.

Mapping tools are adequate for this task. They enable the user to map up to millions of tags fast on the human genome, and scale up well. Nevertheless, differences in speed can be important: e.g., an order of magnitude for mapping 2M tags between MPSCAN and SOAP-v2. If most algorithms are similar, from the user viewpoint the programs are not equivalent: neither in flexibility, ease of use, speed, options, nor in accuracy.

From the algorithmic viewpoint, our results suggest that indexing is not required to perform exact mapping of tags on long sequences. In the class of similarity search tools, the superiority in speed of SSAHA compared to BLAT and MEGABLAST is due to its index, but also to its lack of verification, which induces a poor specificity. In our comparison of seven programs (the largest we are aware of), BOWTIE seems the fastest among mapping tools, but never beats the performance of MPSCAN for exact mapping.

ZOOM, which exploits spaced seeds in its filtration scheme, compares favorably in speed to tools using the splitting strategy or PEX filter, such as SEQMAP, RMAP, SOAP. This suggests the superiority of spaced seeds. However, this superiority has a price in terms of flexibility: sets of spaced seeds are specifically designed for a certain tag length and maximum number of mismatches, and different sets corresponding to different parameter combinations are hard coded in ZOOM. For instance, a set of 4 spaced seeds of weight 13 was manually designed to search for 33 bp tags [20]. Hence, adaptation of ZOOM to a new setup requires the design of specific seeds, which is a theoretically hard and practically difficult problem [25,26,27]. The present limitation of ZOOM to patterns up to 64 bp is certainly due to this bottleneck.

In conclusion, we presented an exact set pattern matching program, MPSCAN, which is based on a filtration scheme that had never been applied to read mapping. Our current implementation has pushed the limit on the number of tags by two orders of magnitude compared to previous pattern matching algorithms [8,22]. We conducted thorough comparisons with similarity search algorithms and mapping tools in term of speed and scalability. Our experiments revealed that BLAT-like tools are inadequate for short read mapping both in terms of scalability and of sensitivity, which, to our knowledge, has never been reported before. From the algorithmic viewpoint, we demonstrated the average running time optimality of MPSCAN, which turns out to be very efficient in practice. Compared to mapping tools for exact mapping, MPSCAN runs faster and scales well: it can even compete in efficiency with programs using a sophisticated genome index, like BOWTIE. Since it uses no index, MPSCAN combines flexibility, low memory footprint, and high efficiency, while avoiding a time consuming index precomputation (cf. building times in [12]). Finally, we provide evidence that exact matching approaches can be relevant for read mapping applications, especially in the perspective of longer reads. It remains open to find filtration strategies that achieve efficient "near exact" mapping.

With future generation of sequencers, which promise further increases in sequencing capacity, read mapping may become a bottleneck. Further research in theoretical and practical pattern matching will be needed to tackle this challenging question.

Acknowledgments. This work was supported by Université Montpellier II [grant BIO-MIPS], Academy of Finland [grant 111060]. We gratefully thank L. Duret, A. Boureux, T. Commes, L. Bréhèlin for helpful comments.

References

1. Kim, J., Porreca, G., Song, L., Greenway, S., Gorham, J., Church, G., Seidman, C., Seidman, J.: Polony Multiplex Analysis of Gene Expression (PMAGE) in Mouse Hypertrophic Cardiomyopathy. Science 316(5830), 1481–1484 (2007)
2. Johnson, D., Mortazavi, A., Myers, R., Wold, B.: Genome-Wide Mapping of in Vivo Protein-DNA Interactions. Science 316(5830), 1497–1502 (2007)
3. Boyle, A.P., Davis, S., Shulha, H.P., Meltzer, P., Margulies, E.H., Weng, Z., Furey, T.S., Crawford, G.E.: High-Resolution Mapping and Characterization of Open Chromatin across the Genome. Cell 132, 311–322 (2008)
4. Schones, D., Zhao, K.: Genome-wide approaches to studying chromatin modifications. Nat. Rev. Genet. 9(3), 179–191 (2008)
5. Mardis, E.R.: ChIP-seq: welcome to the new frontier. Nat. Methods 4(8), 613–614 (2007)
6. Sultan, M., Schulz, M.H., Richard, H., Magen, A., Klingenhoff, A., Scherf, M., Seifert, M., Borodina, T., Soldatov, A., Parkhomchuk, D., Schmidt, D., O'Keeffe, S., Haas, S., Vingron, M., Lehrach, H., Yaspo, M.L.: A Global View of Gene Activity and Alternative Splicing by Deep Sequencing of the Human Transcriptome. Science 321(5891), 956–960 (2008)

7. Barski, A., Cuddapah, S., Cui, K., Roh, T.Y., Schones, D.E., Wang, Z., Wei, G., Chepelev, I., Zhao, K.: High-Resolution Profiling of Histone Methylations in the Human Genome. Cell 129(4), 823–837 (2007)

8. Navarro, G., Raffinot, M.: Flexible Pattern Matching in Strings - Practical on-line search algorithms for texts and biological sequences. Cambridge Univ. Press, Cambridge (2002)

9. Li, H., Ruan, J., Durbin, R.: Mapping short DNA sequencing reads and calling variants using mapping quality scores. Genome Res. 18, 1851–1858 (2008) (in press)

10. Li, R., Li, Y., Kristiansen, K., Wang, J.: SOAP: short oligonucleotide alignment program. Bioinformatics 24(5), 713–714 (2008)

11. Smith, A., Xuan, Z., Zhang, M.: Using quality scores and longer reads improves accuracy of solexa read mapping. BMC Bioinformatics 9(1), 128 (2008)

12. Langmead, B., Trapnell, C., Pop, M., Salzberg, S.: Ultrafast and memory-efficient alignment of short dna sequences to the human genome. Genome Biology 10(3), R25 (2009)

13. Jiang, H., Wong, W.H.: Seqmap: mapping massive amount of oligonucleotides to the genome. Bioinformatics 24(20), 2395–2396 (2008)

14. Saha, S., Sparks, A., Rago, C., Akmaev, V., Wang, C., Vogelstein, B., Kinzler, K., Velculescu, V.: Using the transcriptome to annotate the genome. Nat. Biotech. 20(5), 508–512 (2002)

15. Philippe, N., Boureux, A., Tarhio, J., Bréhélin, L., Commes, T., Rivals, E.: Using reads to annotate the genome: influence of length, background distribution, and sequence errors on prediction capacity. Nucleic Acids Research (2009) doi:10.1093/nar/gkp492

16. Kent, J.W.: BLAT—The BLAST-Like Alignment Tool. Genome Res. 12(4), 656–664 (2002)

17. Zhang, Z., Schwartz, S., Wagner, L., Miller, W.: A greedy algorithm for aligning DNA sequences. J. of Computational Biology 7(1-2), 203–214 (2000)

18. Ning, Z., Cox, A., Mulikin, J.: SSAHA: A Fast Search Method for large DNA Databases. Genome Res. 11, 1725–1729 (2001)

19. Iseli, C., Ambrosini, G., Bucher, P., Jongeneel, C.: Indexing Strategies for Rapid Searches of Short Words in Genome Sequences. PLoS ONE 2(6), e579 (2007)

20. Lin, H., Zhang, Z., Zhang, M.Q., Ma, B., Li, M.: ZOOM! Zillions of oligos mapped. Bioinformatics 24(21), 2431–2437 (2008)

21. Kharchenko, P., Tolstorukov, M.Y., Park, P.J.: Design and analysis of ChIP-seq experiments for DNA-binding proteins. Nat. Biotech. 26(12), 1351–1359 (2008)

22. Salmela, L., Tarhio, J., Kytöjoki, J.: Multipattern string matching with q-grams. ACM Journal of Experimental Algorithmics 11(1) (2006)

23. Navarro, G., Fredriksson, K.: Average complexity of exact and approximate multiple string matching. Theoretical Computer Science 321(2-3), 283–290 (2004)

24. Faulkner, G., Forrest, A., Chalk, A., Schroder, K., Hayashizaki, Y., Carninci, P., Hume, D., Grimmond, S.: A rescue strategy for multimapping short sequence tags refines surveys of transcriptional activity by CAGE. Genomics 91, 281–288 (2008)

25. Kucherov, G., Noé, L., Roytberg, M.: Multiseed Lossless Filtration. IEEE/ACM Transactions on Computational Biology and Bioinformatics 2(1), 51–61 (2005)

26. Ma, B., Li, M.: On the complexity of the spaced seeds. J. of Computer and System Sciences 73(7), 1024–1034 (2007)

27. Nicolas, F., Rivals, E.: Hardness of optimal spaced seed design. J. of Computer and System Sciences 74, 831–849 (2008)

Fast Prediction of RNA-RNA Interaction

Raheleh Salari[1], Rolf Backofen[2], and S. Cenk Sahinalp[1]

[1] School of Computing Science, Simon Fraser University, Burnaby, Canada
[2] Bioinformatics Group, Albert-Ludwigs-University, Freiburg, Germany

Abstract. Regulatory antisense RNAs are a class of ncRNAs that regulate gene expression by prohibiting the translation of an mRNA by establishing stable interactions with a target sequence. There is great demand for efficient computational methods to predict the specific interaction between an ncRNA and its target mRNA(s). There are a number of algorithms in the literature which can predict a variety of such interactions - unfortunately at a very high computational cost. Although some existing target prediction approaches are much faster, they are specialized for interactions with a single binding site.

In this paper we present a novel algorithm to accurately predict the minimum free energy structure of RNA-RNA interaction under the most general type of interactions studied in the literature. Moreover, we introduce a fast heuristic method to predict the specific (multiple) binding sites of two interacting RNAs. We verify the performance of our algorithms for joint structure and binding site prediction on a set of known interacting RNA pairs. Experimental results show our algorithms are highly accurate and outperform all competitive approaches.

1 Introduction

Regulatory non-coding RNAs (ncRNAs) play an important role in gene regulation. Studies on both prokaryotic and eukaryotic cells show that such ncRNAs usually bind to their target mRNA to regulate the translation of corresponding genes. Many regulatory RNAs such as microRNAs and small interfering RNAs (miRNAs/siRNAs) are very short and have full sequence complementarity to the targets. However some of the regulatory antisense RNAs are relatively long and are not fully complementary to their target sequences. They exhibit their regulatory functions by establishing stable joint structures with target mRNA initiated by one or more loop-loop interactions.

In this paper we present an efficient method for RNA-RNA interaction prediction (RIP) problem with multiple binding domains. Alkan et al. [1] proved that RIP, in its general form, is an NP-complete problem and provided algorithms for predicting specific types of interactions and two relatively simple energy models - under which RIP is polynomial time solvable. We focus on the same type of interactions, which to the best of our knowledge, are the most general type of interactions considered in the literature; however the energy model we use is the joint structure energy model recently presented by Chitsaz et al. [5] which is more general than the one used by Alkan et al.

In what follows below, we first describe a combinatorial algorithm to compute the minimum free energy joint structure formed by two interacting RNAs. This algorithm has a running time of $O(n^6)$ and uses $O(n^4)$ space - which makes it impractical for long RNA molecules. Then we present a fast heuristic algorithm to predict the joint structure

S.L. Salzberg and T. Warnow (Eds.): WABI 2009, LNBI 5724, pp. 261–272, 2009.

formed by interacting RNA pairs. This method provides a significant speedup over our combinatorial method, which it achieves by exploiting the observation that the independent secondary structure of an RNA molecule is mostly preserved even after it forms a joint structure with another RNA. In fact there is strong evidence [7,11] suggesting that the probability of an ncRNA binding to an mRNA target is proportional to the probability of the binding site having an unpaired conformation. The above observation has been used by different methods for target prediction in the literature (see below for an overview). However, most of these methods focus on predicting interactions involving only a single binding site, and are not able to predict interactions involving multiple binding sites. In contrast, our heuristic approach can predict interactions involving multiple binding sites by: (1) identifying the collection of accessible regions for both input RNA sequences, (2) using a matching algorithm, computing a set of "non-conflicting" interactions between the accessible regions which have the highest overall probability of occurrence.

Note that an accessible region is a subsequence in an RNA sequence which, with "high" probability, remain unpaired in its secondary structure. Our method considers the possibility of interactions being formed between one such accessible region from an RNA sequence with more than one such region from the other RNA sequence. Thus, in step (1), it extends the algorithm by Mückstein et al. for computing the probability of a specific region being unpaired [12] to compute the joint probability of two (or more) regions remaining unpaired. Because an accessible region from an RNA typically interacts with no more than two accessible regions from the other RNA, we focus on calculating the probability of at most two regions remaining unpaired: within a given an RNA sequence of length n, our method can calculate the probability of any pair of regions of length $\leq w$ each, in $O(n^4.w)$ time and $O(n^2)$ space. In step (2), on two input RNA sequences of length n and m ($n \leq m$), our method computes the most probable non-conflicting matching of accessible regions in $O(n^2.w^4 + n^3/w^3)$ time and $O(w^4 + n^2/w^2)$ space.

Related Work. Early attempts to compute the joint structure of interacting RNAs started by concatenating the two interacting RNA sequences and treated them as a single sequence PairFold [2] and RNAcofold [3]. As these methods typically use secondary structure prediction methods that do not allow pseudoknots, they fail to predict joint structures formed by non-trivial interactions between a pair of RNAs. Another set of methods ignore internal base-pairing in both RNAs, and compute the minimum free energy secondary structure for their hybridization (RNAhybrid [15], UNAFold [6,9], and RNAduplex from Vienna package [3]). These approaches work only for simple cases involving typically very short strands. A further set of studies aim to compute the minimum free energy joint structure between two interacting RNAs. For example Pervouchine [14] devised a dynamic programming algorithm to maximize the number of base pairs among interacting strands. A follow up work by Kato et al. [8] proposed a grammar based approach to RNA-RNA interaction prediction. More generally Alkan et al. [1] studied the joint secondary structure prediction problem under three different models: 1) base pair counting, 2) stacked pair energy model, and 3) loop energy model. The resulting algorithms compute the optimum structure among all possible joint secondary structures that do not contain pseudoknots, crossing interactions, and *zigzags* (please see [1] for the exact definition). In fact the last set of algorithms above are the only methods that have the capability to predict joint secondary structures with multiple loop-loop interactions. However, these algorithms all requires significant computational resources ($O(n^6)$ time and $O(n^4)$ spaces) and thus are impractical for sequences of even modest length. A final group of methods are based on the observation that interaction is a multi-step process that involves: 1) unfolding of the two RNA structures to expose the bases needed for hybridization, 2) the hybridization at the binding site, and 3) restructuring of the complex to a new minimum free energy conformation. The main aim of these methods is to identify the potential binding sites which are going to be unfolded in order to form interactions. One such method presented by Alkan et al. [1], extends existing loop regions in independent structures to find potential binding sites. RNAup [13] presents an extension of the standard partition function approach to compute the probabilities that a sequence interval remains unpaired. IntaRNA [4] considers not only accessibility of a binding sites but also the existence of a seed to predict potential binding sites. All of these methods achieve reasonably high accuracy in predicting interactions involving single binding sites; however, their accuracy levels are not very high when dealing with interactions involving multiple binding sites.

2 Methods

We address the RNA-RNA Interaction Problem (RIP) based on the energy model of Chitsaz et al. [5] over the interactions considered by Alkan et al. [1]. Our algorithm computes the minimum free energy secondary structure among all possible joint secondary structures that do not contain pseudoknots, crossing interactions, and zigzags.

2.1 RNA-RNA Joint Structure Prediction

Recently Chitsaz et al. [5] present an energy model for joint structure of two nucleic acids over the type of interactions introduced by Alkan et al. [1]. Based on the presented energy model they propose an algorithm that consider all possible cases of joint structures to compute the partition function. The specified algorithm with some minor changes can be used to compute the minimum free energy joint structure of two interacting nucleic acids. Following we shortly describe the algorithm.

We are given two RNA sequences $\mathbf{R}$ and $\mathbf{S}$ of lengths n and m. We refer to the ith nucleotide in $\mathbf{R}$ and $\mathbf{S}$ by i_R and i_S respectively. The subsequence from the ith nucleotide to the jth nucleotide in one strand is denoted by $[i, j]$. We denote a base pair between the nucleotides i and j by $i \cdot j$. $MFE(i, j)$ denotes the minimum free energy structure of $[i, j]$, and $MFE(i_R, j_R, i_S, j_S)$ denotes the minimum free energy joint structure of $[i_R, j_R]$ and $[i_S, j_S]$.

Fig. 1 shows the recursion diagram of MFE for the joint structure of $[i_R, j_R]$ and $[i_S, j_S]$. In this figure a horizontal line indicates the phosphate backbone, a dashed curved line encloses a subsequence and denotes its two terminal bases which may be paired or unpaired. A solid vertical line indicates an interaction base pair, a dashed vertical line denotes two terminal bases which may be base paired or unpaired, and a dotted vertical line denotes two terminal bases which are assumed to be unpaired. Grey regions indicate a reference to the substructure of single sequences.

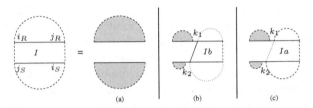

Fig. 1. Recursion for MFE joint structure of subsequences $[i_R, j_R]$ and $[i_S, j_S]$. Case a constitutes no interaction. In case b, the leftmost interaction bond is not closed by any base pair. In case c, the leftmost interaction bond is covered by base pair in at least one subsequence.

The joint structure of two subsequences derived from one of the following cases. The first case is when there is no interaction between the two subsequences. If there are some interaction bonds, the structure has two cases: either the leftmost bond is closed by base pair in at least one of the subsequences or not. If the joint structure starts with a bond which is not closed by any base pair we denote the case by Ib, otherwise the structure starts with a bond which is closed by base pair in at least one subsequence and the case is denoted by Ia. Therefore, $MFE(i_R, j_R, i_S, j_S)$ is calculated by the following dynamic programming:

$$
MFE(i_R, j_R, i_S, j_S) = \min \left\{
\begin{array}{ll}
MFE(i_R, j_R) + MFE(i_S, j_S) & (a), \\[2mm]
\min_{\substack{i_R \le k_1 < j_R \\ i_S < k_2 \le j_S}} \left\{
\begin{array}{l}
MFE(i_R, k_1 - 1) + \\
MFE(k_2 + 1, j_S) + \\
MFE^{Ib}(k_1, j_R, i_S, k_2)
\end{array}
\right\} & (b), \\[6mm]
\min_{\substack{i_R \le k_1 < j_R \\ i_S < k_2 \le j_S}} \left\{
\begin{array}{l}
MFE(i_R, k_1 - 1) + \\
MFE(k_2 + 1, j_S) + \\
MFE^{Ia}(k_1, j_R, i_S, k_2)
\end{array}
\right\} & (c),
\end{array}
\right\}
$$
(1)

in which $MFE^{Ib}(k_1, j_R, i_S, k_2)$ is the minimum free energy for the joint structure of $[k_1, j_R]$ and $[i_S, k_2]$ assuming $k_1 \cdot k_2$ is an interaction bond, and $MFE^{Ia}(k_1, j_R, i_S, k_2)$ is the minimum free energy for the joint structure of $[k_1, j_R]$ and $[i_S, k_2]$ assuming the leftmost interaction bond is covered by a base pair in at least one subsequence. The corresponding dynamic programing for computing the MFE^{Ib} and MFE^{Ia} can be derived from the cases explained in [5] in a similar fashion.

Similar to the partition function algorithm, the minimum free energy joint structure prediction algorithm has $O(n^6)$ running time and $O(n^4)$ space requirements. However the algorithm performs highly accurate (see section 3.2), but it requires substantial computational resources. This could be prohibitive for predicting the joint secondary structure of sufficiently long RNA molecules. Therefore, in the next section we present a fast heuristic algorithm to predict RNA-RNA interaction.

2.2   RNA-RNA Binding Sites Prediction

Our heuristic algorithm for RNA-RNA interaction prediction problem is based on the idea that the external interactions mostly occur between pairs of unpaired regions of single structures. We aim to predict interactions of multiple binding sites as long as they have no crossing. The heuristic algorithm contains the following steps:

- Predict the highly accessible regions in each strands. These regions include the loop regions in native structure of RNA strand. In order to predict accessible regions we chose all the regions which remain unpaired with high probability.
- Predict the optimal non-conflicting interactions between the accessible regions. For every pair of accessible regions of two interacting RNAs a cost of interaction is calculated. Then a matching algorithm runs to find the minimum cost non-conflicting subset of interactions.

Accessible Regions. For a single RNA sequence an accessible region is a subsequence that remains unpaired in equilibrium with high probability. The probability of an unpaired region can be calculated based on the algorithm presented in [12]. Here, we are interested in multiple unpaired regions. For this purpose one should compute the joint probabilities for any subset of possible intervals. Since the computation of all joint probabilities needs substantial time and space, in this paper we only consider the joint probability of two unpaired subsequences.

Denoting the set of secondary structures in which the sequence interval $[k, l]$ remains unpaired by $S^{u[k,l]}$, the corresponding partition function is

$$Q^{u[k,l]}(T) = \sum_{s \in S^{u[k,l]}} e^{-G_s/RT}, \tag{2}$$

where R is the universal gas constant and T is the temperature. In order to compute the $Q^{u[k,l]}$, the standard recursion for the partition function folding algorithm [10] can be extended as:

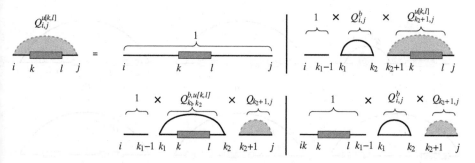

where $i \le k \le l \le j$ and $k_1 \cdot k_2$ is the leftmost base pair. Partition functions $Q^{b,u[k,l]}_{i,j}$ (where $i \cdot j$) and $Q^{m,u[k,l]}_{i,j}$ (where $[i, j]$ is inside a multiloop and constitutes at least one base pair) while the interval $[k, l]$ remains unpaired are derived from the standard algorithm in a similar way. Furthermore, probability of a base pair $p \cdot q$ while $[k, l]$ remains unpaired, $\mathbb{P}(p \cdot q | u[k, l])$, can be calculated by applying the McCaskill algorithm [10] for computing the base pair probability on $Q^{u[k,l]}$. It is easy to see that the desired partition function $Q^{u[k,l]}$ and base pair probability $\mathbb{P}(p \cdot q | u[k, l])$ are computed in same time and space complexity as the standard algorithm by McCaskill ($O(n^3)$ and $O(n^2)$ respectively).

Mückstein et al. [12] introduce an algorithm to compute the probability of unpaired region $\mathbb{P}(u[i,j])$ for a given sequence interval $[i,j]$. Here, we extend the specified algorithm to compute $\mathbb{P}(u[i,j]|u[k,l])$ which is the probability of unpaired region $[i,j]$ while $[k,l]$ remains unpaired. Clearly if some part of $[i,j]$ is within the interval $[k,l]$, the corresponding probability for that part is equal to one. Hence, for computing the probability only the parts of $[i,j]$ which are exterior to $[k,l]$ should be considered. Here, without loss of generality we assume $k \leq l < i \leq j$.

For unpaired interval $[i,j]$ there are two general cases: either it is not closed by any base pair, or it is part of a loop. Fig. 2 summarizes the cases of unpaired interval $[i,j]$ as a part of the loop enclosed by base pair $p \cdot q$ while interval $[k,l]$ remains unpaired. In case x interval $[p,q]$ does not contain interval $[k,l]$, and in the other cases $(a-e)$ interval $[k,l]$ lies in interval $[p,q]$. Probability $\mathbb{P}(u[i,j]|u[k,l])$ can be calculated as follows:

$$
\mathbb{P}(u[i,j]|u[k,l]) = \frac{Q^{u[k,l]}_{1,i-1} \times 1 \times Q_{j+1,n}}{Q^{u[k,l]}}
$$

$$
+ \sum_{l<p<i\leq j<q} \mathbb{P}(p \cdot q|u[k,l]) \times \frac{Q^{pq}_{i,j}}{Q^{b}_{p,q}} \qquad (x) \qquad (3)
$$

$$
+ \sum_{p<k\leq l<i\leq j<q} \mathbb{P}(p \cdot q|u[k,l]) \times \frac{Q^{pq,u[k,l]}[i,j]}{Q^{b,u[k,l]}_{p,q}} \qquad (a-e)
$$

$Q^{pq}[i,j]$ which is introduced by Mückstein et al., counts all structures on $[p,q]$ that $[i,j]$ is part of the loop closed by base pair $p \cdot q$. The quantity $Q^{pq,u[k,l]}[i,j]$ is a variant of $Q^{pq}[i,j]$ while $[k,l]$ lies in $[p,q]$. Recursion of $Q^{pq,u[k,l]}[i,j]$ on cases $(a-e)$ displayed in Fig. 2, is based on different types of loop and position of $[k,l]$. Therefore, we have

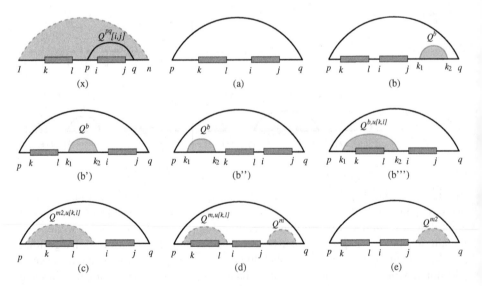

Fig. 2. Cases of unpaired interval $[i,j]$ within a loop enclosed by $p \cdot q$ while $[k,l]$ remains unpaired

$$Q^{pq,u[k,l]}[i,j] = e^{-G_{p,q}^{\text{hairpin}}/RT} \qquad\qquad (a)$$

$$+ \sum_{\substack{j<k_1<k_2<q| \\ l<k_1<k_2<i|p<k_1<k_2<k}} e^{-G_{i,k_1,k_2,j}^{\text{interior}}/RT} Q_{k_1,k_2}^{b} \qquad (b,b',b'')$$

$$+ \sum_{i<k_1<k\le l<k_2<i} e^{-G_{i,k_1,k_2,j}^{\text{interior}}/RT} Q_{k_1,k_2}^{b,u[k,l]} \qquad (b''') \qquad (4)$$

$$+ Q_{p+1,i-1}^{m2,u[k,l]} \; e^{-(a+b+c(q-i))/RT} \qquad (c)$$

$$+ Q_{p+1,i-1}^{m,u[k,l]} Q_{j+1,q-1}^{m} \; e^{-(a+b+c(j-i-1))/RT} \qquad (d)$$

$$+ Q_{j+1,q-1}^{m2} \; e^{-(a+b+c(j-p))/RT} \qquad (e)$$

where Q^{m2} is the partition function of a subsequence inside a multiloop that constitutes at least two base pairs. Q^{m2} which is introduced in Mückstein et al. algorithm can be extended to calculate $Q^{m2,u[k,l]}$. Therefore, the joint probability of two unpaired regions is obtained using

$$\mathbb{P}(u[i,j], u[k,l]) = \mathbb{P}(u[i,j] \mid u[k,l]) \times \mathbb{P}(u[k,l]). \qquad (5)$$

Mückstein et al. algorithm requires $O(n^3)$ running time and $O(n^2)$ space complexity to compute the probability of unpaired region $\mathbb{P}(u[i,j])$ for every possible interval $[i,j]$ assuming the interval length is limited to size w. Using the the extended algorithm, given sequence interval $[k,l]$ computing $\mathbb{P}(u[i,j], u[k,l])$ for every possible interval $[i,j]$ requires the same time and space complexity. Note that for each interval $[k,l]$, $Q^{u[k,l]}$ should be computed separately. Since there are $O(n.w)$ different intervals for a limited interval length w, with $O(n^4.w)$ running time and $O(n^2)$ space complexity we are able to compute the joint probability for all pairs of unpaired regions. The same idea can be used to compute the joint probability of multiple unpaired regions. However, considering each extra interval increases the running time by a factor of $O(n.w)$.

Interaction Matching Algorithm. We are given two lists of non-overlapping accessible regions $T_{\mathbf{R}} = \{r_1, r_2, ..., r_{n'}\}$ and $T_{\mathbf{S}} = \{s_1, s_2, ..., s_{m'}\}$ sorted according to their orders in interacting sequences $\mathbf{R}$ and $\mathbf{S}$. We aim to calculate the optimal set of interaction bonds between the accessible regions under the following constraints: (1) Each accessible region can interact with at most two accessible regions from the other sequence. (2) There is no crossing interaction.

Let Q_{r_i,s_j} be the partition function of all possible joint structures of two interacting sequence r_i and s_j, which can be calculated by piRNA [5]. Define $Q_{r_i,s_j}^{I} = Q_{r_i,s_j} - Q_{r_i}Q_{s_j}$ as the partition function for the set of joint structures that contain some interactions. We denote the interaction between two accessible regions r_i and s_j by $r_i \circ s_j$ which is considered if and only if $\mathbb{P}(r_i \circ s_j) = \dfrac{Q_{r_i,s_j}^{I}}{Q_{r_i,s_j}} > 1/2$. The cost of interaction between two accessible regions r_i and s_j, $C(r_i, s_j)$, is the sum of the following terms:

- $E_u(r_i)$ and $E_u(s_j)$: the energy difference between the complete ensemble and the ensemble in which the interacting subsequences are left unpaired for both accessible regions. We have $E_u(r_i) = (-RT)(\ln(Q_R^{u[r_i]}) - \ln(Q_R)) = (-RT)\ln(\mathbb{P}(u[r_i]))$. Similar equation can be used to calculate $E_u(s_j)$.

- $E_I(r_i, s_j)$: the ensemble energy of interacting joint structure for the two accessible regions where $E_I(r_i, s_j) = (-RT)\ln(\mathbb{P}(r_i \circ s_j))$.

Cost of interaction between an accessible region r_i and two other accessible regions s_k and s_j is defined as $C(r_i, s_k s_j) = E_u(r_i) + E_u(s_k, s_j) + E_I(r_i, s_k s_j)$, where $s_k s_j$ is the concatenation of two subsequences, and $E_u(s_k, s_j) = (-RT)\ln(\mathbb{P}(u[s_k], u[s_j]))$. Similarly the cost of interaction between two accessible regions from **R** and one accessible region from **S** is defined.

As an option, one can use minimum free energy (MFE) instead of ensemble energy (E_I) to define the cost of interaction. Accessible regions r_i and s_j are considered to be able to interact if and only if $MFE(r_i, s_j) < MFE(r_i) + MFE(s_j)$, i.e. there are some interaction bonds in the minimum free energy joint structure. Therefore, we have $C(r_i, s_j) = E_u(r_i) + E_u(s_j) + MFE(r_i, s_j)$. The cost of interaction of an accessible region r_i with two other accessible regions s_k and s_j is defined as $C(r_i, s_k s_j) = E_u(r_i) + E_u(s_k, s_j) + MFE(r_i, s_k s_j)$.

With $H(i, j)$, we denote the minimum cost non-conflicting set of interactions between the accessible regions $\{r_1, ..., r_i\}$ and $\{s_j, ..., s_{m'}\}$. The following dynamic programming computes $H(i, j)$:

$$H(i, j) = min \begin{cases} H(i-1, j+1) + C(r_i, s_j) & (i) \\ \min_{j<k\leq m'}\{H(i-1, k+1) + C(r_i, s_k s_j)\} & (ii) \\ \min_{1\leq k<i}\{H(k-1, j+1) + C(r_k r_i, s_j)\} & (iii) \\ H(i-1, j) & (iv) \\ H(i, j+1) & (v) \\ \infty & (vi) \end{cases} \qquad (6)$$

where $1 \leq i \leq n'$ and $1 \leq j \leq m'$. The algorithm starts by calculating $H(1, m')$ and explores all $H(i, j)$ by increasing i and decreasing j until $i = n'$ and $j = 1$. The

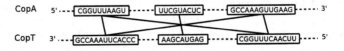

Fig. 3. Interaction between accessible regions of CopA-CopT: a simple example for interaction matching algorithm

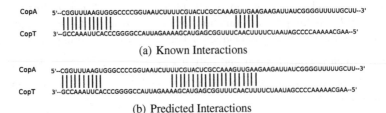

(a) Known Interactions

(b) Predicted Interactions

Fig. 4. Interaction between CopA and CopT. (a) Natural interactions. (b) Predicted interactions.

DP algorithm has $O(n'^2.m' + n'.m'^2)$ time and $O(n'.m')$ space requirements. Also we need $O(n'.m'.w^6)$ time and $O(w^4)$ space to compute the cost of interaction for every pair of accessible regions. Assuming $n' \geq m'$ and $n' \leq n/w$, we can conclude that this step of the algorithm requires $O(n^2.w^4 + n^3/w^3)$ time and $O(w^4 + n^2/w^2)$ space.

CopA-CopT is a well known antisense RNA-target complex observed in E.coli [16]. The joint structure of CopA-CopT contains two disjoint binding sites. Fig. 3 shows the identified accessible regions in CopA and CopT. Two regions connected by an edge are able to interact. Fig. 4 shows the known and predicted interaction bonds between CopA and CopT. Note that internal bonds of both RNAs are not displayed in this figure.

3   Results

3.1   Dataset

In our experiments we used a dataset of 23 known RNA-RNA interactions which includes two recently used test sets. Table 2 contains the list of these RNA pairs. The first 18 sRNA-target pairs are compiled and used as test set by IntaRNA [4]. Next 5 pairs of RNAs which are known to have loop-loop interactions have been used by Kato et al. [8] to evaluate the proposed grammatical parsing approach for RNA-RNA joint structure prediction.

3.2   Joint Structure Prediction

In our first experiment, we assessed the performance of our prediction algorithm for minimum free energy joint structure. For this purpose we used the 5 RNA-RNA complexes from Kato et al. [8] test set. We compared our results with two state-of-the-art methods for joint structure prediction: (1) the grammatical approach by Kato et al. [8] (denoted by EBM as energy-based model), and (2) the DP methods for two models presented by Alkan et al. [1] (denoted by SPM as stacked-pair model and LM as loop model).

In order to estimate the accuracy of prediction, we measured the sensitivity and PPV defined as follows:

$$sensitivity = \frac{number\ of\ correctly\ predicted\ base\ pairs}{number\ of\ true\ base\ pairs}, \qquad (7)$$

$$PPV = \frac{number\ of\ correctly\ predicted\ base\ pairs}{number\ of\ predicted\ base\ pairs}. \qquad (8)$$

As another measure of accuracy we calculated F-measure which considers both sensitivity and PPV. F-measure is the harmonic mean of sensitivity and PPV, and its formula is as follows:

$$F = \frac{2 \times sensitivity \times PPV}{sensitivity + PPV}. \qquad (9)$$

Table 1 shows comparison between the accuracy of our method and other competitors. We referred to our method by inRNAs as an algorithm for prediction the

Table 1. Prediction accuracy of competitive RNA-RNA joint structure prediction methods. Dataset is compiled by Kato et al. [8].

RNA-RNA interaction pairs	Sensitivity				PPV				F-measure			
	inRNAs	EBM	SPM	LM	inRNAs	EBM	SPM	LM	inRNAs	EBM	SPM	LM
CopA-CopT	1.000	0.909	0.955	0.864	0.846	0.800	0.778	0.760	0.917	0.851	0.857	0.809
DIS-DIS	1.000	0.786	0.786	0.786	1.000	0.786	0.786	0.786	1.000	0.786	0.786	0.786
lncRNA$_{54}$-RepZ	0.875	0.917	0.875	0.875	0.792	0.830	0.778	0.778	0.831	0.871	0.824	0.824
R1inv-R2inv	0.900	0.900	1.000	1.000	0.900	0.947	1.000	1.000	0.900	0.923	1.000	1.000
Tar-Tar*	1.000	1.000	1.000	1.000	0.875	0.933	0.875	0.875	0.933	0.965	0.933	0.933
Average	0.955	0.902	0.923	0.905	0.883	0.859	0.843	0.840	0.916	0.879	0.880	0.870

interactions between RNAs. As it can be seen, our method based on the three accuracy measures outperformed the competitors. For Tar-Tar* and R1inv-R2inv pairs that both RNAs are relatively short ($\sim$ 20nt), all methods were accurate enough. However, for DIS-DIS which is not still long (35nt), only our method was able to predict the interaction while the other approaches returned no interaction. CopA-CopT and lncRNA$_{54}$-RepZ are a bit longer ($\sim$ 60nt); CopA-CopT has two disjoint binding sites and lncRNA$_{54}$-RepZ has a continuous binding site. Our method outperformed the others in predicting the joint structure of CopA-CopT, while lncRNA$_{54}$-RepZ was predicted more accurately by EBM. We did not compare the running time between these methods due to the fact that each one uses different platform and hardware. Our method on one Sun Fire processor X4600 2.6 GHz with 64 GB RAM has been running $\sim$ 4000(sec) to predict the joint structures of CopA-CopT and lncRNA$_{54}$-RepZ.

3.3   Binding Sites Prediction

In another experiment, we focused on testing the performance of our heuristic algorithm for interaction prediction. For assessing the predictive power of our algorithm, we compared our algorithm with IntaRNA [4] and RNAup [13]. Based on the experimental results presented by IntaRNA, both IntaRNA and RNAup which incorporate accessibility of target regions, performed better than the other competitive programs.

The results of these two programs for the first 18 RNA pairs are as presented in [4]. For the next 5 RNA pairs, we run IntaRNA with its default settings and RNAup with the same setting that has been used by the experiment in [4][1]. In order to estimate the accuracy of programs, we measured the sensitivity, PPV and F-measure such that only interacting base pairs are considered.

Table 2 shows the results of our programs as well as IntaRNA and RNAup. In this dataset OxyS-fhlA and CopA-CopT are the only ones that have two disjoint binding sites, and our methods outperformed IntaRNA and RNAup by up to 30% improvement in F-measure. Both RNAup and IntaRNA could not predict any correct bond for GcvB-gltI, since they missed the binding site. However, IntaRNA could get 80% accuracy by considering the suboptimal prediction which is close to the accuracy that we have achieved. In overall, the results demonstrate that our method predicted RNA-RNA interactions more accurately in compare to competitive methods.

---

[1] RNAup has been run using parameter -b which considers the probability of unpaired regions in both RNAs and the maximal length of interaction to 80.

Table 2. Prediction accuracy of competitive RNA-RNA interaction prediction methods. Dataset is compiled by Busch et al. [4] and Kato et al. [8].

RNA-RNA interaction pairs	Sensitivity			PPV			F-measure		
	inRNAs	IntaRNA	RNAup	inRNAs	IntaRNA	RNAup	inRNAs	IntaRNA	RNAup
DsrA-RpoS	0.808	0.808	0.808	0.778	0.778	0.778	0.793	0.793	0.793
GcvB-argT	0.950	0.950	0.900	0.864	0.950	0.947	0.905	0.950	0.923
GcvB-dppA	1.000	1.000	1.000	0.850	0.586	0.459	0.919	0.739	0.629
GcvB-gltI	0.750	0.000	0.000	0.500	0.000	0.000	0.600	0.000	0.000
GcvB-livJ	0.634	0.955	0.955	0.824	0.955	0.955	0.717	0.955	0.955
GcvB-livK	0.540	0.542	0.542	0.570	0.565	0.565	0.555	0.553	0.553
GcvB-oppA	1.000	1.000	1.000	0.733	0.957	0.957	0.846	0.978	0.978
GcvB-STM4351	0.760	0.760	0.880	1.000	0.905	0.957	0.864	0.826	0.917
IstR-tisAB	0.722	0.879	0.667	1.000	0.960	1.000	0.839	0.918	0.800
MicA-ompA	1.000	1.000	1.000	1.000	1.000	1.000	1.000	1.000	1.000
MicA-lamB	1.000	1.000	0.826	1.000	0.821	0.704	1.000	0.902	0.760
MicC-ompC	1.000	1.000	0.727	1.000	0.537	0.410	1.000	0.699	0.524
MicF-ompF	0.960	0.960	0.800	0.960	0.960	0.952	0.960	0.960	0.869
OxyS-fhlA	0.813	0.500	0.375	1.000	1.000	1.000	0.897	0.667	0.545
RyhB-sdhD	0.618	0.588	0.794	0.955	1.000	0.794	0.750	0.741	0.794
RyhB-sodB	1.000	1.000	1.000	0.818	0.900	1.000	0.900	0.947
SgrS-ptsG	0.566	0.739	0.739	0.765	1.000	1.000	0.651	0.850	0.850
Spot42-galK	0.432	0.409	0.523	0.760	0.643	0.523	0.551	0.500	0.523
CopA-CopT	0.889	1.000	0.556	0.828	0.391	0.652	0.857	0.562	0.600
DIS-DIS	1.000	1.000	1.000	1.000	1.000	1.000	1.000	1.000	1.000
lncRNA$_{54}$-RepZ	1.000	0.738	0.750	0.889	0.850	0.857	0.941	0.790	0.800
R1inv-R2inv	1.000	1.000	1.000	0.778	1.000	0.778	0.875	1.000	0.875
Tar-Tar*	1.000	1.000	1.000	0.833	0.833	0.833	0.909	0.909	0.909
Average	0.845	0.819	0.776	0.865	0.805	0.784	0.845	0.791	0.763

4   Conclusion

In this work, we introduced a fast algorithm for RNA-RNA interaction prediction. Our heuristic algorithm for RNA-RNA interaction prediction problem incorporates the accessibility of multiple unpaired regions, and a matching algorithm to compute the optimal set of interactions between the target regions. The algorithm requires $O(n^4.w)$ running time and $O(n^2)$ space complexity. The main advantage of our method is its ability to predict multiple binding sites which has been predictable only by expensive algorithms [1,8] so far. On a set of several known RNA-RNA complexes, our proposed algorithm showed a reliable accuracy. Especially, for complexes with multiple binding sites our approach was able to outperform the competitive methods.

It would be interesting to design a method to efficiently compute the joint probability of multiple unpaired regions. Furthermore, the improvement of IntaRNA which got some benefit by considering seed features in comparison to RNAup, encourages us to take into account the existence of seed in the follow up work.

Acknowledgement

R. Salari was supported by Mitacs Research Grant. R. Backofen received funding from the German Research Foundation (DFG grant BA 2168/2-1 SPP 1258), and from the German Federal Ministry of Education and Research (BMBF grant 0313921 FRISYS). S.C. Sahinalp was supported by Michael Smith Foundation for Health Research Career Award.

References

1. Alkan, C., Karakoc, E., Nadeau, J.H., Sahinalp, S.C., Zhang, K.: RNA-RNA interaction prediction and antisense RNA target search. Journal of Computational Biology 13(2), 267–282 (2006)
2. Andronescu, M., Zhang, Z.C., Condon, A.: Secondary structure prediction of interacting RNA molecules. J. Mol. Biol. 345, 987–1001 (2005)
3. Bernhart, S.H., Tafer, H., Mückstein, U., Flamm, C., Stadler, P.F., Hofacker, I.L.: Partition function and base pairing probabilities of RNA heterodimers. Algorithms Mol. Biol. 1, 3 (2006)
4. Busch, A., Richter, A.S., Backofen, R.: IntaRNA: Efficient prediction of bacterial sRNA targets incorporating target site accessibility and seed regions. Bioinformatics, page btn544 (2008)
5. Chitsaz, H., Salari, R., Sahinalp, S.C., Backofen, R.: A partition function algorithm for two interacting nucleic acid strands. Bioinformatics 25(12) (2009)
6. Dimitrov, R.A., Zuker, M.: Prediction of hybridization and melting for double-stranded nucleic acids. Biophysical Journal 87, 215–226 (2004)
7. Hackermller, J., Meisner, N.C., Auer, M., Jaritz, M., Stadler, P.F.: The effect of RNA secondary structures on RNA-ligand binding and the modifier RNA mechanism: a quantitative model. Gene 345, 3–12 (2005)
8. Kato, Y., Akutsu, T., Seki, H.: A grammatical approach to rna-rna interaction prediction. Pattern Recogn. 42(4), 531–538 (2009)
9. Markham, N.R., Zuker, M.: UNAFold: software for nucleic acid folding and hybridization. Methods Mol. Biol. 453, 3–31 (2008)
10. McCaskill, J.S.: The equilibrium partition function and base pair binding probabilities for RNA secondary structure. Biopolymers 29, 1105–1119 (1990)
11. Meisner, N.C., Hackermller, J., Uhl, V., Aszdi, A., Jaritz, M., Auer, M.: mRNA openers and closers: modulating AU-rich element-controlled mRNA stability by a molecular switch in mRNA secondary structure. Chembiochem. 5, 1432–1447 (2004)
12. Mückstein, U., Tafer, H., Hackermüller, J., Bernhart, S.H., Hernandez-Rosales, M., Vogel, J., Stadler, P.F., Hofacker, I.L.: Translational control by RNA-RNA interaction: Improved computation of RNA-RNA binding thermodynamics. Bioinformatics Research and Development 13, 114–127 (2008)
13. Mückstein, U., Tafer, H., Hackermüller, J., Bernhart, S.H., Stadler, P.F., Hofacker, I.L.: Thermodynamics of RNA-RNA binding. Bioinformatics 22, 1177–1182 (2006)
14. Pervouchine, D.D.: IRIS: intermolecular RNA interaction search. Genome Inform. 15, 92–101 (2004)
15. Rehmsmeier, M., Steffen, P., Hochsmann, M., Giegerich, R.: Fast and effective prediction of microRNA/target duplexes. RNA 10, 1507–1517 (2004)
16. Wagner, E.G., Flrdh, K.: Antisense RNAs everywhere? Trends Genet. 18, 223–226 (2002)

FlexSnap: Flexible Non-sequential Protein Structure Alignment[*]

Saeed Salem[1], Mohammed J. Zaki[1], and Chris Bystroff[1,2]

[1] Department of Computer Science
salems@cs.rpi.edu, zaki@cs.rpi.edu
[2] Department of Biology,
Rensselaer Polytechnic Institute, Troy NY 12180, USA
bystrc@rpi.edu

Abstract. Proteins have evolved subject to energetic selection pressure for stability and flexibility. Structural similarity between proteins which have gone through conformational changes can be captured effectively if flexibility is considered. Topologically unrelated proteins that preserve secondary structure packing interactions can be detected if both flexibility and sequence permutations are considered. We propose the FlexSnap algorithm for flexible non-topological protein structural alignment. The effectiveness of FlexSnap is demonstrated by measuring the agreement of its alignments with manually curated non-sequential structural alignments. FlexSnap showed competitive results against state-of-the-art algorithms, like DALI, SARF2, MultiProt, FlexProt, and FATCAT.

1   Background

The wide spectrum of functions performed by proteins are enabled by their intrinsic flexibility [1]. It is known that proteins go through conformational changes to perform their functions. Homologous proteins have evolved to adopt conformational changes in their structure. Therefore, similarity between two proteins which have similar structures with one of them having undergone a conformational change will not be captured unless flexibility is considered.

The problem of flexible protein structural alignment has not received much attention. Even though there are a plethora of methods for protein structure comparison [2, 3, 4, 5, 6, 7, 8], the majority of the existing methods report only sequential alignments and thus cannot capture non-sequential alignments. Non-sequential similarity can occur naturally due to circular permutations [9] or convergent evolution [10]. The case is even harder for flexible alignment since only two methods, FlexProt [11], and FATCAT [12] report flexible alignments. Nevertheless, both methods are inherently limited to *sequential* flexible structural alignment because both methods employ sequential chaining techniques. The complexity of protein structural alignment depends on how the similarity is

---

[*] This work was supported in part by NSF Grants EMT-0829835, and CNS-0103708, and NIH Grant 1R01EB0080161-01A1.

S.L. Salzberg and T. Warnow (Eds.): WABI 2009, LNBI 5724, pp. 273–285, 2009.

assessed. [13] showed that the problem is NP-hard if the similarity score is distance matrix based. Therefore, over the years, a number of heuristic approaches have been proposed, which can mainly be classified into two main categories, dynamic programming and clustering.

Dynamic Programming (DP) is a general paradigm to solve problems that exhibit the optimal substructure property [14]. DP-based methods, STRUCTAL [15] and SSAP [16], construct a scoring matrix S, where each entry, S_{ij}, corresponds to the score of matching the i-th residue in protein A and the j-th residue in protein B. Given a scoring scheme between residues in the two proteins, dynamic programming finds the global alignment that maximizes the score. DP-based methods suffer from two main limitations: first, the alignment is sequential and thus non-topological similarity cannot be detected, and second, it is difficult to design a scoring function that is globally optimal [13]. In fact, structure alignment does not have the optimal substructure property, therefore DP-based methods can find only a suboptimal solution [17].

The other category of alignment methods, the Clustering-based methods, DALI [2], SARF2 [4], CE [5], SCALI [7], and FATCAT [12], seek to assemble the alignment out of smaller compatible (similar) element pairs such that the score of the alignment is as high as possible [18]. Two compatible element pairs are consistent (can be assembled together) if the substructures obtained by elements of the pairs are similar. The clustering problem is NP-hard [19], thus several heuristics have been proposed. The approaches differ in how the set of compatible element pairs is constructed and how the consistency is measured. Both SARF2 and SCALI produce non-sequential alignments.

The two main flexible alignment methods, FlexProt [11] and FATCAT [12], work by clustering (chaining) aligned fragment pairs (AFPs) and allowing flexibility while chaining, by introducing hinges (twists). FlexProt searches for the longest set of AFPs that allow different number of hinges. It then reports different alignments with different number of hinges. The FATCAT method works by chaining AFPs using dynamic programming. The score of an alignment ending with a given AFP is computed as the maximum score of connecting the AFP with any of alignments that end before the AFP. A penalty is applied to the score to compensate for gaps, root mean squared deviation ($rmsd$), and hinges. A third method, which can handle flexible alignments, is the HingeProt [20] method. HingeProt first partitions one of the two proteins into rigid parts using a Gaussian-Network-Model-based (GNM) approach and then aligns each rigid region with the other protein using the MultiProt [6] method. HingeProt uses the MultiProt algorithm in the sequential mode and thus does not report flexible non-sequential alignments. Therefore, the accuracy of the HingeProt approach depends on the accuracy of identifying the rigid domains which is a hard problem as the best known method ,HingeMaster [21], has a sensitivity of only 50%.

In this paper, we propose FlexSnap[1], a greedy algorithm for flexible non-sequential protein alignment. The algorithm assembles the alignment from the

---

[1] A non-sequential permutation of the bold letters in **Flex**ible **n**on-**S**equential **p**rotein **A**lignment

set of AFPs and allows non-sequential alignments and hinges. We demonstrate the effectiveness of FlexSnap by evaluating its alignments' agreement with manually curated non-sequential alignments.

2 Methods

The main idea of the FlexSnap approach is to assemble the alignment from short well-aligned fragment pairs, which are called AFPs. As we assemble the alignment by adding AFPs, we introduce hinges when necessary. Figure 1 shows how the alignment is constructed from smaller aligned fragment pairs. When chaining a fragment pair to the alignment, we choose the fragment that has the highest score when joined with the last rigid region in the alignment. The score rewards longer alignments with small $rmsd$ and penalizes large $rmsd$, gaps, and the introduction of hinges. In the next subsections, we provide a detailed discussion of the FlexSnap algorithm.

2.1 AFPs Extraction

Let $A = \{A_1, A_2, \ldots, A_n\}$ and $B = \{B_1, B_2, \ldots, B_n\}$ be two proteins with n and m residues respectively, and $A_i \in \Re^{3 \times 1}$ (similarly B_i) represents the 3D coordinates of the C_α atom of the i-th residue in protein A. The first step in FlexSnap is to generate a list of aligned fragment pairs (AFPs):

$$AFPs = \{(i, j, l) | \ rmsd(i, j, l) \leq \epsilon\}$$

Each AFP, (i, j, l), is a fragment that starts at the i-th residue in A and j-th residue in B and it has a length of l residues. An AFP is formally represented as a set of l equivalenced pairs between the two proteins, and given as:

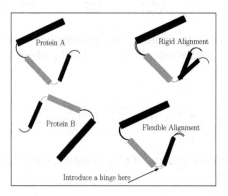

Fig. 1. Flexible Structural Alignment by chaining. The figure shows protein A and protein B which have 3 similar structure fragments. A rigid alignment (top right) is not able to align the blue fragment, but a flexible alignment (bottom right) can do this easily by introducing a hinge between the rigid block (the black and green fragments) and the blue fragment. As we assemble the alignment from well-aligned pairs, we introduce hinges to get a longer alignment and smaller $rmsd$.

$$(i, j, l) = \{(A_i, B_j), (A_{i+1}, B_{j+1}), \cdots, (A_{i+l-1}, B_{j+l-1})\}$$

where (A_i, B_j) indicates that the ith residue of protein A is paired with the jth residue of protein B, and l is AFP's length. Each AFP must satisfy a user-defined similarity constraint. In FlexSnap, approach, we employ the root mean square deviation as the similarity measure, i.e., $rmsd(i, j, l) \leq \epsilon$. Moreover, we require that the length of the AFP be at least L, i.e., $3 \leq L \leq l$. Furthermore, we define b_k^P and e_k^P to be the beginning and end of the AFP$_k$ along the backbone of protein B. e.g, for a triplet AFP$_k = (i, j, l)$ and protein A, $b_k^A = i$ and $e_k^A = i + l - 1$.

The number of possible AFPs can be as large as $O(n^3)$. The set of all AFPs can be obtained by iterating over all the triplets (i, j, l),

$$\text{where} \begin{bmatrix} i & = & 1 \cdots n - L \\ j & = & 1 \cdots m - L \\ L \leq l \leq min(n - i + 1, m - j + 1) \end{bmatrix}$$

and for each triplet checking if the $rmsd(i, j, l) \leq \epsilon$. The $rmsd$ of a fragment can be obtained in $O(l)$ [22]. A naive implementation that iterates over all the triplets (i, j, l) to obtain the set of all the AFPs would have an $O(n^4)$ time complexity. However, by observing that the $rmsd$ of the AFP $(i, j, l + 1)$ can be computed incrementally from the $rmsd$ of AFP (i, j, l) in constant time, the set of aligned fragment pairs (AFPs) can be obtained in $O(n^3)$ time complexity [11].

The main idea to incrementally compute the $rmsd$ is to simplify the $rmsd$ formula. Given two sets, A and B, of N points each, the root mean square deviation ($rmsd$) is calculated as [23]:

$$rmsd^2 = \frac{1}{N} \times \left(\sum_{i=1}^{N} A_i'^2 + \sum_{i=1}^{N} B_i'^2 - 2 \sum_{i=1}^{3} d_i \right) \tag{1}$$

where A' (similarly B') is the points after recentering, i.e., $A_i' = A_i - \frac{\sum_{i=1}^{N} A_i}{N}$, and the d_i's are the singular values of $C = A'B'^T$, which is a 3×3 covariance matrix given as:

$$C = \sum_{i=1}^{n} A_i B_i^T - \frac{\sum_{i=1}^{N} A_i \sum_{i=1}^{n} B_i^T}{N} \tag{2}$$

In rare cases when the determinant of C is negative, then $d_3 = -1 * d_3$. Equation 1 can be simplified as:

$$rmsd^2 = \frac{1}{N} \times \left(\sum_{i=1}^{N} A_i^2 - \frac{1}{N} \left(\sum_{i=1}^{N} A_i \right)^2 + \sum_{i=1}^{N} B_i^2 - \frac{1}{N} \left(\sum_{i=1}^{N} B_i \right)^2 - 2 \sum_{i=1}^{3} d_i \right)$$

It is clear that all the terms used in the $rmsd$ computation can be updated in constant time and thus computing the $rmsd$ for $N + 1$ points requires constant time if we have all the terms evaluated for the first N points. Therefore computing the *rmsd* for $\text{AFP}(i, j, l)$ for all values of l's requires only $O(n)$ time. Thus, the total time complexity for the seeds extraction step is $O(n^3)$.

2.2 Flexible Chaining

The second step in FlexSnap is to construct the alignment by selecting a subset of the AFPs. Given a set of AFPs, P, obtained in the AFPs extraction step, we are interested in finding a subset of AFPs, $R \subseteq P$, such that all the AFPs in R are mutually non-overlapping and the score of the selected AFPs in R is as large as possible. At one hand, we want to get as large an alignment as possible, while on the other hand, we want to minimize the number of hinges and gaps. Therefore, our goal is to optimize a score that rewards long alignments with small *rmsd* and penalizes the introduction of hinges and gaps.

The set of AFPs can be thought of as runs in an $n \times m$ matrix S, where n and m are the sizes of proteins A and B, respectively (see Figure 2). We define a precedence relation, $\prec$, between two AFPs such that $P_i \prec P_j$ if P_i appears either in the upper or lower left quadrant of P_j, i.e. $b_j^A > e_i^A$ and $b_j^B > e_i^B$, or $e_j^A < b_i^A$ and $b_j^B > e_i^B$ (recall that b_i^A (e_i^A) denote beginning (end) of AFP P_i in protein A). Generally speaking, we say that two AFPs, P_i and P_j, can be chained if they do not overlap, i.e., $P_i \prec P_j$ or $P_j \prec P_i$. In Figure 2, P_7 and P_8 can be chained to P_1. For sequential chaining, we define a sequential precedence relation, $\prec_s$, such that P_i precedes P_j (written as $P_i \prec_s P_j$) if P_i appears strictly in the upper left quadrant with respect to P_j, i.e. $b_j^A > e_i^A$ and $b_j^B > e_i^B$. Two AFPs P_i and P_j

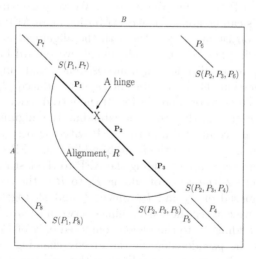

Fig. 2. Flexible Structural Alignment by chaining AFPs. When extending the alignment $R = \{P_1, P_2, P_3\}$, the score of adding each AFP is computed computed and we extend the alignment by the AFP that gives the best score. The score $S(P_4, P_2, P_3))$ indicates the score of adding P_4 to the region composed of P_2 and P_3.

can be sequentially chained together if $P_i \prec_S P_j$ or $P_j \prec_S P_i$. In Figure 2, P_7 and P_2 can be sequentially chained to P_1.

An AFP, P_i, can be chained to an alignment R, denoted as $(R \rightarrow P_i)$, if it does not overlap with any AFP in R. In Figure 2, P_7, P_4, and P_5 can be sequentially chained to R which consists of AFPs $\{P_1, P_2, P_3\}$; and both P_6 and P_8 can be non-sequentially chained to R. Next, we shall introduce our solution for the general flexible chaining problem.

2.3   The FlexSnap Approach

The goal of chaining is to find the highest scoring subset of AFPs, i.e., $R \subseteq P$, such that all the AFPs in R are mutually consistent and non-overlapping. The problem of finding the highest scoring subset of AFPs is essentially the same as finding the maximum weighted clique in a graph $G = (V, E, w)$ where the set of vertices V represent the set of AFPs, each vertex v_i has a weight equal to the score of the AFP, $w(v_i) = S(P_i)$, where the score of an AFP P_i, $S(P_i)$, could be its length or some other combination of length and $rmsd$. There is an edge $(v_i, v_j) \in E$ if the AFPs P_i and P_j do not overlap and are consistent (can be joined with small $rmsd$ or have similar rotation matrices). The problem of finding the maximum weighted clique in a graph is computationally expensive; it is NP-hard [19].

We propose a greedy algorithm to find an approximate solution for the chaining problem. The main idea is to start building the alignment from an initial AFP and add AFPs to the alignment. We start the alignment by selecting the longest AFP, then we iteratively add new AFPs to the alignment as long as the newly added AFP improves the score of the alignment. Given an alignment, R, we add to it the AFP that contributes most. We keep growing the alignment until no more AFPs can be added. The contribution of an AFP to the alignment is scored by how consistent the AFP is with the alignment and how good the AFP is. When adding an AFP to an alignment, we reward longer AFPs with smaller $rmsd$, and we penalize for gaps, inconsistency, and hinges. The penalty takes into consideration: 1) the number of gaps introduced; 2) the increase in $rmsd$ when combining two or more AFPs; 3) the introduction of new hinges.

As depicted in Figure 2, the scores of extending the alignment, R, with P_4, P_5, P_6, P_7, or P_8 are computed and the AFP with the best score is added to the alignment. When measuring the score of adding an AFP to the alignment, we actually measure the score of adding the AFP to the last rigid region in the alignment. In Figure 2, the score of adding P_4 to R is the score of adding P_4 to the region composed of P_2 and P_3. Since P_2 and P_3 together form a rigid sub-alignment (as we can see there is no hinge between them). When adding P_7 to R, the score of adding P_7 to the region composed only of P_1 is computed.

Figure 3 shows the pseudo-code for the greedy chaining algorithm used in FlexSnap. Since the chaining is a greedy algorithm, we run it K times starting from the longest non-overlapping K AFPs and we report the longest alignment. Next, we will discuss how we extend the alignment with the best AFP. More specifically, given an alignment R, the next AFP to chain to the alignment is the one that maximizes the following scoring function:

```
GreedyChaining(A, B, L, H, ε, Dc, Mᵣ, M_g)
A, B: the two proteins to be aligned
L: the minimum length of an AFP, L ≥ 3
ε: the maximum rmsd for an AFP
Dc: the rmsd for introducing a hinge
Mᵣ: the penalty for a hinge
M_g: the penalty for a gap
H: the maximum number of hinges allowed
1.P= seedExtraction(A, B, L, ε)
2.P′= longest AFP in P
3.R = P′
4.While(R can be extended)
5.P′ ← maxₚᵢ(S(R, Pᵢ))
6.R ← R ∪ {P′}
7.End
```

Fig. 3. A greedy algorithm for AFP chaining. The algorithm iteratively chooses an AFP to add to R (lines 5-6) until no more AFPs can be added, or the best score of adding an AFP to R is negative.

$$P' = \max_{\forall P_i, s.t. R \to P_i} (S(R, P_i)) \qquad (3)$$

where $R \to P_i$ indicates that P_i does not overlap with R, and $S(R, P_i)$ is the score of chaining P_i to R. The score, $S(R, P_i)$, is a combination of the weight of the AFP, $W(P_i)$, and the penalty of extending R with P_i, $C(R \to P_i)$. The score is defined as follows:

$$S(R, P_i) = W(P_i) + C(R \to P_i) \qquad (4)$$

where $C(R \to P_i)$ is the penalty incurred when connecting P_i to R, and $W(P_i)$ is the score of the AFP itself. The scoring function rewards longer AFPs with small rmsd and penalize gaps and hinges. If the addition of an AFP P_i to the alignment results in a large rmsd, then we introduce a hinge only if $W(P_i)$ is large enough to compensate for the penalty incurred. A similar approach for penalizing gaps and hinges was used in the FATCAT method [12]. Though their score and cost functions are different, and they do not consider rigid regions as we do in FlexSnap when connecting an AFP to the alignment. The score of connecting P_i to R is defined as follows:

$$C(R \to P_i) = M_r * Z(D_{RP_i}) + M_g * gap \qquad (5)$$

$$\text{where } Z(D_{RP_i}) = \begin{cases} 1 & if D_{RP_i} > D_c \\ \left(\frac{D_{RP_i} - \epsilon}{D_c - \epsilon}\right)^2 & if \epsilon < D_{RP_i} < D_c \\ 0 & otherwise \end{cases}$$

where M_g is the penalty for a gap, M_r is the maximum penalty for a hinge, and D_{RP_i} is the rmsd of connecting P_i to the last rigid region in R. If D_{RP_i} increases above a user-defined threshold, D_c, we introduce a hinge and the penalty is maximum; if not, the penalty is proportional to how far the $rmsd$ value is from ϵ (maximum $rmsd$ for an AFP). Moreover, we allow only a maximum number of H hinges. The score for an AFP is a function of its length and $rmsd$. The score is the length of the AFP, $L(P_i)$, plus a contribution of the $rmsd$ of the AFP, $rmsd(P_i)$, to the score, and is given as:

$$W(P_i) = L(P_i) + \alpha * L(P_i) * \left(\frac{\epsilon - rmsd(P_i)}{\epsilon} \right)^2 \qquad (6)$$

The complexity of the chaining algorithm depends on the number of AFPs, M, that two structures have. In the worst case, M could be close to n^3, but in practice it is much less, i.e., $M \leq n^2$. The complexity of the algorithm is $Mlog(M) + k * M * n$, where k is the number of AFPs in the final solution and n is the size of the larger protein.

2.4   Sequential Flexible Chaining

The above general chaining algorithm reports both sequential and non-sequential alignments. In the results section, we show the quality of its alignments when compared to state-of-the-art non-sequential alignment methods. However, for sequential flexible alignment, there are more efficient chaining algorithms, namely FlexProt and FATCAT. The FATCAT algorithm follows a dynamic programing approach for chaining the AFPs. In FATCAT, the score of an alignment ending with AFP P_i is defined in terms of the score of P_j's and the connection cost of P_i with these P_j's such that $P_j \prec_s P_i$. More specifically, FATCAT defines the score of the alignment that ends with P_i as follows:

$$S(P_i) = W(P_i) + \max_{\forall P_j, s.t. P_j \prec_s P_i} \{ max(S(P_j) + C(P_j \rightarrow P_i), 0) \}$$

where $C(P_j \rightarrow P_i)$ is the penalty incurred when connecting P_i to the alignment that ends with P_j and it is similar to the penalty function used in the general chaining and $W(P_i)$ is the score of the AFP itself. In FATCAT, $C(P_j \rightarrow P_i)$ is the connection cost of P_i and P_j. If P_j belongs to a rigid region and the connection cost of P_i with P_j is small, we will add P_i to the same rigid region as P_j even though P_i might not be consistent with other AFPs in the same region. In Figure 2, if we were connecting P_4 to P_3, FATCAT would compute the connection cost $C(P_3, P_4)$ while it makes more sense to compute $C((P_2, P_3) \rightarrow P_4)$. Moreover, $W(P_i)$ is a function of the length of P_i and its $rmsd$ and thus $S(P_i)$ cannot be optimal because we do not know of a scoring function that involves the $rmsd$ value that is additive and optimal ($rmsd$ score is not a metric since it does not satisfy the triangle inequality property). Therefore, the optimality of FATCAT alignments is not guaranteed since the sub-optimality property of the dynamic programming does not hold if the score incorporates an $rmsd$ term.

   Our approach for sequential chaining is essentially the same as the FATCAT algorithm with the exception that we consider connecting an AFP to the last rigid region in the alignment, not to the last AFP in the alignment as is the case in FATCAT. Moreover, we use a simpler function for scoring an AFP, e.g., $W(P_i) = length(P_i)$. In the results section, we investigate how these two modifications to the FATCAT algorithm would affect the performance of sequential chaining. Since we do not have the exact FATCAT scoring function because some terms are not adequately defined in the paper, we implemented our own and compared the results when including all the AFPs in the last rigid region or not including them.

3   Results and Discussion

To assess the quality of FlexSnap alignment compared to other structural alignment methods, we evaluated the agreement of the methods' alignments with reference manually-curated alignments. We compared our FlexSnap against sequential methods (DALI [2] and CE [5]), non-sequential methods (SARF2 [4], MultiProt [6], and SCALI [7]), and flexible sequential alignment methods (FlexProt [11] and FATCAT [12]). All the experiments were run on a 1.66 GHz Intel Core Duo machine with 1 GB of main memory running Ubuntu Linux. The chaining algorithm is efficient and its running time varies from 1 second to a minute depending on the size of the proteins. We used the corresponding web server for most of the other alignment methods. The optimal values for the different parameters were found empirically such that they give the best agreement with manually curated alignments; we used $L = 8$, $\epsilon = 2\text{Å}$, $D_c = 3\text{Å}$, $\alpha = 0.3$, $M_r = -10$, $M_g = -1$, and $H = 3$ (see Figure 3).

3.1   Non-sequential Alignments

We used the reference alignments for the structure pairs which have circular permutation in the RIPC dataset [24]. The RIPC set contains 40 structurally related protein pairs which are challenging to align because they have indels, repetitions, circular permutations, and show conformational flexibility [24]. There are 10 pairs in the RIPC dataset that have circular permutation. Since the structure pairs have non-sequential alignments, to be fair, we only compare with algorithms that can handle non-sequentiality. However, we report the average agreement for some sequential methods as well. The agreement of a given alignment, S, with the reference alignment, R, is defined as the percentage of the residue pairs in the alignment which are identically aligned as in the reference alignment (I_S) relative to the reference alignment's length (L_R), i.e., $A(S, R) = (I_S/L_R) \times 100$. Table 1 shows the agreements of four different methods with the reference alignments in the RIPC dataset. The results show that FlexSnap is competitive to state-of-the-art methods in non-sequential alignment. In fact, it has the highest average agreement (79%) among the methods shown. The average agreement of most of the sequential alignment methods, we compared with, were drastically lower: DALI [2] (40%), CE [4](36%), FATCAT [12](28%), and LGA [25](38%).

FlexSnap alignments have 100 percent agreement on four structure pairs. One such pair is the alignment of NK-lysin (1nkl, 78 residues) with prophytepsin (1qdm,

Table 1. Comparison of SARF, MultiProt, and FlexSnap on the RIPC dataset. Three values are reported for each alignment: its length, its *rmsd*, and A which is its agreement with the reference alignment in the RIPC dataset.

SCOPID		SARF			MultiProt			SCALI			FlexSnap		
Pro1	Pro2	size	*rmsd*	A	size	*rmsd*	A	size	*rmsd*	A	size	*rmsd*	A
d1nkl__	d1qdma1	67	2.21	92	67	1.82	68	62	1.94	69	73	2.39	100
d1nls__	d2bqpa_	212	1.50	83	213	1.03	100	195	1.62	83	210	2.81	83
d1qasa2	d1rsy__	109	2.27	65	107	1.24	93	98	1.92	82	111	1.73	100
d1b5ta_	d1k87a2	171	2.63	63	144	2.04	0	159	3.38	0	177	2.99	50
d1jwyb_	d1puja_	115	2.43	83	108	1.81	92	110	4.60	83	116	2.61	92
d1jwyb_	d1u0la2	97	2.02	100	103	1.86	91	91	4.52	90	96	2.82	100
d1nw5a_	d2adma_	129	2.52	85	130	2.11	92	132	3.73	84	128	2.91	100
d1gsa_1	d2hgsa1	73	2.59	20	74	1.56	40	69	3.23	40	73	2.81	20
d1qq5a_	d3chy__	88	2.39	67	82	1.97	67	52	2.08	66	93	2.94	67
d1kiaa_	d1nw5a_	146	2.48	83	153	1.85	75	138	3.99	75	141	2.69	75
Avg.	Agreement			74			72			67			79

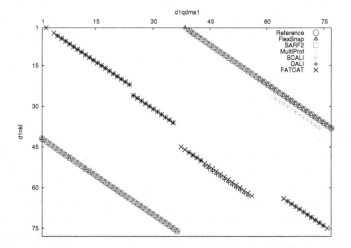

Fig. 4. Comparison of the agreement between the reference alignment and 6 other alignment methods on the structure pair of prophytepsin(d1qdma1) and nk-lysin(d1nkl__). Residue positions of d1qdma1 and d1nkl__ are plotted on the x-axis and y-axis, respectively. Note: The reference alignment pairs are shown in circles. The SARF, MultiProt, SCALI, and FlexSnap plots overlap with the reference alignment. FlexSnap has 100 percent coverage of the reference alignment; there is a triangle in every circle.

chain A, 77 residues). On this pair, all the sequential alignment methods(CE, DALI, FATCAT, and LGA) returned zero agreements. For the non-sequential ones: SARF returned 92%, MultiProt got 68%, and SCALI returned 69%. The reference

Table 2. Comparison of FlexProt, FATCAT, FlexSnap^F, and FlexSnap^{F_2}. Each alignment is reported in the following format: length, *rmsd*, and T which is the number of hinges introduced.

Pro1	Pro2	FlexProt size *rmsd* T			FATCAT size *rmsd* T			FlexSnap^F size *rmsd* T			FlexSnap^{F_2} size *rmsd* T		
1wdnA(223)	1gggA(220)	218	0.94	2	220	1.01	2	220	0.96	2	220	0.96	2
1hpbP(238)	1gggA(220)	220	2.34	2	213	1.59	2	211	1.67	2	210	3.88	1
2bbmA(148)	1clL_(144)	139	2.22	1	144	2.28	1	138	1.8	1	138	1.80	1
2bbmA(148)	1top_(162)	147	2.40	3	145	2.28	3	137	1.78	3	137	1.78	3
1akeA(214)	2ak3A(226)	200	2.44	2	202	1.54	2	207	2.05	2	206	6.72	1
2ak3A(226)	1uke_(193)	182	2.90	2	188	2.97	0	184	2.36	1	184	3.08	0
1mcpL(220)	4fabL(219)	218	1.93	1	217	1.40	1	217	1.49	1	217	1.49	1
1mcpL(220)	1tcrB(237)	212	2.33	1	213	2.20	1	202	2.3	1	200	2.38	1
1lfh_(691)	1lfg_(691)	691	1.41	2	686	0.89	2	688	0.99	2	688	0.99	2
1tfd_(294)	1lfh_(691)	291	1.98	2	290	1.37	2	287	1.89	2	283	1.41	2
1b9wA(91)	1danL(142)	75	2.78	1	80	2.39	2	82	2.25	2	83	2.7	2
1qf6A(641)	1adjA(420)	323	4.43	1	351	2.68	1	326	2.45	3	320	2.47	2
2clrA(275)	3fruA(269)	253	2.71	2	245	3.06	0	254	2.57	3	252	4.31	0
1fmk_(438)	1qcfA(450)	424	1.25	2	433	2.27	0	413	2.71	0	413	2.44	1
1fmk_(438)	1tkiA(321)	231	3.28	2	238	3.07	0	241	2.58	3	242	3.14	2
1a21A(194)	1hwgC(191)	163	2.75	4	153	3.16	1	156	2.35	3	155	3.79	2

alignment had 72 aligned pairs. As shown in Figure 4, the sequential alignment methods (only DALI and FATCAT shown) have their alignment paths along the diagonal and do not agree with with the reference alignment (shown as circles).

3.2 Sequential Flexible Alignments

Table 2 shows the alignments of different methods on the FlexProt dataset [11] which is obtained from the database of macromolecular motions [26]. FlexSnap^F is our implementation of FATCAT with a simpler function for the score of AFP and a different function for the connection cost of two AFPs. In this version, $C(P_j \to P_i)$ calculates the connection cost of P_i with the rigid region to which P_j belongs. In the second version, FlexSnap^{F_2}, $C(P_j \to P_i)$ calculates the connection cost of P_i with only P_j. It is obvious that when considering the entire rigid region, we get much better alignments. Moreover, FlexSnap^F gives comparable results to the FATCAT method. In few cases it got slightly shorter alignments with much better *rmsd* as in the case of the third and fourth alignment pairs.

4 Conclusion

We have introduced FlexSnap, a chaining algorithm that reports both sequential and non-sequential alignments and allows twists (hinges). We assessed the quality of the FlexSnap alignments by measuring its agreements with manually curated non-sequential alignments. Moreover, we employed the scoring function devised in FlexSnap in a FATCAT-like algorithm for sequential flexible alignments. The new algorithm for flexible sequential alignment, FlexSnap^F, gave competitive results against state-of-the-art flexible sequential alignment methods: FlexProt and FATCAT. Our future goal is to compile a list of manually curated flexible non-sequential alignments and measure the agreement of FlexSnap alignments with this dataset. Moreover, we would like to apply the algorithm on SCOP [27] and CATH [28] classifications to investigate how the introduction of flexibility would change the classification of some proteins.

References

[1] Wriggers, W., Schulten, K.: Protein domain movements: detection of rigid domains and visualization of hinges in comparisons of atomic coordinates. Proteins: Structure, Function, and Genetics 29, 1–14 (1997)

[2] Holm, L., Sander, C.: Protein structure comparison by alignment of distance matrices. J. Mol. Biol. 233(1), 123–138 (1993)

[3] Subbiah, S., Laurents, D.V., Levitt, M.: Structural similarity of dna-binding domains of bacteriophage repressors and the globin core. Curr. Biol. 3, 141–148 (1993)

[4] Alexandrov, N.N.: Sarfing the pdb. Protein Engineering 50(9), 727–732 (1996)

[5] Shindyalov, I.N., Bourn, P.E.: Protein structure alignment by incremental combinatorial extension (ce) of the optimal path. Protein Eng. 11, 739–747 (1998)

[6] Shatsky, M., Nussinov, R., Wolfson, H.J.: A method for simultaneous alignment of multiple protein structures. Proteins: Structure, Function, and Bioinformatics 56(1), 143–156 (2004)

[7] Yuan, X., Bystroff, C.: Non-sequential structure-based alignments reveal topology-independent core packing arrangements in proteins. Bioinformatics 21(7), 1010–1019 (2003)

[8] Zhu, J., Weng, Z.: Fast: A novel protein structure alignment algorithm. Proteins:Structure, Function and Bioinformatics 14, 417–423 (2005)

[9] Lindqvist, Y., Schneider, G.: Circular permutations of natural protein sequences: structural evidence. Curr. Opin. Struct. Biol. 7(3), 422–427 (1997)

[10] Milik, M., Szalma, S., Olszewski, K.A.: Common structural cliques: a tool for protein structure and function analysis. Protein Engineering 16(8), 543–552 (2003)

[11] Shatsky, M., Nussinov, R., Wolfson, H.J.: Flexible protein alignment and hinge detection. Proteins: Structure, Function, and Bioinformatics 48, 242–256 (2002)

[12] Ye, Y., Godzik, A.: Flexible structure alignment by chaining aligned fragment pairs allowing twists. Bioinformatics 19, II246–II255 (2003)

[13] Kolodny, R., Linial, N.: Approximate protein structural alignment in polynomial time. PNAS 101, 12201–12206 (2004)

[14] Needleman, S.B., Wunsch, C.D.: A general method applicable to the search for similarities in the amino acid sequence of two proteins. J. Mol. Biol. 48, 443–453 (1970)

[15] Gerstein, M., Levitt, M.: Using iterative dynamic programming to obtain accurate pairwise and multiple alignments of protein structures. In: Proc. Int. Conf. Intell. Syst. Mol. Biol., vol. 4, pp. 59–67 (1996)

[16] Orengo, C.A., Taylor, W.R.: Ssap: sequential structure alignment program for protein structure comparison. Methods Enzymol. 266, 617–635 (1996)

[17] Eidhammer, I., Jonassen, I., Taylor, W.R.: Protein Bioinformatics: An algorithmic Approach to Sequence and Structure Analysis. John Wiley & Sons Ltd., UK (2004)

[18] Eidhammer, I., Jonassen, I., Taylor, W.R.: Structure comparison and structure patterns. J. Comput. Biol. 7(5), 685–716 (2000)

[19] Garey, M.R., Johnson, D.S.: Computers and intractability: A guide to the theory of np-completeness. W.H. Freeman, San Francisco (1979)

[20] Emekli, U., Schneidman-Duhovny, D., Wolfson, H.J., Nussinov, R., Haliloglu, T.: Hingeprot: Automated prediction of hinges in protein structures. Proteins 70(4), 1219–1227 (2008)

[21] Flores, S.C., Keating, K.S., Painter, J., Morcos, F., Nguyen, K., Merritt, E.A., Kuhn, L.A., Gerstein, M.B.: Hingemaster: normal mode hinge prediction approach and integration of complementary predictors. Proteins 73, 299–319 (2008)

[22] Kabsch, W.: A solution for the best rotation to relate two sets of vectors. Acta Crystallogr. A32, 922–923 (1976)

[23] Chwartz, J.T., Sharir, M.: Identification of partially obscured objects in two dimensions by matching of noisy characteristic curves. Int. J. Robotics Res. 6, 29–44 (1987)

[24] Mayr, G., Dominques, F., Lackner, P.: Comparative analysis of protein structure alignments. BMC Structural Biol. 7(50), 564–577 (2007)

[25] Zemla, A.: Lga - a method for finding 3d similarities in protein structures. Nucleic Acids Research 31(13), 3370–3374 (2003)

[26] Gerstein, M., Krebs, W.: A database of macromolecular motions. Nucleic Acids Res. 26(18), 4280–4290 (1998)

[27] Murzin, A., Brenner, S.E., Hubbard, T., Chothia, C.: Scop: A structural classification of proteins for the investigation of sequences and structures. J. Mol. Biol. 247, 536–540 (1995)

[28] Orengo, C.A., Michie, A.D., Jones, S., Jones, D.T., Swindells, M.B., Thornton, J.M.: Cath- a hierarchic classification of protein domain structures. structure 5(8), 1093–1108 (1997)

A Non-parametric Bayesian Approach for Predicting RNA Secondary Structures

Kengo Sato[1,2,3], Michiaki Hamada[4,2], Toutai Mituyama[2], Kiyoshi Asai[5,2], and Yasubumi Sakakibara[3,2]

[1] Japan Biological Informatics Consortium (JBIC),
2-45 Aomi, Koto-ku, Tokyo 135-8073, Japan
[2] Computational Biology Research Center (CBRC),
National Institute of Advanced Industrial Science and Technology (AIST),
2-41-6, Aomi, Koto-ku, Tokyo 135-0064, Japan
{sato-kengo,hamada-michiaki,mituyama-toutai}@aist.go.jp
[3] Department of Biosciences and Informatics, Keio University,
3-14-1 Hiyoshi, Kohoku-ku, Yokohama, Kanagawa 223-8522, Japan
yasu@bio.keio.ac.jp
[4] Mizuho Information & Research Institute, Inc,
2-3 Kanda-Nishikicho, Chiyoda-ku, Tokyo 101-8443, Japan
[5] Graduate School of Frontier Sciences, University of Tokyo,
5-1-5 Kashiwanoha, Kashiwa 277-8562, Japan
asai@k.u-tokyo.ac.jp

Abstract. Since many functional RNAs form stable secondary structures which are related to their functions, RNA secondary structure prediction is a crucial problem in bioinformatics. We propose a novel model for generating RNA secondary structures based on a non-parametric Bayesian approach, called hierarchical Dirichlet processes for stochastic context-free grammars (HDP-SCFGs). Here *non-parametric* means that some meta-parameters, such as the number of non-terminal symbols and production rules, do not have to be fixed. Instead their distributions are inferred in order to be adapted (in the Bayesian sense) to the training sequences provided. The results of our RNA secondary structure predictions show that HDP-SCFGs are more accurate than the MFE-based and other generative models.

1 Introduction

Many functional RNAs form stable secondary structures which are related to their functions such as gene regulation or maturation of mRNAs, rRNAs and tRNAs. Currently, experimental determination of base-pairs of RNA secondary structures is still difficult and expensive, especially for high-throughput assays. Therefore, computational prediction of RNA secondary structures is widely used instead of experimental assays.

The most successful methods for predicting RNA secondary structures of single sequences are based on minimum free energy (MFE) such as Mfold [1] and RNAfold [2]. The energy parameters used in these methods were determined by

S.L. Salzberg and T. Warnow (Eds.): WABI 2009, LNBI 5724, pp. 286–297, 2009.
© Springer-Verlag Berlin Heidelberg 2009

empirical studies of RNA structural analysis. However, some parameters cannot be reliably measured due to the lack of efficient experimental techniques.

Alternative methods are based on probabilistic frameworks, including stochastic context-free grammars (SCFGs) which can model RNA secondary structures without pseudoknots. Several computational methods based on SCFGs have been developed for modeling and analyzing functional non-coding RNA sequences [3,4,5,6,7,8,9,10,11,12]. We can consider grammars of varying complexity for generating RNA sequences with secondary structures. For example, Dowell *et al.* have introduced nine different lightweight SCFGs as shown in Figure 2, and demonstrated that the performances of some of these simple grammars are close to that of the MFE-based models [10]. The reduced accuracy of these SCFGs when compared to the MFE approach is due to the grammars being too simple to represent RNA secondary structures. Rivas *et al.* have proposed a more complicated SCFG which simulates the standard thermodynamics model [13]. However, this grammar also underperforms compared to the MFE-based models because of overfitting in the estimation of its parameters. We conclude that the acquisition of an appropriate grammar is a crucial problem for the use of SCFGs in RNA analysis.

In this paper, we propose a novel model for generating RNA secondary structures based on a non-parametric Bayesian approach, called hierarchical Dirichlet processes for stochastic context-free grammars (HDP-SCFGs) [14]. Here *non-parametric* means that some meta-parameters such as the number of non-terminal symbols and production rules do not have to be fixed. Instead their distributions are inferred in order to be adapted (in the Bayesian sense) to the training sequences. The results of our RNA secondary structure prediction show that HDP-SCFGs are more accurate than the MFE-based and other generative models, and comparable with CONTRAfold. Furthermore, HDP-SCFGs can be used as a prior distribution of more complicated models because HDP-SCFGs are generative models.

2   Methods

In this section, we describe a novel generative model for structured RNAs, and its inference algorithm. Then, a decoding method using generalized centroid estimators are described. More details are provided as the supplemental material at http://www.ncrna.org/software/npbfold/.

2.1   Stochastic Context-Free Grammars for Structured RNAs

Many functional RNAs form secondary structures which consist of hydrogen-bonded base-pairs including the Watson-Crick base-pairs (A-U and G-C), the Wobble base-pairs (G-U), and the other non-canonical base-pairs. Stochastic context-free grammars (SCFGs) can provide a joint probability distribution over RNA sequences and their secondary structures. A context-free grammar is defined by $G = \{V, T, P, S\}$, where V is a set of non-terminal symbols, T is a finite set of terminal symbols (for RNA: $T = \{\mathtt{A}, \mathtt{C}, \mathtt{G}, \mathtt{U}\}$), P is a set of production rules, and $S \in V$ is an initial non-terminal symbol. In language theory, every context-free grammar can be transformed into an equivalent grammar in the Chomsky normal form (CNF) which allows only bifurcation rules and emission rules. However, CNF is not so suitable for parsing RNAs, so we employ a variant of the RNA normal form [15], in which every production rule belongs to one of the following rule types:

$$ B : X \to YZ, \quad P : X \to aYb, \quad L : X \to aY, \quad R : X \to Ya, \quad E : X \to \epsilon, $$

where $X, Y, Z \in V$, $a, b \in T$ and ϵ is the empty string. Rule type P generates base-pairs, and rule types L and R generate loop regions. Given a parameter θ, each production rule is associated with a probability and these probabilities satisfy the condition

$$ \sum_{\alpha \in (V \cup T)^* \cup \{\epsilon\}} P(X \to \alpha | \theta) = 1 \quad \text{for } \forall X \in V. $$

The probability $P(\boldsymbol{x}, \boldsymbol{z} | \theta)$ is the product of all the probabilities of the production rules in a derivation tree $\boldsymbol{z}$ for generating a sequence $\boldsymbol{x}$. We assume that each production rule can be factorized into a product of a transition probability (to go from one non-terminal symbol to zero, one or two non-terminal symbol(s)) and an emission probability (to emit no base, an unpaired base or a base-pair), and these probabilities are given by the multinomial distributions. For a production rule $z \in P$, we denote a non-terminal symbol on the left-hand side of z by $\sigma(z)$, a set of non-terminal symbols on the right-hand side of z by $\tau(z)$, and a set of terminal symbols emitted on z by $\eta(z)$.

2.2   A Model for Generating RNA Sequences with Secondary Structures

We develop *HDP-SCFGs for structured RNAs* which can infer an optimal grammar whose complexity is adaptive and determined by the training sequences provided. Our model can employ an infinite set of non-terminal symbols thanks to the hierarchical Dirichlet process [16,17].

Multinomial parameters for each emission probability distribution on a non-terminal symbol $v \in V$, denoted by $\boldsymbol{\mu}_v$, are drawn from a Dirichlet prior $Dir(\boldsymbol{\alpha}^\mu)$ which contains priors for emitting base-pairs and unpaired bases. Multinomial parameters for each transition probability distribution on a non-terminal symbol v, denoted by $\boldsymbol{\pi}_v$, are drawn from a Dirichlet process (DP) because non-terminal symbols are chosen from the infinite set of non-terminal symbols. We assume that these DP priors hierarchically depend on a top-level prior with the stick-breaking distribution GEM with a concentration parameter α. We write $\boldsymbol{\beta} \sim GEM(\alpha)$ for $\boldsymbol{\beta} = (\beta_1, \beta_2, \ldots)$ when each β_i is given by

$$ \beta_i = v_i \prod_{j=1}^{i-1} (1 - v_j), \quad v_i \sim Beta(1, \alpha), $$

where *Beta* is the beta distribution. The stick breaking distribution $\boldsymbol{\beta}$ plays the important role of governing the distribution of non-terminal symbols and can be optimized in a training procedure (see below) to be adapted to the training data provided. Then, we can obtain multinomial parameters for each transition probability distribution $\boldsymbol{\pi}_v \sim DP(\alpha^\pi, \boldsymbol{\beta}\boldsymbol{\beta}^T)$, where α^π is a concentration parameter for the transition. Theoretically, $\boldsymbol{\beta} \sim GEM(\alpha)$ is a countably infinite set. However, we can truncate it at a large K because (β_i) will decay exponentially in expectation so that $\beta_i \simeq 0$ for $i > K$. Therefore, we can approximate the Dirichlet process by

$$DP(\alpha', \boldsymbol{\beta}) \simeq Dir(\alpha'\beta_1, \ldots, \alpha'\beta_K).$$

Let $\boldsymbol{Z} = (z_1, z_2, \ldots, z_N) \in P^*$ be a series of production rules which generates RNA sequences $\boldsymbol{X} = (x_1, x_2, \ldots, x_L) \in T^*$. The joint probability distribution of all of the random variables can be written as:

$$p(\boldsymbol{X}, \boldsymbol{Z}, \boldsymbol{\pi}, \boldsymbol{\mu}, \boldsymbol{\beta}) = p(\boldsymbol{X}|\boldsymbol{Z}, \boldsymbol{\mu})p(\boldsymbol{Z}|\boldsymbol{\pi})p(\boldsymbol{\mu})p(\boldsymbol{\pi}|\boldsymbol{\beta})p(\boldsymbol{\beta}) \quad (1)$$
$$p(\boldsymbol{X}|\boldsymbol{Z}, \boldsymbol{\mu}) = \prod_{z_i \in \boldsymbol{Z}} Mult(\eta(z_i); \boldsymbol{\mu}_{\sigma(z_i)})$$
$$p(\boldsymbol{Z}|\boldsymbol{\pi}) = \prod_{z_i \in \boldsymbol{Z}} Mult(\tau(z_i); \boldsymbol{\pi}_{\sigma(z_i)})$$
$$p(\boldsymbol{\mu}) = \prod_{v \in V} Dir(\boldsymbol{\mu}_v; \boldsymbol{\alpha}^\mu)$$
$$p(\boldsymbol{\pi}|\boldsymbol{\beta}) = \prod_{v \in V} DP(\boldsymbol{\pi}_v; \alpha^\pi, \boldsymbol{\beta}\boldsymbol{\beta}^T)$$
$$p(\boldsymbol{\beta}) = GEM(\boldsymbol{\beta}; \alpha),$$

where *Mult* is the multinomial distribution. Figure 1 (left) shows the graphical model of the HDP-SCFGs for generating RNA sequences with secondary structures. We can randomly sample RNA sequences in accordance with this procedure.

We can extend the above-mentioned model in which a given grammar is used as a template for HDP-SCFGs. Every non-terminal symbol of the template grammar has its stick-breaking weights, and is divided into an infinite number of sub-symbols. These sub-symbols inherit production rules from the template grammar as shown in Fig. 1 (right). The resulting grammar can be more adapted to the training sequences than the template grammar, and can avoid overfitting due to the stick-breaking priors associated with the sub-symbols.

2.3   Variational Bayesian Inference for HDP-SCFGs

Based on the procedures to generate RNA sequences and their secondary structures described in the previous section, we can calculate the posterior distribution $p(\boldsymbol{Z}, \boldsymbol{\theta}|\boldsymbol{X})$, where $\boldsymbol{\theta} = \{\boldsymbol{\mu}, \boldsymbol{\pi}, \boldsymbol{\beta}\}$, by Markov chain Monte Carlo (MCMC) methods with no approximation. However, the MCMC methods require generating a huge number of samples to obtain a highly accurate posterior. Therefore, we

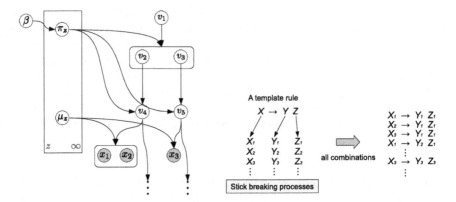

Fig. 1. (left) A graphical model of HDP-SCFGs for structured RNAs. (right) Symbol-associated stick-breaking weights.

approximate this intractable posterior with a tractable approximation using the mean-field approximation, so-called variational Bayesian (VB) inference [17].

We will find a $q \in \mathcal{Q}$ which is close to the true posterior $p(\boldsymbol{Z}, \boldsymbol{\theta}|\boldsymbol{X})$ in the sense of the Kullback-Leibler distance:

$$q^* = \arg\min_{q \in \mathcal{Q}} KL(q(\boldsymbol{Z}, \boldsymbol{\theta})\|p(\boldsymbol{Z}, \boldsymbol{\theta}|\boldsymbol{X})) = \arg\min_{q \in \mathcal{Q}} KL(q(\boldsymbol{Z}, \boldsymbol{\theta})\|p(\boldsymbol{X}, \boldsymbol{Z}, \boldsymbol{\theta})). \quad (2)$$

An approximation q is picked from $\mathcal{Q}$ whose elements are factorized into their components:

$$\mathcal{Q} = \left\{ q(\boldsymbol{Z})q(\boldsymbol{\beta}) \prod_{v=1}^{K} q(\boldsymbol{\pi}_v)q(\boldsymbol{\mu}_v) \right\},$$

where K is the number of non-terminal symbols at which the stick-breaking weight is truncated. We can find a q as one of local minima of (2) by a simple coordinate-ascent algorithm which iteratively optimizes each factor of q in turn while holding the others fixed. This algorithm is similar to the standard expectation-maximization (EM) algorithm: optimizing $q(\boldsymbol{Z})$ corresponds to the E-step, and optimizing $q(\boldsymbol{\pi}) = \prod_v q(\boldsymbol{\pi}_v)$ and $q(\boldsymbol{\mu}) = \prod_v q(\boldsymbol{\mu}_v)$ corresponds to the M-step. However, there is neither a closed form for optimizing $q(\boldsymbol{\beta})$ nor an analogue for this optimization in the standard EM algorithm. Therefore, $q(\boldsymbol{\beta})$ is degenerated into a pointwise function $\delta_{\boldsymbol{\beta}^*}(\boldsymbol{\beta}) = I(\boldsymbol{\beta}^* = \boldsymbol{\beta})$, where $I(condition)$ is an indicator function which takes a value of 1 or 0 depending on whether the $condition$ is true or false.

Let q_i be a factor of $q \in \mathcal{Q}$. We can optimize each q_i holding the other factors q_{-i} fixed as follows:

$$\arg\min_{q_i} KL(q(\boldsymbol{Z}, \boldsymbol{\theta})\|p(\boldsymbol{X}, \boldsymbol{Z}, \boldsymbol{\theta})) \propto \exp E_{q_{-i}}[\log p(\boldsymbol{X}, \boldsymbol{Z}, \boldsymbol{\theta})]. \quad (3)$$

Optimizing $q(\boldsymbol{Z})$ in accordance with (3) for all latent variables, i.e., parse trees in this case, is equivalent to calculating expected rule counts for $\boldsymbol{z}$ under some kinds of weight counts that summarize the uncertainties in the rule probabilities:

$$q^*(\boldsymbol{Z}) \propto \exp\left(E_{q(\boldsymbol{\beta})q(\boldsymbol{\pi})q(\boldsymbol{\mu})}[\log p(\boldsymbol{X}, \boldsymbol{Z}, \boldsymbol{\pi}, \boldsymbol{\mu}, \boldsymbol{\beta})]\right)$$

$$\propto \prod_{z_i \in \boldsymbol{Z}} W^{\mu}_{\sigma(z_i)}(\eta(z_i)) W^{\pi}_{\sigma(z_i)}(\tau(z_i)), \tag{4}$$

where

$$W^{\mu}_v(\eta) = \exp E_{q(\boldsymbol{\mu}_v)}[\log \mu_{v:\eta}], \quad W^{\pi}_v(\tau) = \exp E_{q(\boldsymbol{\pi}_v)}[\log \pi_{v:\tau}]. \tag{5}$$

Here, $\mu_{v:\eta}$ is an element of $\boldsymbol{\mu}_v$ corresponding to the emission of $\eta \in (T \cup \{\epsilon\}) \times (T \cup \{\epsilon\})$, and $\pi_{v:\tau}$ is an element of $\boldsymbol{\pi}_v$ corresponding to the transition from v to $\tau \in V \times V$.

From (1) and (3), the optimal $q(\boldsymbol{\mu}_v)$ can be recognized as the posterior update of the Dirichlet distribution

$$q^*(\boldsymbol{\mu}_v) = Dir(\boldsymbol{\mu}_v; \boldsymbol{\alpha}^{\mu} + \boldsymbol{C}^{\mu}_v), \tag{6}$$

where $\boldsymbol{C}^{\mu}_v = (C^{\mu}_{v:\eta})$ and

$$C^{\mu}_{v:\eta} = E_{q(\boldsymbol{Z})}\left[\sum_{z_i \in \boldsymbol{Z}} I(v = \sigma(z_i))I(\eta = \eta(z_i))\right] \tag{7}$$

is the expected count for the emission of η on the non-terminal symbol v. Similarly, the optimal $q(\boldsymbol{\pi}_v)$ is given by

$$q^*(\boldsymbol{\pi}_v) = Dir(\boldsymbol{\pi}_v; \alpha^{\pi}\boldsymbol{\beta}\boldsymbol{\beta}^T + \boldsymbol{C}^{\pi}_v), \tag{8}$$

where $\boldsymbol{C}^{\pi}_v = (C^{\pi}_{v:\tau})$ and

$$C^{\pi}_{v:\tau} = E_{q(\boldsymbol{Z})}\left[\sum_{z_i \in \boldsymbol{Z}} I(v = \sigma(z_i))I(\tau = \tau(z_i))\right] \tag{9}$$

is the expected count for the transition to τ from v.

Since we found out that $q(\boldsymbol{\mu}_v)$ and $q(\boldsymbol{\pi}_v)$ are the Dirichlet distributions from (6) and (8), the weight counts (5) can be rewritten in an easier form:

$$W^{\mu}_v(\eta) = \exp\left(\Psi\left(\alpha^{\mu}_{\eta} + C^{\mu}_{v:\eta}\right) - \Psi\left(\sum_{\eta'} \alpha^{\mu}_{\eta'} + C^{\mu}_{v:\eta'}\right)\right) \tag{10}$$

$$W^{\pi}_v(\tau) = \exp\left(\Psi\left(\alpha^{\pi}\beta_{\tau_l}\beta_{\tau_r} + C^{\pi}_{v:\tau}\right) - \Psi\left(\alpha^{\pi} + \sum_{\tau'} C^{\pi}_{v:\tau'}\right)\right), \tag{11}$$

where α^{μ}_{η} is an element of $\boldsymbol{\alpha}^{\mu}$ for the emission of η, and $\tau = (\tau_l, \tau_r)$. As shown in (4), (6) and (8), the optimization of $q(\boldsymbol{Z})$ and the optimization of $q(\boldsymbol{\mu})$ and $q(\boldsymbol{\pi})$ depend on each other. Therefore, these variational posterior distributions are iteratively updated. We compute the expected counts (7) and (9) using the inside-outside algorithm with the multinomial weights (10) and (11).

Unlike $q(\boldsymbol{z})$, $q(\boldsymbol{\mu})$ and $q(\boldsymbol{\pi})$, there is no closed form to calculate the optimal $q(\boldsymbol{\beta})$. So, we find $\boldsymbol{\beta}^*$ by a numerical gradient method minimizing the following objective function derived from (1) and (3) :

$$L(\boldsymbol{\beta}^*) = \log GEM(\boldsymbol{\beta}^*; \alpha) + \sum_{z=1}^K E_q[\log Dir(\boldsymbol{\pi}_z; \alpha^{\pi}\boldsymbol{\beta}^*\boldsymbol{\beta}^{*T})].$$

2.4   Decoding RNA Secondary Structures with the Generalized Centroid Estimators

Using the variational posterior distribution $q(\boldsymbol{\theta})$, the probability of a parse tree $\boldsymbol{z}_{new}$ for a new RNA sequence $\boldsymbol{x}_{new}$ is given by

$$p(\boldsymbol{z}_{new}|\boldsymbol{x}_{new}) = E_{p(\boldsymbol{\theta}|\boldsymbol{X})}[p(\boldsymbol{z}_{new}|\boldsymbol{x}_{new},\boldsymbol{\theta})]$$
$$\simeq E_{q(\boldsymbol{\theta})}[p(\boldsymbol{z}_{new}|\boldsymbol{x}_{new},\boldsymbol{\theta})]. \tag{12}$$

However, since inferred HDP-SCFGs generally have structural ambiguity, i.e., there is more than one possible parse tree for a secondary structure, the most likely parse tree is *not* always exactly equivalent to the most likely secondary structure. Therefore, we predict RNA secondary structures with the γ-centroid estimator [18], a kind of posterior decoding method based on statistical decision theory. Here we approximate the intractable integration with respect to $q(\boldsymbol{\theta})$ by using $\boldsymbol{\theta}^{(PME)} = E_{q(\boldsymbol{\theta})}[\boldsymbol{\theta}]$, the posterior mean estimator of $q(\boldsymbol{\theta})$, which can be obtained by:

$$p_v^\mu(\eta) = \frac{\alpha_\eta^\mu + C_{v:\eta}^\mu}{\sum_{\eta'} \alpha_{\eta'}^\mu + C_{v:\eta'}^\mu}, \quad p_v^\pi(\tau) = \frac{\alpha^\pi \beta_{\tau_l}\beta_{\tau_r} + C_{v:\tau}^\pi}{\alpha^\pi + \sum_{\tau'} C_{v:\tau'}^\pi}. \tag{13}$$

The base-pairing probability $p_{ij} = E_{p(\boldsymbol{y}|\boldsymbol{x},\boldsymbol{\theta})}[y_{ij}]$ is the probability that the i-th and j-th nucleotides form a base-pair in $\boldsymbol{y}$ under (13), and can be calculated using the inside-outside algorithm. Then, we can find $\hat{\boldsymbol{y}}$ using the recursive equations:

$$M_{i,j} = \max \begin{cases} M_{i+1,j} \\ M_{i,j-1} \\ M_{i+1,j-1} + (\gamma+1)p_{ij} - 1 \\ M_{i,k} + M_{k+1,j} \end{cases}, \tag{14}$$

and tracing back from $M_{1,|\boldsymbol{x}|}$, where $|\boldsymbol{x}|$ is the length of $\boldsymbol{x}$. We can control the trade-off between specificity and sensitivity by γ. More details about the γ-centroid estimators are provided as the supplemental material and [18].

3   Experimental Results and Discussion

We carried out several experiments on the S151Rfam dataset [12] which was carefully selected from 151 families in the Rfam seed [19]. The accuracy of predicting secondary structures was evaluated by 2-fold cross-validation based on positive predictive value (PPV), sensitivity (SN) and Matthew's correlation coefficient (MCC), which are defined by

$$PPV = \frac{TP}{TP+FP}, \quad SN = \frac{TP}{TP+FN},$$

$$MCC = \frac{TP \cdot TN - FN \cdot FP}{\sqrt{(TP+FN)(TN+FP)(TP+FP)(TN+FN)}},$$

$$
\begin{aligned}
\text{G1:}\ & S \to aS\hat{a}|aS|Sa|SS|\epsilon \\
\text{G2:}\ & S \to aP^{a\hat{a}}\hat{a}|aS|Sa|SS|\epsilon, \quad P^{b\hat{b}} \to aP^{a\hat{a}}\hat{a}|S \\
\text{G3:}\ & S \to aS\hat{a}|aL|Ra|LS, \quad L \to aS\hat{a}|aL, \quad R \to Ra|\epsilon \\
\text{G4:}\ & S \to aS|T|\epsilon, \quad T \to Ta|aS\hat{a}|TaS\hat{a} \\
\text{G5:}\ & S \to aS|aS\hat{a}S|\epsilon \\
\text{G6:}\ & S \to LS|L, \quad L \to aF\hat{a}|a, \quad F \to aF\hat{a}|LS \\
\text{G6s:}\ & S \to LS|L, \quad L \to aF^{a\hat{a}}\hat{a}|a, \quad F^{b\hat{b}} \to aF^{a\hat{a}}\hat{a}|LS \\
\text{G7:}\ & S \to aP^{a\hat{a}}\hat{a}|aL|Ra|LS, \quad L \to aP^{a\hat{a}}\hat{a}|aL, \quad R \to Ra|\epsilon \\
& P^{b\hat{b}} \to aP^{a\hat{a}}\hat{a}|aN\hat{a}, \quad N \to aL|Ra|LS \\
\text{G8:}\ & S \to aS|T|\epsilon, \quad T \to Ta|aP^{a\hat{a}}\hat{a}|TaP^{a\hat{a}}\hat{a} \\
& P^{b\hat{b}} \to aP^{a\hat{a}}\hat{a}|aN\hat{a}, \quad N \to aS|Ta|TaP^{a\hat{a}}\hat{a}
\end{aligned}
$$

Fig. 2. RNA grammars introduced by [10]

$$
\begin{aligned}
\text{G1':}\ & S \to aP\hat{a}|aS|Sa|SS|\epsilon, \quad P \to aP\hat{a}|S \\
\text{G3':}\ & S \to aP\hat{a}|aL|Ra|LS, \quad L \to aP\hat{a}|aL, \quad R \to Ra|\epsilon \\
& P \to aP\hat{a}|aN\hat{a}, \quad N \to aL|Ra|LS \\
\text{G4':}\ & S \to aS|T|\epsilon, \quad T \to Ta|aP\hat{a}|TaP\hat{a} \\
& P \to aP\hat{a}|aN\hat{a}, \quad N \to aS|Ta|TaP\hat{a}
\end{aligned}
$$

Fig. 3. New developed RNA grammars

where TP is the number of base-pairs appearing in both the reference and predicted sequences, FP is the number of base-pairs that appear in the predicted sequence but not in the reference, TN is the number of base-pairs that appear in neither sequence, and FN is the number of base-pairs that appear in the reference but not in the predicted sequence.

First, the proposed algorithm was compared with the standard SCFGs. We employed the previously proposed grammars [10] shown in Fig.2 and modified grammars shown in Fig. 3. G2 is derived from G1 by dividing the non-terminal symbol S into fully lexicalized symbols $P^{a\hat{a}}$ which can emit only a base-pair $(a, \hat{a})$. G1' is also derived from G1 by dividing the non-terminal symbol S into a symbol P which emits any kinds of base-pairs. Therefore, in terms of grammatical complexity, these grammars can be ranked as G1 < G1' < G2. G1 is too simple to represent the thermodynamics of RNA secondary structures, while G2 can simulate stacking base-pairs, but is prone to overfit for the training data. Analogously, we can say that G3 < G3' < G7 and G4 < G4' < G8.

For the standard SCFGs, we inferred their parameters by the expectation-maximization (EM) algorithm (for G1, G1' and G2 because of their structural ambiguity), or the maximum likelihood (ML) estimation using their observation counts (for the other grammars). For the HDP-SCFGs, we used these grammars as templates of the extended HDP-SCFGs as shown in Fig. 1 (right). The top-level hyper-parameter α for each sub-symbol was set to 4 for the states which can emit base-pairs, and 1 otherwise. All the other hyper-parameters were set to 1.

Table 1. Comparison between the HDP-SCFGs, the standard SCFGs and CRFs through Matthew's correlation coefficient (MCC)

	HDP-SCFG		SCFG		CRF	
	γ-centroid	CYK	γ-centroid	CYK	γ-centroid	CYK
G1	**0.61**	0.43	0.32	0.24	0.34	0.31
G1′	**0.62**	0.41	0.59	0.45	0.58	0.48
G2	**0.60**	0.45	0.59	0.45	**0.60**	0.49
G3	**0.58**	0.45	0.49	0.40	0.44	0.37
G3′	**0.58**	0.46	**0.58**	0.53	0.49	0.45
G7	**0.58**	0.49	0.57	0.52	0.46	0.40
G4	**0.58**	0.53	0.19	0.12	0.34	0.30
G4′	**0.59**	0.50	0.19	0.12	0.49	0.47
G8	**0.59**	0.54	0.56	0.50	0.49	0.47
G6	**0.57**	0.54	0.56	0.52	0.53	0.48
G6s	**0.57**	0.54	0.56	0.53	0.56	0.52
G5	**0.58**	0.50	0.04	0.06	0.36	0.29

We predicted secondary structures for each grammar by two decoding methods: the γ-centroid estimators [18] described in the previous section and the CYK algorithm that is a traditional decoding method for SCFGs. For the CYK algorithm on HDP-SCFGs, we employed the posterior mean estimators of the production rules (13) to approximate Equation (12).

Table 1 shows the accuracy of predicting secondary structures using HDP-SCFGs and the standard SCFGs. We fixed $\gamma = 4.0$ for the γ-centroid estimators. When decoding secondary structures with the γ-centroid estimators, the HDP-SCFG outperformed its standard SCFG counterpart for every grammar. Especially, for G1, G3, G4, G4′ and G5, HDP-SCFGs gave a significant improvement in accuracy. However, when decoding secondary structures with the CYK algorithm, the HDP-SCFGs did not show an improvement in accuracy compared to their counterparts for several grammars. HDP-SCFGs divide non-terminal symbols adaptively to maximize the likelihood of the training sequences. This enables the HDP-SCFGs with the γ-centroid estimators to be more accurate even for poor grammars such as G1, G4 and G5. However, this also makes the model structurally more ambiguous so that the chance of inconsistency between the most likely parse tree and the most likely secondary structure might grow. Therefore, the CYK algorithm that finds only the most likely parse trees is inadequate for the HDP-SCFG models. On the other hand, since the γ-centroid estimator calculates the most likely secondary structures supported by all possible parse trees in the sense of statistical decision theory, structural ambiguity of the HDP-SCFG models does not cause the negative effects for the accuracy.

We also compared the HDP-SCFGs to conditional random fields (CRFs) [20,9,12], a kind of discriminative methods, in Table 1. For every grammar, HDP-SCFGs outperformed its CRF counterparts (except for G2). The most important difference between CRFs and HDP-SCFGs is the adaptivity of grammars:

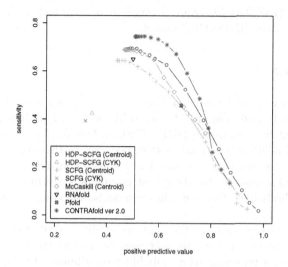

Fig. 4. Accuracy of prediction with various thresholds

CRFs do not employ the stick breaking process to govern the distribution of non-terminal symbols. In our results, the standard SCFGs are almost comparable to CRFs, although Do *et al.* have reported that CRFs are more accurate than the standard SCFGs for all grammars excluding G3 and G4 [12]. This is probably due to different criteria and implementations for evaluation.

We compared our model with existing methods including McCaskill's model [21] with the γ-centroid estimators, RNAfold [2], Pfold model [5] and CON-TRAfold [12]. We inferred the HDP-SCFG and the SCFG for G1$'$, and predicted secondary structures using γ-centroid estimators with $\gamma \in \{2^k | -5 \leq k \leq 10\}$. For McCaskill's model, we predicted secondary structures by the γ-centroid estimators under McCaskill's model, that is, secondary structures are decoded by (14) using McCaskill's base-pairing probabilities. For the Pfold model, we used our implementation with the G6 grammar and the parameters provided by the Pfold web site. We employed CONTRAfold version 2.0 and re-trained the parameters from the S151Rfam dataset. Figure 4 shows the PPV-sensitivity curves for our method and the existing methods. HDP-SCFGs with the γ-centroid estimators performed more accurately than (or were at least comparabe with) the MFE-based models and the other generative models including the Pfold model. Unfortunately, however, our model could not reach the level of performance of CONTRAfold. The difference in the accuracy between CONTRAfold and our model is less than 0.05 points in the MCC (0.67 and 0.62). CONTRAfold employs a rich feature model which mirrors the traditional thermodynamic model of RNA secondary structures and infers their parameters with CLLMs, a generalized CRFs. Although these features contribute to the method's high power for predicting secondary structures, training of the CRF model costs extremely large computational time for calculating partition functions. In fact, only 285 minutes were needed to obtain our HDP-SCFG model on a 2.0 GHz AMD Opteron processor in our experiment, whereas training of the CONTRAfold required 16037 minutes, which is at least 50-fold slower than that of the HDP-SCFG model.

4   Conclusion

We propose a novel model for generating RNA secondary structures based on a non-parametric Bayesian approach, called hierarchical Dirichlet processes for stochastic context-free grammars (HDP-SCFGs). Our results demonstrate that the adaptivity of HDP-SCFGs enables their accuracy in the secondary structure prediction to be more accurate compared with the standard SCFGs and discriminative models. Our results also show that our models are more accurate than the existing methods including MFE-based models, and comparable with CONTRAfold.

CONTRAfold cannot compute the joint probability distribution over RNA sequences and their secondary structures because the feature model violates a number of constraints for representing a probability distribution. On the other hand, since the HDP-SCFGs can calculate the joint probability over sequences and structures, we can build more complicated stochastic models like QRNA [6] and EvoFold [11] using our model as a prior distribution.

Acknowledgements

This work was supported in part by a grant from "Functional RNA Project" funded by the New Energy and Industrial Technology Development Organization (NEDO) of Japan, and was also supported in part by Grant-in-Aid for Scientific Research on Priority Area "Comparative Genomics" No. 17018029 from the Ministry of Education, Culture, Sports, Science and Technology of Japan. We thank Dr. Hisanori Kiryu and our colleagues from the RNA Informatics Team at the Computational Biology Research Center (CBRC) for fruitful discussions.

References

1. Zuker, M., Stiegler, P.: Optimal computer folding of large RNA sequences using thermodynamics and auxiliary information. Nucleic Acids Res. 9(1), 133–148 (1981)
2. Hofacker, I.L.: Vienna RNA secondary structure server. Nucleic Acids Res. 31(13), 3429–3431 (2003)
3. Eddy, S.R., Durbin, R.: RNA sequence analysis using covariance models. Nucleic Acids Res. 22(11), 2079–2088 (1994)
4. Sakakibara, Y., Brown, M., Hughey, R., Mian, I.S., Sjölander, K., Underwood, R.C., Haussler, D.: Stochastic context-free grammars for tRNA modeling. Nucleic Acids Res. 22(23), 5112–5120 (1994)
5. Knudsen, B., Hein, J.: RNA secondary structure prediction using stochastic context-free grammars and evolutionary history. Bioinformatics 15(6), 446–454 (1999)
6. Rivas, E., Eddy, S.R.: Noncoding RNA gene detection using comparative sequence analysis. BMC Bioinformatics 2, 8 (2001)

7. Eddy, S.R.: A memory-efficient dynamic programming algorithm for optimal alignment of a sequence to an RNA secondary structure. BMC Bioinformatics 3, 18
   (2002)
8. Sakakibara, Y.: Pair hidden Markov models on tree structures. Bioinformatics 19(suppl. 1), i232–i240 (2003)
9. Sato, K., Sakakibara, Y.: RNA secondary structural alignment with conditional random fields. Bioinformatics 21(suppl. 2), ii237–ii242 (2005)
10. Dowell, R.D., Eddy, S.R.: Evaluation of several lightweight stochastic context-
    free grammars for RNA secondary structure prediction. BMC Bioinformatics 5, 71
    (2004)
11. Pedersen, J.S., Bejerano, G., Siepel, A., Rosenbloom, K., Lindblad-Toh, K., Lander,
    E.S., Kent, J., Miller, W., Haussler, D.: Identification and classification of conserved
    RNA secondary structures in the human genome. PLoS Comput. Biol. 2(4), e33
    (2006)
12. Do, C.B., Woods, D.A., Batzoglou, S.: CONTRAfold: RNA secondary structure pre-
    diction without physics-based models. Bioinformatics 22(14), e90–e98 (2006)
13. Rivas, E., Eddy, S.R.: Secondary structure alone is generally not statistically signif-
    icant for the detection of noncoding RNAs. Bioinformatics 16(7), 583–605 (2000)
14. Liang, P., Petrov, S., Jordan, M.I., Klein, D.: The infinite PCFG using hierarchi-
    cal Dirichlet processes. In: Proceedings of the 2007 Joint Conference on Empirical
    Methods in Natural Language Processing and Computational Natural Language
    Learning (EMNLP-CoNLL), pp. 688–697 (2007)
15. Durbin, R., Eddy, S., Krogh, A., Mitchison, G.: Biological Sequence Analysis. Cam-
    bridge University Press, Cambridge (1998)
16. Teh, Y.W., Jordan, M.I., Beal, M.J., Blei, D.M.: Hierarchical Dirichlet processes.
    Journal of the American Statistical Association 101, 1566–1581 (2006)
17. Blei, D.M., Jordan, M.I.: Variational inference for Dirichlet process mixtures.
    Bayesian Analysis 1, 121–144 (2005)
18. Hamada, M., Kiryu, H., Sato, K., Mituyama, T., Asai, K.: Prediction of RNA sec-
    ondary structure using generalized centroid estimators. Bioinformatics 25(4), 465–
    473 (2009)
19. Griffiths-Jones, S., Moxon, S., Marshall, M., Khanna, A., Eddy, S.R., Bateman,
    A.: Rfam: annotating non-coding RNAs in complete genomes. Nucleic Acids
    Res. 33(Database issue), D121–D124 (2005)
20. Lafferty, J., McCallum, A., Pereira, F.: Conditional random fields: Probabilistic
    models for segmenting and labeling sequence data. In: Proceedings of the 18th In-
    ternational Conference on Machine Learning, pp. 282–289 (2001)
21. McCaskill, J.S.: The equilibrium partition function and base pair binding probabil-
    ities for RNA secondary structure. Biopolymers 29(6-7), 1105–1119 (1990)

Exact Score Distribution Computation for Similarity Searches in Ontologies

Marcel H. Schulz[1,2], Sebastian Köhler[3,4], Sebastian Bauer[3], Martin Vingron[1], and Peter N. Robinson[3,4]

[1] Max Planck Institute for Molecular Genetics, Ihnestr. 73, 14195 Berlin, Germany
[2] International Max Planck Research School for Computational Biology and Scientific Computing
[3] Institute for Medical Genetics, Charité-Universitätsmedizin Berlin, Augustenburger Platz 1, 13353 Berlin, Germany
[4] Berlin-Brandenburg Center for Regenerative Therapies (BCRT), Charité-Universitätsmedizin Berlin, Berlin, Germany

Abstract. Semantic similarity searches in ontologies are an important component of many bioinformatic algorithms, e.g., protein function prediction with the Gene Ontology. In this paper we consider the exact computation of score distributions for similarity searches in ontologies, and introduce a simple null hypothesis which can be used to compute a P-value for the statistical significance of similarity scores. We concentrate on measures based on Resnik's definition of ontological similarity. A new algorithm is proposed that collapses subgraphs of the ontology graph and thereby allows fast score distribution computation. The new algorithm is several orders of magnitude faster than the naive approach, as we demonstrate by computing score distributions for similarity searches in the Human Phenotype Ontology.

1 Introduction

In this paper we consider the problem of calculating score distributions for similarity searches in ontologies represented by directed acyclic graphs (DAGs). Ontologies are used to describe domains of knowledge, whereby the nodes of the DAG, which are also called *terms* of the ontology, are assigned to items in the domain and the edges between the nodes represent semantic relationships. Ontologies are designed such that terms closer to the root are more general than their descendant terms. For the ontologies we consider in this paper, the *true-path rule* applies, that is, items are annotated to the most specific term possible but are assumed to be implicitly annotated to all ancestors of that term. Examples of current ontologies with an annotated corpus of items are the Gene Ontology (GO) [1], the Mammalian Phenotype Ontology [2], or the Human Phenotype Ontology (HPO) [3]. For example, the HPO provides over 9,000 terms to describe phenotypic abnormalities (signs and symptoms of disease) which have been used to annotate over 4,800 hereditary syndromes in the Online Mendelian Inheritance in Man (OMIM) database [4].

S.L. Salzberg and T. Warnow (Eds.): WABI 2009, LNBI 5724, pp. 298–309, 2009.

Similarity between any two terms within an ontology is based on the annotations to items in the domain and on the structure of the DAG. Different semantic similarity measures have been proposed [5] and the measures have been used in many different applications in computational biology. For example, different studies show that semantic similarity between proteins annotated with GO terms correlate with sequence similarity [6,5,7].

An interesting application of ontologies is for database searches using semantic similarity. Clinical databases such as OMIM are often used by clinical geneticists as a part of the diagnostic process. The geneticist typically enters one or several terms describing the phenotypic features of the patient being evaluated, and a list of candidate diagnoses is returned, which are characterized by some or all of the features. We consider such database searches by computing the similarity between a set of q query terms and a set of target terms annotated to an item from the domain. In the HPO example, each annotated syndrome (item) would represent a target set of HPO terms. When the similarity between the query terms and all annotated syndromes is computed the most similar syndrome to the query can be found. The method we describe in this paper allows us to assign a P-value to the score obtained by such a search corresponding to the probability of obtaining a given similarity score or better by choosing query terms at random.

To date, the comparison of similarity measures in studies such as [6] has been done on an empirical basis by comparing their scores. The use of P-values, instead of the scores themselves, allows the comparison of similarity searches for different similarity measures as well as different query and target set sizes. Exact P-value distributions can not only be used for searching in databases but also provide a new way of comparing the resolution and power of different similarity measures.

For the first time we derive an algorithm to compute the exact score distribution. The algorithm is applicable to similarity measures based on information content [8]. Our algorithm summarises all nodes that contribute the same similarity score to the score distribution to avoid redundant computations and we show that it compares favorably against a naive implementation. Our algorithm thus provides an exact and efficient method for calculating the score distributions for semantic similarity searches in current ontologies such as GO and HPO.

2   Preliminaries

We consider an ontology O composed of a set of *terms* that are linked via an *is-a* or *part-of* relationship. The ontology O can then be represented by a DAG $D = (V, E)$, where every term is a node in V and every link is a directed edge in E. A directed edge going from node n_1 to n_2 is denoted $e_{1,2}$. An *item* i is defined as an abstract entity to which terms of the ontology are annotated. Ontologies such as GO and HPO are used to annotate items of a domain; GO terms are used to annotate genes or proteins, i.e., to assign biological functions or characteristics to the proteins, and HPO terms are used to annotate diseases, i.e., to state that a certain phenotypic feature occurs in persons with the disease.

Let $Anc(n)$ be defined as the ancestors of n, i.e., the nodes that are found on all paths from node n to the root of D, including n. We note that the true-path rule states that if an item is explicitly annotated to a term n, it is implicitly annotated to $Anc(n)$. In order to describe the implicit annotations we define $annot_S$. Let S be the set of terms that has been explicitly annotated to item i, then $annot_S = \{Anc(n)|n \in S\}$, namely all terms that are annotated to item i and all their ancestors in D. Let the set of common ancestors of two nodes n_1 and n_2 be defined as $ComAnc(n_1, n_2) = Anc(n_1) \cap Anc(n_2)$. Let $Desc(n)$ be the set of descendant nodes of n, again including n. Note that in this notation descendant nodes are considered only once, even if there are multiple paths leading to them.

There is an abundant body of literature about similarity measures in ontologies, especially for the GO. We will concentrate in this work on the class of measures that are based on the information content (IC) of a node:

$$IC(n) = -\log \ p(n), \tag{1}$$

where $p(n)$ denotes the frequency of annotations of n among all items in the domain. The information content is a nondecreasing function on the nodes of D as we descend in the hierarchy and therefore it is *monotonic*. The similarity between two nodes was defined by Resnik as the maximum information content among all common ancestors [8]:

$$sim(n_1, n_2) = max\{IC(a)|a \in ComAnc(n_1, n_2)\}. \tag{2}$$

Recent studies have shown that this measure, applied to the GO, is highly correlated with sequence similarity and outperforms many alternative measures [5,7,9], and it produced meaningful clusters of OMIM diseases when applied to the HPO [3]. In the following we introduce two definitions of similarity measures between two sets of terms I_1 and I_2, which are widely used [3,5,9]:

$$sim^{max}(I_1, I_2) = \max_{n_1 \in I_1, n_2 \in I_2} sim(n_1, n_2) \tag{3}$$

$$sim^{avg}(I_1, I_2) = \frac{1}{|I_1|} \sum_{n_1 \in I_1} \max_{n_2 \in I_2} sim(n_1, n_2) \tag{4}$$

Note that Eq. (4) is not *symmetric* [5], i.e., it must not be true that $sim^{avg}(I_1, I_2) = sim^{avg}(I_2, I_1)$. See Fig. 1 for an example computation of sim^{avg}.

In what follows we need to compute the similarity also between a *multiset* and a set of terms. The concept of multisets [10] is a generalization of the concept of sets. In contrast to sets, in which elements can only have a single membership, the elements of multisets may appear more than once. Therefore, all sets are multisets, while a multiset is not a set if an element appears more than once.

Formally, a multiset M is a set of pairs, $M = \{(u_1, M_1), \ldots, (u_d, M_d)\}$, in which $u_i \in U = \{u_1, \ldots, u_d\}$ are the elements of the *underlying set* U, with $U \subseteq V$. M_i defines the *multiplicity* of u_i in the multiset and the sum of the

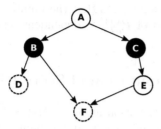

Fig. 1. Computation of sim^{avg} on a DAG with five nodes. The target set $TS = \{B, C\}$ is shown as black nodes and the query set $Q = \{D, F\}$ is shown as nodes surrounded by dashed lines. $sim^{avg}(Q, TS) = \frac{sim(D,B)+sim(F,C)}{2} = \frac{IC(B)+IC(C)}{2}$, under the assumption that $IC(C) \geq IC(B)$.

multiplicities of M is called the *cardinality* of M, denoted $|M|$. Note that we only consider multiplicities in the domain of positive integers, i.e., $M_i \in \mathbb{N}^+$.

Denote the similarity for a multiset M compared to a set of terms I_1, with a slight abuse of notation, as (similar equation for sim^{max} not shown)

$$sim^{avg}(M, I_1) = \frac{1}{|M|} \sum_{(u_i, M_i) \in M} M_i \cdot \max_{n_j \in I_1} sim(u_i, n_j). \tag{5}$$

The *multiset coefficient* $M(n, q) = \binom{n+q-1}{q}$ denotes the number of distinct multisets of cardinality q, with elements taken from a finite set of cardinality n. It describes how many ways there are to choose q elements from a set of n elements if repetitions are allowed.

In this paper we will present methods for analytically calculating the probability distribution of similarity scores for comparisons between a query set Q with q terms against an item that has been annotated with a *target set* TS of nodes. For example, if a clinician chooses a set Q of HPO terms describing abnormalities seen in a patient and uses Eq. (3) or (4) to calculate a score S to a disease that has been annotated with terms of the HPO, we would like to know the probability of a randomly chosen set of q nodes achieving a score of S or greater.

In other words, our methods will be used to calculate a P-value for the null hypothesis that a similarity score of S or greater for a set of q query terms Q and a target set TS has been observed by chance (the alternative hypothesis in the medical example given above would mean that the disease (item) annotated by TS would be at least a reasonable differential diagnosis). We refer to this null hypothesis as the *independence hypothesis*. The P-value for the *independence hypothesis* is defined as:

$$P^{sim}_{q,TS}(S \geq s) = \frac{|\{Q|sim(Q, TS) \geq s, Q = \{n_1, \dots, n_q\} \subseteq V\}|}{\binom{|V|}{q}}. \tag{6}$$

A first result of this consideration is that the number of distinct scores for the complete score distribution of $P_{q,\mathcal{TS}}^{sim}$ is dependent on q and $\mathcal{TS}$, as well as on the similarity measure used.

3   Exhaustive Computation of Score Distributions

We represent the score distribution as $\mathcal{SD} = \{(s_1, f_1), \ldots, (s_k, f_k)\}$. Every pair $(s_i, f_i) \in \mathcal{SD}$ contains a unique score s_i and a count f_i that defines its frequency within the distribution.

A naive approach to calculating the complete score distribution is to determine the similarity of each possible term combination $Q \subseteq V$ of size q with the fixed target set $\mathcal{TS}$. The complete procedure is outlined in Algorithm 1. It requires two basic operations that are applied to the set $\mathcal{SD}$. The first operation called *getScorePair* returns the pair that represents the given score or *nil* in case the entry does not exist. The second operation denoted *putScorePair* puts the given pair into the set $\mathcal{SD}$, overwriting any previously added pair with the same score. For further analyses we assume that both operations have constant running time.

Algorithm 1. Naive score distribution computation for sim^{avg}

    Input: $V, q, \mathcal{TS}$
    Output: The score distribution, i.e., $\mathcal{SD} = \{(s_1, f_1), \ldots, (s_k, f_k)\}$
1 $\mathcal{SD} = \emptyset$;
2 **foreach** $Q = \{n_1, n_2, \ldots, n_q\} \subseteq V$ **do**
3      $s_{new} \leftarrow sim^{avg}(Q, \mathcal{TS})$
4      $(s, f) \leftarrow getScorePair(\mathcal{SD}, s_{new})$
5   **if** $(s, f) \neq nil$ **then**
6         $putScorePair(\mathcal{SD}, (s_{new}, f + 1))$
7   **else**
8         $putScorePair(\mathcal{SD}, (s_{new}, 1))$
9 **return** $\mathcal{SD}$

As the number of possible term combinations is $\binom{|V|}{q}$ and each similarity computation (line 1) costs $\mathcal{O}(q \cdot |\mathcal{TS}|)$ operations for Eq. (4) the Algorithm 1 runs in $\mathcal{O}(|V|^q \cdot q \cdot |\mathcal{TS}|)$ time. A typical size of $|V| = 9,000$ as for the HPO demonstrates that the naive approach is impractical for values $q > 2$. The naive approach neglects the relationships of the nodes in D and $\mathcal{TS}$. We will exploit these relationships in the next section and collapse nodes into subsets with the same contribution to the score distribution computation.

4   Preprocessing of the DAG

We outline now how we reduce the complexity of a DAG D to a tree by exploiting the maximization step in Resnik's formulation of similarity, see Eq. (2). Recall

---

Algorithm 2. Compute *Children* attribute

---

1 for $n_i \in V$ *in topological order* **do**
2   **for** j *in* $e_{i,j} \in E$ **do**   /* set weights of outgoing edges */
3   **if** $n_i \in annot_{TS}$ **then**
4 $w_{i,j} \leftarrow IC(n_i)$
5   **else**
6 $w_{i,j} \leftarrow \max\{w_{h,i}|e_{h,i} \in E\}$

7 convert directed edges in E in undirected edges
8 MST = maximum spanning tree of D
9 for *edge* $e_{i,j} \in$ MST **do**   /* compute MST' */
10   **if** $n_i, n_j \in annot_{TS}$ **then**
11   remove $e_{i,j}$ from MST

12 for $n_i \in annot_{TS}$ **do**
13 $Children(n_i) \leftarrow$ size of connected component in MST' containing n_i

14 return

---

that all nodes that are annotated with terms from the target set TS are contained in $annot_{TS}$. We will prove now that only the IC values of nodes in $annot_{TS}$ are relevant for the score distribution computation, cf. Fig. 1.

Lemma 1. *Given a DAG $D = (V, E)$ and a target set $TS = \{n_1, \ldots, n_k\} \subseteq V$ all scores in the score distribution of similarity measures (3) and (4) are derived from IC values of the nodes in $annot_{TS}$.*

Proof. The first step for the computation of Eq. (3) and Eq. (4) is to maximize $sim(n_1, n_2)$ for each node $n_1 \in Q$ compared to nodes $n_2 \in TS$. Independent of Q, the maximum IC value for $sim(n_1, n_2)$ is taken from a node in $annot_{TS}$. This is easy to see as the nodes in $annot_{TS}$ are the only nodes that could be ancestors of n_1 and n_2, because by definition $Anc(n_2) \subseteq annot_{TS}$.   □

Put differently, Lemma 1 states that D can be collapsed into subtrees that summarize all nodes in V which contribute the same score to the score distribution. In order to explain the property of these subtrees we define the *most informative annotated ancestor (MIA)* of a node $n_1 \in V$ as:

$$MIA(n_1) = \{n_2 | IC(n_2) = \max_{n_2 \in annot_{TS}} sim(n_1, n_2)\}. \tag{7}$$

In words, we consider for each node in V which annotated node in $annot_{TS}$ will be selected in the maximization step over $n_2 \in I_2$ in Eq. (3) and Eq. (4). If several nodes $n_2 \in annot_{TS}$ share the maximum information content with $sim(n_1, n_2)$ in Eq. 7, we arbitrarily select one node n_2, i.e., always $|MIA(n_1)| = 1$. We also define that $MIA(a) = a, \forall a \in annot_{TS}$.

     The definition resolves the problems with nodes that have more than one parent and is the basic property exploited in our algorithm. We can now summarize all nodes in V that have the same MIA. For all nodes $a \in annot_{TS}$ we define the set of *associated descendants* (AD) of a as:

$$AD(a) = \{n | n \in V : MIA(n) = a\}. \tag{8}$$

Example 1. If we assume that $IC(C) \geq IC(B)$ for the DAG in Figure 1, then $AD(A) = \{A\}$, $AD(B) = \{B, D\}$ and $AD(C) = \{C, E, F\}$. Node F belongs to $AD(C)$, because $IC(C) = \max_{n_1 \in \{A,B,C\}} sim(n_1, F)$, see Eq. (7). Note that all nodes in $annot_{TS}$ belong to their own set of ADs.

We will now show how to compute the subtrees in which all nodes are connected to their MIA by computing a *maximum spanning tree* [11]. First, we introduce edge weights $w_{i,j}$ to the edges of D. Let

$$w_{i,j} = \begin{cases} IC(n_i), & \text{if } n_i \in annot_{TS} \\ \max\{w_{h,i} | e_{h,i} \in E\}, & \text{otherwise} \end{cases}. \tag{9}$$

The edge weights are defined in a recursive manner. First, all weights of edges emerging from nodes in $annot_{TS}$ are set. Then the maximum edge weight of all incoming edges for each node not in $annot_{TS}$ are propagated to all outgoing edges of the node, and as such propagated throughout the graph. Computing the edge weights is efficiently done after the nodes of D have been sorted in topological order, see Alg. 2.

We can now explain how we compute $AD(n_i)$ for every $n_i \in annot_{TS}$. To do so, we will convert the directed edges in E into undirected edges and introduce symmetric edge weights, i.e., $w_{i,j} = w_{j,i}$. We denote as $D' = (V', E')$ the modified graph of D with undirected edges and edge weights.

Let MST be the *maximum spanning tree* [11] of D'. Let MST' be the graph when we cut all edges $e_{i,j} \in MST$ for which $n_i, n_j \in annot_{TS}$ holds. MST' has exactly $|annot_{TS}|$ connected components. We apply the MST algorithm in order to remove edges for every node $n \in V$ with multiple parent terms in such a way to keep only the edge to the parent term that is connected to $MIA(n)$. The next lemma establishes the connection between the connected components of MST' and the sets of ADs:

Lemma 2. *Given a DAG $D = (V, E)$ and a target set $TS = \{n_1, \ldots, n_k\} \subseteq V$. For each $n_i \in annot_{TS}$ let C_i be the connected component in MST' to which n_i belongs. It holds that $AD(n_i) = C_i$.*

Proof. By construction of MST' no two nodes in $annot_{TS}$ are in the same connected component. For each $n_j \in C_i$, $n_j \neq n_i$, the following holds: (i) $\max\{w_{h,j} | e_{h,j} \in E'\} = IC(n_i)$ and (ii) $n_j \in Desc(n_i)$. That means for all nodes n_j, n_i is among the ancestors with the highest IC value of nodes in $annot_{TS}$. □

The size of the sets $AD(n_i)$ for each $n_i \in annot_{TS}$ are the important quantities that we are seeking, that is the size of the different connected components of

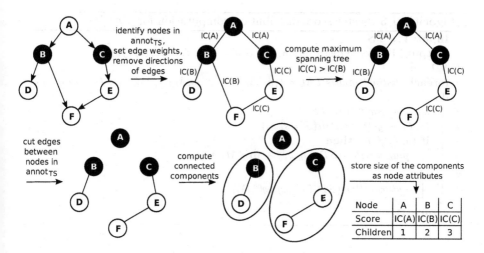

Fig. 2. The general steps of Alg. 2 are shown on the DAG and $\mathcal{TS}$ of Fig. 1. Nodes in $annot_{\mathcal{TS}}$ are colored in black.

MST'. For each $n_i \in annot_{\mathcal{TS}}$ let C_i be the connected component in MST' to which n_i belongs. We define:

$$Children(n_i) = |C_i|. \tag{10}$$

All these steps are summarized in Alg. 2 and Fig. 2, see Example 1.

5   Faster Score Distribution Computation

In this section we present a modified variant of the exhaustive enumeration in Algorithm 3 with greatly improved running time. The intuition of the algorithm is that instead of selecting certain combinations of all nodes of V and constructing the score distribution one by one, we focus on the combinations of nodes in $annot_{\mathcal{TS}}$ with identical scores, and calculate the score frequencies by taking advantage of the nodes' *Children* attribute.

In order to compute the score distribution using only nodes in $annot_{\mathcal{TS}}$ for a fixed q, we need to consider combinations in which the same collapsed node can appear multiple times, regardless of the order. The total count of nodes with repetition has to be q. The properties of these types of combinations have been formalized in the theory of *multisets* [10], see Section 2.

Instead of considering sets of nodes in V we will now consider multisets M of nodes in $annot_{\mathcal{TS}}$. Clearly, $U \subseteq annot_{\mathcal{TS}}$ for any such multiset M with underlying set U. The similarity score computed for M is the same for all of the sets of nodes that correspond to M (see Example 2). Therefore, if we can calculate the number of such sets as well as the score corresponding to each multiset M of nodes in $annot_{\mathcal{TS}}$, we can easily determine the distribution of

Algorithm 3. Faster score distribution computation for sim^{avg}

Input: $V, q, \mathcal{TS}$
Output: The score distribution, i.e., $\mathcal{SD} = \{(s_1, f_1), \dots, (s_k, f_k)\}$
1 $\mathcal{SD} = \emptyset$
2 **foreach** *distinct multiset* $M = \{(u_1, M_1), \dots, (u_d, M_d)\}, |M| = q, U \subseteq annot_{\mathcal{TS}}$
   do
3 $\quad$ $s_{new} \leftarrow sim^{avg}(M, \mathcal{TS})$
4 $\quad$ $(s, f) \leftarrow getScorePair(\mathcal{SD}, s_{new})$
5 $\quad$ **if** $(s, f) \neq nil$ **then**
6 $\quad\quad$ $putScorePair(\mathcal{SD}, (s_{new}, f + freq(M)))$
7 $\quad$ **else**
8 $\quad\quad$ $putScorePair(\mathcal{SD}, (s_{new}, freq(M)))$
9 **return** D

similarity scores over the original graph. The number of ways of drawing M_i nodes from a component of size $Children(u_i)$ can be calculated using the binomial coefficient. The total number of combinations is then simply the product of all binomial coefficients, denoted as the *multiset frequency* for a multiset M:

$$freq(M) = \prod_{(u_i, M_i) \in M} \binom{Children(u_i)}{M_i}. \tag{11}$$

Example 2. We continue example 1 and Fig. 2. Let us assume the scores $IC(A) = 0$, $IC(B) = 0.6$, and $IC(C) = 1.3$. For the query $Q = \{D, E, F\}$ the resulting similarity to $\mathcal{TS} = \{B, C\}$ is $sim^{avg}(Q, \mathcal{TS}) = (0.6 + 1.3 + 1.3)/2 = 1.6$. In total there are 6 query sets with similarity 1.6, namely $\{B, C, E\}, \{B, C, F\}$, $\{B, E, F\}, \{D, C, E\}, \{D, C, F\}, \{D, E, F\}$. After preprocessing, the correspoding multiset M with $sim^{avg}(M, \mathcal{TS}) = 1.6$ is $M = \{(B, 1), (C, 2)\}$. The multiset frequency of M gives the same result as before, $freq(M) = \binom{Children(B)}{1} \cdot \binom{Children(C)}{2} = 2 \cdot 3 = 6$. Instead of iterating over 6 sets we consider one multiset.

The complete procedure is shown in Algorithm 3. We enumerate all distinct multisets instead of iterating over all sets of size q in Alg. 1. Thereby we reduce the number of operations as the next theorem states. In order to compute the score distribution under Eq. (4) we just need to replace the similarity measure in line 3 of Alg. 3.

Theorem 1. *Given a DAG $D = (V, E)$ and a target set $\mathcal{TS} = \{n_1, \dots, n_k\} \subseteq V$ the score distribution of Eq. (3) and Eq. (4) is correctly computed by Alg. 2 and Alg. 3 in $\mathcal{O}(|E| + |V| \log |V| + q \cdot M(|annot_{\mathcal{TS}}|, q))$ time and space.*

Proof. Alg. 2 and Alg. 3 correctly compute $\mathcal{SD} = \{(s_1, f_1), \dots, (s_k, f_k)\}$. All scores s_i are computed from nodes in $annot_{\mathcal{TS}}$ (Lemma 1) and all possible combinations of node scores are considered by enumerating the distinct multisets of cardinality q. The counts f_i are computed correctly by summing $freq(M)$ for all multisets M that have the same score $sim(M, \mathcal{TS}) = s_i$, due to Lemma 2, Eq. (10), and Eq. (11).

The preprocessing of the DAG in Alg. 2, takes time $\mathcal{O}(|E| + |V|\log|V|)$. The topological ordering of V, introducing edge weights to E, removing edges in E, and computing the connected components of MST' can be done with DFS traversals of D and MST' in $\mathcal{O}(|E| + |V|)$ time and space. Computing the maximum spanning tree can be done in $\mathcal{O}(|E| + |V|\log|V|)$ time and space [11].

Algorithm 3 runs in $\mathcal{O}(q \cdot M(|annot_{\mathcal{TS}}|, q))$ time and space. The outer *foreach* loop runs over all distinct multisets with cardinality q given by the multiset coefficient $M(|annot_{\mathcal{TS}}|, q)$. In each loop the computation of the similarity score (line 3) and the multiset frequency, $freq(M)$, cost $\mathcal{O}(q)$ time with a preprocessed lookup table for binomial coefficients (see Section 6). In total, Alg. 2 and Alg. 3 run in $\mathcal{O}(|E| + |V|\log|V| + q \cdot M(|annot_{\mathcal{TS}}|, q))$ time and space.   □

The improvement to the naive algorithm is obvious, for example on average $|annot_{\mathcal{TS}}| = 29$ for all diseases currently annotated with terms of the HPO.

6   Implementation Details

We implemented all algorithms in Java and plan to make them available as open-source software. In order to avoid numerical overflow, we performed the computations in log space, with corresponding minor changes to the pseudocode. See Bejerano et al. [12] for details on how to make fast and accurate additions in log space and how to tabulate calculations of binomial coefficients such that each lookup takes constant time.

Further improvements can be made to Alg. 3. Nodes in $annot_{\mathcal{TS}}$ with identical IC values can be merged into the same node and their *Children* values added up. In our implementation of Alg 3 we enumerate the distinct multisets using a recursive algorithm that stores common partial sums (line 3) and products (Eq. 11) and reduces thus the complexity to $\mathcal{O}(M(|annot_{\mathcal{TS}}|, q))$.

7   Experiments

We have investigated the runtime of both algorithms for score distribution computation on a desktop 2.4 GHz Intel Core Duo with 2GB Ram. All experiments were conducted using the *Organ Abnormality* ontology graph of the HPO [3]. This graph contains 9,019 terms which are annotated to a total of 4,780 OMIM diseases [4]. Our own interest in the algorithm is to compute the score distribution for all annotated OMIM diseases, in order to compute a P-value for a random query against the database. We investigated thus, how much time the naive and new algorithm spend to compute the score distribution for every of the 4,780 diseases for different sizes q of the query set. Figure 3 shows the runtime in *ms* plotted against $|annot_{\mathcal{TS}}|$ for each of the OMIM diseases. We show the results for the Alg. 3 without the time of the preprocessing (on average 40 ms), i.e. Alg. 2, as it only needs to be done once for every $\mathcal{TS}$ (disease).

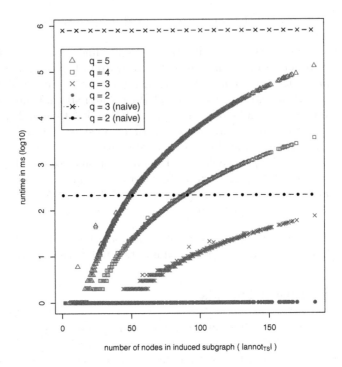

Fig. 3. The figure shows the runtime for complete score distribution computation for all 4,780 OMIM diseases. The number of nodes in $annot_{TS}$ (x-axis) is compared to the runtime in ms (y-axis, log10). The naive algorithm with $q = 2, 3$ (black) is compared to the new algorithm with $q = 2, \ldots, 5$ (gray) without preprocessing, see text.

First, we see that the runtime of the naive algorithm (Alg 1) is independent of $|annot_{TS}|$. Second, we observe that the new algorithm scales exponentially with q, but all score distributions ($q \leq 4$) are computed in less than 10 seconds. The naive algorithm is clearly slower independent of the parameter setting, for example for $q{=}3$ the naive computation of every score distribution took on average 790,556 ms, whereas the new algorithm took on average 42 ms (with preprocessing), that is an 18,000-fold improvement.

8   Discussion

In this work we have tackled the unstudied problem of computing the score distribution for similarity searches with ontologies. We have devised an efficient preprocessing of the underlying DAG of the ontology that reduces the complexity for similarity measures based on Resnik's popular definition of similarity [8]. We have introduced a new algorithm based on multiset enumeration, which can be applied to score distribution computation for Eq. (4) and Eq. (3). In experiments with the HPO, as well as in theory, we show that the new algorithm is much faster than exhaustive enumeration of the score distribution and that it is applicable to current ontologies.

The algorithm we describe here can be used as a component of a procedure to find the best hit in a database, i.e., we need to calculate the score for each entry in the database and rank the results according to P-value. The P-values could be precomputed for each entry in the database (such as all the diseases in OMIM or each protein in the human proteome). For small q, the P-values could be calculated dynamically. Clearly, our algorithm could be parallelized.

We believe that the preprocessing we propose will prove useful for algorithms that compute the score distribution of other semantic similarity measures [5]. Analysis of the shape of the distributions for different input parameters and further algorithmic improvements will be the subject of future work.

References

1. The Gene Ontology Consortium: Gene Ontology: tool for the unification of biology. Nature Genet. 25, 25–29 (2000)
2. Smith, C.L., Goldsmith, C.A.W., Eppig, J.T.: The Mammalian Phenotype Ontology as a tool for annotating, analyzing and comparing phenotypic information. Genome Biol. 6(1), R7 (2005)
3. Robinson, P.N., Köhler, S., Bauer, S., Seelow, D., Horn, D., Mundlos, S.: The Human Phenotype Ontology: a tool for annotating and analyzing human hereditary disease. Am. J. Hum. Genet. 83(5), 610–615 (2008)
4. Amberger, J., Bocchini, C.A., Scott, A.F., Hamosh, A.: McKusick's Online Mendelian Inheritance in Man (OMIM). Nucleic Acids Res. 37(Database issue), D793–D796 (2009)
5. Couto, F., Silva, M.J., Coutinho, P.M.: Measuring semantic similarity between Gene Ontology terms. Data and Knowledge Engineering 61(1) (April 2007)
6. Lord, P.W., Stevens, R.D., Brass, A., Goble, C.A.: Investigating semantic similarity measures across the Gene Ontology: the relationship between sequence and annotation. Bioinformatics 19(10), 1275–1283 (2003)
7. Joshi, T., Xu, D.: Quantitative assessment of relationship between sequence similarity and function similarity. BMC Genomics 9(8), 222 (2007)
8. Resnik, P.: Semantic similarity in a taxonomy: an information-based measure and its application to problems of ambiguity in natural language. Artificial Intelligence Research 11, 95–130 (1999)
9. Pesquita, C., Faria, D., Bastos, H., Ferreira, A., Falco, A., Couto, F.: Metrics for GO based protein semantic similarity: a systematic evaluation. BMC Bioinformatics 9(suppl. 5), S4 (2008)
10. Blizard, W.D.: Multiset theory. Notre Dame Journal of Formal Logic 30(1), 36–66 (1989)
11. Thomas, H., Cormen, C.E.L., Rivest, R.L.: Introduction to Algorithms, 2nd edn. McGraw-Hill Science / Engineering / Math. (December 2003)
12. Bejerano, G., Friedman, N., Tishby, N.: Efficient exact p-value computation for small sample, sparse, and surprising categorical data. Journal of Computational Biology 11(5), 867–886 (2004)

Linear-Time Protein 3-D Structure Searching with Insertions and Deletions

Tetsuo Shibuya[1], Jesper Jansson[2], and Kunihiko Sadakane[3]

[1] Human Genome Center, Institute of Medical Science, University of Tokyo
4-6-1 Shirokanedai, Minato-ku, Tokyo 108-8639, Japan
[2] Ochanomizu University, 2-1-1 Ohtsuka, Bunkyo-ku, Tokyo 112-8610, Japan
[3] National Institute of Informatics, 2-1-2 Hitotsubashi, Chiyoda-ku, Tokyo 101-8430, Japan
tshibuya@hgc.jp, Jesper.Jansson@ocha.ac.jp, sada@nii.ac.jp

Abstract. It becomes more and more important to search for similar structures from molecular 3-D structure databases in the structural biology of the post genomic era. Two molecules are said to be similar if the RMSD (root mean square deviation) of the two molecules is less than or equal to some given constant bound. In this paper, we consider an important, fundamental problem of finding all the similar substructures from 3-D structure databases of chain molecules (such as proteins), with consideration of indels (*i.e.*, insertions and deletions). The problem has been believed to be very difficult, but its computational difficulty has not been well known. In this paper, we first show that the same problem in arbitrary dimension is NP-hard. Moreover, we also propose a new algorithm that dramatically improves the average-case time complexity for the problem, in case the number of indels k is bounded by some constant. Our algorithm solves the above problem in average $O(N)$ time, while the time complexity of the best known algorithm was $O(Nm^{k+1})$, for a query of size m and a database of size N.

1 Introduction

Molecules with similar 3-D structures are said to have similar functions. It means that we can predict the molecular function by searching for molecules with similar structures in the databases. Thus, finding similar 3-D structures from 3-D databases is very important [2,10,12,17]. Due to recent technological evolution of molecular structure determination methods, more and more structures of biomolecules, especially proteins, are solved, as shown in the PDB (Protein Structure Data Bank) database [3]. Moreover, a huge number of molecular structures are predicted with various computational techniques recently. Hence, faster searching techniques against these molecular structure databases are seriously needed.

A protein is a chain of amino acids, and its structure is often represented by a sequence of 3-D coordinates that represents the positions of amino acids. Usually, the 3-D coordinates of the C_α atom in each amino acid is used as the representative position of that amino acid. Note that there are also other important chain molecules in living cells, such as DNAs, RNAs, and glycans. In this

S.L. Salzberg and T. Warnow (Eds.): WABI 2009, LNBI 5724, pp. 310–320, 2009.

paper, we consider a problem of searching for similar structures from a structure database of chain molecules, which consists of sequences of 3-D coordinates that represent molecular structures.

The RMSD (Root Mean Square Deviation) [1,9,14,15,17,20,21] is the most widely-used similarity measure between molecular structures, which is also used in various other fields, such as robotics and computer vision. It determines geometric similarity between two same-length sequences of 3-D coordinates. It is defined as the square root of the minimum value of the average squared distance between each pair of corresponding atoms, over all the possible rotations and translations. (See section 2.2 for more details.) The RMSD measure corresponds to the Hamming distance in the textual pattern matching, from the viewpoint that it does not consider any indels (*i.e.*, insertions and deletions) between them. In the case of textual string comparison, especially comparison of two textual strings of bio-molecules (such as proteins and DNA), we often prefer to use the string alignment score that considers indels to compare two bio-sequences, rather than the Hamming distance. Likewise, it is also important to consider indels when we compare two molecular 3-D structures. But it is much harder than the textual string cases to compare two 3-D structures with consideration of indels, though an ordinary pair-wise alignment algorithm for textual strings requires only quadratic time.

In this paper, we consider a problem of searching for substructures of database structures whose RMSDs to a given query is within some constant, permitting indels. It is widely known that the contact map problem [13] is NP-hard and the structure alignment problems are believed to be very difficult. But the difficulty of our problem is unknown, as our problem is different from the contact map problem. We show in this paper that our problem is also NP-hard if the dimension of the problem is arbitrary. But it does not mean that our problem is always difficult. If the number of indels is at most some constant, the problem can be solved in polynomial time, though the time complexity of known algorithms is still very large. The best-known algorithm for the problem is a straight-forward algorithm that requires $O(Nm^{k+1})$ time for a database of size N and a query of size m, where k is the maximum number of indels. It is the worst-case time complexity, but the average-case time complexity of the algorithm is still all the same $O(Nm^{k+1})$. We propose in this paper a much faster algorithm that runs in average-case $O(N)$ time, assuming that the database structures can be considered as random walks. The model under this assumption is called the 'random-walk model' (It is also called the 'freely-jointed chain model' or just the 'ideal chain model'. See section 2.3 for more details.), and is very often used in molecular physics [4,8,11,18]. It is also used in the analysis of algorithms for protein structure comparison [22]. As demonstrated in [22], theoretical analyses based on the random-walk model have high consistency with the actual experimental results on the PDB database.

The organization of this paper is as follows. Section 2 describes the notations used in this paper and previous related work as preliminaries. Section 3 describes the problem that we solve. Section 4 describes the NP-hardness of our problem.

Section 5 describes our new algorithm and the computational time analysis of the algorithm. Section 6 concludes our results.

2   Preliminaries

2.1   Notations and Definitions

A chain molecule $\mathbf{S}$ whose i-th 3-D coordinates (vector) is s_i is noted as $\mathbf{S} = (s_1, s_2, \ldots, s_n)$. The length n of $\mathbf{S}$ is denoted by $|\mathbf{S}|$. A structure $\mathbf{S}[i..j] = (s_i, s_{i+1}, \ldots, s_j)$ $(1 \le i \le j \le n)$ is called a *substructure* of $\mathbf{S}$. A structure $\mathbf{S}' = (s_{a_1}, s_{a_2}, \ldots, s_{a_\ell})$ $(1 \le a_1 < a_2 < \ldots < a_\ell \le n)$ is called a *subsequence structure* of $\mathbf{S}$. $\mathbf{S}'$ is also called a *k-reduced subsequence structure* of $\mathbf{S}$, where $k = |\mathbf{S}| - |\mathbf{S}'|$. $R \cdot \mathbf{S}$ denotes the structure $\mathbf{S}$ rotated by the rotation matrix R, i.e., $R \cdot \mathbf{S} = (Rs_1, Rs_2, \ldots, Rs_n)$. $|v|$ denotes the norm of the vector v. $\mathbf{0}$ denotes the zero vector. $\langle x \rangle$ denotes the expected value of x. $Prob(\mathcal{X})$ denotes the probability of the event $\mathcal{X}$.

2.2   RMSD: Root Mean Square Deviation

The RMSD (root mean square deviation) [1,9,14,15,20,21] is the most widely-used geometric similarity measure between two sequences of 3-D coordinates. The RMSD between two 3-D coordinates sequences $\mathbf{S} = (s_1, s_2, \ldots, s_n)$ and $\mathbf{T} = (t_1, t_2, \ldots, t_n)$ is defined as the minimum value of $\sqrt{\frac{1}{n} \sum_{i=1}^{n} |s_i - (R \cdot t_i + v)|^2}$ over all the possible rotation matrices R and translation vectors v. Let $RMSD(\mathbf{S}, \mathbf{T})$ denote the minimum value. $RMSD(\mathbf{S}, \mathbf{T})$ can be computed in $O(n)$ time [1,9,14,15]. Note that the RMSD can be defined in any other dimensions, by considering the above vectors and matrices are in any d-dimensions.

2.3   Random-Walk Model for Chain Molecules

The *random-walk model* (also called the *freely-jointed chain model*, or just the *ideal chain model*), is a very widely used simple model for analyzing behavior of chain molecules in molecular physics [4,8,11,18]. The model is also used for analyzing the computational time complexities of algorithms for protein structures [22]. In the model, we assume that the chain molecules can be considered as random walks. The model ignores many physical/chemical constraints, but it is known to reflect the behavior of real molecules very well. In fact, experiments in [22] showed high consistency between the experimental results obtained from the PDB database and the theoretical results deduced from the random-walk model. Consider a chain molecule $\mathbf{S} = (s_0, s_2, \ldots, s_n)$ of length $n + 1$, in which the distance between any two adjacent atoms is fixed to some constant r.[1] In the random-walk model, a bond between two adjacent atoms, i.e., $b_i = s_{i+1} - s_i$, is

---

[1] In the case of proteins, the distance between two adjacent C_α atoms is fixed to 3.8Å. We can let $r = 1$ by considering the distance between two adjacent atoms as the unit of distance.

considered as a random vector that satisfies $|\boldsymbol{b}_i| = r$, and $\boldsymbol{b}_i$ is considered to be independent from any other bond $\boldsymbol{b}_j$ $(j \neq i)$.

2.4 Shibuya's Lower Bound of the RMSD [22]

Let $\mathbf{U}^{left}$ denote $(\boldsymbol{u}_1, \boldsymbol{u}_2, \ldots, \boldsymbol{u}_{\lfloor \ell/2 \rfloor})$ and $\mathbf{U}^{right}$ denote $(\boldsymbol{u}_{\lfloor \ell/2 \rfloor+1}, \boldsymbol{u}_{\lfloor \ell/2 \rfloor+2}, \ldots, \boldsymbol{u}_{2 \cdot \lfloor \ell/2 \rfloor})$ for a structure $\mathbf{U} = (\boldsymbol{u}_1, \boldsymbol{u}_2, \ldots, \boldsymbol{u}_\ell)$. Let $G(\mathbf{U})$ denote the centroid of the structure $\mathbf{U}$, i.e., $G(\mathbf{U}) = \frac{1}{\ell} \sum_{i=1}^{\ell} \boldsymbol{u}_i$. Let $F(\mathbf{U})$ denote $|G(\mathbf{U}^{left}) - G(\mathbf{U}^{right})|/2$, and let $D(\mathbf{S}, \mathbf{T})$ denote $\sqrt{2 \cdot |\mathbf{S}^{left}|/|\mathbf{S}|} \cdot |F(\mathbf{S}) - F(\mathbf{T})|$ for two structures such that $|\mathbf{S}| = |\mathbf{T}|$. Shibuya proved in [22] that $D(\mathbf{S}, \mathbf{T})$ is always smaller than or equal to $RMSD(\mathbf{S}, \mathbf{T})$. In [22], he also proved the following lemma:

Lemma 1 (Shibuya [22]). *The probability $Prob(D(\mathbf{S}, \mathbf{T}) < c)$ is in $O(c/\sqrt{n})$, where $n = |\mathbf{S}| = |\mathbf{T}|$, under the assumption that either $\mathbf{S}$ or $\mathbf{T}$ follows the random-walk model.*

3 The k-Indel 3-D Substructure Search Problem

From now on, we deal with the following problem.

k-**Indel 3-D Substructure Search Problem:** We are given a text structure $\mathbf{P}$ of size N and a query structure $\mathbf{Q}$ of size m $(1 < m \leq N)$, both of which are represented by 3-D coordinates sequences of the residues. We are also given a constant positive real c and a small constant positive integer k $(k \ll m)$. The problem is to find all the positions i $(1 \leq i \leq N - m + k + 1)$ such that the RMSD between some k'-reduced subsequence structure of $\mathbf{Q}$ and some k''-reduced subsequence structure of $\mathbf{P}[i..i - k' + k'' + m - 1]$ is at most c, for some non-negative integers k' and k'' $(k' + k'' \leq k, k'' - k' \leq N - m - i + 1)$.

If there exists some triple set $\{i, k', k''\}$ that satisfies the above condition, we say that $\mathbf{Q}$ matches with $\mathbf{P}[i..i - k' + k'' + m - 1]$ with threshold c and (at most) $k' + k''$ indels. Usually, c is set to a constant proportional to the distance between two adjacent residue coordinates in the molecular structures. In the case of protein structures, c is often set to 1–2Å, while the distance between two adjacent C_α atoms is 3.8Å. Usual structure databases may contain more than 1 structures, but problems against the databases with multiple structures can be reduced to the above single-text problem by just concatenating all the structures into a single long text structure and ignoring matches that cross over the boundaries of two concatenated structures.

The same problem without indels, i.e., the problem in case $k = 0$, is studied very well. If we directly apply the Kabsch's algorithm [14,15] introduced in section 2.2, the problem without indels can be solved in $O(Nm)$ time. For the problem, Schwartz and Sharir [20] proposed an algorithm based on the fast Fourier transform technique that runs in $O(N \log m)$ time.[2] Recently, Shibuya [22]

---

[2] The original algorithm runs in $O(N \log N)$ time. See [22] for the technique to improve it into $O(N \log m)$.

proposed a breakthrough average-case (expected) linear-time algorithm, assuming that the text structures follow the random-walk model. He showed that his algorithm is much faster than other algorithms also in practice. Moreover he showed that the experimental results on the whole PDB database agrees with the theoretical analysis based on the random-walk model. But none of these algorithms considers any indels.

On the other hand, there have been almost no algorithmic study for cases $k > 0$, due to the difficulty of the problem, though the problem is very important. The difficulty of the problem is not well known, though the problem is similar to the famous contact map problem, which is known to be NP-hard [13]. In section 4, we will show that the problem is NP-hard, in case the dimension of the problem is arbitrary.

According to section 2.2, the RMSD between two structures of size m can be computed in $O(m)$ time. The possible number of subsequence structures to be compared in the k-indel 3-D substructure search problem is less than $_{2m+k}C_k \cdot N$, which is in $O(Nm^k)$. Thus, our problem can be computed in $O(Nm^{k+1})$ time, either in the worst-case analysis or in the average-case analysis. As far as we know, it is the best-known time complexity, and there have been known no algorithms other than the above straight-forward algorithm. But it also means that the problem can be computed in polynomial time, in case the number of indels is bounded by some constant. In section 5, we will propose the first algorithm with better average-case time complexity, $i.e.$, $O(N)$, for the above problem in case the number of the indels is at most some constant, which is a substantial improvement for the problem. Note that the worst-case time complexity of our algorithm is still the same as the above straight-forward algorithm. Note also that our analysis of the average-case time complexity is based on the assumption that the text structure follows the random-walk model,[3] like the analyses in [22].

4   An NP-Hardness Result

Consider the following variant of the k-indel 3-D substructure search problem.

k-Indel Structure Comparison Problem: We are given two structures **P** and **Q**, both of whose lengths are n. Find a k-reduced subsequence structure **P′** of **P** and a k-reduced subsequence structure **Q′** of **Q**, such that the RMSD between **P′** and **Q′** is at most some given threshold c.

It is trivial that the k-indel structure comparison problem is in the class NP, as the correctness of any instance can be checked in linear time. Moreover, it is also trivial that the k-indel 3-D substructure search problem is at least as difficult as the k-indel comparison problem in 3-D, and the k-indel 3-D substructure search problem is NP-hard if the k-indel structure comparison problem in 3-D is NP-complete. The two problems can be extended to the problems in any

---

[3] We give this random-walk assumption only on the database structures, and we give no assumption on the query structures.

dimensional space. From now on, we show the k-indel structure comparison problem in arbitrary dimension is NP-complete, by reduction from the following k-cluster problem (or the densest k-subgraph problem), whose decision problem is known to be NP-complete [6].

k-Cluster Problem (Densest k-Subgraph Problem): Given a graph $G = (V, E)$, find a size k subset of V such that the number of edges induced by the subset is the largest.

Let $V = \{v_1, v_2, \ldots, v_n\}$. Consider an arbitrary subset $V' = \{v_{g_1}, v_{g_2}, \ldots, v_{g_k}\}$ of V, where $g_1 < g_2 < \ldots < g_k$, and let x be the number of edges induced by V'.

There must exist a sequence of points $\mathbf{P} = (\mathbf{p}_1, \mathbf{p}_2, \ldots, \mathbf{p}_n)$ in $n-1$ dimensional space, such that $|\mathbf{p}_i - \mathbf{p}_j| = \alpha$ if $\{v_i, v_j\} \in E$ and $|\mathbf{p}_i - \mathbf{p}_j| = \beta$ if $\{v_i, v_j\} \notin E$, where α and β are any constants that satisfy $0 < \alpha < \beta < 2\alpha$. Let $\mathbf{Q}$ be a sequence of n zero vectors $(\mathbf{0}, \ldots, \mathbf{0})$ in the same $n - 1$ dimensional space. Let $\mathbf{P}_{V'} = (\mathbf{p}_{g_1}, \mathbf{p}_{g_2}, \ldots, \mathbf{p}_{g_k})$, and $\mathbf{Q}_{V'}$ be a sequence of k zero vectors $(\mathbf{0}, \ldots, \mathbf{0})$ in the $n - 1$ dimensional space.

It is well known that the translation of the two structures in 3-D is optimized when the centroids of the two structures are placed at the same position (e.g., at the origin of the coordinates) [1,14], in computing the RMSD. It is also true in any dimensions d, which can be easily proved as follows. Consider two arbitrary d-dimensional structures $\mathbf{S} = (\mathbf{s}_1, \mathbf{s}_2, \ldots, \mathbf{s}_n)$ and $\mathbf{T} = (\mathbf{t}_1, \mathbf{t}_2, \ldots, \mathbf{t}_n)$, and an arbitrary d-dimensional translation vector $\mathbf{v}$. Then the following equation holds:

$$\sum_{i=1}^{n}(\mathbf{s}_i - \mathbf{t}_i + \mathbf{v})^2 = n\{\mathbf{v} + \frac{\sum_{i=1}^{n}(\mathbf{s}_i - \mathbf{t}_i)}{n}\}^2 + \sum_{i=1}^{n}(\mathbf{s}_i - \mathbf{t}_i)^2 - \frac{\{\sum_{i=1}^{n}(\mathbf{s}_i - \mathbf{t}_i)\}^2}{n}. \quad (1)$$

Thus the translation is optimized when $\mathbf{v} = -\frac{\sum_{i=1}^{n}(\mathbf{s}_i - \mathbf{t}_i)}{n}$. It means that the translation is optimized when the two structures are moved so that the centroids of the two structures are at the same position.

From now on, we consider computing the RMSD between $\mathbf{P}_{V'}$ and $\mathbf{Q}_{V'}$. It is trivial that the centroid of $\mathbf{Q}_{V'}$ is at the origin of the coordinates, and moreover $\mathbf{Q}_{V'}$ does not change its shape by any rotation, as all the vectors in $\mathbf{Q}_{V'}$ are zero vectors. Hence, we do not have to consider the optimization of the rotation for computing the RMSD between the two structures. Therfore we obtain the following equation:

$$RMSD(\mathbf{P}_{V'}, \mathbf{Q}_{V'}) = \{\sum_{i=1}^{k}(\mathbf{p}_{g_i} - \frac{\sum_{j=1}^{k}\mathbf{p}_{g_j}}{k})^2 / k\}^{1/2}$$

$$= \{\sum_{i=1}^{k-1}\sum_{j=i+1}^{k}(\mathbf{p}_{g_i} - \mathbf{p}_{g_j})^2 / k\}^{1/2}$$

$$= \{(\alpha^2 \cdot x + \beta^2 \cdot (\frac{k(k-1)}{2} - x)) / k\}^{1/2} \quad (2)$$

It means that $RMSD(\mathbf{P}_{V'}, \mathbf{Q}_{V'})$ is smaller if x is larger, as $0 < \alpha < \beta$. Thus we can obtain the answer of the decision problem of the k-cluster problem by solving the $(n - k)$-indel $n - 1$ dimensional structure comparison problem on the two structures $\mathbf{P}$ and $\mathbf{Q}$. Hence the k-indel structure comparison problem in arbitrary dimensional space is NP-complete, and consequently we conclude that the k-indel substructure search problem in arbitrary dimensional space is NP-hard:

Theorem 1. *The k-indel substructure search problem in arbitrary dimensional space is NP-hard.*

5   The New Linear Expected Time Algorithm

5.1   The Algorithm

To improve the performance of the algorithms for approximate matching of ordinary textual strings, we often divide the query into several parts to improve the query performance [19]. For example, in case we want to search for textual strings with k indels, we can efficiently enumerate candidates for the matches by dividing the query into $k + 1$ substrings and finding the exact matches of these substrings, as at least one of the divided substrings must exactly match somewhere in the text. Similarly, we also divide the query 3-D structure into several substructures in our algorithm.

In our algorithm, we first divide the query $\mathbf{Q}$ of size m into $3k+2$ equal-length substructures of size $m' = \lfloor m/(3k + 2) \rfloor$. Note that k is the number of maximum indels defined in section 3, which is considered to be a small constant. We call each substructure a 'divided substructure'. Let $\mathbf{Q}_j$ denote the j-th divided substructure, *i.e.*, $\mathbf{Q}[(j - 1)m' + 1..j \cdot m']$. We ignore the remaining part (*i.e.*, $\mathbf{Q}[(3k + 2)m' + 1..m]$) in case m is not a multiple of $3k + 2$.

Consider the case that $\mathbf{Q}$ matches with $\mathbf{P}_i = \mathbf{P}[i..i - k' + k'' + m - 1]$ with threshold c and (at most) $k = k' + k''$ indels. Let $\mathbf{Q}'$ and $\mathbf{P}'_i$ denote the k'-reduced subsequence structure of $\mathbf{Q}$ and the k''-reduced subsequence structure of $\mathbf{P}[i..i - k' + k'' + m - 1]$ respectively, such that $RMSD(\mathbf{Q}', \mathbf{P}'_i) \leq c$. Let $\mathbf{Q}'_j$ denote the largest substructure of $\mathbf{Q}'$ such that $\mathbf{Q}'_j$ is a subsequence structure of $\mathbf{Q}_j$. Let h_j denote the first index of $\mathbf{Q}'_j$ in $\mathbf{Q}'$, *i.e.*, $\mathbf{Q}'_j = \mathbf{Q}'[h_j..h_{j+1} - 1]$ $(h_{3k+2} = m - k' + 1)$. Let $\mathbf{P}'_{i,j} = \mathbf{P}'[h_j..h_{j+1} - 1]$. It is easy to see that there are at least $2k + 2$ pairs of subsequence structures $\mathbf{Q}'_j$ and $\mathbf{P}'_{i,j}$ such that $\mathbf{Q}'_j = \mathbf{Q}_j$ and $\mathbf{P}'_{i,j}$ is a substructure of $\mathbf{P}_i$. We call these (at least $2k + 2$ pairs of) substructures 'ungapped substructures'. Notice that the length of the ungapped substructures is m'. Let the index of an ungapped structure $\mathbf{P}'_{i,j}$ in $\mathbf{P}_i$ be p_j, *i.e.*, $\mathbf{P}'_{i,j} = \mathbf{P}_i[p_j..p_j + m' - 1]$. It is easy to see that $|(j - 1) \cdot m' + 1 - p_j| \leq k$, as we allow only at most k indels. Then, an inequality $RMSD(\mathbf{Q}'_j, \mathbf{P}'_{i,j}) \leq c \cdot \sqrt{m/m'}$ holds for ungapped substructures $\mathbf{Q}'_j$ and $\mathbf{P}'_{i,j}$, according to the following lemma:

Lemma 2. *Consider a pair of two structures $\mathbf{S} = (s_1, s_2, \ldots, s_n)$ and $\mathbf{T} = (t_1, t_2, \ldots, t_n)$, both of whose length is n. Let $\mathbf{S}' = (s_{a_1}, s_{a_2}, \ldots, s_{a_{n'}})$ be some subsequence structure of $\mathbf{S}$, and let $\mathbf{T}' = (t_{a_1}, t_{a_2}, \ldots, t_{a_{n'}})$. Then, $RMSD(\mathbf{S}', \mathbf{T}') \leq \sqrt{n/n'} \cdot RMSD(\mathbf{S}, \mathbf{T})$.*

Proof. According to the definition of the RMSD, the following inequality holds:

$$RMSD(\mathbf{S'}, \mathbf{T'}) = \min_{R,v} \sqrt{\frac{1}{n'} \sum_{i=1}^{n'} |\mathbf{s}_{a_i} - (R \cdot \mathbf{t}_{a_i} + \mathbf{v})|^2}$$

$$\leq \min_{R,v} \sqrt{\frac{1}{n'} \sum_{i=1}^{n} |\mathbf{s}_i - (R \cdot \mathbf{t}_i + \mathbf{v})|^2}$$

$$= \sqrt{n/n'} \cdot RMSD(\mathbf{S}, \mathbf{T}). \qquad (3)$$

$\square$

In summary, at least $2k + 2$ divided substructures $\mathbf{Q}_j = \mathbf{Q}[(j-1)m' + 1..j \cdot m']$ (among the $3k + 2$ divided substructures) must satisfy the following constraint:

- There must be some substructure $\mathbf{P}[\ell..\ell + m' - 1]$ of $\mathbf{P}$ such that $RMSD(\mathbf{Q'_j}, \mathbf{P}[\ell..\ell + m' - 1]) \leq c \cdot \sqrt{m/m'}$ and $i + (j-1)m' - k \leq \ell \leq i + (j-1)m' + k$.

These $2k + 2$ (or more) divided substructures must also satisfy the following weaker constraint, as an inequality $D(\mathbf{S}, \mathbf{T}) \leq RMSD(\mathbf{S}, \mathbf{T})$ holds for any pair of same-length structures $\mathbf{S}$ and $\mathbf{T}$ (see section 2.4 for the definition of $D(\mathbf{S}, \mathbf{T})$).

- There must be some substructure $\mathbf{P}[\ell..\ell + m' - 1]$ of $\mathbf{P}$ such that $D(\mathbf{Q'_j}, \mathbf{P}[\ell..\ell+m'-1]) \leq c \cdot \sqrt{m/m'}$ and $i+(j-1)m'-k \leq \ell \leq i+(j-1)m'+k$.

We call the divided substructures that satisfy the latter weaker constraint 'hit substructures' for the position i.

Based on the above discussions, we propose the following simple algorithm for the k-indel 3-D substructure problem.

Algorithm

1. We first enumerate all the positions i in $\mathbf{P}$ such that there are at least $2k+2$ hit substructures for the position i, by computing all the $D(\mathbf{Q}_j, \mathbf{P}[i..i+m'-1])$ values for all the pairs of i ($1 \leq i \leq N - m' + 1$) and j ($1 \leq j \leq 3k + 2$).
2. For each position i found in step 1, we check the RMSDs between all the pairs of k'-reduced subsequence structure of $\mathbf{Q}$ and k''-reduced subsequence substructure of $\mathbf{P}[i..i + m - k' + k'' + m - 1]$ such that $k' + k'' \leq k$ and $k'' - k' \leq N - m - i + 1$. If any one of the checked RMSDs is smaller or equal to c, output i as the position of a substructure similar to the query $\mathbf{Q}$.

In the next section, we analyze the average-case time complexity of the algorithm.

5.2   The Average-Case Time Complexity of the Algorithm

For each $\mathbf{Q}_j$ (whether it is a hit substructure or not), we can compute $D(\mathbf{Q}_j, \mathbf{P}[i..i + m' - 1])$ for all i ($1 \leq i \leq N - m' + 1$) in total $O(N)$ time, as $G(\mathbf{P}[i..i + m' - 1])$ (i.e., the centroid of $\mathbf{P}[i..i + m' - 1]$) can be computed in $O(N)$ time for all i. Thus, we can execute the step 1 of the algorithm in section 5.1 in $O(k^2 \cdot N)$ time, which is in $O(N)$ as we consider k is a small fixed constant. Let N' denote the number of candidates enumerated in step 1 of the algorithm in section 5.1. As the number of pairs to check in step 2 for each position is less than $_{2m+k}C_k$ (which is in $O(m^k)$), and each RMSD can be computed in $O(m)$ time, the computational complexity of step 2 is $O(N'm^{k+1})$. In total, the computational complexity of the algorithm is $O(N + N'm^{k+1})$. In the worst case, the algorithm could be as bad as the naive $O(Nm^{k+1})$-time algorithm, as N' could be N at worst.

But, in the following, we show that $\langle N' \rangle$ is only in $O(N/m^{k+1})$ and consequently the average-case (expected) time complexity of the algorithm is astonishingly $O(N)$, under the assumption that $\mathbf{P}$ follows the random-walk model. According to Lemma 1 in section 2.4, the probability that a divided substructure $\mathbf{Q}_i$ is a hit substructure for the position i is in $O(k \cdot c \cdot \sqrt{m/m'}/\sqrt{m'}) = O(c \cdot k^2/\sqrt{m})$, under the random-walk assumption. Consider that the above probability can be bounded by $a \cdot c \cdot k^2/\sqrt{m}$ if m is large enough, where a is an appropriate constant. Then, the probability that $2k + 2$ of the $3k + 2$ divided substructures are hit substructures is $O((a \cdot c \cdot k^2/\sqrt{m})^{2k+2} \cdot _{3k+2}C_{2k+2})$, which is in $O(c^{2k+2} \cdot k^{5k+4}/m^{k+1})$. Thus $\langle N' \rangle$ is in $O(N \cdot c^{2k+2} \cdot k^{5k+4}/m^{k+1})$, and the following lemma holds, considering that both c and k are small fixed constants.

Lemma 3. $\langle N' \rangle$ is in $O(N/m^{k+1})$.

Consequently the expected time complexity of the step 2 of the above algorithm is only in $O(N)$. (More precisely, it is $O(c^{2k+2} \cdot k^{5k+4} \cdot N)$, but we consider that both c and k are small fixed constants.) In conclusion, the total expected time complexity of the algorithm in section 5.1 is only $O(N)$, under the assumption that $\mathbf{P}$ follows the random walk model:[4]

Theorem 2. *The total expected time complexity of our algorithm is $O(N)$, under the assumption that $\mathbf{P}$ follows the random walk model.*

6   Concluding Remarks

We considered the k-indel 3-D substructure search problem, in which we search for similar 3-D substructures from molecular 3-D structure databases, with consideration of indels. We showed that the same problem in arbitrary dimensional space is NP-hard. Moreover, we proposed a linear expected time algorithm, under the assumption that the number of indels is bounded by a constant and the database structures follow the random-walk model. There are several open problems. First of all, the difficulty of our problem restricted to 3-D space is not known. As for our algorithm, its expected time complexity is $O(N)$ for a database of size N, but its coefficient, i.e., $c^{2k+2} \cdot k^{5k+4}$, is very large (c is the

---

[4] The same discussion can be done if the query $\mathbf{Q}$ follows the random walk model, instead of $\mathbf{P}$.

threshold of the RMSD and k is the maximum number of indels, both of which we consider as constant numbers). It would be more practical if we can design algorithms with better coefficients. Another open problem is whether we can design a worst-case (deterministically) linear-time algorithm for our problem, though no worst-case linear-time algorithm is known even for the no-indel case.

Acknowledgement. This work was partially supported by the Grant-in-Aid from the Ministry of Education, Culture, Sports, Science and Technology of Japan. Jesper Jansson was supported by the Special Coordination Funds for Promoting Science and Technology.

References

1. Arun, K.S., Huang, T.S., Blostein, S.D.: Least-squares fitting of two 3-D point sets. IEEE Trans Pattern Anal. Machine Intell. 9, 698–700 (1987)
2. Aung, Z., Tan, K.-L.: Rapid retrieval of protein structures from databases. Drug Discovery Today 12, 732–739 (2007)
3. Berman, H.M., Westbrook, J., Feng, Z., Gilliland, G., Bhat, T.N., Weissig, H., Shindyalov, I.N., Bourne, P.E.: The protein data bank. Nucl. Acids Res. 28, 235–242 (2000)
4. Boyd, R.H., Phillips, P.J.: The Science of Polymer Molecules: An Introduction Concerning the Synthesis, Structure and Properties of the Individual Molecules That Constitute Polymeric Materials. Cambridge University Press, Cambridge (1996)
5. Cooley, J.W., Tukey, J.W.: An algorithm for the machine calculation of complex Fourier series. Math. Comput. 19, 297–301 (1965)
6. Corneil, D.G., Perl, Y.: Clustering and domination in perfect graphs. Discrete Appllied Mathematics 9(1), 27–39 (1984)
7. Dayantis, J., Palierne, J.-F.: Monte Carlo precise determination of the end-to-end distribution function of self-avoiding walks on the simple-cubic lattice. J. Chem. Phys. 95, 6088–6099 (1991)
8. de Gennes, P.-G.: Scaling Concepts in Polymer Physics. Cornell University Press (1979)
9. Eggert, D.W., Lorusso, A., Fisher, R.B.: Estimating 3-D rigid body transformations: a comparison of four major algorithms. Machine Vision and Applications 9, 272–290 (1997)
10. Eidhammer, I., Jonassen, I., Taylor, W.R.: Structure comparison and structure patterns. J. Computational Biology 7(5), 685–716 (2000)
11. Flory, P.J.: Statistical Mechanics of Chain Molecules. Interscience, New York (1969)
12. Gerstein, M.: Integrative database analysis in structural genomics. Nat. Struct. Biol., Suppl., 960–963 (2000)
13. Goldman, D., Istrail, S., Papadimitriou, C.H.: Algorithmic aspects of protein structure similarity. In: Proc. 40th Annual Symposium on Foundations of Computer Science, pp. 512–522 (1999)
14. Kabsch, W.: A solution for the best rotation to relate two sets of vectors. Acta Cryst. A32, 922–923 (1976)
15. Kabsch, W.: A discussion of the solution for the best rotation to relate two sets of vectors. Acta Cryst. A34, 827–828 (1978)
16. Kallenberg, O.: Foundations of Modern Probability. Springer, Heidelberg (1997)

17. Koehl, P.: Protein structure similarities. Current Opinion in Structural Biology 11, 348–353 (2001)
18. Kramers, H.A.: The behavior of macromolecules in inhomogeneous flow. J. Chem. Phys. 14(7), 415–424 (1946)
19. Navarro, G.: A guided tour to approximate string matching. ACM Computing Surveys 33(1), 31–88 (2001)
20. Schwartz, J.T., Sharir, M.: Identification of partially obscured objects in two and three dimensions by matching noisy characteristic curves. Intl. J. of Robotics Res. 6, 29–44 (1987)
21. Shibuya, T.: Efficient substructure RMSD query algorithms. J. Comput. Biol. 14(9), 1201–1207 (2007)
22. Shibuya, T.: Searching protein 3-D structures in linear time. In: Batzoglou, S. (ed.) RECOMB 2009. LNCS (LNBI), vol. 5541, pp. 1–15. Springer, Heidelberg (2009)

Visualizing Phylogenetic Treespace Using Cartographic Projections

Kenneth Sundberg, Mark Clement, and Quinn Snell

Department of Computer Science
Brigham Young University

Abstract. Phylogenetic analysis is becoming an increasingly important tool for biological research. Applications include epidemiological studies, drug development, and evolutionary analysis. Phylogenetic search is a known NP-Hard problem. The size of the data sets which can be analyzed is limited by the exponential growth in the number of trees that must be considered as the problem size increases. A better understanding of the problem space could lead to better methods, which in turn could lead to the feasible analysis of more data sets. We present a definition of phylogenetic tree space and a visualization of this space that shows significant exploitable structure. This structure can be used to develop search methods capable of handling much larger datasets.

1 Introduction

Phylogenetic analysis has become an integral part of many biological research programs. These include such diverse areas as human epidemiology [7,20], viral transmission [8,13], and biogeography [10]. With the advent of new automated sequencing technologies, the ability to generate data for inferring evolutionary histories (phylogenies) for a great diversity of organisms has increased dramatically. Researchers are now commonly generating many sequences from many individuals. However, our ability to analyze the data has not kept pace with data generation.

Phylogenetic search is a difficult problem. When parsimony is used as the optimality criterion the problem is known to be NP-complete [9]. The search problem itself, independent of scoring, is known to be NP-Hard [6]. This means that optimal phylogenetic searches on even hundreds of taxa will take years to complete and heuristic searches for near optimal trees must be used.

A variety of heuristic search methods have been used to find optimal trees within a treespace. The most common method is to search treespace using tree rearrangements [21,18,23,12]. Other methods such as those based on Bayesian inference [19], or genetic algorithms [24] also exist. However all of these methods rely only on local information to guide the phylogenetic search. This limitation arises because no global exploitable structures have been previously observed in treespace.

Greater understanding of the problem space may allow more sophisticated search techniques to be applied, with a consequent improvement in the effectiveness of the search. One technique that can be used to better understand the

S.L. Salzberg and T. Warnow (Eds.): WABI 2009, LNBI 5724, pp. 321–332, 2009.

space of phylogenetic search, and the behavior of search algorithms within this space is visualization. This includes two separate activities; first, defining the search space of phylogenetic trees, or treespace, and second, developing methods to display treespace in a way that is exploitable in search techniques.

This visualization must have the following properties to be useful.

- Each tree should map to a single deterministic position. Otherwise the method is restricted to post-processing, and can not be used to guide a search.
- Distance between trees should be easy to calculate. If it is not the visualization will not be able to be used in real time to guide a search.
- The visualization should reveal exploitable structure. This is important because if a visualization shows no structure it provides no guidance for a search.
- This mapping should be reversible, meaning that there should be a method of turning a position into a tree. This is necessary as structure suggests a space where good trees might be found, to be useful in searching it must be possible to quickly find trees in the suggested space.

This work presents an elegant linear projection of trees. This projection can be computed much faster than current alternatives and is better at preserving structural continuity between trees after the projection. Furthermore this projection is deterministic, allowing it to be used as an inline rather than a post-process analysis. This property coupled with the structural preservation allows the consideration of novel search strategies in the new projected space.

Section 3 presents a definition of treespace and section 3.3 presents an elegant projection of that space that has all four of these desirable properties. This projection is then used to visualize the treespace and expose structure that can be exploited to guide the searches of common, but computationally expensive, methods.

2   Related Work

Treespace consists of all of the possible phylogenetic trees for a given set of taxa and their relationships with each other. This space is the domain of whatever search strategy is employed. Previous search strategies have not explicitly defined this domain, and the treespace that implicitly arises from these strategies is very cumbersome to work with. Figure 1 contains a visual comparison of three treespaces that have been used previously and are discussed in the following sections.

2.1   Subtree Transfer Induced Spaces

The most common treespaces used in phylogenetic search are the spaces implicitly defined by the subtree transfer operations, such as Nearest Neighbor Interchange (NNI), Subtree Prune and Regraft (SPR), Tree Bisection and Reconnection (TBR), or p-Edge Contraction and Refinement (p-ECR) [11], used

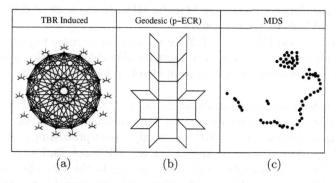

TBR Induced	Geodesic (p–ECR)	MDS

(a)   (b)   (c)

Fig. 1. A visual comparison of three treespaces previously used. The graph structure induced by TBR moves is highly connected. The geodesic structure consists of tiles of Euclidean space each consisting of one topology with all its possible branch lengths, joined together along valid p-ECR moves. Finally Multidimensional Scaling (MDS) plots trees in locations that preserves some distance metric.

during the search. These operations in turn induce distances between trees [1]. These treespaces take the form of graphs where each node is a specific tree. Each pair of trees that can reach each other with a single subtree transfer operation is connected with an edge of the graph. This is also the space used by Keith *et al.* [17] to build their generalized Gibbs sampler.

Unfortunately, though this space has been commonly used for searching, it is not easily visualized. For example using TBR, a very popular subtree transfer operation, the graph that represents this treespace has $O(n!!)$ nodes and each node is degree $O(n^3)$. Displaying this graph is clearly not practical for any problem of significant size. Worse, as this treespace is essentially a graph, there is no significant meaning to position, violating the first two criteria for a useful visualization. Also, calculating TBR distance is NP-Hard [1], which violates the third criterion. Finally this graph structure shown in Figure 1a does not exhibit exploitable structure, the fourth criterion, as trees of similar score are not grouped together.

2.2 Geodesic Tree Space

Billera *et al.* [3] introduced a new description of treespace, which has been further refined by Hultman [15]. The structure of this space, shown in Figure 1b is essentially the graph induced by all p-ECR transitions. However, it also differentiates trees by branch length within a topology. As a p-ECR induced space, this treespace fails to meet the criteria for the same reasons as other subtree transfer induced spaces.

2.3 Multidimensional Scaling

Multidimensional Scaling (MDS) has also been used to visualize treespace [2,14]. This method does not directly define a treespace, rather it uses the space induced by the distance metric used for the MDS. In the work of Hillis *et al.* [14] Robinson-Foulds distance was used.

Using this method Hillis *et al.* [14] were able to show some important characteristics of phylogenetic search. The most notable characteristic visualized was the presence of plateaus, large groups of closely related trees, that tend to slow down the search.

There are however some significant limitations to their use of MDS, which may not apply to the many variants of Multidimensional Scaling or to other manifold based methods. First, MDS is strictly a post-processing step. All of the points to be projected must be known beforehand, which limits the method to analysis of a search. Secondly there is no meaning to the space between points. It is not possible under MDS to determine a tree that would map to a specific point. Third, the axes of the new space have no consistent meaning. The only thing that MDS tries to preserve is some sense of distance, direction does not have any meaning after MDS is performed. As a result of these limitations, while MDS is a good visualization technique it does not meet the criteria of this work.

3   The Hypersphere of Trees in Split Space

Another treespace is one defined in terms of partitions of taxa. A projection can be defined from this space which both deterministically maps trees to single points and is reversible. These properties give us the first three criteria for a good treespace and visualization. In the results section we show that this space also displays exploitable structure.

3.1   Split Space

There are several varieties of trees that can be used in phylogenetics. Both candidate scoring metrics (likelihood and parsimony) work with unrooted trees so the space is further constrained to contain only unrooted and fully resolved trees.

Definition 1. *An n-tree is a graph in which all vertices have degree one or three, with exactly n vertices of degree one.*

Every branch in an n-tree divides the taxa on the tree into two sets. Thus every branch is a partition of the taxa. Some of these branches, those that connect to the leaves, are common to all n-trees, and are called trivial.

We define a space, called *split space*, where every possible nontrivial partition is associated with a unique dimension. We denote the split space associated with trees of n taxa as $\mathbb{T}_n$. The location of a tree in $\mathbb{T}_n$ is a vector, where each element corresponding to a partition in the tree is 1 and all others are 0.

There is a one-to-one mapping between vectors in split space and n-trees. The mapping from an n-tree to a vector in $\mathbb{T}_n$ is simple. Every element in the vector associated with a partition in the tree is set to 1, all others to 0. This mapping is one-to-one but not onto, as there are more possible vectors than

n-trees. Building an n-tree from a vector in $\mathbb{T}_n$ is also possible. As there is a mapping from an n-tree to a vector in $\mathbb{T}_n$ and the reverse mapping also exists, these trees and vectors are equivalent.

3.2   The Hypersphere of Trees

A hypersphere consists of the set of all points which are equidistant from a given center point. The set of all vectors in $\mathbb{T}_n$ which correspond to valid n-trees has this structure as shown in theorem 1.

Lemma 1. *All n-trees have $2n - 3$ branches, $n - 3$ of these are nontrivial.*

Theorem 1. *All n-trees lie on a hypersphere in $\mathbb{T}_n$.*

Proof. By Definition 1, n-trees are fully resolved. All fully resolved trees on n taxa have $n - 3$ nontrivial branches by Lemma 1. The vector representing any tree has exactly $n - 3$ elements which are 1 and all others are 0. The magnitude of this vector is $\sqrt{n - 3}$, which is the same for all n-trees. Thus all n-trees lie on a hypersphere.

3.3   Projecting the Sphere

Directly visualizing the hypersphere model is clearly infeasible as the number of dimensions that would need to be included quickly exceeds the number of dimensions that we can conveniently visualize. Therefore some form of dimension reduction is needed.

3.4   Sphere to Plane Projections

Cartographic projections are particularly apt at sphere to plane transformations. The basic cartographic projection takes a hypersphere in n dimensions and projects it onto a hyperplane of m dimensions. This is done by selecting m vectors. The inner product of each point to be projected with each of the selected vectors is computed. These inner products become the coordinates of the projected point. In this work three dimensions are used because it is well known how to display 3-D data.

3.5   Implementation Details

The extremely high dimensionality of $\mathbb{T}_n$ makes explicit storage of the three reference vectors needed for the cartographic projection infeasible. Likewise, due to the size of these vectors the typical calculations used for computing inner products require infeasible amounts of time. A naïve implementation of cartographic projections is adequate for very small numbers of taxa, but more sophisticated techniques are required for most data sets.

3.6   Hash Table Vector Representations

The memory usage of a straightforward implementation of cartographic projections is exponential in the number of taxa. Rather than explicitly storing the very large reference vectors a hash table representation is chosen. This representation has a fixed memory size, which can be arbitrarily chosen independently of the number of taxa.

To construct this table, a hash function and three representative vectors, one for each reference vector, of a feasible dimensionality are chosen. The hash function chosen must have a range equal to the set of natural numbers up to the dimensionality of the reference vectors and a domain equal to the set of natural numbers up to the dimensionality of the representative vectors.

Together these representative vectors and the hash function are used to compute the elements of the reference vectors as needed. The ith element of each reference vector is defined to be the element of the corresponding representative vector with the hashed value of i as follows:

$$X_i \leftarrow X'_{h(i)}$$

This representation allows a fixed amount of memory to be adequate for data sets of any number of taxa. This bound on memory usage is critical for the visualization of large data sets.

3.7   Orthogonality and Normalization of the Reference Vectors

It is desirable that the three reference vectors be orthonormal as they form the basis for the visualization. It is not practical to directly enforce this constraint, as each vector is implicitly defined by its representative vector and the hash function. Yet, it is still possible to make the reference vectors mutually linearly independent and give bounds on their normality and orthogonality [22]. These bounds do not depend on the hash function, so any good hash function should be adequate. Bob Jenkins' one at a time hash function [16] was used for the results in section 4.

If the representative vectors are made to be orthogonal then regardless of the choice of hash function, the true reference vectors are linearly independent. Furthermore, given the size of the representative vectors used (65535 elements) and only 20 taxa the reference vectors must be within 7.32×10^{-5} degrees of orthogonal. As the number of taxa increases this bound becomes even tighter.

Normalizing the reference vectors is more difficult. As the vectors have a very high dimensionality, normalization tends to make each individual element too small to be represented with finite precision arithmetic. As an alternative, each representative vector is made to have the same length as the others, without constraining this length to be one.

3.8   Calculating the Inner Product

The naïve method of calculating an inner product grows linearly with the dimension of the two vectors involved. Unfortunately, in this case the size of those vectors grows as the combinations of taxa. This method therefore gives worse than exponential performance with respect to number of taxa. However, for any given tree of n taxa, the vector representing that tree will have exactly $n - 3$ non-zero components by lemma 1. Furthermore, each of these will be exactly 1. Exploiting these properties gives an algorithm that computes the needed inner products in time $O(n)$, where n is the number of taxa.

This method begins with a hash table. Each element of the hash table contains one element from each of the reference vectors. The keys into the hash table are partition sets. The mapping of a tree is accomplished with the following steps.

1. A list of the $n - 3$ partition sets is built : $O(n)$
2. Each partition is used to lookup a set of x,y, and z values in the hash table : $O(1) * O(n)$
3. The $n - 3$ values are summed giving the final mapping : $O(n)$

These steps give an overall runtime execution of $O(n)$.

This method has two main advantages. First the time needed to compute the inner product scales with the number of taxa, not the dimensionality of split space. Second only a fixed amount of storage for the hash table is required. This upper bound on necessary storage makes the visualization of larger data sets feasible.

4   Results

The definitions of $\mathbb{T}_n$ and the cartographic projection are deterministic, reversible and have an easily calculated distance metric, fulfilling three of the four criteria for a useful visualization. The fourth criterion, exploitable structure, is the most important. The cartographic projection places similarly scored trees together in the data sets examined. This creates a gradient, an exploitable structure, which allows future work to develop a gradient descent strategy, which would be an improvement over current hill climbing techniques.

4.1   Locality of Structure

To have any exploitable structure there must be some correlation between position in the projected space and the topology of the trees near that position. Three methods will be considered: first, the method of Cartographic Projections, second, Multidimensional Scaling in two dimensions as in Tree Set Vis [14], and finally Multidimensional Scaling in three dimensions to account for any affects from the extra degree of freedom. The test case will be the exhaustive set of all trees of 7 taxa, with each method run 100 times as they all have random elements. Once each projection is calculated, the nearest n neighbors for every tree are found, with n ranging from 0 to 25. A majority rule consensus tree is then constructed for each of these neighborhoods. The resolution of these trees is reported, with a value of 1 indicating that the tree was fully resolved and a value of 0 indicating that the tree was fully unresolved.

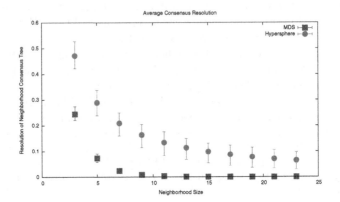

Fig. 2. The average degree of consensus across near neighbors among all trees with 7 taxa, note that higher values are better. As both cartographic projections and MDS have a random component each point consists of 100 projections with the average, minimum and maximum values for the consensus across all neighborhoods of the given size plotted. MDS was run both in the two dimensional case as in TreeSetVis and in a three dimensional case as the chosen cartographic projection resulted in a three dimensional result.

Figure 2 shows the results of this test. The points are plotted with the minimum, average and maximum values for the resolution. Note that cartographic projections are superior to both two and three dimensional MDS in every case. Not only are close trees more structurally similar, but also the neighborhoods over which some degree of topological similarity is found are much larger. It is thus concluded that cartographic projections produce, in terms of topology, a smoother mapping of treespace. Further this superiority is not due to the added flexibility of projecting onto three dimensions rather than two.

4.2   Results from Nine Taxa Set Exhaustive Searches

To explore the inherent structure of the maximum parsimony problem, several nine taxa data sets were fully analyzed. The size of nine taxa was selected because with only $135,135$ possible solution trees, it was very feasible to exhaustively enumerate all solutions for many different data sets of this size and to plot all of the points. Each set was exhaustively enumerated and scored using PAUP* [23]. The three reference points for the projection were chosen at random. Under this projection each of the possible trees mapped to a unique point in the new three dimensional space. The same projection was used for all of the data sets. These points were then colored according to the parsimony score of the corresponding tree, with white indicating a poor score and black indicating a good score.

In all of the data sets, there is significant exploitable structure. In some, such as that shown on the right in Figure 3a, a clear nearly linear gradient was visible throughout the entire cloud of possible trees. While in others such as that shown

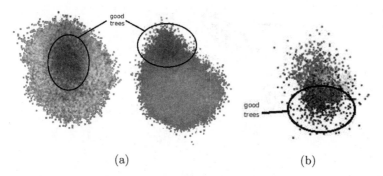

(a)   (b)

Fig. 3. Various datasets under cartographic projection. Dark points represent trees with better scores. (a) shows two distinct 9 taxa datasets, (b) shows a search through the Zilla 500 dataset, showing every tree that improved the best score that the search had found and every 100th tree considered that did not have a better score.

on the left in Figure 3a, clustering of scores is clear. Even though the gradient was much more complex, it would still be possible to use gradient descent.

4.3   Visualizing an Exhaustive Search with MDS

The tool Tree Set Viz was used to produce a visualization of a complete data set for comparison with our cartographic projections. Due to the very high memory requirements of multi-dimensional scaling, it was not possible to use a nine taxa data set. An eight taxa subset was used instead. The program was run overnight to allow the program adequate time to converge to the mapping shown in figure 4.

A few features are noteworthy. First, as all of the trees lie on the surface of a specific sphere, the best MDS solutions are circular. Also the MDS clustering, like the cartographic projection, has a large concentration of good trees. Although it is not clear that the clustering of scores caused by MDS is inferior to that of cartographic projections, it is crucial to note that MDS is a post process step and can not be used to guide a search. Therefore any structure is inherently not exploitable structure.

4.4   Results from Large Data Set Searches

It is not practical to exhaustively search the tree space associated with a large data set. Instead the phylogenetic search program PSODA [4] was modified to output every tree that it was going to perform a TBR rearrangement on, and every 100th rearrangement so produced. This gives not only the path of best trees found by the search as it progressed, but also a sampling of the trees that were rejected.

Figure 3b shows a projection of a TBR search with the Zilla data set [5] using cartographic projections. Again, a clustering of scores is apparent among the trees considered by the search, revealing exploitable structure in this difficult dataset.

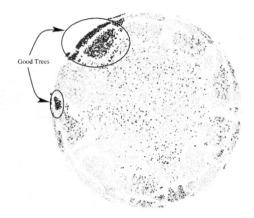

Fig. 4. A multi-dimensional scaling (MDS) visualization of an exhaustive search of 8 Taxa. Dark points represent trees with better scores.

5 Future Work

The cartographic projection from the hypersphere of trees has revealed significant structure to the problem of phylogenetic search. Further contributions can be made in improving our understanding of the revealed structure. More importantly new search techniques can be developed that can exploit this structure.

The current projection from split space to the 3-D visualization is based on the random selection of the points in split space. These points are guaranteed to result in linearly independent reference vectors and are very likely to result in vectors which are orthonormal as well. Although the initial random selection provides encouraging results, a more intelligent selection of basis vectors could improve the quality of the visualization.

There are two directions in which to take this work with respect to improving phylogenetic searches by utilizing the structure seen in the visualization. The first is to create a human guided search. As the projection from split space to the visualization is a simple linear transformation, it is possible to select a point in the visualized space and calculate the subspace of split space that corresponds to that point. A tree or trees in that subspace would then be generated and added to the list of trees used in a typical TBR based search, thereby restarting the search from the desired location. The second approach is to calculate and directly use the apparent gradient seen in the visualization to find better trees.

6 Conclusions

This cartographic projection from $\mathbb{T}_n$ fulfills all defined criteria for a good visualization. First the mapping from n-trees to $\mathbb{T}_n$ is one-to-one and further the cartographic projection for $\mathbb{T}_n$ to $\mathbb{R}^3$ is linear. This means that each tree maps to exactly one point, and this point is not affected by any outside influences. Also because the mapping is linear, it is reversible, which meets the second criterion. Euclidean distance in $\mathbb{T}_n$ is easy to calculate. Robinson-Foulds distance is also closely related to $\mathbb{T}_n$ as both definitions are based on the partition sets of trees. Either of these distance metrics are easily calculated and meet our third criterion.

More importantly, the use of a cartographic inspired projection has revealed significant structure to the problem of phylogenetic search. The visualization shows a general clustering of trees with similar scores, and in some data sets a clear gradient structure is observed. This promises to be useful in furthering our understanding of the problem of phylogenetic search and for informing the development of new methods in the field. These new methods will expand our ability to perform phylogenetic analysis which has implications for many biological fields.

References

1. Allen, B.L., Steel, M.: Subtree Transfer Operations and Their Induced Metrics on Evolutionary Trees. Annals of Combinatorics 5(1), 1–15 (2001)
2. Amenta, N., Klingner, J.: Case study: visualizing sets of evolutionary trees. In: IEEE Symposium on Information Visualization, INFOVIS 2002, pp. 71–74 (2002)
3. Billera, L.J., Homes, S.P., Vogtmann, K.: Geometry of the space of phylogenetic trees. Advances in Applied Mathematics 27(4), 733–767 (2001)
4. Carroll, H., Ebbert, M., Clement, M., Snell, Q.: PSODA: Better tasting and less filling than PAUP. In: Proceedings of the 4th Biotechnology and Bioinformatics Symposium (BIOT 2007), October 2007, pp. 74–78 (2007)
5. Chase, M.W., Soltis, D.E., Olmstead, R.G., Morgan, D., Les, D.H., Mishler, B.D., Duvall, M.R., Price, R.A., Hills, H.G., Qiu, Y.L., et al.: Phylogenetics of Seed Plants: An Analysis of Nucleotide Sequences from the Plastid Gene rbcL. Annals of the Missouri Botanical Garden 80(3), 528–580 (1993)
6. Chor, B., Tuller, T.: Maximum likelihood of evolutionary trees is hard. In: Miyano, S., Mesirov, J., Kasif, S., Istrail, S., Pevzner, P.A., Waterman, M. (eds.) RECOMB 2005. LNCS (LNBI), vol. 3500, pp. 296–310. Springer, Heidelberg (2005)
7. Clark, A.G., Weiss, K.M., Nickerson, D.A., Taylor, S.L., Buchanan, A., Stengard, J., Salomaa, V., Vartiainen, E., Perola, M., Boerwinkle, E., Sing, C.F.: Haplotype structure and population genetic inferences from nucleotide-sequence variation in human lipoprotein lipase. American Journal of Human Genetics 63, 595–612 (1998)
8. Crandall, K.A.: Multiple interspecies transmissions of human and simian t-cell leukemia/lymphoma virus type i sequences. Molecular Biology and Evolution 13, 115–131 (1996)
9. Day, W.H.E., Johnson, D.S., Sankoff, D.: The computational complexity of inferring rooted phylogenies by parsimony. Mathematical Biosciences 81(33-42), 299 (1986)
10. DeSalle, R.: Molecular approaches to biogeographic analysis of Hawaiian Drosophilidae. In: Wagner, W.L., Funk, V. (eds.) Hawaiian Biogeography, pp. 72–89 (1995)
11. Ganapathy, G., Ramachandran, V., Warnow, T.: Better Hill-Climbing Searches for Parsimony. In: Workshop on Algorithms in Bioinformatics (2003)

12. Guindon, S., Gascuel, O.: A Simple, Fast, and Accurate Algorithm to Estimate Large Phylogenies by Maximum Likelihood. Systematic Biology 52(5), 696–704 (2003)
13. Herring, B.L., Bernardin, F., Caglioti, S., Stramer, S., Tobler, L., Andrews, W., Cheng, L., Rampersad, S., Cameron, C., Saldanha, J., et al.: Phylogenetic analysis of WNV in North American blood donors during the 2003-2004 epidemic seasons. Virology (2007)
14. Hillis, D.M., Heath, T.A., St. John, K.: Analysis and Visualization of Tree Space. Systematic Biology 54(3), 471–482 (2005)
15. Hultman, A.: The topology of spaces of phylogenetic trees with symmetry. Discrete Mathematics 307(14), 1825–1832 (2007)
16. Jenkins, B.: A new hash function for hash table lookup. Dr. Dobb's Journal (1997)
17. Keith, J.M., Adams, P., Ragan, M.A., Bryant, D.: Sampling phylogenetic tree space with the generalized Gibbs sampler. Mol. Phylogenet Evol. 34(3), 459–468 (2005)
18. Meier, R., Ali, F.B.: Software Review. The newest kid on the parsimony block: TNT (Tree analysis using new technology). Systematic Entomology 30(1), 179 (2005)
19. Ronquist, F., Huelsenbeck, J.P.: MrBayes 3: Bayesian phylogenetic inference under mixed models. Bioinformatics 19(12), 1572–1574 (2003)
20. Sing, C.F., Haviland, M.B., Zerba, K.E., Templeton, A.R.: Application of cladistics to the analysis of genotype-phenotype relationships. European Journal of Epidemiology 8, 3–9 (1992)
21. Stamatakis, A.: RAxML-VI-HPC: maximum likelihood-based phylogenetic analyses with thousands of taxa and mixed models. Bioinformatics 22(21), 2688 (2006)
22. Sundberg, K., Clement, M., Snell, Q.: On the use of cartographic projections in visualizing phylogenetic treespace. Technical report, Brigham Young University (2009), http://dna.cs.byu.edu/papers/pdf/TR_BYU_CSL-2009-1.pdf
23. Swofford, D.L.: PAUP*. Phylogenetic Analysis Using Parsimony (* and Other Methods). Version 4. Sinauer Associates, Sunderland (2003)
24. Zwickl, D.J.: Genetic algorithm approaches for the phylogenetic analysis of large biological sequence datasets under the maximum likelihood criterion. PhD thesis, The University of Texas at Austin (2006)

A Simulation Study Comparing Supertree and Combined Analysis Methods Using SMIDGen

M. Shel Swenson[1], François Barbançon[2], C. Randal Linder[3], and Tandy Warnow[4]

[1] Department of Mathematics, The University of Texas at Austin
[2] Microsoft, Redmond, WA, USA
[3] Section of Integrative Biology, The University of Texas at Austin
[4] Department of Computer Sciences, The University of Texas at Austin

Abstract. Supertree methods comprise one approach to reconstructing large molecular phylogenies given estimated source trees for overlapping subsets of the entire set of taxa. These source trees are combined into a single supertree on the full set of taxa using various algorithmic techniques, with the most common being matrix representation with parsimony (MRP). When the data allow, the competing approach is a combined analysis (also known as a "supermatrix" or "total evidence" approach) whereby the different sequence data matrices for each of the different subsets of taxa are concatenated into a single supermatrix, and a tree is estimated on that supermatrix. In this paper, we report an extensive simulation study comparing the supertree methods MRP and weighted MRP against combined analysis methods on large model trees, using a novel simulation methodology (Super-Method Input Data Generator, or *SMIDGen*), which better reflects biological processes and the practices of systematists. This study shows that combined analysis based upon maximum likelihood outperforms all the other methods, giving especially big improvements when the largest subtree does not contain most of the taxa.

Supertree methods – methods that, given a set of trees with overlapping sets of taxa, return a tree on the combined taxon set – are used when a combined analysis (also known as a supermatrix or total evidence approach) is impractical or impossible (1; 2; 3). There is substantial interest in supertree methods (4).

Matrix representation with parsimony (MRP) (5; 6) is currently the most widely used supertree method. It operates by encoding the set of source trees as a matrix of partial binary characters, one character for each branch of each source tree, and then analyzing that matrix using a parsimony heuristic. Weighted MRP (7) is a variant of MRP in which the partial binary characters are weighted, and this weighted matrix representation is then analyzed using weighted parsimony. In the case of weighted MRP, the source trees could be majority rule or greedy consensus trees obtained during a bootstrap analysis, or a tree obtained on the original dataset whose branches have been weighted with bootstrap values or posterior probabilities.

Several studies (mostly based upon simulation) have evaluated the performance of different supertree methods in terms of the topological accuracy of the resultant phylogenies, and have investigated how different properties of the input–in particular, the percentage of missing data–and phylogenetic analysis impact final phylogenetic accuracy (8; 9; 10; 11; 12; 13; 14; 15; 16). Most of the supertree methods require that the

S.L. Salzberg and T. Warnow (Eds.): WABI 2009, LNBI 5724, pp. 333–344, 2009.
© Springer-Verlag Berlin Heidelberg 2009

input trees be rooted, a property that is not true of all systematic studies, and potentially problematic because accurate rooting is itself a nontrivial issue. Of the supertree methods that do not require rooted input trees, MRP and weighted MRP are the most promising of the current supertree methods.

Only two studies have evaluated the relative performance of supertree methods and combined analysis methods. Bininda-Emonds and Sanderson (8), who evaluated combined analyses based upon maximum parsimony (MP), and Criscuolo et al. (16), who evaluated combined analyses based upon maximum likelihood (ML). Both found that combined analysis provided a somewhat more topologically accurate reconstruction than MRP. Specifically, Criscuolo et al. found that combined analysis based upon ML improved slightly on weighted MRP based upon ML source trees, while Bininda-Emonds and Sanderson found weighted MRP provided a slight improvement in tree accuracy over combined analysis when MP was used for both the source trees and the combined analysis. However, the experimental methodology of these earlier studies included elements that were neither biologically accurate nor reflective of systematic practice, so their conclusions regarding the relative performance of supertree methods and combined analysis need to be revisited. In particular, both studies examined datasets with small numbers of taxa (while supertree analyses are primarily used for very large numbers of taxa), and for their source trees, they chose taxa at random from their model trees (while most source tree datasets are generally based upon clades – sets of taxa densely sampled within one subtree of the tree). Given these methodological issues, the relative topological accuracy of supertree methods and combined analysis should still be considered open.

This paper introduces a novel simulation methodology, SMIDGen, which better reflects both biological processes and systematic practice. We used SMIDGen to compare MRP and weighted MRP supertree methods and combined analysis on datasets with up to 1000 sequences. Under the conditions of our study, combined analysis using maximum likelihood consistently outperformed all other methods with respect to topological accuracy, suggesting that with more realistic simulations, MRP and weighted MRP supertree methods do not provide an acceptable alternative to combined analysis based upon maximum likelihood.

1   Methods

Experimental Design: Estimating a species phylogeny from a combined dataset involving multiple markers requires either a combined analysis, or the use of a supertree method. In a combined analysis, the alignments of individual markers are concatenated, and then a phylogenetic estimation method is applied to the concatenated alignments. Alternatively, for a supertree analysis, source trees are estimated individually on each aligned marker, and the source trees are then combined into a tree on the full dataset using a supertree method. (Supertree methods may also be employed when combined analysis is impossible, e.g., when only source trees are available or when the source trees are derived from data types that cannot be used as input to a combined analysis. These cases for supertree construction are not the subject of this paper.) The availability of sequence data for each marker depends upon a number of factors, including biological processes (the novel gain of a gene within the evolutionary history and its loss in some lineages due to silencing, and whether the marker evolves at the right rate for the taxa being reconstructed), and technical and practical issues (e.g., difficulties obtaining tissue samples for some taxa and difficulty successfully producing the sequence from some taxa). The consequence of these issues is that the pattern of missing data in both the supertree and combined analyses is not random.

Second, the source trees for a supertree method are produced by analyses of datasets selected by systematists, typically with the intent of estimating the phylogeny for a lower level taxonomic group (genera, families, and sometimes orders). We refer to these as "clade-based" studies. Clade-based studies usually have dense taxonomic sampling within the desired group (the ingroup), which tends to improve the ingroup's phylogenetic accuracy. For the taxa used as outgroups, sampling is almost always less dense. In addition to clade-based studies, systematists also produce "scaffold" phylogenies for higher level taxonomic groups, e.g., angiosperms (17) and metazoa (18; 19). Scaffold phylogenies sample taxa widely distributed across the group to provide a broad-scale sense of the relationships of the lower-level groups contained within the higher level group. In the context of a supertree analysis, scaffold phylogenies can provide the necessary "glue" for connecting phylogenies. In the absence of scaffold phylogenies, often the only overlapping taxa between source trees will be the small number of taxa used as outgroups for the clades of interest. Thus, supertree efforts will often consist of a large number of clade-based phylogenies that are densely sampled within their clades of interest and one or a small number of broadly distributed scaffold phylogenies based upon markers that evolve slowly (see Table 2 in the online supplementary material).

The simulation methodology we developed (SMIDGen) has six basic steps. In many cases we follow standard protocols, but with a few significant changes to increase simulation realism. The changes are indicated below.

Step 1: Generate model trees. We followed standard methodology here, generating trees under a pure birth process, and deviating these from ultrametricity (the molecular clock hypothesis). Unlike previous studies, we focused on large trees with 100 to 1000 taxa. We generated 30 replicates for each model tree size. (See the online supplementary material.) However, we report the results for only 10 replicates for the 1000-taxon datasets due to the very long running times for the ML analyses of these datasets.

Step 2: Evolve gene sequences down the model tree. We modified the standard methodology for this step. We first determined the subtree within the model tree for which each gene would be present, using a gene "birth-death" process; this produced missing data patterns that reflect biological data. Each gene was then evolved down its subtree under a GTR+Gamma+Invariable process (i.e., General Time Reversible process, with rates for sites drawn from a Gamma plus Invariable distribution (20)).

Step 3: Dataset production. This step was also performed in a novel way, producing datasets for both combined analysis and source tree estimation that reflect the practice of systematists. We produced datasets of genes selected to estimate trees on specific clades (rooted subtrees) within the tree, and also datasets of genes selected to form the scaffold tree. For each clade dataset, we selected three genes (each evolving on the same tree, but almost always under a different set of parameters); the scaffold dataset was based on 1-4 genes (again, each evolving on the same tree, but often under a different set of parameters). Thus, each dataset provided to the phylogeny estimation routine was a combined dataset.

Step 4: Estimation of source trees and the combined analysis trees. We followed standard practice here, using RAxML (21) for maximum likelihood (ML) and PAUP* (22) for maximum parsimony (MP) to estimate trees. However, we did not use a partitioned analysis within RAxML, thus somewhat hampering the accuracy of the maximum likelihood analysis. See the online supplementary materials for further details.

Step 5: Estimation of the supertrees. We used both standard and weighted MRP. See the online supplementary materials for further details.

Step 6: Performance evaluation. We mostly followed standard practice here, evaluating topological accuracy (with respect to false negative and false positive rates) and running time. We also explored the impact of dataset parameters on topological accuracy.

Steps 1, 4, and 5 were handled using standard methods, and are described in detail in the online supplementary materials; the other steps are described in detail here.

Step 2: Evolve Gene Sequences: For each model tree, we generated a large suite of genes for use in inferring the source trees. Genes for inferring scaffold trees always appeared at the root of the tree and did not go extinct; these are termed universal genes. Five universal genes were evolved for each model tree. The genes we created for the clade-based source trees did not occupy the entire tree and are called non-universal genes.

We simulated 100 non-universal genes as follows. We selected a single birth node for each gene by randomly selecting the gene's birth point using the model tree topology and branch lengths (see online supplementary materials for details). Once born, a gene could go extinct on later branches. As with the birth process, extinction was governed by both the model tree topology and branch lengths. Unlike the birth process, a gene could go extinct at multiple independent points on the tree following its first appearance (see online supplementary materials for details). Thus, the process for determining birth and extinction points of the genes produced a connected set of nodes in the tree containing the gene. This subtree constituted the model tree topology for the given gene.

Following generation of the birth-death patterns for the genes, gene sequences, each of length 500, were evolved under the GTR+Gamma+I model, where the parameters of the model were chosen with equal probability from a pool of parameter sets estimated by Ganesh Ganapathy (23) on three biological datasets: (a) the Angiosperm data set – 288 aligned DNA sequences of a group of Angiosperms, each of length 4811 (17; 24); (b) the Nematode data set – 682 aligned small subunit rRNA sequences, consisting of 678 species of Nematodes and four outgroups, each of length 1808 (25); and (c) the rbcL data set – 500 aligned rbcL DNA sequences each of length 1398 (26). Genes were evolved at a fast, medium or slow rate, implemented by rescaling the model tree branch lengths by a factor of 2.0, 1.0, or 0.1, respectively. Universal genes were always slow – reflecting the practice in systematics of using slower evolving genes for higher taxonomic groups. Twenty-five of the non-universal genes were fast, 50 were medium, and

25 slow. We evolved each gene independently down its model tree using the program Seq-Gen (27) (see the online supplementary materials for the Seq-Gen commands and for the parameters of the evolutionary processes).

Step 3: Data Set Production: For each model tree, we created DNA sequence data sets for phylogenetic analyses. Data sets differed in the taxa and genes used and whether they were scaffold or clade-based, in order to mimic taxon-sampling strategies used by systematists. The result of this process was a collection of data matrices, which we then used for the combined and supertree analyses.

For each clade-based data set, we selected a clade of interest from the model tree using a process similar to that used to find a birth node for each non-universal gene (see the online supplementary materials for details), restricting clade selection by setting bounds on the number of extant taxa in a clade to avoid selection of either very small or very large clades. For each 100-taxon model tree, we selected five clades with a clade size of at least 20. For each 500-taxon model tree, we selected 15 clades with a clade size of at least 30, and for each 1000-taxon model tree, we selected 25 clades ranging in size between 30 and 500. We never created more than one clade-based source tree for any clade in the model tree.

For each clade chosen, we selected the three non-universal genes that covered the largest number of taxa in the clade, breaking ties randomly. Once the three non-universal genes were chosen, we restricted the taxa in the clade to only those that had all three of the genes. This process produced clade-based datasets without any missing sequence data, but that may not have contained all the taxa of the specified clade. Since this process could produce datasets with small numbers of taxa, we excluded any clade-based data set that had fewer than ten taxa.

For the scaffold data sets, we used the same technique as in Bininda-Emonds and Sanderson (2001), and selected a subset of taxa uniformly at random from the model tree, with a fixed probability p, which we called the "scaffold-factor." By design, the scaffold datasets generated by this process had on average $p \times n$ taxa, where n is the number of taxa in the model tree. We generated scaffold data sets with a scaffold factor of either 0.20, 0.5, 0.75 or 1.0, for either one, two or four universal genes. The larger scaffold factors were chosen to ensure some model conditions had the taxonomic overlap necessary to potentially reconstruct an accurate supertree. For the smaller scaffold factors, we detected a handful of datasets with such low taxon overlap that it would have been inappropriate to apply either a supertree or a supermatrix analysis (see (28, pg. 257) for a description of the conditions needed to apply a supertree analysis). These datasets were discarded from our study. Because of the combination of scaffold and clade-based source trees, and because all were larger than some minimum size, we were able to achieve good coverage of most taxa in the model tree.

Step 6: Performance Evaluation: Steps 1 through 5 resulted in four supertree methods and two combined analysis methods. The supertree methods were MRP based upon MP trees (MRP-MP), MRP based upon ML trees (MRP-ML), weighted MRP based upon MP trees (wMRP-MP), and weighted MRP based upon ML trees (wMRP-ML)), and the combined analysis methods were based upon either MP (CA-MP) or upon ML (CA-ML). We calculated topological error using the false negative (FN) rate–the number of edges present in the model tree but not in the estimated tree, divided by n-3–and the false positive (FP) rate–the number of edges present in the estimated tree but not in the model tree, divided by n-3. We calculated the average error rates and standard deviations for each model condition.

We recorded the running time of each method on each dataset. Because the analyses were run under Condor (a distributed software environment (29)), the running times (for the larger datasets, especially) are inexact, and larger than they would be if run on a dedicated processor. Running times are provided to give an approximate estimation of the time needed to perform these analysis. We report the maximum and minimum running time for each model condition.

Finally, we explored the impact of the topological error of the source trees, the scaffold factor, the number of scaffold genes, and the number of taxa, on the topological error of the resultant supertrees.

2   Results and Discussion

Relative Performance of Methods: Interestingly, the six methods we studied had roughly the same relative performance (with respect to topological accuracy, measured with respect to both FN and FP rates) under most model conditions. CA-ML was consistently the best method, with *much* lower topological error than the other methods for most model conditions. Following CA-ML were the other ML-based methods–wMRP-ML, and MRP-ML, in that order–and then the three MP-based methods–CA-MP, MRP-MP, and wMRP-MP, usually in that order (Figs. 1 and 2). CA-ML's advantage was substantial for cases where the scaffold factor was less than 100% (with the biggest advantage for the smallest scaffold factors), and this advantage increased slightly with the number of taxa and decreased with the number of scaffold genes (Fig. 3).

In comparing the performance of the different algorithms (combined analysis or supertree method, based upon either maximum parsimony or maximum likelihood), we discovered that certain algorithm design choices had a large impact on the topological accuracy of the trees that were constructed. In particular, the choice of optimization

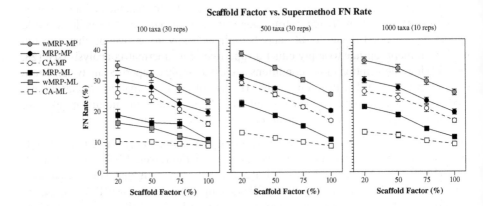

Fig. 1. False Negative (FN) rates (means with standard error bars) for supertree and supermatrix reconstructions as a function of the scaffold factor, for datasets with four scaffold genes

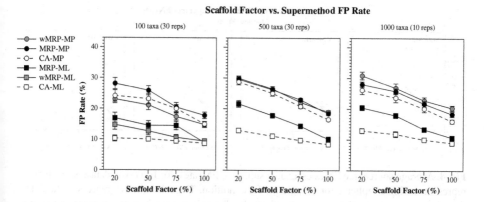

Fig. 2. False Positive (FP) rates (means with standard error bars) for supertree and supermatrix reconstructions as a function of the scaffold factor, for datasets with four scaffold genes

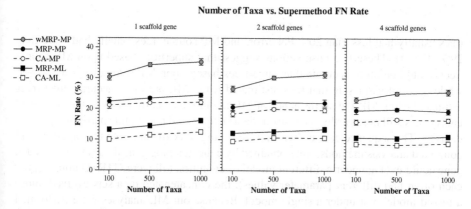

Fig. 3. FN rates (mean with standard error bars) for supertree and supermatrix reconstructions as a function of the number of taxa in the model tree and the number of scaffold genes. Only data sets with 100% scaffold factors are presented.

problem, i.e. whether we used maximum likelihood or maximum parsimony, had the largest impact on the final accuracy, with methods that used maximum likelihood (CA-ML, wMRP-ML, and MRP-ML) as a group more accurate than the methods based upon maximum parsimony (CA-MP, wMRP-MP, and MRP-MP). The second most significant algorithmic choice was whether we performed a combined or a supertree analysis, with CA-MP more accurate than MRP-MP and wMRP-MP, and similarly CA-ML more accurate than MRP-ML and wMRP-ML.

The first of these observations (that ML-based analyses were more accurate than MP-based analyses) is in some ways not surprising. Supertree methods are sensitive to error in their source trees, so improving the accuracy of the source trees will improve the accuracy of the supertrees. Furthermore, our study showed that source trees based upon ML were on average more accurate than source trees based upon MP. ML source

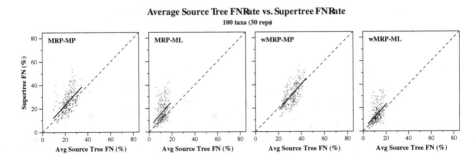

Fig. 4. Average source tree FN rates against supertree FN rates, for 100 taxon datasets. Each point represents a single replicate for a single model condition. The solid line is a regression line. The dotted line represents supertree constructions that have the same FN as the average source tree given as input. Points above the dotted line correspond to supertrees that are less topologically accurate than the average source tree, while points below the line correspond to supertrees that improved upon the average accuracy of the source trees.

trees usually had less than 20% FN error, and MP source trees usually had more than 20% (Fig. 4). These two observations suggest that supertrees based upon ML source trees should be more accurate than supertrees based upon MP source trees. However, while it is not surprising that trees estimated using ML are more accurate than trees estimated using MP, the performance of ML in our study, particularly in the combined analyses, is still noteworthy because our data analyses did not include partitioning of the datasets. The sequence datasets we analyzed using RAxML, for both the supertree and combined analysis methods, were obtained by concatenating gene data sets. Since each gene could evolve under a different model of evolution (all were GTR+Gamma+I, but each often had different parameter values), the concatenated data sets evolved under a mixed model, not under a single model. Because our ML analyses were performed under a simpler model than the one that generated the data, our ML-estimated source trees may have been less accurate than they would have been if they had been estimated under a partitioned analysis for each gene. That ML still outperformed MP under these conditions shows that even without partitioning the data by gene, maximum likelihood was still able to recover a more accurate topology than MP.

Running Time: Table 1 provides information on the running times of the methods studied here. Since these methods were run under Condor (a distributed system lacking dedicated processors), these numbers should be considered approximate and are given only as an indication of the general trends. Running times for wMRP-ML were prohibitively high for the 500 and 1000 taxon datasets, making it infeasible to use wMRP-ML on datasets of these sizes. Almost as problematic is wMRP-MP, which can be extremely slow on some 1000 taxon datasets. Combined analysis using maximum likelihood takes more time than combined analysis using maximum parsimony or MRP-ML and MRP-MP, but is still acceptable in its computational requirements (although it requires more than a day of analysis on the 1000 taxon datasets).

Table 1. Minimum and maximum of the average running times for each of the six methods over each model condition. For the four supertree methods this table displays the full running time, which includes the time taken to compute source trees. Running times are given in hours:minutes:seconds. N/A indicates the runs were not completed due to excessive running time.

Method	100 taxa		500 taxa		1000 taxa	
	Min.	Max.	Min.	Max.	Min.	Max.
CA-MP	0:00:19	0:00:22	0:26:27	0:37:44	4:49:07	11:13:18
CA-ML	0:09:22	0:24:19	7:02:24	11:11:52	24:33:39	34:14:48
MRP-MP	0:02:26	0:03:14	0:35:50	1:04:27	3:15:25	5:29:37
MRP-ML	0:04:12	0:05:25	0:42:23	1:09:41	3:14:56	5:53:37
wMRP-MP	0:01:59	0:02:27	3:40:18	12:33:24	30:34:14	295:10:35
wMRP-ML	7:00:43	16:23:16	N/A	N/A	N/A	N/A

3 Comparison with Earlier Studies

The most appropriate studies for comparison with the work presented here are those by Bininda-Emonds and Sanderson (2001) and Criscuolo et al. (2006) since they both performed simulation studies using MRP and combined analysis approaches. However, Bininda-Emonds and Sanderson only examined CA-MP, wMRP-MP, and MRP-MP, while Criscuolo et al. only examined CA-ML and MRP-ML; thus, our study is the first to consider wMRP-ML as a supertree method. Direct, quantitative comparisons with their studies are somewhat complicated because they used different metrics for assessing the topological accuracy of estimated trees relative to the true trees: Criscuolo et al. (2006) used quartet distances and Bininda-Emonds and Sanderson (2001) used the consensus fork index (CFI). We, therefore, restrict our comparisons to qualitative differences.

The result that is common among all three studies is that combined analysis is more accurate than MRP, whether based upon MP (Bininda-Emonds and Sanderson (2001)) or ML (Criscuolo et al. (2006)). Beyond this, the three studies come to different conclusions. Bininda-Emonds and Sanderson found that wMRP-MP was more accurate than CA-MP, whereas across all model conditions, our study found that CA-MP was more accurate than wMRP, having much lower FN error rates and comparable FP error rates. Since neither earlier study evaluated wMRP-ML as a supertree method, our findings with respect to its performance cannot be compared to their findings, except insofar as our findings are helpful in understanding wMRP as a generic supertree method. We found that wMRP, when applied to ML-based source trees, was not as accurate as combined analysis using maximum likelihood.

The differences between our findings and those of Bininda-Emonds and Sanderson (2001) and Criscuolo et al. (2006) could be due to several factors. Since our results show increasing the number of taxa increases the relative advantage of combined analysis over MRP and wMRP, we suspect that a major factor is likely to be the number of taxa in the experiments. Whereas we examined datasets with at least 100 taxa and up to 1000 taxa, Bininda-Emonds and Sanderson (2001) explored datasets with at most 32 taxa, and Criscuolo et al. (2006) explored 48 and 96 taxon datasets. Since most recent empirical supertree studies have included upwards of 200 taxa (see the online supplementary materials) and it is likely that future empirical supertree analyses will also tend to be in the range of our analyses, our results may be a better indicator of the relative performance of MRP supertree and combined analysis methods (and in particular, of wMRP-MP and CA-MP) for current uses of these methods. Another factor that is likely to be relevant is the methods we used to select taxa for the source tree datasets. For our experiments, we designed the taxon selection for the source trees to more nearly replicate systematic practice, producing two types of source datasets: scaffold datasets, and clade-based datasets. Our scaffold datasets were a random sample of the taxa and the clade-based datasets were obtained by identifying particular clades, then picking the genes that covered the clades best. The technique used by both Bininda-Emonds and Sanderson (2001) and Criscuolo et al. (2006) produced only scaffold datasets, because their taxa were always randomly selected from the full dataset. While the comprehensiveness of taxon sampling in in-groups in biological studies varies depending on the purpose of the study, the resources available to the researchers, and the ability to collect or access source material, there is almost always a clear non-random distribution of taxon-sampling effort in most of the individual trees that would be used as input for a supertree method or for producing a supermatrix. Thus, while we do not know how these differences would impact the relative performance of the methods studied, it is clear that their technique for producing the source datasets is not as consistent with systematic practice as ours.

4 Summary

Our study has two contributions. First, for simulating supertree and combined analysis datasets, we provide a new experimental methodology (SMIDGen) that reflects both biological processes and systematic practice. The SMIDGen software is available upon request from the first author. All of the datasets and model trees generated by SMIDGen and used in this study are available as benchmarks in our online supplementary material, at http://www.cs.utexas.edu/users/mswenson/pubs/. Second, we show that combined analysis using maximum likelihood produces more accurate trees than the other methods we tested, with the degree of improvement increasing with the number of taxa, and as the density of the scaffold tree decreases. This result clearly shows that MRP supertree methods do not produce sufficiently good analyses to substitute for combined analyses. Furthermore, with the availability of fast and highly accurate software for maximum likelihood, such combined analyses should not pose a substantial computational problem. However, we did not test any other supertree methods (many of which require the input source trees to be rooted), leaving open the possibility that some of these may outperform combined analysis using ML.

Acknowledgments

This research was supported in part by the US National Science Foundation under grants DEB 0733029, 0331453 (CIPRES), 0121680, and DGE 0114387. We also wish to thank the referees for their helpful and detailed comments.

References

[1] Sanderson, M.J., Purvis, A., Henze, C.: Phylogenetic supertrees: Assembling the trees of life. Trends Ecol. Evol. 13, 105–109 (1998)

[2] Bininda-Emonds, O.R.P., Gittleman, J.L., Steel, M.A.: The (super) tree of life: Procedures, problems, and prospects. Annu. Rev. Ecol. Syst. 33, 265–289 (2002)

[3] Bininda-Emonds, O.R.P.: The evolution of supertrees. Trends in Ecology and Evolution 19, 315–322 (2004)

[4] Bininda-Emonds, O.R.P.: Phylogenetic Supertrees: Combining Information To Reveal The Tree Of Life. Computational Biology (2004)

[5] Baum, B.R.: Combining trees as a way of combining data sets for phylogenetic inference, and the desirability of combining gene trees. Taxon 41, 3–10 (1992)

[6] Ragan, M.A.: Phylogenetic inference based on matrix representation of trees. Mol. Phylo. Evol. 1, 53–58 (1992)

[7] Ronquist, F.: Matrix representation of trees, redundancy, and weighting. Syst. Biol. 45, 247–253 (1996)

[8] Bininda-Emonds, O.R.P., Sanderson, M.J.: Assessment of the accuracy of matrix representation with parsimony analysis supertree construction. Syst. Biol. 50, 565–579 (2001)

[9] Chen, D., Diao, L., Eulenstein, O., Fernández-Baca, D., Sanderson, M.J.: Flipping: A supertree construction method. In: Bioconsensus. DIMACS: Series in Discrete Mathematics and Theoretical Computer Science, vol. 61, pp. 135–160. American Mathematical Society-DIMACS, Providence (2003)

[10] Burleigh, J.G., Eulenstein, O., Fernández-Baca, D., Sanderson, M.J.: MRF supertrees. In: Bininda-Emonds, O.R.P. (ed.) Phylogenetic Supertrees: Combining Information To Reveal The Tree Of Life, pp. 65–86. Kluwer Academic, Dordrecht (2004)

[11] Eulenstein, O., Chen, D., Burleigh, J.G., Fernández-Baca, D., Sanderson, M.J.: Performance of flip supertree construction with a heuristic algorithm. Syst. Biol. 53, 299–308 (2004)

[12] Lapointe, F.J., Levasseur, C.: Everything you always wanted to know about average consensus and more. In: Bininda-Emonds, O.R.P. (ed.) Phylogenetic Supertrees: Combining Information To Reveal The Tree Of Life, pp. 87–106. Kluwer Academic, Dordrecht (2004)

[13] Piaggio-Talice, R., Burleigh, J.G., Eulenstein, O.: Quartet supertrees. In: Bininda-Emonds, O.R.P. (ed.) Phylogenetic Supertrees: Combining Information to Reveal the Tree of Life, pp. 173–191. Kluwer Academic, Dordrecht (2004)

[14] Ross, H.A., Rodrigo, A.G.: An assessment of matrix representation with compatibility in supertree construction. In: Bininda-Emonds, O.R.P. (ed.) Phylogenetic Supertrees: Combining Information To Reveal The Tree Of Life, pp. 35–64. Kluwer Academic, Dordrecht (2004)

[15] Chen, D., Eulenstein, O., Fernández-Baca, D., Burleigh, J.G.: Improved heuristics for minimum-flip supertree construction. Evol. Bioinform. 2, 401–410 (2006)

[16] Criscuolo, A., Berry, V., Douzery, E., Gascuel, O.: SDM: A fast distance-based approach for (super) tree building in phylogenomics. Syst. Biol. 55, 740–755 (2006)

[17] Soltis, D.E., Soltis, P.S., Nickrent, D.L., Johnson, L.A., Hahn, W.J., Hoot, S.B., Sweere, J.A., Kuzoff, R.K., Kron, K.A., Chase, M.W.: Angiosperm phylogeny inferred from 18S Ribosomal DNA sequences. Ann. Mo. Bot. Garden 84, 1–49 (1997)

[18] Glenner, H., Hansen, A.J., Sørensen, M.V., Ronquist, F., Huelsenbeck, J.P., Willerslev, E.: Bayesian inference of the metazoan phylogeny: A combined molecular and morphological approach. Current Biology 14, 1644–1649 (2004)

[19] Dunn, C.W., Hejnol, A., Matus, D.Q., Pang, K., Browne, W.E., Smith, S.A., Seaver, E., Rouse, G.W., Obst, M., Edgecombe, G.D., Sorensen, M.V., Haddock, S.H.D., Schmidt-Rhaesa, A., Okusu, A., Kristensen, R.M., Wheeler, W.C., Martindale, M.Q., Giribet, G.: Broad phylogenomic sampling improves resolution of the animal tree of life. Nature 452(7188), 745–749 (2008)

[20] Swofford, D.L., Olson, G.J., Waddell, P.J., Hillis, D.M.: Phylogenetic Inferrence, 2nd edn., pp. 407–425. Sinauer Associates, Sunderland (1996)

[21] Stamatakis, A.: RAxML-NI-HPC: Maximum likelihood-based phylogenetic analyses with thousands of taxa and mixed models. Bioinformatics 22, 2688–2690 (2006)

[22] Swofford, D.L.: PAUP*: Phylogenetic analysis using parsimony (* and other methods). Ver. 4. Sinauer Associates, Sunderland (2002)

[23] Ganapathy, G.: Algorithms and Heuristics for Combinatorial Optimization in Phylogeny. PhD thesis, University of Texas at Austin (2006)

[24] Brauer, M.J., Holder, M.T., Dries, L.A., Zwickl, D.J., Lewis, P.O., Hillis, D.M.: Genetic algorithms and parallel processing in maximum-likelihood phylogeny inference. Mol. Biol. Evol. 19, 1717–1726 (2002)

[25] Baldwin, J., Fitch, D., De Ley, P., Nadler, S.: The Nematode Branch of the Assembling the Tree of Life Project: NemATOL (2008), http://nematol.unh.edu

[26] Chase, M.W., Soltis, D.E., Olmstead, R.G., Morgan, D., Les, D.H., Mishler, B.D., Duvall, M.R., Price, R.A., Hills, H.G., Qiu, Y.L.: Phylogenetics of seed plants: An analysis of nucleotide sequences from the plastid gene rbcL. Ann. Mo. Bot. Garden 80, 528–580 (1993)

[27] Rambaut, A., Grassly, N.C.: Seq-Gen: An application for the Monte Carlo simulation of DNA sequence evolution along phylogenetic trees. Comput. Appl. Biosci. 13, 235–238 (1997)

[28] Page, R.D.M.: Taxonomy, supertrees, and the tree of life. In: Bininda-Emonds, O.R.P. (ed.) Phylogenetic Supertrees: Combining Information to Reveal the Tree of Life, pp. 247–265. Kluwer Academic, Dordrecht (2004)

[29] Thain, D., Tannenbaum, T., Livny, M.: Distributed computing in practice: the condor experience. Concurrency and Computation: Practice and Experience 17, 323–356 (2005)

Aligning Biomolecular Networks Using Modular Graph Kernels

Fadi Towfic[1,2,*], M. Heather West Greenlee[1,3], and Vasant Honavar[1,2]

[1]Bioinformatics and Computational Biology Graduate Program
[2]Department of Computer Science
[3]Department of Biomedical Sciences
Iowa State University, Ames, IA
{ftowfic,mheather,honavar}@iastate.edu

Abstract. Comparative analysis of biomolecular networks constructed using measurements from different conditions, tissues, and organisms offer a powerful approach to understanding the structure, function, dynamics, and evolution of complex biological systems. We explore a class of algorithms for aligning large biomolecular networks by breaking down such networks into subgraphs and computing the alignment of the networks based on the alignment of their subgraphs. The resulting subnetworks are compared using graph kernels as scoring functions. We provide implementations of the resulting algorithms as part of BiNA, an open source biomolecular network alignment toolkit. Our experiments using *Drosophila melanogaster*, *Saccharomyces cerevisiae*, *Mus musculus* and *Homo sapiens* protein-protein interaction networks extracted from the DIP repository of protein-protein interaction data demonstrate that the performance of the proposed algorithms (as measured by % GO term enrichment of subnetworks identified by the alignment) is competitive with some of the state-of-the-art algorithms for pair-wise alignment of large protein-protein interaction networks. Our results also show that the inter-species similarity scores computed based on graph kernels can be used to cluster the species into a species tree that is consistent with the known phylogenetic relationships among the species.

1 Introduction

The rapidly advancing field of systems biology aims to understand the structure, function, dynamics, and evolution of complex biological systems [9]. Such an understanding may be gained in terms of the underlying networks of interactions among the large number of molecular participants involved including genes, proteins, and metabolites [47,16]. Of particular interest in this context is the problem of comparing and aligning multiple networks e.g., those generated from measurements taken under different conditions, different tissues, or different organisms [40]. Network alignment methods present a powerful approach for detecting conserved modules across several networks constructed from different

---

* Corresponding author.

S.L. Salzberg and T. Warnow (Eds.): WABI 2009, LNBI 5724, pp. 345–361, 2009.
© Springer-Verlag Berlin Heidelberg 2009

species, conditions or timepoints. The detection of conserved network modules may allow the discovery of disease pathways, proteins/genes critical to basic biological functions, and the prediction of protein functions.

The problem of aligning two networks, in the absence of the knowledge of how each node in one network maps to one or more nodes in the other network, requires solving the subgraph isomorphism problem, which is known to be computationally intractable (NP-Hard) [15]. However, in practice, it is possible to establish correspondence between nodes in the two networks to be aligned and to design heuristics that strike a balance between the speed, accuracy and robustness of the alignment of large biological networks. For instance, MaWISh [29] is a pairwise network alignment algorithm with a runtime complexity of $O(mn)$ (where m and n are the number of vertices in the two networks being compared) that relies on a scoring function that takes into account protein duplication events as well as interaction loss/gain events between pairs of proteins to detect conserved protein clusters. Hopemap [44] is an iterative clustering-based alignment algorithm for Protein-Protein Interaction networks. HopeMap starts by clustering homologs based on their sequence similarity and already known KEGG/InParanoid Orthology status. The algorithm then proceeds to search for strongly connected components and outputs the conserved components that statisfy a predefined user threshold [44]. Graemlin 2.0 is a linear time algorithm that relies on a feature-based scoring function to perform an approximate global alignment of multiple networks. The scoring function for Graemlin 2.0 takes into account protein deletion, duplication, mutation, presence and count as well as edge/paralog deletion across the different networks being aligned [13]. NetworkBLAST-M [23] is a progressive multiple network alignment algorithm that constructs a layered alignment graph, where each layer corresponds to a network and edges between layers connect homologs across different networks. Highly conserved subnetworks from networks from different species are first aligned based on highly conserved orthologous clusters, then the clusters are expanded using an iterative greedy local search algorithm [23].

Against this background, we explore a class of algorithms for aligning large biomolecular networks using a *divide and conquer* strategy that takes advantage of the *modular* substructure of biological networks [17,36,19]. The basic idea behind our approach is to align a pair of networks based on the optimal alignments of the subnetworks of one network with the subnetworks of the other. Different ways of decomposing a network into subnetworks in combination with different choices of measures of *similarity* between a pair of subnetworks yield different algorithms for aligning biomolecular networks.

We utilize variants of state-of-the-art *graph kernels* [6,7], first developed for use in training support vector machines for classification of graph-structured patterns, to compute the *similarity* between two subgraphs. The use of graph kernels to align networks offers several advantages: It is easy to substitute one graph kernel for another (to incorporate different application-specific criteria) without changing the overall approach to aligning networks; it is possible to combine multiple graph kernels to create more complex kernels [7] as needed. Our experiments with the fly, yeast, mouse and human protein-protein interaction networks extracted from DIP (Database of Interacting Proteins) [38] demonstrate the feasibility of the proposed approach for aligning large biomolecular networks.

The rest of the paper is organized as follows: Section 2 precisely formulates the problem of aligning two biomolecular networks and describes the key elements of our proposed solution. Section 3 describes the experimental setup and experimental results. Section 4 concludes with a summary of the main contributions of the paper in the broader context of related literature and a brief outline of some directions for further research.

2   Aligning Protein-Protein Interaction Networks

2.1   Problem Formulation

We consider the problem of pair-wise alignment of protein-protein interaction networks. We model protein-protein interaction networks as undirected and unweighted graphs. In a protein-protein interaction network, the vertices in the graph correspond to proteins and the edges denote interactions between the two proteins. Let the graphs $G_1(V_1, E_1)$ and $G_2(V_2, E_2)$ denote two protein-protein interaction networks where $V_1 = \{v_1^1, v_2^1, v_3^1, ...v_n^1\}$ and $V_2 = \{v_1^2, v_2^2, v_3^2, ...v_m^2\}$, respectively, denote the vertices of G_1 and G_2; and E_1 and E_2 denote the edges of G_1 and G_2 respectively. Let a matrix $\mathbf{P}$ with $|V_1|$ rows and $|V_2|$ columns (i.e, $n \times m$ matrix) denote a set of matches between the vertices of G_1 and G_2. The mapping matrix $\mathbf{P}$ is defined such that for any two vertices v_x^1 and v_y^2 (where $1 \leq x \leq n$ and $1 \leq y \leq m$) from graphs G_1 and G_2, respectively, $P_{v_x^1 v_y^2} = 1$ if v_x^1 from G_1 is matched to v_y^2 from G_2 and $P_{v_x^1 v_y^2} = 0$ if v_x^1 in G_1 is not a match to v_y^2 in G_2. For example, the matches between nodes may be based on homology between the sequences of the corresponding proteins. Thus, each node in G_1 is matched to 0 or more nodes of G_2 and vice versa. Note that the number of such matches for any node in G_1 is much smaller than the total number of nodes in G_2 and vice versa.

$C_1(L_1, O_1)$ is said to be a subgraph of $G_1(V_1, E_1)$ if $L_1 \subset V_1$ and $O_1 \subset E_1$ where O_1 consists only of edges whose end points are in L_1. We associate with the graphs $G_1(V_1, E_1)$ and $G_2(V_2, E_2)$ sets of subgraphs $S_1 = \{C_1, C_2, C_3, ...C_l\}$ and $S_2 = \{Z_1, Z_2, Z_3, .., Z_w\}$ (respectively), where $C_i(L_i, O_i)$ $1 \leq i \leq l$ is a subgraph of G_1 and $Z_j(W_j, Q_j)$ $1 \leq j \leq w$ is a subgraph of G_2. Our basic strategy is to find a best match for each subgraph in S_1 from S_2 by optimizing a scoring function, $K(C_i, Z_j)$, such that we obtain: (i) a set of vertices that satisfy $P_{v_x^1 v_y^2} = 1$, where $v_x^1 \in L_i$ and $v_y^2 \in W_j$ and (ii) a set of edges where: if (v_x^1, v_d^1) is an edge in O_i, then (v_y^2, v_g^2) is an edge in Q_j where $P_{v_x^1 v_y^2} = 1$ and $P_{v_d^1 v_g^2} = 1$. The resulting solution to the network alignment problem satisfies the condition that each subgraph in S_1 has at most one matching subgraph in S_2. Thus, a pairwise alignment of the networks $G_1(V_1, E_1)$ and $G_2(V_2, E_2)$ is expressed in terms of an optimal alignment among the sets of the corresponding sets of subgraphs in S_1 and S_2.

2.2 Divide-and-Conquer Approach to Aligning Protein-Protein Interaction Networks

As noted earlier, our basic approach to aligning a pair of protein-protein interaction networks involves (a) decomposing each network into a collection of smaller subnetworks; (b) compute the alignment of the two networks in terms of the optimal alignments of the subnetworks of one network with the subnetworks of the other. Different choices of methods for decomposing a network into subnetworks in combination with different choices of measures of *similarity* between a pair of subnetworks yield different algorithms for aligning protein-protein interaction networks. In our current implementation, we establish the matches between nodes in the two protein-protein interaction networks to be aligned based on reciprocal BLASTp [2] hits between the corresponding protein sequences. Thus, $P_{v_x^1 v_y^2} = 1$ if and only if the correponding protein sequences of v_x^1 and v_y^2 are reciprocal BLASTp hits [21] for each other (at some chosen user-specified threshold). Alternatively, the mapping can be established based on known homologies (e.g between the human WNT1 and mouse Wnt1 proteins) [28,10].

Decomposing Networks into k-hop Neighborhoods. A k-hop neighborhood-based approach to alignment uses the notion of k-hop neighborhood. The k-hop neighborhood of a vertex $v_x^1 \in V_1$ of the graph $G_1(V_1, E_1)$ is simply a subgraph of G_1 that connects v_x^1 with the vertices in V_1 that are reachable in k hops from v_x^1 using the edges in E_1. Given two graphs $G_1(V_1, E_1)$ and $G_2(V_2, E_2)$, a mapping matrix $\mathbf{P}$ that associates each vertex in V_1 with zero or more vertices in V_2 and a user-specified parameter k, we construct for each vertex $v_x^1 \in V_1$ its corresponding k-hop neighborhood C_x in G_1. We then use the mapping matrix $\mathbf{P}$ to obtain the set of matches for vertex v_x^1 among the vertices in V_2; and construct the k-hop neighborhood Z_y for each matching vertex v_y^2 in G_2 and $P_{v_x^1 v_y^2} = 1$. Let $S(v_x^1, G_2)$ be the resulting collection of k-hop neighborhoods in G_2 associated with the vertex v_x^1 in G_1. We compare each k-hop subgraph C_x in G_1 with each member of the corresponding collection $S(v_x^1, G_2)$ to identify the k-hop subgraph of G_2 that is the best match for C_x (based on a chosen similarity measure). This process is illustrated in figure 1. The runtime complexity of the k-hop neighborhood based network alignment algorithm is $O(bmg)$ where m is the number of nodes in the query network G_1, b is the maximum number of matches in the target network G_2 for any node in the query network, and g is the running time of the similarity measure or scoring function used to compare a pair of k-hop subnetworks.

Decomposing Networks Into Clusters. A graph clustering based alignment algorithm works as follows: Given two node-labeled graphs $G_1(V_1, E_1)$ and $G_2(V_2, E_2)$, and a mapping matrix $\mathbf{P}$ that associates each vertex in V_1 with zero or more vertices in V_2, we first extract collections of subgraphs $H_1 = \{C_1, C_2, C_3, ...C_l\}$ and $H_2 = \{Z_1, Z_2, Z_3, ...Z_w\}$ from G_1 and G_2 respectively. In principle, any graph clustering algorithm may be used to construct the subgraph sets H_1 and H_2. In our experiments, we used the bicomponent clusterer as implemented in the JUNG (Java Universal Network/Graph) framework

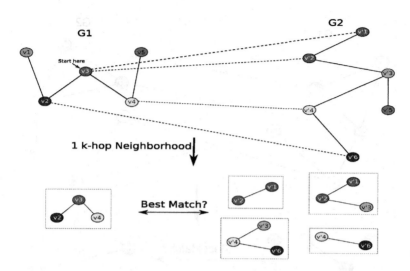

Fig. 1. General schematic of the k-hop neighborhood alignment algorithm. The input to the algorithm are two graphs (G_1 and G_2) with corresponding relationships among their nodes using mapping matrix **P** (similarly colored nodes are sequence homologous according to a BLAST search, for example $P_{v_2 v_6'} = 1$). The algorithm starts at an arbitrary vertex in G_1 (red vertex in the figure) and constructs a k-hop neighborhood around the starting vertex (1-hop neighborhood in the figure). The algorithm then matches each of the nodes in the 1-hop neighborhood subgraph from G_1 to nodes in G_2 using mapping matrix **P**. 1-hop subgraphs are then constructed around each of the matching vertices. The 1-hop subgraphs from G_2 are then compared using a scoring function (e.g. a graph kernel) to the 1-hop subgraph from G_1 and the maximum scoring match is returned.

[35,46] to extract H_1 and H_2. Briefly, the bicomponent clusterer searches for all biconnected components (graphs that cannot be disconnected by removing a single node/vertex [18]) by traversing a graph in a depth-first manner (please see [32] for more details). Once the subgraph sets H_1 and H_2 of the biconnected subgraphs of G_1 and G_2 (respectively) are extracted, an all vs. all comparison is conducted to identify for each subgraph in H_1, the best matching subgraph in H_2 using a scoring function (e.g. a graph kernel, see figure 2). The running time complexity of this algorithm is $O(lwg)$ where l is the number of clusters extracted from the query network G_1, w is the number of clusters extracted from the target network G_2, and g is the running time of the scoring function used to compare a pair of clusters (subgraphs).

2.3 Scoring Functions

We now proceed to describe the similarity measures or scoring functions used to compare a pair of subgraphs (e.g., a pair of k-hop subgraphs or a pair of bi-component clusters described above).

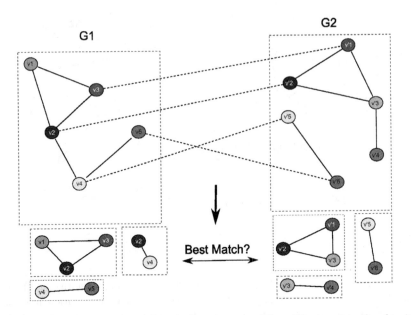

Fig. 2. Schematic for the cluster-based alignment algorithm. The input to the algorithm are two graphs (G_1 and G_2) with corresponding relationships among their nodes using mapping matrix **P** (similarly colored nodes are sequence homologous according to a BLAST search, for example $P_{v_2v_2'} = 1$). Subgraphs are generated from G_1 and G_2 using a graph clustering algorithm (e.g. bicomponent clusterer that finds biconnected subgraphs) and the subgraphs from G_1 are compared against the subgraphs from G_2 to find the best matching subgraphs using an appropriate scoring function.

Modified Shortest Path Distance Graph Kernel. The shortest path graph kernel was first described by Borgwardt and Kriegel [6]. As the name implies, the kernel compares the length of the shortest paths between any two nodes in a graph based on a pre-computed shortest-path distance. The shortest path distances for each graph may be computed using the Floyd-Warshall algorithm as implemented in the CDK (Chemistry Development Kit) package [41]. We modified the Shortest-Path Graph Kernel to take into account the sequence homology of nodes being compared as computed by BLAST [2]. The shortest path graph kernel for subgraphs Z_{G_1} and Z_{G_2} (e.g., k-hop subgraphs, bicomponent clusters extracted from G_1 and G_2 respectively) is given by:

$$K(Z_{G_1}, Z_{G_2}) = \log \left[\sum_{v_i^1, v_j^1 \in Z_{G_1}} \sum_{v_k^2, v_p^2 \in Z_{G_2}} \delta(v_i^1, v_k^2) \times \delta(v_j^1, v_p^2) \times d(v_i^1, v_j^1) \times d(v_k^2, v_p^2) \right]$$

(1)

where $\delta(v_x^1, v_y^2) = \frac{BlastScore(v_x^1, v_y^2) + BlastScore(v_y^2, v_x^1)}{2}$. $d(v_i^1, v_j^1)$ and $d(v_k^2, v_p^2)$ are the lengths of the shortest paths between v_i^1, v_j^1 and v_k^2, v_p^2 computed by the Floyd-Warshall algorithm. The runtime of the Floyd-Warshall Algorithm is $O(n^3)$. The

Graph 1   **Graph 2**

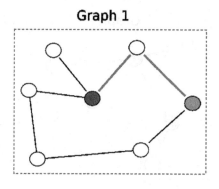
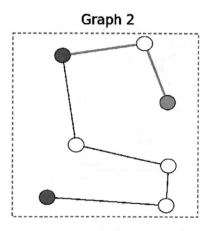

Fig. 3. An example of the graph matching conducted by the shortest path graph kernel. Similarly colored nodes are sequence homologous according to a BLAST search. As can be seen from the figure, the graph kernel compares the lengths of the shortest paths around homologous vertices across the two graphs. The red edges show the matching shortest path in both graphs as computed by the graph kernel. The shortest path distance graph kernel takes into account the sequence homology score for the matching vertices across the two graphs as well as the distances between the two matched vertices within the graphs.

shortest path graph kernel has a runtime of $O(n^4)$ (where n is the maximum number of nodes in larger of the two graphs being compared). Please see figure 3 for a general outline of the comparison technique used by the shortest-path graph kernel.

Modified Random Walk Graph Kernel. The random walk graph kernel [45] has been previously utilized by Borgwardt et al. [7] to compare protein-protein interaction networks. The random walk graph kernel for subgraphs Z_{G_1} and Z_{G_2} (e.g., k-hop subgraphs, bicomponent clusters extracted from G_1 and G_2 respectively) is given by:

$$K(Z_{G_1}, Z_{G_2}) = p \times (\mathbf{I} - \lambda K_x)^{-1} \times q \qquad (2)$$

where $\mathbf{I}$ is the identity matrix, λ is a user-specified variable controlling the length of the random walks (a value of 0.01 was used for the experiments in this paper), K_x is an $nm \times nm$ matrix (where n is the number of vertices in Z_{G_1} and m is the number of vertices in Z_{G_2} resulting from the Kronecker product $K_x = Z_{G_1} \otimes Z_{G_2}$, specifically,

$$K_{\alpha\beta} = \delta(Z_{G_{1_{ij}}}, Z_{G_{2_{kl}}}), \alpha \equiv m(i-1) + k, \beta \equiv m(j-1) + l \qquad (3)$$

Where $\delta(Z_{G_{1_{ij}}}, Z_{G_{2_{kl}}}) = \frac{BlastScore(Z_{G_{1_{ij}}}, Z_{G_{2_{kl}}}) + BlastScore(Z_{G_{2_{kl}}}, Z_{G_{1_{ij}}})}{2}$; p and q are $1 \times nm$ and $nm \times 1$ vectors used to obtain the sum of all the entries of the inverse expression $((\mathbf{I} - \lambda K_x)^{-1})$.

Graph 1

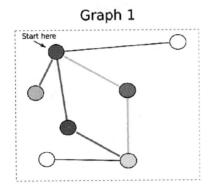

Graph 2

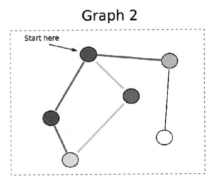

Fig. 4. An example of the graph matching conducted by the random walk graph kernel. Similarly colored vertices are sequence homologous according to a BLAST search. As can be seen from the figure, the graph kernel compares the neighborhood around the starting vertices in each graph using random walks. Colored edges indicate matching random walks across the two graphs of up to length 2. The random walk graph kernel takes into account the sequence homology of the vertices visited in the random walks across the two graphs as well as the general topology of the neighborhood around the starting vertex.

We adapted the random walk graph kernel to align protein-protein interaction networks by taking advantage of the reciprocal BLAST hits (RBH) among the proteins in the networks from different species [21]. Naive implementation of our modified random-walk graph kernel, like the original random-walk graph kernel [45], has a runtime complexity of $O(r^6)$ (where $r = max(n, m)$). This is due to the fact that the product graph's adjacency matrix is $nm \times nm$, and the matrix inverse operation takes $O(h^3)$ time, where h is the number of rows in the matrix being inverted (thus, the total runtime is $O((rm)^3)$ or $O(r^6)$ where $r = max(n, m)$). However, runtime complexity of the random walk graph kernel (and hence our modified random walk graph kernel) can be improved to $O(r^3)$ by making use of the Sylvester equations as proposed by Borgwardt et al. [7]. Figure 4 illustrates the computation of the random walk graph kernel.

2.4   Implementation

The the k-hop network neighborhood and bicomponent clustering based protein-protein interaction network alignment algorithms are implemented in BiNA (http://www.cs.iastate.edu/~ftowfic), an open source Biomolecular Network Alignment toolkit. The current implementation includes variants of the shortest path and random walk graph kernels for computing similarity between pairs of subnetworks. The modular design of BiNA allows the incorporation of alternative strategies for decomposing networks into subnetworks and alternative similarity measures (e.g., kernel functions) for computing the similarity between subnetworks.

3   Experiments and Results

We conducted experiments using k-hop subgraph and bi-component cluster based strategies for decomposing graphs into collections of subgraphs. In each case, both modified shortest path and modified random walk graph kernels were used to compute similarity between pairs of subgraphs. We compare the performance of the resulting algorithms with variants of NetworkBLAST [24] and HopeMap [44], which are among the state-of-the-art algorithms for pair-wise alignment of protein-protein interaction networks, using metrics proposed by Kalaev et al [24]. The NetworkBLAST algorithm uses BLASTp [2] to match the nodes across the different networks being aligned, whereas HopeMap uses In-Paranoid [34] orthology groups to match the nodes across different networks. The HopeMap and NetworkBLAST algorithm were adapted by Tian and Samatova to utilize KEGG Ortholog (KO) groups to match nodes across different species (NetworkBLAST-ko and HopeMap-ko) [44]. The results in table 1 for Network-BLAST, HopeMap, NetworkBLAST-ko and HopeMap-ko are taken from Tian and Samatova's HopeMap paper [44].

In addition, we used our network alignment algorithms to generate phylogenetic trees from protein-protein interaction networks. We now proceed to describe the experimental setup and the results of this study.

3.1   Datasets

The yeast, fly, mouse and human protein-protein interaction networks were obtained from the Database of Interacting Proteins (DIP) release $1/26/2009$ [38]. The sequences for each dataset were obtained from uniprot release 14 [4]. The DIP sequence ids were matched against their uniprot counterparts using a mapping table provided on the DIP website. All proteins from DIP that had obsolete uniprot IDs or were otherwise not available in release 14 of the uniprot database were removed from the dataset. The fly, yeast, mouse and human protein-protein interaction networks consisted of $6,645$, $4,953$, 424 and $1,321$ nodes and $20,010$, $17,590$, 384 and $1,716$ edges, respectively. The protein sequences for each dataset were downloaded from uniprot [4]. BLASTp [2] with a cutoff of 1×10^{-10} was used to match protein sequences across species.

3.2   Comparison with NetworkBLAST and HopeMap

To evaluate the alignments, Kalaev et al.'s approach was implemented as described in the NetworkBLAST [24] and the HopeMap [44] papers. Recall from section 2.1 that the output of the alignment algorithm is a set of subgraphs S_1 and S_2 (corresponding to the query and target networks, respectively). The set of subgraphs $S_2 = \{Z_1, Z_2, Z_3, ..., Z_w\}$ in the target network is evaluated by searching for overrepresented Gene Ontology (GO) categories from the biological process annotation [3]. The GOTermFinder [8] tool was utilized to compute enrichment p-values (p-value significance cutoff $= 0.05$) that have been corrected for multiple testing using the false discovery rate. Briefly, GOTermFinder computes p-values for a set of GO annotations for the set of proteins in subgraphs

$Z_{1..w}$ based on the number of proteins in the subgraph Z_x (where $1 \leq x \leq w$, and the number of vertices in Z_x is r) and the number of proteins in the genome of the target network (n) and their respective GO annotation. The p-value is computed based on the hypergeometric distribution as the probability of k or more out of r proteins being assigned a given annotation (where k is the number of proteins in the subgraph Z_x possessing the GO category of interest), given that y of n proteins possess such an annotation in the genome in general. The number of subgraphs, f, that had one or more GO categories overrepresented is computed (where $f \leq w$) and the fraction of subgraphs from the target network that had a significant number of GO categories overrepresented is then computed ($\frac{f}{w} \times 100$, % coherent subnetworks). The specificity of the alignment method is measured by the percent of coherent subnetworks discovered for each species. The sensitivity of the methods is indicated by the number of distinct GO categories covered by the functionally coherent subnetworks. The purpose of this evaluation approach is to determine whether or not the matching subgraphs found in the target network represent a functional module/pathway (functionally coherent subgraphs) based on the GO annotation of the proteins in the subgraph.

k-hop Neighborhood based Alignment. The results in table 1 show a comparison of the performance of the k-hop neighborhood based alignment with variants of NetworkBLAST [24] and HopeMap [44].

As can be seen from the results in table 1, the performance measured in terms of % GO enrichment observed when the fly protein-protein interaction network is aligned with the yeast protein-protein interaction network, and vice-versa, using k-hop neighborhood based alignment (with k, the number of hops set equal to 1 and 2) is comparable to that of of variants of NetworkBLAST and Hopemap algorithms (as reported in [44]). The modified random walk graph kernel (RWKernel) yields higher % GO enrichment than the modified shortest path graph kernel (SPKernel) in the case of the fly dataset. The effectiveness of

Table 1. Comparison of the k-hop neighborhood based protein-protein interaction network alignment algorithm (using the SPKernel and RWKernels) with variants of NetworkBLAST and HopeMap (as reported in [44]) using the functional coherence measure: As can be seen from the table, the k-hop neighborhood based algorithm with $k=2$ is competitive with variants of NetworkBLAST and Hopemap that use sequence identity or KEGG ortholog groups (NetworkBLAST-ko and HopeMap-ko, respectively) for node-matching

Method	% GO enrichment in Yeast	# GO enriched in Yeast	% GO enrichment in Fly	# GO enriched in Fly
NetworkBLAST	94.87	67	84.62	62
HopeMap	98.73	65	78.48	46
NetworkBLAST-ko	100	9	100	8
HopeMap-ko	100	24	92.31	24
SPKernel-1Hop	100 (Score cutoff = 800)	51	78 (Score cutoff = 900)	22
SPKernel-2Hop	100 (Score cutoff = 1800)	46	76 (Score cutoff = 2400)	9
RWKernel-1Hop	100 (Score cutoff = 500)	71	85 (Score cutoff = 900)	19
RWKernel-2Hop	100 (Score cutoff = 800)	107	100 (Score cutoff = 3100)	1

Table 2. Sample comparison of the k-hop alignment algorithm with the SPKernel and RWKernel on the mouse and human DIP datasets

Method	% GO enrichment in Mouse	# GO enriched in Mouse	% GO enrichment in Human	# GO enriched in Human
SPKernel-1Hop	53 (Score cutoff = 40)	19	85 (Score cutoff = 70)	70
RWKernel-1Hop	100 (Score cutoff = 80)	1	100 (Score cutoff = 50)	8
SPKernel-2Hop	94 (Score cutoff = 450)	4	100 (Score cutoff = 200)	13
RWKernel-2Hop	94 (Score cutoff = 110)	4	100 (Score cutoff = 80)	17

k-hop neighborhood based network alignment algorithm is further confirmed by results of aligning the human and mouse protein-protein interaction networks shown in table 2.

It is worth noting that the k-hop based network algorithms which use only BLASTp hits to match nodes across networks are competitive with variants of NetworkBLAST and HopeMap (including those that use other evidence for orthology: InParanoid orthology groups in the case of HopeMap, phylogeny in the case of NetworkBLAST, and KEGG orthologs in the case of NetworkBLAST-ko, and HopeMap-ko) or utilize GO annotations as part of their scoring functions (in case of HopeMap).

Bicomponent Cluster based Alignment. The results in table 3 show the performance of the bicomponent cluster-based alignment algorithm using "bicomponent clusterer", as implemented in JUNG [35] (Java Universal Network/Graph Framework). The clustering algorithm produced 1,236, 2,110, 579 and 1,893 clusters, with 5, 4, 2 and 2.5 proteins per cluster, respectively, on the yeast, fly, mouse and human datasets extracted from DIP. As can be seen from table 3, the performance of the bicomponent clustering based alignment is comparable to that of k-hop neighborhood based alignment algorithm (see 1) when the modified random walk graph kernel is used for comparing subgraphs. However, the performance of the bicomponent clustering based alignment using the modified shortest path graph kernel is substantially worse than that obtained using the modified random walk kernel. This is consistent with the observation that random walk graph kernel is more sensitive to differences between the graphs being compared than the shortest path kernel.

Table 3. Performance for the cluster-based alignment algorithm with the Shortest Path graph kernel (SPKernel) and the Random Walk graph kernel (RWKernel) using the bicomponent clustering algorithm on the fly and yeast DIP datasets

Method	% GO enrichment in Yeast	# GO enriched in Yeast	% GO enrichment in Fly	# GO enriched in Fly
SPKernel with Bicomponent Clusterer	67 (Score cutoff = 5)	2	50	1
RWKernel with Bicomponent Clusterer	100 (Score cutoff = 4)	2	100 (Score cutoff = 4)	1

Table 4. Performance for the cluster-based alignment algorithm with the Shortest
Path graph kernel (SPKernel) and the Random Walk graph kernel (RWKernel) using
the bicomponent clustering algorithm on the mouse and human DIP datasets

Method	% GO enrichment in Mouse	# GO enriched in Mouse	% GO enrichment in Human	# GO enriched in Human
SPKernel with Bicomponent Clusterer	70 (Score cutoff = 16)	4	33 (Score cutoff = 20)	1
RWKernel with Bicomponent Clusterer	96 (Score cutoff = 4)	6	83 (Score cutoff 15)	4

3.3 Reconstructing Phylogenetic Relationships from Network Alignments

The accuracy with which known phylogenetic relationships between species can
be recovered by a protein-protein interaction network alignment algorithm serves
as an additional measure of the quality of network alignments produced by the algorithm. The pairwise similarity scores associated with a pairwise network alignment can be used to construct an *inter-species similarity graph* where the nodes
denote the species and the weight on the links connecting pairs of nodes denote the
pairwise alignment scores output by a network alignment algorithm. The resulting inter-species similarity graph can be partitioned (or alternatively, the nodes
of the graph can be clustered) hierarchically to produce a phylogenetic tree.

We constructed the inter-species similarity graph using all possible pair-wise
alignments of protein-protein interaction networks from yeast, fly, human and
mouse obtained with k-hop neighborhood based alignment using the modified
random walk kernel. The resulting inter-species similarity network is shown in
figure 5 (left). We used spectral clustering algorithm [33], to recursively partition

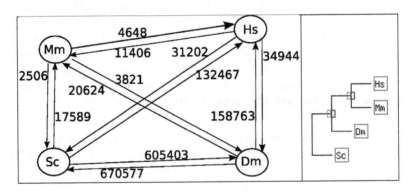

Fig. 5. (Left) Graph representation of the relationships between the mouse (Mm),
human (Hs), yeast (Sc) and fly (Dm) networks compared using the 1-hop RWKernel
algorithm. Higher scores indicate greater similarity between the two connected networks. **(Right)** The tree constructed using a hierarchical spectral clustering algorithm
on the graph shown to the left. The tree figure was created using proWeb Tree viewer
[43].

the inter-species similarity graph to obtain a hierarchical clustering of species which is shown in figure 5 (right). We observe that the tree shown in figure 5 is consistent with the generally accepted phylogenetic relationships among yeast, fly, human and mouse [20].

4   Summary and Discussion

Aligning biomolecular networks from different species, tissues and conditions allows offers a powerful approach to discover shared components that can help explain the observed phenotypes. Specifically, applications of network alignment allow the discovery of conserved pathways among different species [25,42], finding protein groups that are relevant to disease [22,31], discovery of the chemical mechanism of metabolic reactions [37,26] and more [48,27,39,5,1]. We have explored a novel class of graph kernel based polynomial time algorithms for aligning biomolecular networks. The proposed algorithms align large biomolecular networks by decomposing them into easy to compare substructures. The resulting subnetworks are compared using graph kernels as scoring functions. The modularity of kernels [11] offers the possibility of constructing composite kernel functions using existing kernel functions that capture different but complementary notions of similarity between graphs [7].

The runtime complexity of the k-hop neighborhood based alignment algorithm is $O(bmg)$ where m is the number of nodes in the query network G_1, b is the maximum number of matches in the target network G_2 for any node in the query network, and g is the running time of the similarity measure or scoring function used to compare a pair of k-hop subnetworks. The running time complexity of this algorithm is $O(lwg)$ where l is the number of clusters extracted from the query network G_1 , w is the number of clusters extracted from the target network G_2 , and g is the running time of the scoring function used to compare a pair of clusters (subgraphs). In comparison, the run-time complexity of NetworkBLAST-M $(O((np)^d s3^s))$, where n is the number of nodes in each of the networks, s the number of networks, p an upper bound on the node degree and d the number of seed spines used to generate the alignment. In the special case of pairwise network alignment $(s=2)$, the run-time complexity of Network-BLAST reduces to $O((np)^d)$. The runtime complexity of HopeMap is linear in terms of the total number of nodes and edges in the alignment graph [44], which is $O(2n+2n^2)$ in terms of the input graphs (where each input graph has at most n nodes).

The k-hop network neighborhood based and bicomponent clustering based protein-protein interaction network alignment algorithms are implemented in BiNA (http://www.cs.iastate.edu/~ftowfic), an open source Biomolecular Network Alignment toolkit. The current implementation includes variants of the shortest path and random walk graph kernels for computing similarity between pairs of subnetworks. The modular design of BiNA allows the incorporation of alternative strategies for decomposing networks into subnetworks and alternative similarity measures (e.g., kernel functions) for computing the similarity between subnetworks.

Our experiments with the fly, yeast, mouse and human protein-protein interaction networks extracted from DIP (Database of Interacting Proteins) [38] demonstrate that the performance of the proposed algorithms (as measured by % GO term enrichment of subnetworks identified by the alignment) is competitive with variants of the NetworkBLAST and HopeMap, which are among the state-of-the-art algorithms for pair-wise alignment of large protein-protein interaction networks [24,44].

Our results show that the inter-species similarity scores computed on the basis of pair-wise protein-protein interaction network alignments can be used to cluster the species into a species tree that is consistent with the known phylogenetic relationships among the species. Taken together with the results reported by Frost et al. [14] and Kuchaiev et al. [30] on reconstruction of phylogenetic relationships by comparing metabolic networks, we conjecture that (a) the accuracy with which a network alignment algorithm can be used to recover known phylogenetic relationships among species can be used as useful metric for evaluating the algorithm and (b) protein-protein interaction networks can be used as a useful source of information in reconstructing phylogenies. As this evaluation approach works at a global level (it only considers the total alignment score between two species, not the specific alignment scores for the subnetworks), new evaluation approaches would need to be considered to determine the feasibility of the specific alignments/matches generated by network alignment algorithms.

Some interesting directions for further work on the biomolecular network alignment algorithms include:

- Design of alternative measures of performance for assessing the quality of the generated network alignments.
- Algorithms for aligning networks that contain directed links, such as transcriptional regulatory networks, multiple types of nodes (proteins, DNA, RNA) and multiple types of links.
- Extensions that allow the alignment of multiple networks.
- The use of more sophisticated graph-clustering algorithms (such as MCL [12]).
- Automated tuning of parameters (e.g λ for the random walk kernel) using parameter learning techniques [13].
- Optimizations that reduce the runtime memory requirements of the algorithm.

Acknowledgments. This research was supported in part by an Integrative Graduate Education and Research Training (IGERT) fellowship to Fadi Towfic, funded by the National Science Foundation grant (DGE 0504304) to Iowa State University and a National Science Foundation Research Grant (IIS 0711356) to Vasant Honavar. The authors are grateful to the WABI-09 anonymous referees for their helpful comments on the manuscript.

References

1. Aittokallio, T., Schwikowski, B.: Graph-based methods for analysing networks in cell biology. Briefings in Bioinformatics 7(3), 243 (2006)
2. Altschul, S.F., Madden, T.L., Schäffer, A.A., Zhang, J., Zhang, Z., Miller, W., Lipman, D.J.: Gapped BLAST and PSI-BLAST: a new generation of protein database search programs. Nucleic Acids Research 25(17), 3390 (1997)
3. Ashburner, M., Ball, C.A., Blake, J.A., Botstein, D., Butler, H., Cherry, J.M., Davis, A.P., Dolinski, K., Dwight, S.S., Eppig, J.T., et al.: Gene ontology: Tool for the unification of biology. The Gene Ontology Consortium. Nature genetics 25(1), 25 (2000)
4. Bairoch, A., Apweiler, R., Wu, C.H., Barker, W.C., Boeckmann, B., Ferro, S., Gasteiger, E., Huang, H., Lopez, R., Magrane, M., et al.: The Universal Protein Resource (UniProt). Nucleic Acids Research 33, D154 (2005)
5. Barabasi, A.L., Oltvai, Z.N.: Network biology: understanding the cell's functional organization. Nature Reviews Genetics 5(2), 101–113 (2004)
6. Borgwardt, K.M., Kriegel, H.P.: Shortest-Path Kernels on Graphs. In: Proceedings of the Fifth IEEE International Conference on Data Mining, pp. 74–81 (2005)
7. Borgwardt, K.M., Kriegel, H.P., Vishwanathan, S.V.N., Schraudolph, N.N.: Graph Kernels For Disease Outcome Prediction From Protein-Protein Interaction Networks. In: Proceedings of the Pacific Symposium of Biocomputing (2007)
8. Boyle, E.I., Weng, S., Gollub, J., Jin, H., Botstein, D., Cherry, J.M., Sherlock, G.O.: GO: TermFinder–open source software for accessing Gene Ontology information and finding significantly enriched Gene Ontology terms associated with a list of genes. Bioinformatics (Oxford, England) 20(18), 3710 (2004)
9. Bruggeman, F.J., Westerhoff, H.V.: The nature of systems biology. Trends Microbiol. 15(1), 45–50 (2007)
10. Burrus, L.W., McMahon, A.P.: Biochemical analysis of murine Wnt proteins reveals both shared and distinct properties. Experimental cell research 220(2), 363–373 (1995)
11. Cristianini, N., Shawe-Taylor, J.: An introduction to support vector machines. Cambridge University Press, Cambridge (2000)
12. Enright, A.J., Van Dongen, S., Ouzounis, C.A.: An efficient algorithm for large-scale detection of protein families. Nucleic Acids Research 30(7), 1575 (2002)
13. Flannick, J., Novak, A., Do, C.B., Srinivasan, B.S., Batzoglou, S.: Automatic parameter learning for multiple network alignment. In: Vingron, M., Wong, L. (eds.) RECOMB 2008. LNCS (LNBI), vol. 4955, pp. 214–231. Springer, Heidelberg (2008)
14. Forst, C.V., Flamm, C., Hofacker, I.L., Stadler, P.F.: Algebraic comparison of metabolic networks, phylogenetic inference, and metabolic innovation. BMC Bioinformatics 7(1), 67 (2006)
15. Garey, M.R., Johnson, D.S.: Computers and Intractability: A Guide to the Theory of NP-Completeness. WH Freeman & Co., New York (1979)
16. Ge, H., Walhout, A.J.M., Vidal, M.: Integrating 'omic' information: a bridge between genomics and systems biology. Trends in Genetics 19(10), 551–560 (2003)
17. Han, J.D., Bertin, N., Hao, T., Goldberg, D.S., Berriz, G.F., Zhang, L.V., Dupuy, D., Walhout, A.J., Cusick, M.E., Roth, F.P., Vidal, M.: Evidence for dynamically organized modularity in the yeast protein-protein interaction network. Nature 430(6995), 88–93 (2004)
18. Harary, F.: Graph theory (1969)

19. Hartwell, L.H., Hopfield, J.J., Leibler, S., Murray, A.W.: From molecular to modular cell biology. Nature 402(6761 suppl.), C47–C52 (1999)
20. Hedges, S.B.: The origin and evolution of model organisms. Nature Reviews Genetics 3(11), 838–849 (2002)
21. Hirsh, A.E., Fraser, H.B.: Protein dispensability and rate of evolution. Nature 411(6841), 1046–1049 (2001)
22. Ideker, T., Sharan, R.: Protein networks in disease. Genome Research 18(4), 644 (2008)
23. Kalaev, M., Bafna, V., Sharan, R.: Fast and accurate alignment of multiple protein networks. In: Vingron, M., Wong, L. (eds.) RECOMB 2008. LNCS (LNBI), vol. 4955, pp. 246–256. Springer, Heidelberg (2008)
24. Kalaev, M., Smoot, M., Ideker, T., Sharan, R.: NetworkBLAST: comparative analysis of protein networks. Bioinformatics 24(4), 594 (2008)
25. Kelley, B.P., Sharan, R., Karp, R., Sittler, E.T., Root, D.E., Stockwell, B.R., Ideker, T.: Conserved pathways within bacteria and yeast as revealed by global protein network alignment. Proc. Natl. Acad. Sci. 100, 11394–11399 (2003)
26. Kharchenko, P., Church, G.M., Vitkup, D.: Expression dynamics of a cellular metabolic network. Molecular Systems Biology 1 (2005)
27. Kirac, M., Ozsoyoglu, G.: Protein Function Prediction Based on Patterns in Biological Networks. In: Vingron, M., Wong, L. (eds.) RECOMB 2008. LNCS (LNBI), vol. 4955, pp. 197–213. Springer, Heidelberg (2008)
28. Koonin, E.: Orthologs, paralogs and evolutionary genomics. Annu. Rev. Genet. 39, 309–338 (2005)
29. Koyuturk, M., Kim, Y., Topkara, U., Subramaniam, S., Szpankowski, W., Grama, A.: Pairwise alignment of protein interaction networks. Journal of Computational Biology 13(2), 182–199 (2006)
30. Kuchaiev, O., Milenkovic, T., Memisevic, V., Hayes, W., Przulj, N.: Topological network alignment uncovers biological function and phylogeny. Arxiv, 0810.3280v2 (2009)
31. Lim, J., Hao, T., Shaw, C., Patel, A.J., Szabó, G., Rual, J.F., Fisk, C.J., Li, N., Smolyar, A., Hill, D.E., et al.: A Protein–Protein Interaction Network for Human Inherited Ataxias and Disorders of Purkinje Cell Degeneration. Cell 125(4), 801–814 (2006)
32. Manber, U.: Introduction to algorithms: a creative approach. Addison-Wesley Longman Publishing Co., Inc., Boston (1989)
33. Ng, A., Jordan, M., Weiss, Y.: On spectral clustering: Analysis and an algorithm. In: Advances in Neural Information Processing Systems 14: Proceedings of the 2002 [sic] Conference, p. 849. MIT Press, Cambridge (2002)
34. O'Brien, K.P., Remm, M., Sonnhammer, E.L.L.: Inparanoid: a comprehensive database of eukaryotic orthologs. Nucleic Acids Research 33(Database issue), D476 (2005)
35. O'Madadhain, J., Fisher, D., White, S., Boey, Y.: The JUNG (Java Universal Network/Graph) Framework. University of California, California (2003)
36. Ravasz, E., Somera, A.L., Mongru, D.A., Oltvai, Z.N., Barabasi, A.L.: Hierarchical organization of modularity in metabolic networks. Science 297(5586), 1551–1555 (2002)
37. Ross, J., Schreiber, I., Vlad, M.O.: Determination of Complex Reaction Mechanisms: Analysis of Chemical, Biological, and Genetic Networks. Oxford University Press, USA (2006)

38. Salwinski, L., Miller, C.S., Smith, A.J., Pettit, F.K., Bowie, J.U., Eisenberg, D.: The database of interacting proteins: 2004 update. Nucleic Acids Research 32(Database issue), D449 (2004)
39. Scott, J., Ideker, T., Karp, R.M., Sharan, R.: Efficient Algorithms for Detecting Signaling Pathways in Protein Interaction Networks. Journal of Computational Biology 13(2), 133–144 (2006)
40. Sharan, R., Ideker, T.: Modeling cellular machinery through biological network comparison. Nature Biotechnology 24, 427–433 (2006)
41. Steinbeck, C., Hoppe, C., Kuhn, S., Floris, M., Guha, R., Willighagen, E.L.: Recent Developments of the Chemistry Development Kit (CDK)-An Open-Source Java Library for Chemo-and Bioinformatics. Current Pharmaceutical Design 12(17), 2111–2120 (2006)
42. Stuart, J.M., Segal, E., Koller, D., Kim, S.K.: A Gene-Coexpression Network for Global Discovery of Conserved Genetic Modules. Science 302(5643), 249–255 (2003)
43. Taylor, N.: proWeb Tree Viewer, http://www.proweb.org/treeviewer/
44. Tian, W., Samatova, N.F.: Pairwise alignment of interaction networks by fast identification of maximal conserved patterns. In: Proc. of the Pacific Symposium on Biocomputing (2009)
45. Vishwanathan, S.V.N., Borgwardt, K.M., Schraudolph, N.N.: Fast Computation of Graph Kernels. Technical report, NICTA (2006)
46. White, S., Smyth, P.: Algorithms for estimating relative importance in networks. In: Proceedings of the ninth ACM SIGKDD international conference on Knowledge discovery and data mining, pp. 266–275. ACM, New York (2003)
47. Wong, S.L., Zhang, L.V., Tong, A.H.Y., Li, Z., Goldberg, D.S., King, O.D., Lesage, G., Vidal, M., Andrews, B., Bussey, H., et al.: Combining biological networks to predict genetic interactions. Proceedings of the National Academy of Sciences 101(44), 15682–15687 (2004)
48. Zhou, X., Kao, M.C.J., Wong, W.H.: Transitive functional annotation by shortest-path analysis of gene expression data. Proceedings of the National Academy of Sciences 99(20), 12783–12788 (2002)

MADMX: A Novel Strategy for Maximal Dense Motif Extraction

Roberto Grossi[1], Andrea Pietracaprina[2,*], Nadia Pisanti[1], Geppino Pucci[2,*],
Eli Upfal[3,*,**], and Fabio Vandin[2,*,* * *]

[1] Dipartimento di Informatica, Università di Pisa, Pisa, Italy
{grossi,pisanti}@di.unipi.it
[2] Dipartimento di Ingegneria dell'Informazione, Università di Padova, Padova, Italy
{capri,geppo,vandinfa}@dei.unipd.it
[3] Department of Computer Science, Brown University, Providence RI, USA
eli@cs.brown.edu

Abstract. We develop, analyze and experiment with a new tool, called
MADMX, which extracts frequent motifs, possibly including don't care
characters, from biological sequences. We introduce *density*, a simple
and flexible measure for bounding the number of don't cares in a motif,
defined as the ratio of solid (i.e., different from don't care) characters to
the total length of the motif. By extracting only *maximal dense motifs*,
MADMX reduces the output size and improves performance, while enhanc-
ing the quality of the discoveries. The efficiency of our approach relies
on a newly defined combining operation, dubbed *fusion*, which allows
for the construction of maximal dense motifs in a bottom-up fashion,
while avoiding the generation of nonmaximal ones. We provide experimental evidence of the efficiency and the quality of the motifs returned
by MADMX.

1 Introduction

The discovery of frequent patterns (*motifs*) in biological sequences has attracted
wide interest in recent years, due to the understanding that sequence similarity is
often a necessary condition for functional correlation. Among other applications,
motif discovery proves an important tool for identifying regulatory regions and
binding sites in the study of functional genomics. From a computational point
of view, a major complication for the discovery of motifs is that they may feature some sequence variation without loss of function. The discovery process
must therefore target *approximate motifs*, whose occurrences are similar but not
necessarily identical. Approximate motifs are often modeled through the use of

---

* Support for these authors was provided, in part, by the European Union under
the FP6-IST/IP Project AEOLUS.
** Supported in part by NSF awards IIS-0325838 and DMI-0600384, and ONR Award
N000140610607.
* * * Contact Author. This work was done, in part, while the author was visiting the
Department of Computer Science of Brown University.

S.L. Salzberg and T. Warnow (Eds.): WABI 2009, LNBI 5724, pp. 362–374, 2009.

the *don't care* character in certain positions, which is a wild card matching all characters of the alphabet, called *solid characters* [10].

Finding interesting approximate motifs is computationally challenging. As the number of don't cares increases and/or the minimum frequency threshold decreases, the output may explode combinatorially, even if the discovery targets only maximal motifs—a subset of the motifs which implicitly represents the complete set. Moreover, even when the final output is not too large, partial data during the inference of target motifs might lead to memory saturation or to extensive computation during the intermediate steps.

A large body of literature in the last decade has dealt with efficient motif discovery [9,3,12,4,16,8,6,5,2], and an excellent survey of known results can be found in the book [10]. In order to alleviate the computational burden of motif extraction and to limit the output to the most promising or interesting discoveries, some works combine the traditional use of a frequency threshold with restrictions on the flexibility of the extracted motifs, often captured by limitations on the number of occurring don't cares.

In a recent work, Apostolico et al. [2] study the extraction of *extensible motifs*, comprising standard don't cares and extensible wild cards. The latter are spacers of variable length that can take different size (within pre-specified limits) in each occurrence of the motif. An efficient tool, called VARUN, is devised in [2] for extracting all maximal extensible motifs (according to a suitable notion of maximality defined in the paper) which occur with frequency above a given threshold σ and with upper limits D on the length of the spacers. VARUN returns the extracted motifs sorted by decreasing z-score, a widely adopted statistical measure of interestingness. The authors demonstrate the effectiveness of their approach both theoretically, by proving that each maximal motif features the highest z-score within the class of motifs it represents, and experimentally, by showing that the returned top-scored motifs comprise biologically relevant ones when run on protein families and DNA sequences.

A slightly more general way of limiting the number of don't cares in a motif has been explored in [13]. The authors define $\langle L, W \rangle$ motifs, for $L \leq W$, where at least L solid characters must occur in each substring of length W of the motif. They propose a strategy for extracting $\langle L, W \rangle$ motifs which are also maximal, although their notion of maximality is not internal to the class of $\langle L, W \rangle$ motifs. As a consequence, the algorithm is not complete, since it disregards all those $\langle L, W \rangle$ motifs that are subsumed by a maximal non-$\langle L, W \rangle$ one.

Our results. Our work focuses on the discovery of *rigid motifs*, which contain blocks of solid characters (solid blocks) separated by one or more don't cares. We propose a more general approach for controlling the number of don't cares in rigid motifs. Specifically, we introduce the notion of *dense motif*, a frequent pattern where the fraction of solid characters is above a given threshold. Our density notion is more flexible and general than the one considered in [10,2], since it allows for arbitrarily long runs of don't cares as long as the fraction of solid characters in the pattern is above the threshold. We define a natural notion of *maximality* for dense patterns and devise an efficient algorithm, called

MADMX (pronounced *Mad Max*), which performs complete MAximal Dense Motif eXtraction from an input sequence, with respect to user-specified frequency and density thresholds.

The key technical result at the core of our extraction strategy is a closure property which affords the complete generation of all maximal dense motifs in a breadth-first fashion, through an *apriori*-like strategy [1], starting from a relatively small set of solid blocks, and then repeatedly applying a suitable combining operator, called *fusion*, to pairs of previously generated motifs. In this fashion, our strategy avoids the generation and consequent storage of intermediate patterns which are not in the output set, which ensures time and space complexities polynomial in the combined size of the input and the output.

We performed a number of experiments on MADMX to assess the biological significance of maximal dense motifs and to compare MADMX against its most recent and close competitor VARUN. For the first objective, we used MADMX to extract maximal dense motifs from a number of human DNA fragments. We compared the output set against those in RepBase [7], the largest repository of repetitive patterns for eukaryotic species, using REPEATMASKER [15], a popular tool for masking repetitive DNA. The experiments show that all of our returned motifs are occurrences of patterns in RepBase, and *fully* characterize the family of SINE/ALU repeats (and partially the LINE/L1 family). This provides evidence that the notion of density, when applied to rigid motifs, captures biological significance.

Next we compared the z-score performance of MADMX and VARUN. We ran both algorithms on several families of DNA fragments, limiting VARUN to the generation of rigid motifs and setting the parameters so as to obtain comparable output sizes, with motifs listed by decreasing z-score. The experiments show that the top-m highest-ranking motifs returned by MADMX almost always feature higher z-scores than the corresponding top-m ones returned by VARUN, even for large values of m, with only a modest increase in running time, which may be partly due to the fact that coding of MADMX is yet to be optimized. In fairness, we must remark that VARUN deals also with extensible motifs while MADMX only targets rigid motifs.

The paper is organized as follows. In Section 2 several technical definitions and properties of motifs with don't cares are given. Section 3 proves the closure property at the base of MADMX and provides a high-level description of the algorithm. In Section 4, the experimental validation of MADMX is presented.

2   Preliminary Definitions and Properties

Let Σ be an alphabet of m characters and let $s = s[0]s[1]\ldots s[n-1]$ be a string of length n over Σ. We use $s[i\ldots j]$ to denote the substring $s[i]\,s[i+1]\cdots s[j]$ of s, for $i \leq j$. Characters in Σ are also called *solid characters*. We use $\circ \notin \Sigma$ to denote a distinguished character called *wild card* or *don't care* character. Let ϵ denote the empty string. A *pattern* x is a string in $\{\epsilon\} \cup \Sigma \cup \Sigma(\Sigma \cup \{\circ\})^*\Sigma$.

However, whenever necessary, we will assume that patterns are implicitly padded to their left and right with arbitrary sequences of don't care characters.

Given two patterns x, y we say that y is *more specific* than x, and write $x \preceq y$, iff for every $i \geq 0$ either $x[i] = y[i]$ or $x[i] = \circ$. Given two patterns x, y we say that x *occurs in* y *at position* ℓ iff $x \preceq y[\ell \ldots \ell + |x| - 1]$: we also say that y *contains* x. For a string s, the *location list* $\mathcal{L}_x$ of a pattern x in s is the complete set of positions at which x occurs in s. We refer to $f(x) = |\mathcal{L}_x|$ as the *frequency* of pattern x in s. (Note that $f(\epsilon) = n$.) As in [16], the *translated representation* of the location list $\mathcal{L}_x = \{l_0, l_1, l_2, \ldots, l_k\}$ is $\tau(\mathcal{L}_x) = \{l_1 - l_0, l_2 - l_0, \ldots, l_k - l_0\}$. Given two patterns x, y, we say that y *subsumes* x in s if $f(x) = f(y)$ and y contains x. As a consequence, if y subsumes x then $\tau(\mathcal{L}_x) = \tau(\mathcal{L}_y)$. A pattern x is *maximal* if it is not subsumed by any other pattern y. (We observe that this notion of maximality coincides with that of [12].) Given a pattern x, its *maximal extension* $\mathcal{M}(x)$ is the maximal pattern that subsumes x, which can be shown to be unique [12].

In what follows, we call *solid block* a string in Σ^+ and a *don't care block* a string in $\circ^+$. Furthermore, given a pattern x, $dc(x)$ denotes the number of don't care characters contained in x.

Definition 1. *The* density $\delta(x)$ *of* x *is:* $\delta(x) = 1 - dc(x)/|x|$. *Given a (density) threshold* ρ, $0 < \rho \leq 1$, *we say that a pattern* x *is* dense *if* $\delta(x) \geq \rho$.

Note that a solid block is a dense pattern with respect to every threshold ρ.

It is reasonable to concentrate the attention on dense patterns that are not subsumed by any other dense pattern, since they are the most interesting dense representatives in the equivalence classes induced by "sharing" the same translated representation; these representatives are defined below.

Definition 2. *A dense pattern* x *is a* maximal dense pattern *in* s *if it is not subsumed by any other dense pattern* $x' \neq x$.

Observe that a maximal dense pattern x needs not be a maximal pattern in the general sense, since $\mathcal{M}(x)$ might be a nondense pattern. However, every dense pattern x is subsumed by *at least* one maximal dense pattern. In fact, all of the maximal dense patterns that subsume x are dense substrings of $\mathcal{M}(x)$, namely, those that contain x and are not substrings of any other dense substring of $\mathcal{M}(x)$. We want to stress that there might be several maximal dense patterns that subsume x. As an example, for $\rho = 2/3$, the dense pattern $x = \mathtt{B}$ in the string $S = \mathtt{AdBeCfAgBhC}$ is subsumed by maximal dense patterns $\mathtt{A} \circ \mathtt{B}$ and $\mathtt{B} \circ \mathtt{C}$, while $\mathcal{M}(x) = \mathtt{A} \circ \mathtt{B} \circ \mathtt{C}$ is not dense.

Definition 3. *Given a frequency threshold* σ *and a density threshold* ρ, *a pattern* x *is a* dense maximal motif *in* s *if* x *is a maximal dense pattern in* s *with respect to* ρ, *and* $f(x) \geq \sigma$. *A dense maximal motif for* $\rho = 1$ *is also referred to as* maximal solid block.

Problem of interest. We are given an input string s, a frequency threshold σ, and a density threshold ρ. Find all the maximal dense motifs in s.

In the rest of the paper, we will omit referencing the input string s when clear from the context. An important property of maximal dense patterns, which we will exploit in our mining strategy, is that all of their solid blocks are maximal solid blocks. This property is stated in the following proposition whose proof, omitted for brevity, extends a similar result holding for arbitrary maximal patterns [16,11].

Proposition 1. *Let x be a maximal dense pattern with respect to a density threshold ρ, and let $b = x[i \ldots j]$ be a solid block in x such that $x[i-1] = x[j+1] = \circ$ and $j \geq i$. Then, b is a maximal solid block.*

3   An Algorithm for MAximal Dense Motif eXtraction

In this section we describe our algorithm, called MADMX (pronounced *Mad Max*), for MAximal Dense Motif eXtraction. The algorithm adopts a breadth-first *apriori*-like strategy [1], similar in spirit to the one developed in [2], using maximal solid blocks as building blocks by Proposition 1. MADMX operates by repeatedly combining together, in a suitable fashion, pairs of maximal dense motifs, and extracting from the combinations less frequent maximal dense motifs.

A key notion for the algorithm, underlying the aforementioned combining operations, is the *fusion* of characters/patterns.

Definition 4. *Given three characters $c, c_1, c_2 \in \Sigma \cup \{\circ\}$, we say that c is the fusion of c_1 and c_2, and write $c = c_1 \bigtriangledown c_2$, if one of the following holds:*

1. *$c = c_1 = c_2$;*
2. *$c_1 = \circ$, $c = c_2 \neq \circ$;*
3. *$c = c_1 \neq \circ$, $c_2 = \circ$.*

The above notion of fusion generalizes to patterns as follows.

Definition 5. *Given three patterns x, y, z and an integer d, we say that z is the d-fusion of x and y, and write $z = x \bigtriangledown_d y$, if z can be obtained by removing the leading and trailing don't care characters from the pattern m defined as $m[i] = x[i+d] \bigtriangledown y[i]$, for all indices i.*

The breadth-first strategy adopted by our algorithm crucially relies on the following theorem, which highlights the structure of dense motifs:

Theorem 1. *Let x be a maximal dense motif with $dc(x) > 0$. Then:*

(a) there exists a maximal solid block b in x such that $\mathcal{M}(x) = \mathcal{M}(b)$, or
(b) there exist two maximal dense motifs y_1, y_2 such that:
   - *$\mathcal{M}(x) = \mathcal{M}(y_1 \bigtriangledown_d y_2)$, for some d;*
   - *there are two maximal solid blocks b_1, b_2 in x and an integer $\hat{d} > 0$ such that b_1 is a maximal solid block in y_1, b_2 is a maximal solid block in y_2, and $b_1 \circ^{\hat{d}} b_2$ is contained in $y_1 \bigtriangledown_d y_2$;*
   - *$f(x) < \min\{f(y_1), f(y_2)\}$;*

For the proof of Theorem 1 we need to define another type of pattern combination, namely the operation of *merge* between two patterns, which is similar to the one introduced in [12]. Given two characters c_1, c_2, we define the operator $\oplus$ between them such that $c_1 \oplus c_2 = \circ$, if $c_1 \neq c_2$, and $c_1 \oplus c_2 = c_1 = c_2$, otherwise.

Definition 6. *Given two patterns x, y and an integer d, the d-merge of x and y is the pattern $z = x \oplus_d y$ which can be obtained by removing all leading and trailing don't cares from the pattern m defined as $m[i] = x[i + d] \oplus y[i]$ for all i.*

We want to stress the difference between the notions of merging and fusion: the merge of two patterns x, y is always well defined and more general than x, y, while the fusion of x, y may not exist and, if it does, is more specific than x, y.

For the proof of Theorem 1 we also need the property established by the following lemma.

Lemma 1. *Let x and y be maximal patterns, and d be an integer such that $z = x \oplus_d y \neq \epsilon$. Then z is a maximal pattern. Moreover, if $z \neq x$ (resp., $z \neq y$) then $f(z) > f(x)$ (resp., $f(z) > f(y)$).*

Proof. First we prove that z is maximal. By contradiction, suppose that this is not the case. Then, there exists a position i such that $z[i] = \circ$ and we can replace the $\circ$ with a solid character c without decreasing the frequency of the pattern. (Note that the position of the substitution can be to the left of the first character in z or to the right of the last character in z.) Since x and y are more specific than z, to every occurrence of x and y in the string corresponds an occurrence of z. Hence, every occurrence of x (resp., y) in the string, contains c in its $i + d$th (resp., ith) position. Therefore, by maximality of x and y, it must be $z[i] = x[i + d] = y[i] = c$, which is a contradiction. The relations between the frequencies of x, y and z follow trivially by their maximality. $\square$

We are now ready to prove the theorem.

Proof (Theorem 1). Given a pattern x and two nonnegative integers $i \leq j$, we let $x^*[i \ldots j]$ denote the pattern obtained by removing all the leading and trailing don't care characters from $x[i \ldots j]$. Since x is a maximal dense pattern and $dc(x) > 0$, it is easy to see that there exist two dense patterns x_1, x_2 and an integer $d > 0$ such that $x = x_1 \circ^d x_2$, hence there exists an index $s_1 > 0$ such that $x^*[0 \ldots s_1 - 1]$ and $x^*[s_1 + 1 \ldots |x| - 1]$ are dense. We call these two patterns the *level-1 decomposition* of x (observe that many such decompositions may exist). Also, we let $\ell_1 = 0$ and $r_1 = |x| - 1$. Now, consider the following iterative process:

1. If in the level-i decomposition of x both $x^*[\ell_i \ldots s_i - 1]$ and $x^*[s_i + 1 \ldots r_i]$ have frequency strictly greater than $f(x)$, *or* at least one of $x^*[\ell_i \ldots s_i - 1]$ and $x^*[s_i + 1 \ldots r_i]$ is a solid block with frequency equal to $f(x)$, then terminate;
2. Otherwise, let $y = x^*[\ell_{i+1} \ldots r_{i+1}]$ be (an arbitrary) one of $x^*[\ell_i \ldots s_i - 1]$ or $x^*[s_i + 1 \ldots r_i]$ which is not a solid block and has frequency equal to $f(x)$. Since y is dense, there exists an index s_{i+1}, $\ell_{i+1} < s_{i+1} < r_{i+1}$ such that $x^*[\ell_{i+1} \ldots s_{i+1} - 1]$ and $x^*[s_{i+1} + 1 \ldots r_{i+1}]$ are both dense. Call these two patterns the level-$(i+1)$ decomposition of x. Set $i = i + 1$ and go to Step 1.

Assume that the decomposition process ends by finding a solid block b that is a solid block in x and has $f(b) = f(x)$. Then, $\mathcal{M}(b) = \mathcal{M}(x)$ and the theorem follows. Otherwise, at the last level j of the decomposition, we have that $f(x) < \min \{f(x^*[\ell_j \ldots s_j - 1]), f(x^*[s_j + 1 \ldots r_j])\}$. In this latter case, as explained in Section 2 (after Definition 2), we can determine two maximal dense patterns y_1, y_2 such that y_1 contains $x^*[\ell_j \ldots s_j - 1]$, y_2 contains $x^*[s_j + 1 \ldots r_j]$, and with $\mathcal{M}(y_1) = \mathcal{M}(x^*[\ell_j \ldots s_j - 1])$ and $\mathcal{M}(y_2) = \mathcal{M}(x^*[s_j + 1 \ldots r_j])$. Since $f(y_1) = f(x^*[\ell_j \ldots s_j - 1])$ and $f(y_2) = f(x^*[s_j + 1 \ldots r_j])$, we have that $f(x) < \min \{f(y_1), f(y_2)\}$. Observe that by construction there must exist two solid blocks b_1, b_2 in x and an integer $\hat{d}$ such that b_1 is a solid block in y_1, b_2 is a solid block in y_2, and $b_1 \circ^{\hat{d}} b_2$ is a sequence of two solid blocks in x. In fact, b_1 (resp., b_2) is the last (resp., the first) solid block of $x^*[\ell_j \ldots s_j - 1]$ (resp., $x^*[s_j + 1 \ldots r_j]$).

Next, we show that there exists a d such that the d-fusion $y_1 \triangledown_d y_2$ is well defined, contains $b_1 \circ^d b_2$, and $\mathcal{M}(y_1 \triangledown_d y_2) = \mathcal{M}(x)$. We proceed as follows. Let us "align" $\mathcal{M}(x)$ and y_1 so to match the occurrences of b_1 in both patterns. Then, for a certain integer p, $\mathcal{M}(x)[i + p]$ corresponds to $y_1[i]$. Assume, for the sake of contradiction, that there exists an index j such that $\mathcal{M}(x)[j + p]$ is not more specific than $y_1[j]$. Then, Lemma 1 implies that $z = \mathcal{M}(x) \oplus_p \mathcal{M}(y_1) \neq \mathcal{M}(y_1)$, which contains $x^*[\ell_j \ldots s_j - 1]$, is maximal and has frequency strictly greater than $f(y_1)$, which is impossible because we have chosen y_1 such that $\mathcal{M}(x^*[\ell_j \ldots s_j - 1]) = \mathcal{M}(y_1)$ and therefore $f(x^*[\ell_j \ldots s_j - 1]) = f(y_1)$. Therefore, $\mathcal{M}(x)$ contains y_1. A similar argument shows that $\mathcal{M}(x)$ contains y_2.

Since y_1 and y_2 are contained in $\mathcal{M}(x)$, there must exist a d such that $y_1 \triangledown_d y_2$ is well defined and can be aligned with $\mathcal{M}(x)$ in such a way to match the blocks b_1 and b_2 of y_1 and y_2 with the corresponding blocks in $\mathcal{M}(x)$. Moreover, $\mathcal{M}(x)$ contains $y_1 \triangledown_d y_2$, hence $f(y_1 \triangledown_d y_2) \geq f(\mathcal{M}(x)) = f(x)$. However, since $y_1 \triangledown_d y_2$ contains both $x^*[\ell_j \ldots s_j - 1]$ and $x^*[s_j + 1 \ldots r_j]$, it contains also $x^*[\ell_j \ldots r_j]$, which, by the decomposition process, has frequency equal to $f(x)$. Therefore, $f(y_1 \triangledown_d y_2) \leq f(x)$, and the theorem follows since $f(y_1 \triangledown_d y_2) = f(x)$. □

In essence, Theorem 1 guarantees that we can find any maximal dense motif x either within $\mathcal{M}(b)$, for some maximal solid block b, or by d-fusing two higher-frequency maximal dense motifs y_1, y_2, for some d, finding $z = \mathcal{M}(y_1 \triangledown_d y_2)$ and then possibly "trimming" z on both sides to obtain x.

Algorithm MADMX, whose pseudocode is reported in Figure 1, implements the strategy inspired by Theorem 1. It employs three (initially empty) sets *previous*, *current*, and *next*. In Line 2, the algorithm first stores the maximal solid blocks b in s for the given frequency in the set *blocks* (see Section 2). Then, it extracts all of the appropriate maximal dense motifs from $\mathcal{M}(b)$ in Lines 3–6, using the function extractMaximalDense, as implied by Theorem 1(a). Finally, Lines 7–16 implement the strategy as implied by Theorem 1(b). (In Line 10 a d-fusion $y_1 \triangledown_d y_2$ is considered *valid* if it satisfies the second property of Theorem 1(b).)

Algorithm MADMX()

Input: String s, frequency threshold σ, density threshold ρ
Output: Maximal dense motifs

1 $previous \leftarrow \emptyset$, $current \leftarrow \emptyset$, $next \leftarrow \emptyset$;
2 $blocks \leftarrow$ maximal solid blocks of s with frequency $\geq \sigma$;
3 **for each** $b \in blocks$ **do**
4   find $\mathcal{M}(b)$;
5      $\mathcal{DM} \leftarrow$ extractMaximalDense($\mathcal{M}(b)$);
6   **for each** $x \in \mathcal{DM}$ **do** $current \leftarrow current \cup \{x\}$;
7 **while** $current \neq \emptyset$ **do**
8   **for each** $x_1 \in current$ **do**
9   **for each** $x_2 \in previous \cup current$ **do**
10   **for each** d s.t. $z = x_1 \bigtriangledown_d x_2$ is a valid fusion **do**
11   find $\mathcal{M}(z)$;
12           $\mathcal{DM} \leftarrow$ extractMaximalDense($\mathcal{M}(z)$);
13   **for each** $x \in \mathcal{DM}$ **do**
14   **if** $f(x) \geq \sigma$ and $x \notin previous \cup current$ **then** $next \leftarrow next \cup \{x\}$;
15     $previous \leftarrow previous \cup current$;
16     $current \leftarrow next$; $next \leftarrow \emptyset$;
17 **return** $previous$;

Fig. 1. Pseudocode of algorithm MADMX

An important issue for the efficiency of MADMX is that it needs to compute the exact frequency of each generated pattern. For what concerns the fusion operation of two patterns x_1, x_2 in Line 10, observe that a simple computation on the pairs $(\ell_1, \ell_2) \in \mathcal{L}_{x_1} \times \mathcal{L}_{x_2}$ is sufficient to yield the frequencies of all the valid fusions of two patterns. However, given $z = x_1 \bigtriangledown_d x_2$, for a maximal dense pattern w which does not contain z in its entirety, we can only conclude that $f(w) \geq f(z)$. We then label the motifs for which the exact frequencies are known as *final*, and those for which only a lower bound to their frequencies is known as *tentative*, and update the lower bounds and the labels during the execution of the algorithm. Whenever the set *current* contains no final motifs, we can label as final the tentative motif in *current* with the highest lower bound to its frequency, and continue with the generation. The proof of the correctness of this assumption and further details on the implementation of the algorithm will be provided in the full version of this extended abstract. A crude upper bound on the running time of MADMX can be derived by observing that, for each pair of dense maximal motifs in output, the time spent during all the operations concerning that pair is (naively) $O\left(n^3\right)$, where n is the length of the input string. If P patterns are produced in output, the overall time complexity is $O\left(n^3 P^2\right)$.

4   Experimental Validation of MADMX

We developed a first, non-optimized, implementation of MADMX in C++ also including an additional feature which eliminates, from the set of initial maximal solid blocks, those shorter than a given threshold min_ℓ. The purpose of this latter heuristics is to speed up motif generation driving it towards the discovery of (possibly) more significant motifs, with the exclusion of spurious, low-complexity ones. (The code is available for download at http://www.dei.unipd.it/wdyn/?IDsezione=4534.)

We performed two classes of experiments to evaluate how significant is the set of motifs found using our approach. The first class of experiments, described in Section 4.1, compares our motifs with the known biological repetitions available in RepBase [7], a very popular genomic database. The second class of experiments, described in Section 4.2, aims at comparing the motifs extracted by MADMX with those extracted by VARUN using the same z-score metric employed in [2] for assessing their relative statistical significance.

4.1   Evaluating Significance by Known Biological Repetitions

RepBase [7] is one of the largest repositories of prototypic sequences representing repetitive DNA from different eukaryotic species, collected in several different ways. RepBase is used as a reference collection for masking and annotation of repetitive DNA through popular tools such as REPEATMASKER [15]. REPEAT-MASKER screens an input DNA sequence s for simple repeats and low complexity portions, and interspersed repeats using RepBase. Sequence comparisons are performed through Smith-Waterman scoring. REPEATMASKER returns a detailed annotation of the repeats occurring in s, and a modified version of s in which all of the annotated repeats are masked by a special symbol (N or X). With the current version of RepBase, on average, almost 50% of a human genomic DNA sequence will be masked by the program [15].

Most of the interspersed repeats found by REPEATMASKER belong to the families called SINE/ALU and LINE/L1: the former are *Short INterspersed Elements* that are repetitive in the DNA of eukaryotic genomes (the Alu family in the human genome); the latter are *Long Interspersed Nucleotide Elements*, which are typically highly repeated sequences of 6K–8K bps, containing RNA polymerase II promoters. The LINE/L1 family forms about 15% of the human genome.

We have conducted an experimental study using MADMX and REPEATMASKER on *Human Glutamate Metabotropic Receptors* HGMR 1 (410277 bps) and HGMR 5 (91243 bps) as input sequences. We have downloaded the sequences from the March 2006 release of the UCSC Genome database (http://genome.ucsc.edu). REPEATMASKER version was open-3.2.7, sensitive mode, with the query species assumed to be homologous; it ran using blastp version 2.0a19MP-WashU, and RepBase update 20090120.

The experiments to assess the biological significance of the maximal dense motifs extracted by MADMX involved three separate stages. In the first stage, we ran REPEATMASKER on the input sequences HGMR 1 and HGMR 5, searching for interspersed repeats using RepBase. One of the output files (.out) of REPEAT-MASKER contains the list of found repeats, and provides, for each occurrence, the substring $s[i \ldots j]$ of the input sequence s which is locally aligned with (a substring of) the repeat.

In the second stage, we ran MADMX on the same DNA sequences, with density threshold $\rho = 0.8$, frequency threshold $\sigma = 4$, and $min_\ell = 15$. In order to filter out simple repeats and low complexity portions, which are dealt with by REPEATMASKER without resorting to RepBase, we modified MADMX eliminating periodic maximal solid blocks (with short periods), which are the seeds of simple repeats. Then, we identified the occurrences of the motifs returned by MADMX in the input sequences, using REPEATMASKER as a pattern matching tool (i.e., replacing RepBase with the set of motifs returned by MADMX as the database of known repeats). The underlying idea behind this use of REPEATMASKER was to employ the same local alignment algorithms, so to make the comparison fairer.

In the third stage, we cross-checked the intervals associated with the occurrences of the RepBase repeats against those associated with the occurrences of our motifs. Surprisingly, MADMX was able to identify and characterize *all* of the intervals of the known SINE/ALU repeats in HGMR 1 and HGMR 5 (respectively, 56 repeats plus an extra unclassified for HGMR 1, and 20 plus an extra unclassified for HGMR 5). The remaining occurrences of the motifs permitted to identify 29 repeats out of 78 of the LINE/L1 family in HGMR 1. (A more detailed account of the whole range of experiments conducted using REPEATMASKER and the data sets by Tompa et el. and Sandve et al. will be provided in the full version.)

4.2 Evaluating Significance by Statistical z-Score Ranking

The z-score is the measure of the distance in standard deviations of the outcome of a random variable from its expectation. Consider a DNA sequence s of length n as if it was generated by a stationary, i.i.d. source with equiprobable symbols; an approximation to the z-score for a motif of length m that contains c solid characters and appears f times in s is given by $Z = \dfrac{f - (n-m+1) \times p}{\sqrt{(n-m+1) \times p \times (1-p)}}$, where $p = (1/4)^c$. This metric was used in [2] to assess the significance of the motifs extracted by VARUN and to rank them in the output.

We employed the code for VARUN provided by the authors to extract the rigid motifs from the DNA sequences analyzed in [2]. We then ran MADMX on the same sequences using the same frequency parameters, and setting the minimum density threshold ρ in such a way to obtain a comparable yet smaller output size. In this fashion, we tested the ability of MADMX to produce a succinct yet significant set of motifs, by virtue of its more flexible notion of density.

The results are shown in Table 1. For VARUN we used $D = 1$, thus allowing at most one don't care between two solid characters, and ran MADMX with $min_\ell = 1$, so to obtain the *complete* family of maximal dense motifs. In the table, there is a row of the table for each sequence (identified in the first column). Each sequence, whose total length is reported in the second column, is obtained as the concatenation of a number of smaller subsequences, reported in the third column.

Table 1. Results of the comparison with VARUN

name	length	#	σ	VARUN output	time	ρ	MADMX output	time	$m=10$	$m=50$	$m=100$	m^*	$\hat{m}$
ace2	500	1	2	1866	3s	0.7	1762	18s	10	50	100	1571	1067
ap1	500	1	2	1555	1s	0.7	1304	5s	10	50	100	392	13
gal4	3000	6	4	9764	12s	0.67	7606	67s	10	49	99	16	16
gal4$^{(*)}$	3000	6	4	9764	12s	0.65	11733	191s	10	50	100	9764	301
uasgaba	1000	2	2	4586	30s	0.70	4194	90s	10	50	100	175	175

On each sequence, both tools were run with the same frequency threshold σ, and the table reports for both the output size in terms of the number of motifs returned and the execution time in seconds. Also, for MADMX, the table reports the density threshold ρ used in each experiment.

For each experiment, we compared the best top-m z-scores, with $m = 10, 50$, and 100, as follows. Note that, in general, the top-m motifs found by MADMX and VARUN differ. Thus, we let z_M^i (resp., z_V^i) be the z-score of the ith motif in decreasing z-score order obtained by MADMX (resp., VARUN). For each m, the table reports how many times it was $z_M^i \geq z_V^i$, for $1 \leq i \leq m$. Also, column m^* (resp., column $\hat{m}$) gives the maximum m such that $z_M^i \geq z_V^i$ (resp., $z_M^i > z_V^i$) for every $1 \leq i \leq m$.

Even when MADMX is calibrated to yield a slightly smaller output, the quality of the motifs extracted, as measured by the z-score, is higher than those output by VARUN. Indeed, for sequences ace2 and uasgaba a very large prefix of the top-ranked motifs extracted by MADMX features strictly greater z-scores of the corresponding top-ranked ones extracted by VARUN. In fact, for all of the four sequences, at least the thirteen top-ranked motifs enjoy this property. To shed light on the slightly worse performance of MADMX on gal4, we re-ran MADMX with a different density threshold, so to obtain a slightly larger output (see row gal4$^{(*)}$). In this case, the top-301 motifs extracted by MADMX have z-score strictly greater than the corresponding motifs extracted by VARUN, while the execution time remains still acceptable.

For all runs, the top z-score of a motif discovered by MADMX is considerably higher than the one returned by VARUN. Specifically, on ace2 our best z-score is $387\,763$ vs. $12\,027$ of VARUN; on ap1, we have $12\,027$ vs. $1\,490$; on gal4 it is 75 vs. 28; on gal4$^{(*)}$ it is 150 vs. 28; on uasgaba we have $134\,532$ vs. $67\,059$. This reflects the high selectivity of MADMX, which is to be attributed mostly to adoption of a more flexible density constraint.

We must remark that MADMX (in its current nonoptimized version) is slower than VARUN, but it still runs in time acceptable from the point of view of a user. To further investigate the tradeoff between execution time and significance of the discovered motifs, we repeated the experiments running MADMX with $\text{min}_\ell = 2$ and $\rho = 0.65$, for all sequences. The running time of MADMX was almost halved, while the small output produced still featured high quality. In fact, for sequences ace2, ap1, and uasgaba the top-100 motifs extracted by MADMX have z-score greater or equal than the corresponding ones returned by VARUN.

We also have attempted a comparison between VARUN and MADMX on longer sequences (such as HGMR 1) at higher frequencies (since, unfortunately, VARUN does not seem to be able to handle low frequencies on very long sequences). Even allowing a higher number of don't cares between solid characters ($D = 2$) for the motifs of VARUN, all of the top-m z-scores featured by the motifs extracted by MADMX are greater than or equal to the corresponding scores in the ranking of VARUN, with m reaching the size of VARUN's output. In fairness, we remark that VARUN was designed to work at its best on protein sequences, while MADMX's main target are DNA sequences. Hence, these two tools should be regarded as complementary. Moreover, VARUN has the advantage of retrieving flexible motifs, while MADMX focuses only on rigid ones.

Acknowledgments. The authors wish to thank Alberto Apostolico and Matteo Comin for providing the code and giving valuable insights on VARUN, Ben Raphael for suggesting the use of REPEATMASKER, and Roberta Mazzucco and Francesco Peruch for coding MADMX.

References

1. Agrawal, R., Srikant, R.: Fast algorithms for mining association rules. In: Proc. of 20th VLDB, pp. 487–499 (1994)
2. Apostolico, A., Comin, M., Parida, L.: VARUN: discovering extensible motifs under saturation constraints. IEEE Trans. on Computational Biology and Bioinformatics (to appear, 2009)
3. Apostolico, A., Parida, L.: Incremental paradigms of motif discovery. Journal of Computational Biology 11(1), 15–25 (2004)
4. Apostolico, A., Tagliacollo, C.: Optimal offline extraction of irredundant motif bases. In: Lin, G. (ed.) COCOON 2007. LNCS, vol. 4598, pp. 360–371. Springer, Heidelberg (2007)
5. Apostolico, A., Tagliacollo, C.: Incremental discovery of the irredundant motif bases for all suffixes of a string in $O(n^2 \log n)$ time. Theoretical Computer Science 408(2-3), 106–115 (2008)
6. Arimura, H., Uno, T.: Mining maximal flexible patterns in a sequence. In: Satoh, K., Inokuchi, A., Nagao, K., Kawamura, T. (eds.) JSAI 2007. LNCS (LNAI), vol. 4914, pp. 307–317. Springer, Heidelberg (2008)
7. Jurka, J., Kapitonov, V.V., Pavlicek, A., Klonowski, P., Kohani, O., Walichiewicz, J.: Repbase Update, a database of eukaryotic repetitive elements. Cytogenet. Genome Res. 110, 462–467 (2005)
8. Morris, M., Nicolas, F., Ukkonen, E.: On the complexity of finding gapped motifs. CoRR, abs/0802.0314 (2008)
9. Parida, L.: Some results on flexible-pattern discovery. In: Giancarlo, R., Sankoff, D. (eds.) CPM 2000. LNCS, vol. 1848, pp. 33–45. Springer, Heidelberg (2000)
10. Parida, L.: Pattern discovery in bioinformatics. Mathematical and Computational Biology Series. Chapman & Hall / CRC, Boca Raton (2008)
11. Pisanti, N.: Segment-based distances and similarities in genomic sequences. PhD thesis, University of Pisa, Italy (2002)
12. Pisanti, N., Crochemore, M., Grossi, R., Sagot, M.F.: Bases of motifs for generating repeated patterns with wild cards. IEEE Trans. on Computational Biology and Bioinformatics 2(1), 40–50 (2005)

13. Rigoutsos, I., Floratos, A.: Combinatorial pattern discovery in biological sequences: the TEIRESIAS algorithm. Bioinformatics 14(1), 55–67 (1998)
14. Saha, S., Bridges, S., Magbanua, Z.V., Peterson, D.G.: Empirical comparison of *ab initio* repeat finding programs. Nucleic Acids Res. 36(7), 2284–2294 (2008)
15. Smit, A.F.A., Hubley, R., Green, P.: RepeatMasker Open-3.0. 1996–2004, http://www.repeatmasker.org
16. Ukkonen, E.: Structural analysis of gapped motifs of a string. In: Kučera, L., Kučera, A. (eds.) MFCS 2007. LNCS, vol. 4708, pp. 681–690. Springer, Heidelberg (2007)

Large-Scale Neighbor-Joining with NINJA

Travis J. Wheeler

Department of Computer Science
The University of Arizona, Tucson AZ 85721, USA
twheeler@cs.arizona.edu

Abstract. Neighbor-joining is a well-established hierarchical clustering algorithm for inferring phylogenies. It begins with observed distances between pairs of sequences, and clustering order depends on a metric related to those distances. The canonical algorithm requires $O(n^3)$ time and $O(n^2)$ space for n sequences, which precludes application to very large sequence families, e.g. those containing 100,000 sequences. Datasets of this size are available today, and such phylogenies will play an increasingly important role in comparative biology studies. Recent algorithmic advances have greatly sped up neighbor-joining for inputs of thousands of sequences, but are limited to fewer than 13,000 sequences on a system with 4GB RAM. In this paper, I describe an algorithm that speeds up neighbor-joining by dramatically reducing the number of distance values that are viewed in each iteration of the clustering procedure, while still computing a correct neighbor-joining tree. This algorithm can scale to inputs larger than 100,000 sequences because of external-memory-efficient data structures. A free implementation may by obtained from http://nimbletwist.com/software/ninja

Keywords: Phylogeny inference, Neighbor joining, external memory.

1 Introduction

The neighbor-joining (NJ) method of Saitou and Nei [1] is a widely-used method for constructing phylogenetic trees, owing its popularity to good speed, generally good accuracy [2], and proven statistical consistency (informally: NJ reconstructs the correct tree given a sufficiently long sequence alignment) [3,4,5].

NJ is a hierarchical clustering algorithm. It begins with a distance matrix, where d_{ij} is the observed distance between clusters i and j, and initially each of the n input sequences forms its own cluster. NJ repeatedly joins a pair of clusters that are closest under a measure, q_{ij}, that is related to the d_{ij} values. The canonical algorithm [6] finds the minimum q_{ij} at each iteration by scanning through the entire current distance matrix, requiring $O(r^2)$ work per iteration, where r is the number of remaining clusters. The result is a $\Theta(n^3)$ run time, using $\Theta(n^2)$ space. Thus, while NJ is quite fast for n in the hundreds or thousands, both time and space balloon for inputs of tens of thousands of sequences.

As a frame of reference, there are 8 families in Pfam [7] containing more than 50,000 sequences, and 3 families in Rfam [8] with more than 100,000 sequences,

S.L. Salzberg and T. Warnow (Eds.): WABI 2009, LNBI 5724, pp. 375–389, 2009.

and since the number of sequences in genbank is growing exponentially [9], these numbers will certainly increase. Phylogenies of such size are applicable, for example, to large-scale questions in comparative biology (e.g. [10]).

Related work. QuickTree [11] is a very efficient implementation of the canonical NJ algorithm. Due to low data-structure overhead, it is able to compute trees up to nearly 40,000 sequences before running out of memory on a 4GB system. QuickJoin [12,13], RapidNJ [14], and the bucket-based method of [15] all produce correct NJ trees, reducing run time by finding the globally smallest q_{ij} without looking at the entire matrix in each iteration. While all methods still suffer from worst-case running time of $O(n^3)$, they offer substantial speed improvements in practice. Unfortunately, the memory overhead of the employed data structures reduces the number of sequences for which a tree can be computed (e.g. on a system with 4GB RAM, RapidNJ scales to 13,000 sequences, and QuickJoin scales to 8000).

The focus of this paper is on exact NJ tools, but I briefly mention other distance-based methods for completeness. Relaxed [16] and fast [17] neighbor joining are NJ heuristics that improve speed by choosing the pair to merge from an incomplete subset of all pairs; they do not guarantee an exact NJ tree in the typical case that pairwise distances are not nearly-additive (very close to the distances induced by the true tree). Minimum-evolution with NNI offers an alternate fast approach with good quality and conjectured consistency [18]. An implementation of the relaxed neighbor joining heuristic, ClearCut [19], is faster than NINJA on very large inputs, and a recent implementation of a minimum-evolution heuristic with NNI, FastTree [20], is notable for constructing accurate trees on datasets of the scale discussed here, with speed at least 10-fold greater than that acheived by NINJA on very large inputs.

Contributions. I present NINJA, an implementation of an algorithm in the spirit of QuickJoin and RapidNJ: it produces a correct NJ phylogeny, and achieves increased speed by restricting its search for the smallest q_{ij} at each iteration to a small portion of the quadratic-sized distance matrix. The key innovations of NINJA are (1) introduction of a search-space filtering scheme that is shown to be consistently effective even in the face of difficult inputs, and (2) inclusion of data structures that efficiently use disk storage as external memory in order to overcome input size limits.

The result is a statistically consistent phylogeny inference tool that is roughly an order of magnitude faster than a very fast implementation of the canonical algorthm, QuickTree (for example, calculating a NJ tree for 60,000 sequences in less than a day on a desktop computer), and is scalable to hundreds of thousands of sequences.

Overview. The next section gives necessary details of the canonical NJ algorithm. Section 3 describes the primary filtering heuristic used to avoid viewing most of the distance matrix at each iteration, called *d-filtering*. Section 4 describes a secondary filtering method, called *q-filtering*, which is of primary value on the kinds of inputs where d-filtering is ineffective. Section 5 gives the full algorithm, and finally section 6 analyzes the impact of these methods, and compares the scalability of NINJA to that of other exact neighbor joining tools.

2   Canonical Neighbor-Joining

NJ [1,6] is a hierarchical clustering algorithm. It begins with a distance matrix, D, where d_{ij} is the observed distance between clusters i and j, and initially each sequence forms its own cluster. NJ forms an unrooted tree by repeatedly joining pairs of clusters until a single cluster remains. At each iteration, the pair of clusters merged are those that are closest under a transformed distance measure

$$q_{ij} = (r-2)\, d_{ij} - t_i - t_j \,, \tag{1}$$

where r is the number of clusters remaining at the time of the merge, and

$$t_i = \sum_k d_{ik} \,. \tag{2}$$

When the $\{i,j\}$ pair with minimum q_{ij} is found, D is updated by inactivating both the rows and columns corresponding to clusters i and j, then adding a new row and column containing the distances to all remaining clusters for the newly formed cluster ij. The new distance $d_{ij|k}$ between the cluster ij and each other cluster k is

$$d_{ij|k} = (d_{i|k} + d_{j|k} - d_{i|j})/2 \,. \tag{3}$$

There are n-1 merges, and in the canonical algorithm each iteration takes time $O(r^2)$ to scan all of D. This results in an overall running time of $O(n^3)$.

3   Restricting Search of the Distance Matrix

3.1   The d-Filter

A valid filter must retain the standard NJ optimization criterion at each iteration: merge a pair $\{i,j\}$ with smallest q_{ij}. To avoid scanning the entire distance matrix D, the pairs can be organized in a way that makes it possible to view only a few values before reaching a bound that ensures that the smallest q_{ij} has been found.

To acheive this we use a bound that represents a slight improvement to that used in RapidNJ [14]. In that work, $(\{i,j\}, d_{ij})$ triples are grouped into sets, sorted in order of increasing d_{ij}, with one set for each cluster. Thus, when there are r remaining clusters, each cluster i has a related set S_i containing $r-1$ triples, storing the distances of i to all other clusters j, sorted by d_{ij}. Then, for each cluster i, S_i is scanned in order of increasing d_{ij}. The value of q_{ij} is calculated (equation 1) for each visited entry, and kept as $q_{\min}$ if it is the smallest yet seen.

To limit the number of triples viewed in each set, a second value is calculated for each visited triple, a lower bound on q-values among the unvisited triples in the current set S_i: $q_{\text{bound}} = (r-2)\, d_{ij} - t_i - t_{\max}$, where $t_{\max} = \max_k\{t_k\}$.

In a single iteration, t_{max} is constant, and for a fixed set S_i, t_i is constant and the sorted d_{ij} values are by construction non-decreasing. Thus, if $q_{bound} \geq q_{min}$, no unvisited entries in S_i can improve q_{min}, and the scan is stopped. After this bounded scan of all sets, it is guaranteed that the correct q_{min} has been found. This is the approach of RapidNJ.

Improving the d-filter. While this method is correct, and provides dramatic speed gains [14], it can be improved. First, observe that the bound is dependent on t_{max}, which may be very loose (see fig. 2a). One way to provide tighter bounds is to abandon the idea of creating one list per cluster. Instead, the interval (t_{min}, t_{max}) is divided into evenly spaced disjoint bins, where each bin B_x covers the interval $[T_x^{min}, T_x^{max})$. For X bins, then, the size of the interval between min and max values will be $(t_{max} - t_{min})/X$ (the default number of bins is 30). Each cluster i is associated with the bin B_x for which $T_x^{min} \leq t_i < T_x^{max}$. Adopt the notation that cluster i's bin $B(i) = x$. Note that bins may contain differing numbers of clusters. Then create a set $S_{\{x,y\}}$ for each bin-pair $\{B_x, B_y\}$.

Now, instead of placing $(\{i,j\}, d_{ij})$ triples into per-cluster sets as before, place them in per-bin-pair sets $S_{\{B(i),B(j)\}}$, still sorting triples within a set by increasing d_{ij}. To find q_{min}, traverse the sets, scanning through each as before, but now calculating the bound based on current triple $(\{i,j\}, d_{ij})$, taken from set $S_{\{x,y\}}$, as

$$q_{bound} = (r-2)\, d_{ij} - T_x^{max} - T_y^{max}. \tag{4}$$

This improves the filter because, for an unvisited pair $\{i',j'\}$ from the same set $S_{\{x,y\}}$, setting $\rho = (r-2)\, d_{i'j'}$, $\rho - T_x^{max} - T_y^{max}$ will usually be a tighter bound on $q_{i'j'}$ than is $\rho - t_{i'} - t_{max}$.

Updating data structures. After merging clusters i and j, the rows and columns associated with those clusters are inactivated in D, and a new row and column are added for the merged cluster ij. Entries in the sets also require update. The new cluster, ij, is associated with the bin $B(ij) = \operatorname{argmin}_x \{T_x^{max} > t_{ij}\}$. Triples $(\{ij,k\}, d_{ij|k})$ for distances to each remaining cluster are added to the appropriate sets, $S_{\{B(ij),B(k)\}}$. Triples for the removed clusters i and j are removed from sets in a lazy fashion: i and j are marked as retired, and when triples involving either i or j are encountered while scanning sets, they are removed.

While this method provides tighter q_{bound} values than the method of keeping one set per cluster, these bounds will tend to relatively loosen over time. Before any merges are performed, the intervals of these sets are non-overlapping, but because the change in t_k after a merge may be different for each cluster k, this non-overlapping property is no longer guaranteed to hold after a merge is performed. The result is a loosening of the value T_x^{max} as a bound for t_i for an arbitrary cluster (i.e. the bound may be greater than $(t_{max} - t_{min})/X$). The loosening of the bound grows as iterations pass, though it is still tighter than the per-cluster bound until the set ranges overlap almost completely.

It may seem appealing to move a cluster to a new bin when that bin could provide a tighter bound, but doing so would incur substantial work to take all corresponding triples out of the various bin-pair sets. The strategy taken by NINJA is to occasionally rebuild the sets from scratch. For a constant $K > 1$ (the default is $K = 2$), the sets are rebuilt after r/K merges have been performed since the last rebuild, where r is the number of clusters remaining at the time of that prior rebuild. Overall runtime for these set constructions is dominated by the time of the first construction, $O(n^2 \log n)$.

3.2 Overcoming Memory Limits

The size of D is quadratic in the number of sequences, as is the size of the sets of triples described above. If these structures grow to exceed available RAM, an application may either abort or store the structures to *external storage* (i.e. disk). If the pattern of disk access is is random, the latter will result in frequent paging. The dramatic difference in latency between disk and RAM access (on the order of 10^6-fold difference [21]) necessitates I/O-efficient algorithms if external storage is to be used. I describe methods for efficiently handling both the sets $S_{\{x,y\}}$ and the distance matrix.

Bin-pair sets in external memory. The set of triples associated with each bin pair set $S_{\{x,y\}}$ has been described as a sorted list. In fact, in order to allow fast insertion of triples for new clusters, such a list would likely be implemented as a data structure such as a binary search tree. Binary search trees have poor I/O behavior when stored to disk, but could be easily replaced by a B-tree [22] or B+ tree, which allow for logarithmic number of disk I/Os for both insertions and reads.

However, since only a small portion of the entries in a set are accessed, the effort of keeping a totally ordered data structure is unnecessary. A min-heap [23] provides the tools necessary to scan through increasing d_{ij}, with less overhead since it only need keep a partial order. NINJA implements an *external memory array heap* [24], keyed on d_{ij}. This heap structure can store more triples than would fill a 1TB hard drive while maintaining a memory footprint smaller than 2MB, and guarantees an ammortized number of I/O operations for *insert* and *extract-min* operations that is logarithmic in the number of inserted triples. One heap is used for each set $S_{\{x,y\}}$.

Distance matrix in external memory. Though the heaps are used to identify the cluster pair $\{i, j\}$ to merge, the distance matrix D should still be maintained. After a merge, $d_{ij|k}$ is calculated for every cluster k. From equation 3, we see that we must view d_{ik} and d_{jk} for every k, which is more efficiently done by traversing the rows and columns for i and j in D than by scanning through the heaps.

Since NJ assumes a symmetric D, an efficient way to store D for in-memory use is to keep only its upper-right triangle: distances for cluster i are spread across row i and column i, such that all reside in the upper triangle. When a pair of clusters $\{i, j\}$ is merged, a new row and column are said to be added, but no additional space is actually required: the distances of the new cluster ij to all remaining clusters k can be stored in the cells previously belonging to one of the retired clusters, say i, so $d_{ij|k}$ is stored in the cell where d_{ik} was stored. Clusters i and j are noted as retired, and the mapping of cluster ij's stored location is simple.

However, when D is stored to disk, this approach will lead to poor disk paging behavior, because values for cluster i are split between row i (which can be accessed efficiently from disk, with many consecutive values per disk block), and column i (which will be spread across the disk, with typically one value per disk block). Therefore, a modification is required. For an input of n sequences, a file F stores a matrix with with $2n$ columns and n rows. The full initial D (i.e. both the upper and lower triangles) is stored to F, filling the first n columns for each row. When a merge is performed, and new distances are calculated, the values d_{ik} and d_{jk} can be gathered by sweeping through rows i and j, allowing the number of distance values that fit in a disk page to be gathered at the cost of a single disk access. The mapping for the storage location of the new $d_{ij|k}$ values will be different for rows and columns: if ij is formed as the result of the pth merge, then it will map to the row in F where i was stored, but will fill a new column $n + p - 1$. Newly calculated distances are not immediately stored to disk, instead waiting until enough values have been calculated to allow for efficient disk I/O. Suppose b distance values fit in a disk block: then d_{ij}s for new clusters are appended to a b x n in-memory matrix M until all b columns of that matrix are full. At that time, each row of M is appended to the same row in F (requiring one disk I/O per row), and each column is translated and written into the mapped row in F (requiring up to $\lceil n/b \rceil$ I/Os).

4   Candidate Handling

Due to the nature of heaps, all viewed $(\{i, j\}, d_{ij})$ triples are removed from their containing heaps during the search for $q_{\min}$; call these the *candidates*. The d-filter method described in section 3 dramatically reduces the number of candidates viewed in most cases, but inputs with relationships like those seen in figure 2a reduce the efficacy of d-filtering, for reasons described in section 6. Examples of the impact on run time are given in table 2c.

Here I describe a second level of filtering, called the *q-filter*. It works by sequestering candidates passing the d-filter, and organizing them in a way that allows a new bound to limit the number of those candidates that are viewed in each iteration.

q-filter on a candidate heap. Let $q_{ij}(p)$, $r(p)$, and $t_i(p)$ correspond to the values of q_{ij}, r, and t_i at a fixed previous iteration p. And let $\delta_i(p) = (r - 2)t_i(p) - (r(p) - 2)t_i$. Then it is easy to show that, for the current iteration,

$$q_{ij} = \frac{(r - 2)\, q_{ij}(p)\ + \delta_i(p) + \delta_j(p)}{r(p) - 2}. \tag{5}$$

Suppose all candidates on hand at iteration p are stored as $(\{i, j\}, q_{ij}(p))$ triples in a candidate set, sorted according to their $q_{ij}(p)$ values. Assign the current

r and t_i as $r(p)$ and $t_i(p)$ for that set. Since relative q-values change by small amounts from one iteration to the next, the $\{i,j\}$ pair with the smallest q_{ij} at a future iteration is likely to be near the front of this sorted list. It can be found by initializing $q_{\min}$ to ∞, then scanning candidates in order of increasing $q_{ij}(p)$, updating $q_{\min}$ when an entry with a smaller q_{ij} is found.

Let \$ be the set of all clusters with at least one representative in the candidate set, and define

$$\Delta_{\max}(p) := \max_{\substack{i,j \in \$ \\ i \neq j}} \{\delta_i(p) + \delta_j(p)\} . \tag{6}$$

Then scanning of this sorted list may be stopped when an element is found with

$$\frac{(r-2)\, q_{ij}(p)\ + \Delta_{\max}(p)}{r(p) - 2} \geq q_{\min} . \tag{7}$$

The candidate set can be large enough to exceed memory for very large inputs, and because only a partial order is required, NINJA stores the contents of the candidate set in an external memory heap array, as described for the d-bound bin-pairs in section 3. The heap formed from such candidates is called a *candidate heap*.

Candidate heap chain. Adding a new candidate to a candidate heap created in a previous iteration p_a (with associated $r(p_a)$ and $t(p_a)$ values) is problematic: (1) if the candidate involves a cluster j that was formed after p_a, then $q_{ij}(p_a)$ and $t_j(p_a)$ are undefined, and (2) even if both clusters existed before p_a, the candidate would need to be stored on the heap with a back-calculated $q_{ij}(p_a)$ (and thus looser than necessary bounds) to retain sensible δ-values.

The response in NINJA is to keep a chain of candidate heaps. At initiation, there are no candidates. In each iteration, newly gathered candidates from the d-filter are placed in a single candidate pool. When the size of that pool exceeds a threshold (default is 50,000; it should be fairly large because of the overhead required to form an external memory array heap), a candidate heap is created and populated with the triples in the pool, and the pool is then emptied. As more candidates are gathered, they are again stored in the pool, until it exceeds threshold, at which time a second candidate heap is formed, filled from the candidate pool, and linked to the first. This is repeated until the tree is complete. This results in a chain of candidate heaps. The chain is destroyed when bin-pair heaps are rebuilt (section 3.1).

At each iteration, these heaps are scanned for elements with small q_{ij} by removing triples until the bound (7) is reached. Those viewed triples with $q_{ij} > q_{\min}$ are placed in the candidate pool, rather than being returned to their source candidate heap, because the δ-bound usually gets looser, so they'd almost always just be pulled back off their original heap on the next iteration. When a candidate heap drops below a certain size (default = 60% of original size), it is liquidated, and all triples placed in the candidate pool.

5   Algorithm Overview

At each iteration p_a, NINJA follows this process, tracking q_{min} at each step:

1. Scan all candidates in the pool, keeping the one with smallest q_{ij}.
2. Sweep through the candidate heap chain, for each heap removing triples until reaching the bound (7), and placing those triples in the candidate pool. Possibly liquidate heaps in the chain if they become too empty. Steps 1 and 2 typically provide a good bound on the best q_{ij} value for the iteration, because they start with a set of previously filtered candidates.
3. Sweep through the bin-pair heaps, for each heap removing triples until reaching bound (4), and placing those triples in the candidate pool.
4. If the size of the candidate pool exceeds threshold, move all candidates into a new heap, storing $q_{ij}(p_a)$ for each candidate, and $t_i(p_a)$-values and $r(p_a)$ for the heap. Append this heap to the candidate heap chain.
5. Having found the q_{ij} with minimum value, merge clusters i and j, update the bin-pair heaps and the in-memory part of the distance matrix M with entries for new cluster ij, and possibly write out to the on-disk distance matrix D. Also occasionally liquidate the candidate heap chain and rebuild the bin-pair heaps (see section 3.1).

6   Results and Discussion

To assess the effectiveness of the two-tiered filtering algorithm, I have implemented it in an application called NINJA. Three variants were used in various tests in the results shown below. The default variant, NINJA, stores the distance matrix on disk, and uses both the d-filter described in section 3 and the q-filter described in section 4, both implemented with external-memory array heaps [24]. The variant labeled NINJA-d-filter is identical to NINJA, except that it implements only the d-filter, not the q-filter. The variant labeled NINJA-InMem also uses only the d-filter, but does so with in-memory data structures - keeping the distance matrix entirely in memory, and using a binary heap in place of the external-memory array heap. NINJA-InMem makes it possible to directly assess the impact of external-memory components of the algorithm. On a machine with 4GB RAM, NINJA-InMem is only able to compute neighbor-joining trees on inputs of fewer than about 7000 sequences, due to overhead memory use.

For comparison purposes, I tested two tools that similarly avoid viewing the entire distance matrix at each iteration, QuickJoin and RapidNJ, and a very fast implementation of the canonical algorithm, QuickTree. To my knowledge, these are the fastest available tools that implement exact NJ. Both of the former tools are unable to handle inputs of more than 13,000 sequences on a machine with 4GB of RAM, but an experimental external-memory version of RapidNJ, called RapidDiskNJ, has been released. A verison of RapidDiskNJ downloaded on 04/24/09 was used as a reference for large inputs. Note that the filter used in RapidNJ and RapidDiskNJ is almost equivalent to that used in NINJA-d-filter and NINJA-InMem. Tree constructing methods that do not form NJ trees are not included in this analysis due to space limits. It is worth noting that `ClearCut` and `FastTree` are both faster than `NINJA`.

`QuickTree` was implemented in C, `QuickJoin` and both `RapidNJ` variants were implemented in $C++$, and the `NINJA` variants were implemented in *Java*.

Environment. Experiments were run on a bank of 8 identical dedicated systems running CentOS 4.5 (kernel 2.6.9-55), with 64 bit 2.33 GHz Xeon processors, 4 GB allocated RAM, and 500 GB 7200 RPM SATA hard drives. `NINJA` used roughly 60 GB of disk space for the largest inputs. The "real time" output from the standard `time` tool was used to measure run time.

Data. `Pfam` [7] families were used as sample input for the tools. Each protein domain family was preprocessed to remove duplicate sequences, and all 415 families with more than 2000 unique sequences were used. Phylip formatted distance matrices, calculated with `QuickTree`, were used as input to all tools.

Effect of filters. At each iteration, the canonical algorithm scans through all $r(r-1)/2$ cells in the distance matrix, where r is the number of remaining

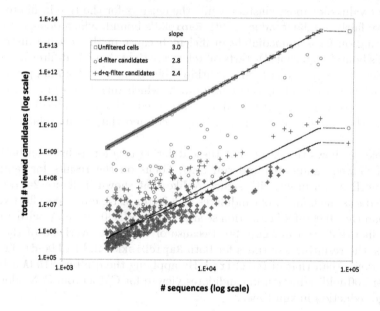

Fig. 1. Number of candidates viewed during tree-building with and without filters, for all 415 Pfam alignments with more than 2000 non-duplicate sequences. Data points are placed on a log-log plot: the slope on such a plot gives the exponent of growth. The canonical algorithm treats all cells as candidates, and the coresponding count of unfiltered cells shows the expected slope of 3 for $\Theta(n^3)$ number of cells. The d-filter often reduces the number of viewed candidates by more than 3 orders of magnitude, but is less effective for some inputs. Addition of the q-filter results in more consistent filtering success across all inputs, and an observed growth rate in number of viewed cells of roughly $O(n^{2.4})$.

clusters. It thus views $\Theta(n^3)$ cells (candidates) over the course of building a complete NJ tree on n sequences. Figure 1 shows the often dramatic reduction in number of candidates passing the d-filter, relative to this total count of cells. It also highlights instances where the d-filter is mostly ineffective, and shows that more consistent success is achieved when the q-filter is used in conjuction with the d-filter. It is important to note that figures 1, 3, and 4 are all log-log plots. Thus, the roughly linear growth observed in all plots corresponds to polynomial growth of both candidates and run time, with the polynomial exponent visible in the log-log slope.

Inputs causing bad d-filtration. Figure 2a shows an example of the kind of input that makes the d-filter fairly ineffective. It contains large clusters of very closely related sequences, and a few relatively long branches. Contrast this to the more evenly-distributed sequences seen in figure 2b, for which the d-filter is quite effective.

The reason for the computational difficulty of trees like the one in figure 2a is that the clusters on the very long branches have very large t values relative to the t values for most clusters, while the clusters for the tree in figure 2b will all have fairly similar t values. With RapidNJ's bound, which depends on $t_{\max}$, d-filtering on 2a is immediately inefficient because of this t-value discrepancy. NINJA's bound (equation 4) starts off relatively tight, but the d-filtering becomes inefficient as the range of t-values within a bin grows. This can happen dramatically when clusters along one long branch, which thus begin in a high t-value bin, are merged (with corresponding relative reduction in t values), while other clusters sharing the same bin are not merged, and thus retain relatively high t values

Table 2c shows the effect that these differing tree forms have on both number of viewed candidates and run time. Focus on the results for family Cytochrom_B_N, an input with structure like that shown in figure 2a: d-filtering only reduces the number of candidates by a factor of 10, much less effective filtering than the 10,000-fold reduction seen in family WD40, an input with structure much like that seen in figure 2b. Because of the extra overhead of their algorithms, the resulting run times for both RapidDiskNJ and NINJA-d-filter are much worse than that of QuickTree. By applying the q-filter, NINJA achieves a further 100-fold reduction in candidates viewed for Cytochrom_B_N, along with a large reduction in run time.

Comparison to other tools. Figures 3 and 4 compare NINJA variants to other neighbor-joining tools. They focus on inputs of more than 2000 sequences, since smaller inputs are solved by the canonical algorithm (implemented in QuickTree) in under 10 seconds.

The orders-of-magnitude reduction in viewed candidates seen in figure 1 does not translate to a similar reduction in run-time because the underlying data structures required to gain this filtering advantage incur a great deal of overhead relative to the simple scanning of a matrix. In addition, the large-scale applications (NINJA and RapidDiskNJ) incur a constant-factor overhead from

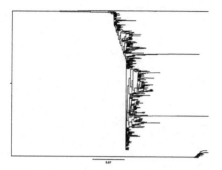

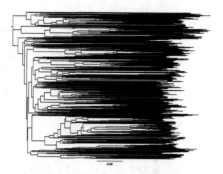

(a) Flu_M1, 707 sequences. Example of a topology for which filtering is ineffective

(b) QRPTase_N, 707 sequences. A topology for which filtering is effective

		Number candidates			Run time (minutes)			
Pfam ID	Sequence number	All cells	d-filter only	d+q filters	QuickTree	RapidDiskNJ	NINJA d-filter	NINJA d+q filters
RuBisCO_large	17,490	9E+11	5E+10	1E+09	64	124	331	25
PPR	18,961	1E+12	2E+08	2E+08	155	20	21	20
Cytochrom_B_N	33,789	6E+12	6E+11	7E+09	539	1,678	6,092	146
WD40	33,327	6E+12	2E+08	2E+08	756	52	121	110
RVT_1	56,822	3E+13	2E+12	1E+10	n/a	>18,500	>18,500	717
ABC_tran	53,116	2E+13	2E+08	2E+08	n/a	159	554	530

(c) Impact of d- and q-filtering at various input sizes

Fig. 2. Trees (a) and (b) are both of 707 sequences, and represent approximately equal evolutionary distance between the most divergent pair of sequences. They are mid-point-rooted, and images were created using FigTree (http://tree.bio.ed.ac.uk/software/figtree/). Pfam datasets with relationships like those shown in (a), with many closely related sequences and a few relatively long branches, cause d-filtering to be ineffective. In table (c), the first of each pair has topology similar to (a), and shows poor d-filtering; the second has topology similar to (b), and shows good d-filtering. Run times for RapidNJ and NINJA-d-filter are very slow when d-filtering is ineffective, but the additional q-filter used by NINJA results in much better filtering, and therefore improves runtimes even for these hard cases. On a system with 4GB RAM, QuickTree crashes on all Pfam families with more than 37,000 sequences. RapidDiskNJ and NINJA-d-filter both took longer than 13 days to compute a tree for RVT_1.

disk accesses. Those factors are mitigated by using algorithms with good disk-paging behavior, but are nevertheless present.

Figure 3 shows run times for a random sample of medium-sized (2000-7000 sequences) inputs from Pfam. A sample is shown, rather than the entire datset, to improve visibility of the chart, and agrees with trends for the full set of similarly-sized inputs. Note that QuickTree's run time grows with a slope of 2.9 on a log-log plot, essentially what is expected of a $\Theta(n^3)$ algorithm. QuickJoin and RapidNJ are in-memory versions of competitor algorithms - both show a reduction in

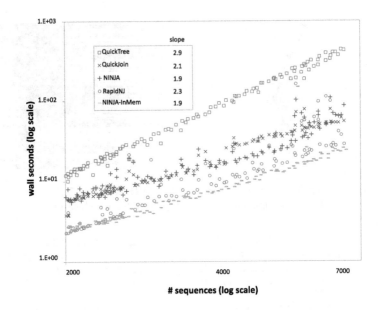

Fig. 3. Performance of NINJA and NINJA-InMem compared to that of QuickTree, QuickJoin, and RapidNJ on a random sample of medium-sized (2000 to 7000 sequences) Pfam inputs

run-time, and a growth rate that is slightly more than quadratic. This is in agreement with results from [14]. Results for NINJA-InMem and NINJA are presented to show their relative performance to each other and the other tools. Both show a roughly quadratic run-time growth on this data set. NINJA-InMem is slightly faster than the fastest other tool, RapidNJ. Since the two tools use essentially the same bounding method for their d-filter methods, this difference is likely explained by the tighter bounds generated by the bin-pair approach of NINJA.

Figure 4 shows run times for all inputs from Pfam with more than 7000 sequences. Results are given for the variant of each tool that best handles these large inputs: QuickTree, RapidDiskNJ, and NINJA. Only NINJA successfully computed NJ trees for all inputs; QuickTree crashed on all inputs with more than 37,000 sequences, while RapidDiskNJ failed to complete within 13 days on the two largest inputs. QuickTree continues to exhibit the expected slope (3.0) on a log-log plot for a $O(n^3)$ algorithm. Interestingly, both RapidDiskNJ and NINJA also show a similar cubic slope for these larger inputs, in conflict with the lower rate of growth observed for smaller inputs in figure 3 and [14]. Inspection of the data suggests that this is due to an increased frequency in these larger datasets of the sort of difficult inputs characterized by figure 2a. Note that the number of viewed candidates was observed in figure 1 as growing with a power of 2.4. The logarithmic overhead of heap data structures is responsible for the observation that run time grows faster than the number of candidates.

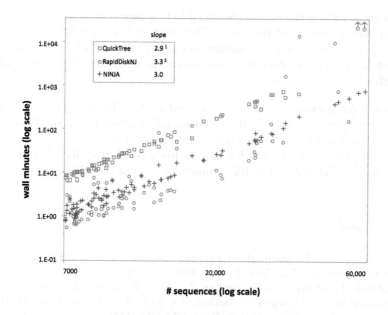

Fig. 4. Performance of NINJA compared to that of QuickTree and RapidDiskNJ on all large (7000 to 60,000 sequences) Pfam inputs. (1) On a system with 4GB RAM, QuickTree crashes on inputs with more than 37,000 sequences. (2) RapidDiskNJ failed to complete within 13 days for the two largest inputs; the uncertain times-to-completion are represented with the arrowed cirles in the upper right corner. The slope for RapidDiskNJ, which shows that its run time is growing faster than n^3, does not include these two points.

7   Conclusion

I have presented a new tool, NINJA, that builds a tree under the traditional optimization criteria of NJ, with the associated guarantee of statistical consistency. NINJA speeds up NJ by employing a two-tiered filtering regime, which greatly reduces the number of viewed candidates in each iteration relative to the complete scan of the distance matrix that is employed in the canonical algorithm. NINJA also overcomes memory constraints seen in earlier filtering-based work by incorporating external-memory-efficient data structures into the algorithm, specifically the external memory array heap [24] and simple on-disk storage of the distance matrix. The latter structure can be trivially co-opted by any NJ tool to overcome memory constraints due to the size of the distance matrix.

Though this method greatly speeds up NJ, and makes it possible to construct extremely large NJ trees, the run time still appears to be in $O(n^3)$ despite the dramatic reduction in viewed candidates. Though this does not represent an improvement in growth rate, the reduced constant factor makes it feasible to construct trees for inputs with well over 100,000 sequences in a matter of a small number of days of computation on a modern desktop.

The accuracy of NINJA is not discussed in this paper, as accuracy of any exact NJ tool is expected to be the same. That said, it is a straightforward exercise to incorporate the variance-minimization calculations of BioNJ [25], which have been show to improve accuracy over canonical neighbor-joining, into NINJAś algorithm.

Acknowledgements. I thank Karen Cranston, John Kececioglu, Morgan Price, and Mike Sanderson for helpful discussions, and Mike Sanderson and Darren Boss for use of, and assistance with, Mike's computing cluster.

I am supported by a PhD Fellowship from the University of Arizona National Science Foundation IGERT Comparative Genomics Initiative Grant DGE-0654435.

References

1. Saitou, N., Nei, M.: The neighbor-joining method: a new method for reconstructing phylogenetic trees. Mol. Biol. Evol. 4, 406–425 (1987)
2. Nakhleh, L., Moret, B.M.E., Roshan, U., John, K.S., Sun, J., Warnow, T.: The accuracy of fast phylogenetic methods for large datasets. In: Proc. 7th Pacific Symp. on Biocomputing, PSB 2002, pp. 211–222 (2002)
3. Atteson, K.: The Performance of Neighbor-Joining Methods of Phylogenetic Reconstruction. Algorithmica 25, 251–278 (1999)
4. Felsenstein, J.: Inferring phylogenies (January 2004)
5. Bryant, D.: On the Uniqueness of the Selection Criterion in Neighbor-Joining. Journal of Classification 22, 3–15 (2005)
6. Studier, J.A., Keppler, K.J.: A note on the neighbor-joining algorithm of Saitou and Nei. Mol. Biol. Evol. 5(6), 729–731 (1988)
7. Finn, R.D., Tate, J., Mistry, J., Coggill, P.C., Sammut, S.J., Hotz, H.R.R., Ceric, G., Forslund, K., Eddy, S.R., Sonnhammer, E.L.L., Bateman, A.: The Pfam protein families database. Nucleic Acids Res. 36(Database issue), D281–D288 (2008)
8. Griffiths Jones, S., Moxon, S., Marshall, M., Khanna, A., Eddy, S.R., Bateman, A.: Rfam: annotating non-coding RNAs in complete genomes. Nucleic Acids Res. 33(Database issue), D121–D124 (2005)
9. Goldman, N., Yang, Z.: Introduction. Statistical and computational challenges in molecular phylogenetics and evolution. Philos. Trans. R Soc. Lond B Biol. Sci. 363(1512), 3889–3892 (2008)
10. Smith, S.A., Beaulieu, J.M., Donoghue, M.J.: Mega-phylogeny approach for comparative biology: an alternative to supertree and supermatrix approaches. BMC Evol. Biol. 9, 37 (2009)
11. Howe, K., Bateman, A., Durbin, R.: QuickTree: building huge Neighbour-Joining trees of protein sequences. Bioinformatics 18(11), 1546–1547 (2002)
12. Mailund, T., Pedersen, C.N.S.: QuickJoin–fast neighbour-joining tree reconstruction. Bioinformatics 20(17), 3261–3262 (2004)
13. Mailund, T., Brodal, G.S., Fagerberg, R., Pedersen, C.N.S., Phillips, D.: Recrafting the neighbor-joining method. BMC Bioinformatics 7, 29 (2006)
14. Simonsen, M., Mailund, T., Pedersen, C.N.S.: Rapid Neighbour-Joining. In: Crandall, K.A., Lagergren, J. (eds.) WABI 2008. LNCS (LNBI), vol. 5251, pp. 113–122. Springer, Heidelberg (2008)

15. Zaslavsky, L., Tatusova, T.: Accelerating the neighbor-joining algorithm using the adaptive bucket data structure. In: Măndoiu, I., Sunderraman, R., Zelikovsky, A. (eds.) ISBRA 2008. LNCS (LNBI), vol. 4983, pp. 122–133. Springer, Heidelberg (2008)

16. Evans, J., Sheneman, L., Foster, J.: Relaxed neighbor joining: a fast distance-based phylogenetic tree construction method. J. Mol. Evol. 62(6), 785–792 (2006)

17. Elias, I., Lagergren, J.: Fast Neighbor Joining. Theor. Comput. Sci. 410, 1993–2000 (2009)

18. Desper, R., Gascuel, O.: Fast and accurate phylogeny reconstruction algorithms based on the minimum-evolution principle. Journal of Computational Biology 9(5), 687–705 (2002)

19. Sheneman, L., Evans, J., Foster, J.A.: Clearcut: a fast implementation of relaxed neighbor joining. Bioinformatics 22(22), 2823–2824 (2006)

20. Price, M.N., Dehal, P.S., and Arkin, A.P.: FastTree: Computing Large Minimum-Evolution Trees with Profiles instead of a Distance Matrix. Molecular Biology and Evolution 26, 1641–1650 (2009) doi:10.1093/molbev/msp077

21. Patterson, D.A.: Latency lags bandwidth. Communications of the ACM 47(10), 71–75 (2004)

22. Bayer, R., McCreight, E.: Organization and Maintenance of Large Ordered Indexes. Acta Informatica 1, 173–189 (1972)

23. Corman, T.H., Leiserson, C.E., Rivest, R.L., Stein, C.: Introduction to algorithms, 2nd edn. MIT Press, Cambridge (2001)

24. Brengel, K., Crauser, A., Ferragina, P., Meyer, U.: An Experimental Study of Priority Queues in External Memory. In: Vitter, J.S., Zaroliagis, C.D. (eds.) WAE 1999. LNCS, vol. 1668, pp. 345–359. Springer, Heidelberg (1999)

25. Gascuel, O.: BIONJ: an improved version of the NJ algorithm based on a simple model of sequence data. Mol. Biol. Evol. 14(7), 685–695 (1997)

A Unifying View on Approximation and FPT of Agreement Forests

(Extended Abstract*)

Chris Whidden** and Norbert Zeh***

Faculty of Computer Science, Dalhousie University, Halifax, Nova Scotia, Canada
{whidden,nzeh}@cs.dal.ca

Abstract. We provide a unifying view on the structure of maximum (acyclic) agreement forests of rooted and unrooted phylogenies. This enables us to obtain linear- or $O(n \log n)$-time 3-approximation and improved fixed-parameter algorithms for the subtree prune and regraft distance between two rooted phylogenies, the tree bisection and reconnection distance between two unrooted phylogenies, and the hybridization number of two rooted phylogenies.

1 Introduction

Phylogenies, or evolutionary trees, are a standard model to represent the evolutionary history of a set of species and are an indispensable tool in evolutionary biology [14]. Since determining the correct phylogeny of a set of species is hard, a host of heuristic methods for computing phylogenies have been proposed. The computation of distances between phylogenies under different metrics has proven essential for assessing the quality of phylogenies proposed by these heuristics, as well as for visualizing tree space (see, e.g., [13]). Of particular interest are metrics that model reticulation events, such as hybridization, lateral gene transfer, and recombination. These events result in species being composites of genes derived from different ancestors, and the analysis of different genes may produce different phylogenies over the same set of species. The *subtree prune and regraft* (rSPR) distance [11] and the *hybridization number* [2] of two rooted trees over the same set of species are important tools that often help to discover such events [15,16]. A related distance measure for unrooted trees is the *tree bisection and reconnection* (TBR) distance [1].

While biologically meaningful, TBR distance, rSPR distance, and hybridization number are NP-hard to compute [1, 6, 8]. As a result, significant efforts have been made to develop approximation and fixed-parameter (FPT) algorithms, as well as heuristic approaches [3, 12], for computing these distances. The best previous approximation algorithm for rSPR distance [17] provides a 3-approximation in $O(n^2)$ time. This algorithm builds on earlier results from [11].

---

* For details, see [18].
** Supported by an NSERC CGS-M graduate scholarship.
*** Supported in part by NSERC and the Canada Research Chairs programme.

S.L. Salzberg and T. Warnow (Eds.): WABI 2009, LNBI 5724, pp. 390–402, 2009.

Another 3-approximation algorithm [5] has running time $O(n^5)$; the "shifting lemma" central to its analysis is also the key to our new results. The same paper presents the best previous FPT algorithm for rSPR distance with running time $O(4^k k^4 + n^3)$, where k is the distance between the two trees. For TBR distance, the best previous approximation algorithm [9] computes an 8-approximation in polynomial time, and the best previous FPT algorithm [10] has running time $O(4^k k^5 + p(n))$, where $p(\cdot)$ is a polynomial function. We are not aware of any approximation algorithms for hybridization number; the best previous FPT algorithm for this problem [7] has running time $O((28k)^k + n^3)$.

The contribution of this paper is to provide a unifying view on the structure of rooted and unrooted agreement forests, using the "shifting lemma" mentioned above. This allows us to show that the framework of the algorithms of [11,17,4] can be used not only to approximate rSPR distances but also to obtain approximation and FPT algorithms for rSPR distance, TBR distance, and hybridization number. Our approximation algorithms provide a 3-approximation of the distance between the two trees and run in linear or, in the case of hybridization number, $O(n \log n)$ time. Our FPT algorithms for rSPR distance, TBR distance, and hybridization number have running times $O(3^k n)$, $O(4^k n)$, and $O(3^k n \log n)$, respectively. Using standard kernelizations, their running times can be reduced to $O(3^k k + n^3)$, $O(4^k k + n^3)$, and $O(3^k k \log k + n^3)$, respectively. All our algorithms represent improvements over the best previous algorithms for the same problems, substantial ones except in the case of the approximation algorithm for rSPR distance and the FPT algorithm for TBR distance. It should be noted here that "our" 3-approximation algorithm for rSPR distance is the algorithm of [17], with a minor modification to reduce its running time to $O(n)$. We believe, however, that the correctness proof obtained using our approach is simpler than the one presented in [17].

The rest of this paper is organized as follows. Section 2 introduces the necessary terminology and notation. Section 3 presents the main structural theorems at the heart of our approximation and FPT algorithms. Section 4 presents the approximation algorithms. Section 5 discusses briefly how to turn the approximation algorithms into FPT algorithms based on bounded search trees.

2 Preliminaries

Throughout this paper, we mostly use the definitions and notation from [1, 4, 6, 5, 17]. An *(unrooted binary phylogenetic) X-tree* is a tree T whose leaves are the elements of a label set X and whose internal nodes each have degree three. For a subset V of X, $T(V)$ is the smallest subtree of T that connects all nodes in V. The *V-tree induced by T* is the tree $T|V$ obtained from $T(V)$ using *forced contractions*, each of which replaces a vertex of degree two and its incident edges with a single edge between its neighbours. Figure 1 illustrates these definitions.

A *rooted X-tree* is obtained from an unrooted one, T, by subdividing one of T's edges, declaring the node this introduces to be the root, and defining parent-child and ancestor-descendant relations accordingly. For a subset V of X, $T(V)$

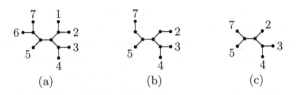

Fig. 1. (a) An X-tree T. (b) The subtree $T(V)$ for $V = \{2, 3, 4, 5, 7\}$. (c) The tree $T|V$ obtained from $T(V)$ using forced contractions.

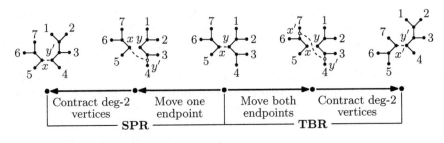

Fig. 2. Illustration of TBR and SPR operations

and $T|V$ are defined as in the unrooted case, but the construction of $T|V$ from $T(V)$ excludes the root of $T(V)$ from forced contractions.

Given an unrooted X-tree T, a *tree bisection and reconnection* (TBR) operation cuts an edge xy, thereby dividing T into two subtrees T_x and T_y containing x and y, respectively. Then it introduces two new vertices x' and y' into T_x and T_y by subdividing one edge of each tree, and adds an edge $x'y'$ to reconnect the two trees. Finally, x and y are removed using forced contractions. A *subtree prune and regraft* (SPR) operation is the one-sided equivalent of a TBR operation in that it introduces a new vertex y' into only T_y, adds an edge xy', and then removes y using a forced contraction. Figure 2 illustrates both operations. For a rooted tree T, a *rooted SPR* (rSPR) operation is an SPR operation in which the endpoints of edge xy are chosen so that T_y contains the root of T. Moreover, if y is the root of T_y, it is removed and its child becomes the new root.

TBR and rSPR operations define distance measures $d_{\mathrm{TBR}}(\cdot, \cdot)$ and $d_{\mathrm{rSPR}}(\cdot, \cdot)$ between X-trees, defined as the number of such operations required to transform one tree into the other. A related distance measure for rooted X-trees is their *hybridization number*, $\mathrm{hyb}(T_1, T_2)$, which is defined in terms of hybrid networks of the two trees. A *hybrid network* of T_1 and T_2 is a directed acyclic graph H such that both T_1 and T_2, with their edges directed away from the root, can be obtained from H by deleting edges and performing forced contractions. For a vertex $x \in H$, let $\deg_{\mathrm{in}}(x)$ be its in-degree and $\deg_{\mathrm{in}}^-(x) = \max(0, \deg_{\mathrm{in}}(x) - 1)$. Then the hybridization number of T_1 and T_2 is $\min_H \sum_{x \in H} \deg_{\mathrm{in}}^-(x)$, where the minimum is taken over all hybrid networks H of T_1 and T_2.

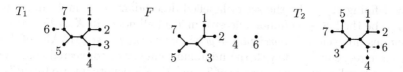

Fig. 3. Two X-trees T_1 and T_2 and an agreement forest F of T_1 and T_2. F is obtained from each tree by cutting the dashed edges.

TBR distance, rSPR distance, and hybridization number are known to be one less than the sizes of appropriately defined maximum agreement forests (MAF's). To define these MAF's, we first introduce some terminology.

Given a forest F and a subset E of its edges, $F - E$ denotes the forest obtained by deleting the edges in E from F. If F has components $T_1, T_2, \ldots, T_k$ with label sets $X_1, X_2, \ldots, X_k$, F *yields* the forest F' whose components $T'_1, T'_2, \ldots, T'_k$ satisfy $T'_i = T_i | X_i$, for all $1 \le i \le k$; if all nodes of a component T_i are unlabelled (that is, $X_i = \emptyset$), we define $T'_i = T_i | X_i = \emptyset$. If $T - E$ yields F, for an X-tree T and a subset E of its edges, we say that F is a *forest of T*.

Given two X-trees T_1 and T_2, a forest F is an *agreement forest* of T_1 and T_2 if there exist edge sets E_1 and E_2 such that $T_1 - E_1$ and $T_2 - E_2$ yield F; see Figure 3. F is a *maximum agreement forest* (MAF) of T_1 and T_2 if there is no agreement forest of T_1 and T_2 with fewer connected components. We use $m(T_1, T_2)$ to denote the number of connected components in an MAF of T_1 and T_2. For a forest F of T_2, we use $e(T_1, T_2, F)$ to denote the size of the smallest edge set E such that $F - E$ yields an agreement forest of T_1 and T_2. Allen and Steel [1] showed that, for two unrooted X-trees, T_1 and T_2, $d_{\text{TBR}}(T_1, T_2) = e(T_1, T_2, T_2) = m(T_1, T_2) - 1$.

In the rooted case, MAF's are similarly related to rSPR distances. This, however, is true only if the MAF is defined with respect to augmented versions of the two trees, obtained by adding a new root node with label ρ to both trees and making it the parent of their original roots. An agreement forest of two rooted X-trees T_1 and T_2 is then defined as a collection $\{T'_\rho, T'_1, T'_2, \ldots, T'_k\}$ of rooted trees with label sets $X_\rho, X_1, X_2, \ldots, X_k$ that satisfy the following conditions [6]:

1. The label sets $X_\rho, X_1, X_2, \ldots, X_k$ partition $X \cup \{\rho\}$, and $\rho \in X_\rho$.
2. For all $i \in \{\rho, 1, 2, \ldots, k\}$, $T'_i = T_1 | X_i = T_2 | X_i$ in the rooted sense.
3. The trees in each of the sets $\{T_1(X_i) \mid i \in \{\rho, 1, 2, \ldots, k\}\}$ and $\{T_2(X_i) \mid i \in \{\rho, 1, 2, \ldots, k\}\}$ are vertex-disjoint.

An MAF is again one with the minimum number of connected components. Bordewich and Semple [6] proved that, for two rooted X-trees T_1 and T_2, $d_{\text{rSPR}}(T_1, T_2) = e(T_1, T_2, T_2) = m(T_1, T_2) - 1$.

The hybridization number of two rooted X-trees T_1 and T_2 corresponds to an MAF of T_1 and T_2 with an additional constraint. An agreement forest F of T_1 and T_2 is said to contain a cycle if there exist two nodes x and y that are roots of trees in F and such that x is an ancestor of y in T_1, while y is an ancestor of x in T_2. (Each node x in F can be mapped to nodes $\phi_1(x)$ in T_1 and $\phi_2(y)$ in

T_2 by defining X_x to be the set of labelled descendants of x in F and defining $\phi_i(x)$ to be the lowest common ancestor in T_i of all nodes in X_x.) F is *acyclic* if it contains no cycles. A *maximum acyclic agreement forest* (MAAF) of T_1 and T_2 is an agreement forest with the minimum number of connected components among all acyclic agreement forests of T_1 and T_2. We denote its size by $\tilde{m}(T_1, T_2)$ and the number of edges in a forest F of T_2 that must be cut to obtain an acyclic agreement forest of T_1 and T_2 by $\tilde{e}(T_1, T_2, F)$. Baroni et al. [2] showed that, for two rooted X-trees T_1 and T_2, $\mathrm{hyb}(T_1, T_2) = \tilde{e}(T_1, T_2, T_2) = \tilde{m}(T_1, T_2) - 1$.

For two nodes a and b of a forest F, we write $a \sim_F b$ to indicate that there exists a path from a to b in F. Two labelled leaves a and c with the same neighbour, denoted r_{ac}, form a *sibling pair* (a, c). Our algorithms rely on the following two lemmas, which were proved in [4, 6] for rooted MAF's but are easily seen to apply also to the unrooted case.

Lemma 1. *Let T_1 and T_2 be X-trees, F a forest of T_2, and (a, c) a sibling pair that exists in T_1 and F. Let T_1', T_2', and F' be obtained from T_1, T_2, and F by relabelling node a as (a, c), removing c, and performing forced contractions to eliminate degree-2 nodes. We refer to this as* contracting the sibling pair (a, c). *Then $e(T_1, T_2, F) = e(T_1', T_2', F')$.*

Lemma 2. *Let T_1 and T_2 be X-trees, F a forest of T_2, and c a singleton in F. Let T_1', T_2', and F' be obtained from T_1, T_2, and F by removing c and performing forced contractions to eliminate degree-2 nodes. Then $e(T_1, T_2, F) = e(T_1', T_2', F')$.*

3   The Structure of Agreement Forests

This section presents results that provide the intuition and correctness proofs for our approximation and FPT algorithms, presented in Sections 4 and 5. All these algorithms start with a pair of trees (T_1, T_2) and then cut edges, remove singletons, and contract sibling pairs in both trees until they are identical. The intermediate state is that T_1 has been reduced to a smaller tree, and T_2 to a forest F. Each iteration has to decide which edges in F to cut next. The results in this section identify small edge sets in F so that at least one edge in each of these sets has the property that cutting it reduces $e(T_1, T_2, F)$ by one. Thus, the approximation algorithm cuts all edges in the identified set, and the size of the edge set cut in each step gives the approximation ratio of the algorithm. The FPT algorithm tries each edge in the set in turn, and the size of the set gives the branching factor for a bounded search tree algorithm. The following lemma by Bordewich et al. [5] is the central tool used in all our proofs.

Lemma 3 (Shifting Lemma). *Let F be a forest of an X-tree, e and f edges in the same component of F, and E a subset of edges of F such that $f \in E$ and $e \notin E$. Let v_f be the end-vertex of f closest to e, and v_e an end-vertex of e. If $v_f \sim_{F-E} v_e$ and $x \nsim_{F-(E \cup \{e\})} v_f$, for all $x \in X$, then $F - E$ and $F - (E \setminus \{f\} \cup \{e\})$ yield the same forest.[1]*

---

[1] In the rooted case, it is assumed that $\rho \in X$.

The other tool we need is an observation that relates incompatible triples and quartets to agreement forests. A *triple* $ab|c$ in a rooted tree T is defined by three leaves a, b, c such that the path from a to b is vertex-disjoint from the path from c to the root. A *quartet* $ab|cd$ in an unrooted tree T is defined by four leaves a, b, c, d such that the two paths from a to b and from c to d are vertex-disjoint. Given a tree T and a forest F, we say a triple $ab|c$ or quartet $ab|cd$ of T is *incompatible* with F if its leaves do not all belong to the same component of F or define a different triple or quartet in F (e.g., $ac|b$ or $ac|bd$).

Observation 1. *(i) Let T_1 and T_2 be rooted X-trees, F a forest of T_2, and E a set of edges such that $F - E$ yields an agreement forest of T_1 and T_2. If $ab|c$ is a triple of T_1 incompatible with F, then $a \nsim_{F-E} b$ or $a \nsim_{F-E} c$.*
(ii) Let T_1 and T_2 be unrooted X-trees, F a forest of T_2, and E a set of edges such that $F - E$ yields an agreement forest of T_1 and T_2. If $ab|cd$ is a quartet of T_1 incompatible with F, then $a \nsim_{F-E} b$, $a \nsim_{F-E} c$ or $c \nsim_{F-E} d$.

Now consider two X-trees T_1 and T_2 and a forest F of T_2, and let (a,c) be a sibling pair of T_1 that does not exist in F and such that neither a nor c is a singleton in F. If a and c belong to the same tree of F, the *sibling* b of a in F is the node adjacent to a's neighbour that is not on the path from a to c in F; otherwise, b is any node at distance two from a in F. Note that b may not be a leaf. We use e_a and e_b to denote the edges connecting a and b to their common neighbour r_{ab}, and B to denote the subtree of F induced by all nodes b' such that e_b belongs to the path from b' to a. The sibling d of c and edges e_c and e_d are defined analogously. In the rooted case, we choose b so that r_{ab} is the parent of a and b; to ensure that $c \notin B$, we assume that the distance from the root to a is no less than the distance from the root to c, which is easily ensured by swapping the roles of a and c if necessary.

With these tools in hand, we are now ready to prove three results characterizing edges that need to be cut in order to make progress towards an M(A)AF. The first result considers rooted MAF's.

Theorem 1. *Let T_1 and T_2 be rooted X-trees, F a forest of T_2, and (a,c) a sibling pair of T_1 that is not a sibling pair of F. Assume that neither a nor c is a singleton in F. Then $e(T_1, T_2, F - \{e_x\}) = e(T_1, T_2, F) - 1$, for some $x \in \{a,b,c\}$. In particular, $e(T_1, T_2, F - \{e_a, e_b, e_c\}) \le e(T_1, T_2, F) - 1$.*

Proof. It suffices to prove that there exists an edge set E of size $e(T_1, T_2, F)$ such that $F - E$ yields an MAF of T_1 and T_2 and $E \cap \{e_a, e_b, e_c\} \ne \emptyset$. So assume that $F - E$ yields an MAF F' of T_1 and T_2 and that $E \cap \{e_a, e_b, e_c\} = \emptyset$. We prove that we can replace an edge $f \in E$ with an edge in $\{e_a, e_b, e_c\}$ without changing the forest yielded by $F - E$.

First assume that $b' \nsim_{F-E} r_{ab}$, for all leaves $b' \in B$. In this case, we choose an arbitrary leaf $b' \in B$ and the first edge $f \in E$ on the path from r_{ab} to b'. Lemma 3 implies that $F - E$ and $F - (E \setminus \{f\} \cup \{e_b\})$ yield the same forest.

Now assume that there exists a leaf $b' \in B$ such that $b' \sim_{F-E} r_{ab}$. We prove that this implies that c is a singleton in F'. Since $e_a \notin E$, we have $a \sim_{F-E} r_{ab}$

and, hence, $a \sim_{F-E} b'$. Since (a, c) is a sibling pair of T_1, $ac|b'$ is a triple of T_1, while $c \notin B$ implies that either $ab'|c$ is a triple of F or $a \nsim_F c$. In either case, the triple $ac|b'$ is incompatible with F. By Observation 1(i), this implies that $a \nsim_{F-E} c$ because $F - E$ yields an agreement forest of T_1 and T_2 and $a \sim_{F-E} b'$. Then, however, either a or c is a singleton of F' because (a, c) is a sibling pair of T_1. Since $a \sim_{F-E} b'$, a cannot be a singleton of F', that is, c must be a singleton.

Now, as c is not a singleton in F, there exists a leaf l such that $c \sim_F l$. Since c is a singleton in F', at least one of the edges on the path from c to l in F belongs to E; let f be the one closest to c. Since c is a singleton in F', edges e_c and f satisfy the conditions of Lemma 3, and $F - E$ and $F - (E \setminus \{f\} \cup \{e_c\})$ yield the same forest.   □

Note that Theorem 1 also holds if we replace $e(\cdot, \cdot, \cdot)$ with $\tilde{e}(\cdot, \cdot, \cdot)$. To see this, it suffices to consider a set E in the proof such that $F - E$ yields an MAAF instead of an MAF. The next theorem provides an analogous result for unrooted MAF's. Its proof is similar to that of Theorem 1 and is omitted.

Theorem 2. *Let T_1 and T_2 be unrooted X-trees, F a forest of T_2, and (a, c) a sibling pair of T_1 that is not a sibling pair of F. Assume that neither a nor c is a singleton in F. Then $e(T_1, T_2, F - \{e_x\}) = e(T_1, T_2, F) - 1$, for some $x \in \{a, b, c, d\}$.*

Similar to Theorem 1, Theorem 2 implies that $e(T_1, T_2, F - \{e_a, e_b, e_c, e_d\}) \leq e(T_1, T_2, F) - 1$. However, we can do a little better.

Theorem 3. *Let T_1 and T_2 be unrooted X-trees, F a forest of T_2, and (a, c) a sibling pair of T_1 that is not a sibling pair of F. Assume that neither a nor c is a singleton in F. Then $e(T_1, T_2, F - \{e_a, e_b, e_c\}) \leq e(T_1, T_2, F) - 1$.*

Proof. Let E be an edge set of size $e(T_1, T_2, F)$ such that $F - E$ yields an MAF of T_1 and T_2. We can again assume that $E \cap \{e_a, e_b, e_c\} = \emptyset$, as otherwise the theorem holds trivially. As in the proof of Theorem 1, we can also assume there exists a leaf $b' \in B$ such that $b' \sim_{F-E} r_{ab}$ and, hence, $b' \sim_{F-E} a$. We prove that this implies that c is a singleton of the forest F' yielded by $F - E_{ab}$, where $E_{ab} = E \cup \{e_a, e_b\}$. As in the proof of Theorem 1, this implies that there exists an edge $f \in E$ such that $F - E_{ab}$ and $F - (E_{ab} \setminus \{f\} \cup \{e_c\})$ yield the same forest, F', which proves the theorem. We distinguish two cases.

If $a \nsim_{F-E} c$, c is a singleton of F' because (a, c) is a sibling pair of T_1 and $a \sim_{F-E} b'$. If $a \sim_{F-E} c$, (a, c) being a sibling pair in T_1 and $c \notin B$ imply that $ac|b'd'$ is a quartet of T_1 incompatible with F, for all $d' \notin X_B \cup \{a, c\}$, where X_B is the label set of B. Hence, by Observation 1(ii), $a \sim_{F-E} b'$ and $a \sim_{F-E} c$ imply that $c \nsim_{F-E} d'$, for each such leaf d'. This, however, implies that $c \nsim_{F-E_{ab}} x$, for all $x \in X$, that is, c is a singleton of F'.   □

While Theorem 1 suffices as a basis for an algorithm to compute or approximate an MAF of two rooted trees, a little extra work is required to obtain an MAAF. As observed after the proof of Theorem 1, we can use this theorem to make progress towards an MAAF until we obtain an agreement forest of the two trees. If this forest is in fact acyclic, we are done. Otherwise, we need to continue cutting edges to remove all cycles that may exist. The next theorem identifies candidate edges to cut. In this theorem, we consider two trees, A and B, of the agreement forest whose roots, a and b, form a cycle. We call (a, b) a *cycle pair* and use e_a to denote any of the two edges in A incident to a, and e_b to denote any of the two edges in B incident to b.

Theorem 4. *Let T_1 and T_2 be two rooted X-trees, F an agreement forest of T_1 and T_2, and (a, b) a cycle pair of F. Then $\tilde{e}(T_1, T_2, F - \{e_x\}) = \tilde{e}(T_1, T_2, F) - 1$, for some $x \in \{a, b\}$. In particular, $\tilde{e}(T_1, T_2, F - \{e_a, e_b\}) \leq \tilde{e}(T_1, T_2, F) - 1$.*

Proof. Once again, our goal is to show that there exists a set E of $\tilde{e}(T_1, T_2, F)$ edges of F such that $F - E$ yields an MAAF of T_1 and T_2 and $E \cap \{e_a, e_b\} \neq \emptyset$. So we choose E to be a set such that $F - E$ yields an MAAF F' of T_1 and T_2, and we show that, if $E \cap \{e_a, e_b\} = \emptyset$, we can find an edge $f \in E$ such that, for some $x \in \{a, b\}$, $F - E$ and $F - (E \setminus \{f\} \cup \{e_x\})$ yield the same forest. Let A_1 and A_2 be the two subtrees of A rooted in a's children, and let B_1 and B_2 be the two subtrees of B rooted in b's children.

First observe that there exists an index i such that either $a' \not\sim_{F-E} a$ for all $a' \in X_{A_i}$ or $b' \not\sim_{F-E} b$ for all $b' \in X_{B_i}$. If this was not the case, there would exist leaves $a_1 \in A_1$, $a_2 \in A_2$, $b_1 \in B_1$, and $b_2 \in B_2$ such that $a_1 \sim_{F-E} a_2$ and $b_1 \sim_{F-E} b_2$, implying that both a and b exist in F', and F' would not be acyclic.

So assume w.l.o.g. that $a' \not\sim_{F-E} a$, for all $a' \in A_1$. In this case, Lemma 3 implies that, if we choose a leaf $a' \in A_1$ and the edge $f \in E$ closest to a on the path from a to a', then $F - E$ and $F - (E \setminus \{f\} \cup \{e_a\})$ yield the same forest. $\square$

4 Approximation Algorithms

Rooted and unrooted MAF. The first algorithm we present is a 3-approximation algorithm for rooted MAF, that is, rSPR distance. This is essentially the algorithm discussed in [17], modified to achieve linear time. We include it here to demonstrate that Theorem 1 proves its correctness, and also as a reference for the other algorithms. The algorithm maintains a triple (T_1, T_2, F) and modifies it through a series of transformations. Initially, $F = T_2$. T_1 and T_2 shrink over time but are always trees with the same label set; F is always a forest of T_2. The algorithm also maintains a counter, D, of the number of edges in F it has cut so far. We use $T_1^{(i)}$, $T_2^{(i)}$, $F^{(i)}$, and $D^{(i)}$ to denote snapshots of T_1, T_2, F, and D after the ith transformation. The algorithm terminates when the label set of $T_1^{(i)}$ and $T_2^{(i)}$ has size at most 2, including the root label ρ, which is never eliminated. The output is the value of $D^{(i)}$ at the time of termination. Each iteration applies one of the following cases, illustrated in Figure 4.

1. As long as F contains a singleton $c \neq \rho$, the algorithm removes c from T_1, T_2, and F and performs forced contractions in T_1 and T_2 to merge the other two edges incident to c's parents in T_1 and T_2. D remains unchanged.

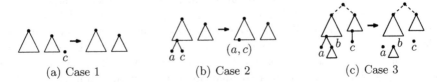

Fig. 4. The three cases of the approximation algorithm for rooted MAF. Only F is shown. In Figure (c), the dashed edges indicate that a and c may or may not belong to the same tree of F in Case 3.

For the other two cases, the algorithm chooses a fixed sibling pair (a, c) of T_1.

2. If (a, c) is also a sibling pair of F, the algorithm contracts the sibling pair as discussed in Lemma 1. D remains unchanged.
3. If (a, c) is not a sibling pair in F, then assume w.l.o.g. that a's distance from the root of T_2 is no less than that of c. Node a must have a sibling b in F because F contains no singletons. In this case, the algorithm cuts edges e_a, e_b, and e_c in F and increases D by three. T_1 and T_2 remain unchanged.

Theorem 5. *Given two rooted X-trees T_1 and T_2, a 3-approximation of $e(T_1, T_2, T_2) = d_{rSPR}(T_1, T_2)$ can be computed in linear time.*

Proof. We use the algorithm above and output the final value of D as the approximation of $e(T_1, T_2, T_2)$. We argue below that the algorithm terminates, in linear time. If the algorithm terminates after k iterations, k' of which change D, then its output is $D^{(k)} = 3k'$. We prove that $e(T_1, T_2, T_2) \le 3k' \le 3e(T_1, T_2, T_2)$, thereby proving that the value $D^{(k)}$ is a 3-approximation of $e(T_1, T_2, T_2) = d_{rSPR}(T_1, T_2)$.

For every iteration that leaves D unchanged (Cases 1 and 2), Lemmas 1 and 2 show that the applied transformations do not alter $e(T_1, T_2, F)$, and thus $e(T_1^{(i)}, T_2^{(i)}, F^{(i)}) = e(T_1^{(i-1)}, T_2^{(i-1)}, F^{(i-1)})$. Every iteration that changes D applies Case 3. Since (a, c) is not a sibling pair of F in this case, and neither a nor c is a singleton in F, Theorem 1 implies that $e(T_1^{(i)}, T_2^{(i)}, F^{(i)}) \le e(T_1^{(i-1)}, T_2^{(i-1)}, F^{(i-1)}) - 1$. Hence, we have $e(T_1, T_2, T_2) \ge k'$, that is, $D^{(k)} = 3k' \le 3e(T_1, T_2, T_2)$. Conversely, since the $3k'$ edges we cut in T_2 yield an agreement forest of T_1 and T_2, we have $e(T_1, T_2, T_2) \le 3k'$.

To bound the running time of the algorithm, we observe that it terminates after $O(n)$ iterations, as each iteration removes at least one vertex or edge from T_1 or F. Using arguments similar to the ones presented in [4], we can show that each iteration takes constant time. (See [18] for details.) □

The 3-approximation algorithm for unrooted MAF and, hence, for TBR distance is the same as for the rooted case, except that the edges e_a, e_b, and e_c in Case 3 are used in their unrooted meaning. Moreover, in the unrooted case, edge e_b is less trivial to identify in constant time. Instead, we cut e_a and one additional edge incident to r_{ab}. It is not hard to see that cutting any two of the three edges incident to r_{ab} has the same effect as cutting e_a and e_b. Hence, Theorem 3 establishes the correctness of this procedure, and we obtain:

Theorem 6. *Given two unrooted X-trees T_1 and T_2, a 3-approximation of $e(T_1, T_2, T_2) = d_{TBR}(T_1, T_2)$ can be computed in linear time.*

Rooted MAAF. The approximation algorithm for rooted MAAF consists of three phases: a preprocessing phase and two cutting phases. The preprocessing phase labels every node in T_1 and T_2 with its preorder number and with the interval of preorder numbers of its descendants, in order to identify cycles in an agreement forest of T_1 and T_2 later. The first cutting phase runs the algorithm for rooted MAF to obtain an agreement forest F of T_1 and T_2. (The described algorithm computes only the number of edges that need to be cut in T_2 to obtain an agreement forest F, but it is easily augmented to compute F itself.) The second cutting phase identifies and breaks cycles in F. Whenever we cut an edge in either of the two cutting phases, we increase D by one. The algorithm terminates when no cycle pair remains in F. The output is the final value of D.

To implement the second cutting phase, we maintain two sets, R_d and R_t, of roots of trees in F. R_d contains a set of roots that do not form any cycles with each other. R_t contains roots that may be involved in cycles. Initially, $R_d = \emptyset$ and R_t contains the roots of all trees in F. Each iteration of the algorithm removes a root a from R_t and tests whether a forms a cycle with any root $b \in R_d$. This is true if and only if a's preorder interval in T_1 contains $b's$, and $b's$ preorder interval in T_2 contains a's (or vice versa). If not, we add a to R_d and move on to the next root in R_t. If there is a root $b \in R_d$ such that (a, b) is a cycle pair, we cut one of the two edges incident to each of a and b and increase D by two. This breaks the two trees in F with roots a and b into two subtrees each; their roots are the children of a and b in F. We add these children to R_t and then move on to the next iteration. The algorithm terminates when $R_t = \emptyset$, at which point F is an acyclic agreement forest of T_1 and T_2.

Each iteration of this algorithm is easily implemented in linear time, resulting in a total running time of $O(n^2)$. In the full paper (see also [18]), we show how to reduce the running time to $O(n \log n)$. Thus, we have the following result.

Theorem 7. *Given two unrooted X-trees T_1 and T_2, a 3-approximation of $\tilde{e}(T_1, T_2, T_2) = hyb(T_1, T_2)$ can be computed in $O(n \log n)$ time.*

Proof. We have already discussed the running time of the algorithm. To prove the approximation bound, consider all iterations over the two cutting phases of the algorithm. An iteration that leaves D unchanged does not increase $\tilde{e}(T_1, T_2, T_2)$. By Theorem 1, every application of Case 3 in the first cutting phase decreases $\tilde{e}(T_1, T_2, F)$ by at least one and cuts three edges in F. By Theorem 4, every time we cut two edges in the current agreement forest F in the second cutting phase, $\tilde{e}(T_1, T_2, F)$ decreases by at least one. Hence, the number k' of iterations that change D is at most $\tilde{e}(T_1, T_2, T_2)$, and each such iteration increases D by at most three. Thus, $D \leq 3k' \leq 3\tilde{e}(T_1, T_2, T_2)$ at the end of the algorithm. On the other hand, once the algorithm terminates, R_t is empty, and the roots in R_d do not form cycles. The resulting agreement forest is therefore acyclic, and we cut D edges to obtain it. Thus, $\tilde{e}(T_1, T_2, T_2) \leq D$. Together with the upper bound, this shows that the final value of D is a 3-approximation of $\tilde{e}(T_1, T_2, T_2)$.   □

5   Fixed-Parameter Algorithms

The approximation algorithms discussed in the previous section are easily modified to obtain fixed-parameter algorithms for the respective problems. As is customary when discussing such algorithms, we focus on the decision version: "Given two X-trees T_1 and T_2, a distance measure $d(\cdot, \cdot)$, and a parameter k, is $d(T_1, T_2) \leq k$?" If this decision version can be solved in $O(c^k \text{poly}(n))$ time, then the exact distance $d = d(T_1, T_2)$ can be found in $O(c^d \text{poly}(n))$ time by iteratively trying larger guesses of k until we obtain the first positive answer.

To obtain such a decision algorithm for (rooted or unrooted) MAF, we modify the approximation algorithm from Section 4. We denote an invocation of the algorithm on trees T_1', T_2', forest F', and distance bound k by $\mathcal{A}(T_1', T_2', F', k)$. If T_1' and T_2' have at most two nodes each, the algorithm returns "yes" if and only if $k \geq 0$. Otherwise, whenever the approximation algorithm applies Case 1 or 2, so does $\mathcal{A}(T_1', T_2', F', k)$. When the approximation algorithm would apply Case 3, $\mathcal{A}(T_1', T_2', F', k)$ recurses. In the rooted case, the algorithm makes three recursive calls $\mathcal{A}(T_1', T_2', F' - \{e_x\}, k - 1)$, for $x \in \{a, b, c\}$, and returns "yes" if and only if one of these recursive calls does. In the unrooted case, the algorithm makes four recursive calls $\mathcal{A}(T_1', T_2', F' - \{e_x\}, k - 1)$, for $x \in \{a, b, c, d\}$, and again returns "yes" if and only if one of these recursive calls does.

Theorem 8. *For two rooted X-trees T_1 and T_2 and a parameter k, it takes $O(3^k n)$ time to decide whether $e(T_1, T_2, T_2) \leq k$. In the unrooted case, it takes $O(4^k n)$ time.*

Proof. The correctness of the algorithm follows immediately from Lemmas 1 and 2, and from Theorems 1 and 2. As for the running time, we can view each recursive call as a truncated invocation of the approximation algorithm from Section 4 that recurses as soon as it would invoke Case 3. Hence, each recursive call takes $O(n)$ time. The recursion depth is k, and each recursive call spawns at most three recursive calls, four in the unrooted case. Hence, the number of recursive calls is $O(3^k)$ in the rooted case and $O(4^k)$ in the unrooted case. This gives the claimed running times of $O(3^k n)$ and $O(4^k n)$, respectively.   □

To obtain an FPT algorithm for MAAF, we augment the above algorithm for rooted MAF as follows. For every recursive call $\mathcal{A}(T_1', T_2', F', k)$ that would output "yes" without recursing, we compute the corresponding agreement forest F of the two original trees T_1 and T_2. Note that F is not necessarily an MAF, as k may be greater than $e(T_1, T_2, T_2)$. If F is acyclic, the algorithm answers "yes". Otherwise, it invokes a second recursive algorithm $\mathcal{B}(T_1, T_2, F, k)$. This invocation returns "yes" if $k \geq 0$ and F contains no cycle, and "no" if $k < 0$. If $k \geq 0$ and F contains a cycle pair (a, b), $\mathcal{B}(T_1, T_2, F, k)$ makes two recursive calls $\mathcal{B}(T_1, T_2, F - \{e_a\}, k - 1)$ and $\mathcal{B}(T_1, T_2, F - \{e_b\}, k - 1)$, and returns "yes" if and only if one of the two calls does. The correctness and running time of the algorithm are established similarly to Theorem 8 (see [18]). Hence, we have:

Theorem 9. *For two rooted X-trees T_1 and T_2 and a parameter k, it takes $O(3^k n \log n)$ time to decide whether $\tilde{e}(T_1, T_2, T_2) \leq k$.*

Known kernelizations [1,6,7] reduce trees T_1 and T_2 to trees T_1' and T_2' in $O(n^3)$ time such that $e(T_1, T_2, T_2) = e(T_1', T_2', T_2') = k$ and T_1' and T_2' have size $O(k)$. The same holds for $\tilde{e}(T_1, T_2, T_2)$. Hence, we obtain the following corollary.

Corollary 1. *For two rooted X-trees T_1 and T_2 and a parameter k, it takes $O(3^k k + n^3)$ time to decide whether $e(T_1, T_2, T_2) \leq k$ and $O(3^k k \log k + n^3)$ time to decide whether $\tilde{e}(T_1, T_2, T_2) \leq k$. In the unrooted case, it takes $O(4^k k + n^3)$ time to decide whether $e(T_1, T_2, T_2) \leq k$.*

References

1. Allen, B.L., Steel, M.: Subtree transfer operations and their induced metrics on evolutionary trees. Annals of Comb. 5, 1–15 (2001)
2. Baroni, M., Grünewald, S., Moulton, V., Semple, C.: Bounding the number of hybridisation events for a consistent evolutionary history. J. Math. Biol. 51, 171–182 (2005)
3. Beiko, R.G., Hamilton, N.: Phylogenetic identification of lateral genetic transfer events. BMC Evol. Biol. 6, 15 (2006)
4. Bonet, M.L., St. John, K., Mahindru, R., Amenta, N.: Approximating subtree distances between phylogenies. J. Comp. Biol. 13, 1419–1434 (2006)
5. Bordewich, M., McCartin, C., Semple, C.: A 3-approximation algorithm for the subtree distance between phylogenies. J. Disc. Alg. 6, 458–471 (2008)
6. Bordewich, M., Semple, C.: On the computational complexity of the rooted subtree prune and regraft distance. Annals of Comb. 8, 409–423 (2005)
7. Bordewich, M., Semple, C.: Computing the hybridization number of two phylogenetic trees is fixed-parameter tractable. IEEE/ACM Trans. on Comp. Biol. and Bioinf. 4, 458–466 (2007)
8. Bordewich, M., Semple, C.: Computing the minimum number of hybridization events for a consistent evolutionary history. Disc. Appl. Math. 155, 914–928 (2007)
9. Chataigner, F.: Approximating the maximum agreement forest on k trees. Inf. Proc. Letters 93, 239–244 (2005)
10. Hallett, M., McCartin, C.: A faster FPT algorithm for the maximum agreement forest problem. Theory of Comp. Sys. 41, 539–550 (2007)
11. Hein, J., Jiang, T., Wang, L., Zhang, K.: On the complexity of comparing evolutionary trees. Disc. Appl. Math. 71, 153–169 (1996)
12. Hickey, G., Dehne, F., Rau-Chaplin, A., Blouin, C.: SPR distance computation for unrooted trees. Evol. Bioinf. 4, 17–27 (2008)
13. Hillis, D.M., Heath, T.A., St. John, K.: Analysis and visualization of tree space. Syst. Biol. 54, 471–482 (2005)
14. Hillis, D.M., Moritz, C., Mable, B.K. (eds.): Molecular Systematics. Sinauer Associates (1996)
15. Maddison, W.P.: Gene trees in species trees. Syst. Biol. 46, 523–536 (1997)

16. Nakhleh, L., Warnow, T., Lindner, C.R., St. John, K.: Reconstructing reticulate evolution in species—theory and practice. J. Comp. Biol. 12, 796–811 (2005)
17. Rodrigues, E.M., Sagot, M.-F., Wakabayashi, Y.: The maximum agreement forest problem: Approximation algorithms and computational experiments. Theor. Comp. Sci. 374, 91–110 (2007)
18. Whidden, C., Zeh, N.: A unifying view on approximation and FPT of agreement forests (2009), http://www.cs.dal.ca/~nzeh/Publications/maf.pdf

Structural Alignment of RNA with Complex Pseudoknot Structure

Thomas K.F. Wong[1], T.W. Lam[1], Wing-Kin Sung[2], and S.M. Yiu[1]

[1] Department of Computer Science, The University of Hong Kong, Hong Kong
{kfwong,twlam,smyiu}@cs.hku.hk
[2] School of Computing, National University of Singapore, Singapore
ksung@comp.nus.edu.sg

Abstract. The secondary structure of an ncRNA molecule is known to play an important role in its biological functions. Aligning a known ncRNA to a target candidate to determine the sequence and structural similarity helps in identifying de novo ncRNA molecules that are in the same family of the known ncRNA. However, existing algorithms cannot handle complex pseudoknot structures which are found in nature. In this paper, we propose algorithms to handle two types of complex pseudoknots: simple non-standard pseudoknots and recursive pseudoknots. Although our methods are not designed for general pseudoknots, it already cover all known ncRNAs in both Rfam and PseudoBase databases. A preliminary evaluation on our algorithms show that it is useful to identify ncRNA molecules in other species which are in the same family of a known ncRNA.

Keywords: Structural alignment, non-coding RNA, pseudoknots.

1   Introduction

A non-coding RNA (ncRNA) is a RNA molecule that does not translate into a protein. It has been shown to be involved in many biological processes [1,2,3]. The number of ncRNAs within the human genome was underestimated before, but recently some databases reveal over 212,000 ncRNAs [4] and more than 1,300 ncRNA families [5]. Large discoveries of ncRNAs and their families show the possibilities that ncRNAs may be as diverse as protein molecules [6]. Identifying ncRNAs is an important problem in biological study.

It is known that the secondary structure of an ncRNA molecule usually plays an important role in its biological functions. Some researches attempted to identify ncRNAs by considering the stability of secondary structures formed by the substrings of a given genome [16]. This method is not effective because a random sequence with high GC composition also allows an energetically favorable secondary structure [10]. A more promising direction is comparative approach which makes use of the idea that if a DNA region from which a RNA is transcribed has similar sequence and structure to a known ncRNA, then this region is likely to be an ncRNA gene whose corresponding ncRNA is in the same family of the known ncRNA. Thus, to locate ncRNAs in a genome, we can use a

S.L. Salzberg and T. Warnow (Eds.): WABI 2009, LNBI 5724, pp. 403–414, 2009.
© Springer-Verlag Berlin Heidelberg 2009

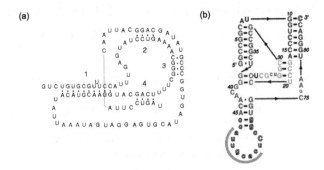

Fig. 1. (a): The secondary structure of RF00140 from Rfam 9.1 database [5]. Consider three base pairs: one from region 1, one from region 2 and one from region 4, they are mutually crossing each other (i.e. any two of them are crossing). (b): The secondary structure of self-cleaving ribozymes of hepatitis delta virus from [14] (i.e. RF00094 from Rfam 9.1 database).

known ncRNA as a query and search along the genome for substrings with similar sequence and structure to the query. The key of this approach is to compute the structural alignment between a query sequence with known structure and a target sequence with unknown structure. The alignment score represents their sequence and structural similarity. RSEARCH [11] and FASTR [12] belong to this category.

However, these tools do not support pseudoknots. Given two base pairs at positions (i,j) and (i',j'), where $i < j$ and $i' < j'$, pseudoknots are base pairs *crossing* each other, i.e. $i < i' < j < j'$ or $i' < i < j' < j$. In some studies, secondary structures including pseudoknots are found involved in some functions such as telomerase [7], catalytic functions [8], and self-splicing introns [9]. The presence of pseudoknots makes the problem computationally harder. Usually the large time complexity and considerable memory required for these algorithms make it impractical to search long pseudoknotted ncRNA along the genome.

Recently, Han et al. [15] developed PAL to solve the problem that supports secondary structures with standard pseudoknot of degree k and their algorithm runs in $O(kmn^k)$ where m is the length of the query sequence and n is the length of the target sequence. Their algorithm cannot handle more complex pseudoknot structures such as one with 3 base pairs mutually crossing each other (i.e. any two of them are crossing) as in Fig. 1(a) or the structure allowing recursive pseudoknots (i.e. pseudoknot/regular structures exist within another pseudoknot structure) as in Fig. 1(b). In Rfam 9.1 database [5], among 71 pseudoknotted families, 18 of them have complex pseudoknot structure. In the PseudoBase database [13], among 304 pseudoknot RNAs, 8 of them have complex pseudoknot structures. The small number may reflect the uncommon situation of having complex pseudoknots, but it may also reflect the difficulty of finding ncRNAs with complex pseudoknots due to the limitation of existing tools.

In this paper, we consider more complex pseudoknot structures which are found in nature. We define a class of pseudoknots called *simple non-standard pseudoknot* which allows some restricted cases with 3 base pairs mutually crossing each other. Our algorithm can apply to this complex structure in the same time complexity as Han's algorithm [15] for standard pseudoknot structure (i.e. $O(kmn^k)$ for degree k). Then, we propose an algorithm to handle recursive pseudoknot structure, our algorithm runs in $O(kmn^{k+1})$ if it is an *odd* structure or $O(kmn^{k+2})$ if it is an *even* structure. The definitions of odd and even structures for recursive pseudoknots will be given in Section 4.

Although our method is not designed for generic pseudoknots, we found that our method already cover all ncRNAs with complex pseudoknots in both Rfam 9.1 and PseudoBase databases. A preliminary experiment shows that our algorithms are useful in identifying ncRNAs from other species which are in the same family of a known ncRNA.

2   Pseudoknot Definitions

Let $A = a_1 a_2 \ldots a_m$ be a length-m ncRNA sequence and M be the secondary structure of A. M is represented as a set of base pair positions. i.e. $M = \{(i,j)|1 \leq i < j \leq m, (a_i, a_j)$ is a base pair$\}$. Let $M_{x,y} \subseteq M$ be the set of base pairs within the subsequence $a_x a_{x+1} \ldots a_y$, $1 \leq x < y \leq m$, i.e., $M_{x,y} = \{(i,j) \in M | x \leq i < j \leq y\}$. Note that $M = M_{1,m}$. We assume that there is no two base pairs sharing the same position, i.e., for any $(i_1, j_1), (i_2, j_2) \in M$, $i_1 \neq j_2$, $i_2 \neq j_1$, and $i_1 = i_2$ if and only if $j_1 = j_2$.

Definition 1. $M_{x,y}$ *is a regular structure if there does not exist two base pairs* $(i,j), (k,l) \in M_{x,y}$ *such that* $i < k < j < l$ *or* $k < i < l < j$. *Note that an empty set is also considered as a regular structure.*

A regular structure is one without pseudoknots. On the other hand, a standard pseudoknot of degree k allows certain types of pseudoknots. A structure is a standard pseudoknot of degree k if the RNA sequence can be divided into k consecutive regions (see Fig. 2(a)) such that base pairs must have end points in adjacent regions and base pairs that are in the same adjacent regions cannot cross each other. The formal definition is as follows.

Definition 2. $M_{x,y}$ *is a standard pseudoknot of degree* $k \geq 3$ *if* $\exists$ *a set of pivot points* $x_1, x_2, \ldots, x_{k-1}$ $(x = x_0 < x_1 < x_2 < \ldots < x_{k-1} < x_k = y)$ *that satisfy the following. Let* $M_w(1 \leq w \leq k-1) = \{(i,j) \in M_{x,y} | x_{w-1} \leq i < x_w \leq j < x_{w+1}\}$. *Note that we allow* $j = x_k$ *for* M_{k-1} *to resolve the boundary case.*
• *For each* $(i,j) \in M_{x,y}$, $(i,j) \in M_w$ *for some* $1 \leq w \leq k-1$.
• $M_w(1 \leq w \leq k-1)$ *is a regular structure.*

Note that a standard pseudoknot of degree 3 usually is referred as a *simple pseudoknot*. Now, we define a simple *non-standard* pseudoknot to include some structures with three base pairs crossing each other. For a simple non-standard

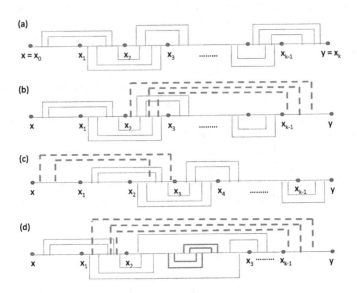

Fig. 2. (a)Standard pseudoknot of degree k. (b) Simple non-standard recursive pseudoknot of degree k (Type I). (c) Simple non-standard recursive pseudoknot of degree k (Type II). (d) Recursive pseudoknot (note the pseudoknot inside $[x_2, x_3]$).

pseudoknot of degree k, similar to a standard pseudoknot, the RNA sequence can be divided into k regions with the region at one of the ends (say, the right end) designated as the special region. Base pairs with both end points in the first $k-1$ regions have the same requirements as in a standard pseudoknot. And there is an extra group of base pairs that can start in one of the first $k-2$ regions and end at the last special region and again these pairs do not cross each other (see Fig. 2(b)). See the formal definition below.

Definition 3. $M_{x,y}$ *is a simple non-standard pseudoknot of degree $k \geq 4$ (Type I) if $\exists x_1, ..., x_{k-1}$ and t where $x = x_0 < x_1 < ... < x_{k-1} < x_k = y$ and $1 \leq t \leq k-2$ that satisfy the following. Let $M_w(1 \leq w \leq k-2) = \{(i,j) \in M_{x,y} | x_{w-1} \leq i < x_w \leq j < x_{w+1}\}$. Let $X = \{(i,j) \in M_{x,y} | x_{t-1} \leq i < x_t, x_{k-1} \leq j \leq y\}$.*
- *For each $(i,j) \in M_{x,y}$, either $(i,j) \in M_w(1 \leq w \leq k-2)$ or $(i,j) \in X$.*
- *M_w and X are regular structures.*

Type II simple non-standard pseudoknots (see Fig. 2(c)) are symmetric to Type I simple non-standard pseudoknots with the special region on the left end. In the rest of the paper, we only consider Type I simple non-standard pseudoknots and simply refer it as simple non-standard pseudoknots.

Lastly, we define what a recursive pseudoknot is (see Fig. 2(d)).

Definition 4. $M_{x,y}$ *is a recursive pseudoknot of degree $k \geq 3$ if $M_{x,y}$ is either regular, standard pseudoknot of degree k or simple non-standard pseudoknot of* *degree k (if $k \geq 4$), or $\exists a_1, b_1, ..., a_s, b_s (x \leq a_1 < b_1 < ... < a_s < b_s \leq y)$ that satisfy the followings. Each M_{a_i,b_i} is called a recursive region.*

- *M_{a_i,b_i}, for $1 \leq i \leq s$, is a recursive pseudoknot of degree $\leq k$.*
- *$(M_{x,y} - \bigcup_{1 \leq i \leq s} M_{a_i,b_i})$ is either regular structure, standard pseudoknot of degree $\leq k$ or simple non-standard-pseudoknot of degree $\leq k$.*

3   Algorithm for Simple Non-standard Pseudoknots

3.1   Structural Alignment

Let $S[1...m]$ be a query sequence with known secondary structure M, and $T[1...n]$ be a target sequence with unknown secondary structure. S and T are both sequences of {A,C,G,U}. A structural alignment between S and T is a pair of sequences $S'[1...r]$ and $T'[1...r]$ where $r \geq m, n$, S' is obtained from S and T' is obtained from T with spaces inserted to make both of the same length. A space cannot appear in the same position of S' and T'. The score of the alignment, which determines the sequence and structural similarity between S' and T', is defined as follows [12].

$$score = \sum_{i=1}^{r} \gamma(S'[i], T'[i]) + \sum_{\substack{i,j \text{ s.t. } \eta(i), \eta(j) \in M, \\ S'[i], S'[j], T'[i], T'[j] \neq '\_'}} \delta(S'[i], S'[j], T'[i], T'[j]) \quad (1)$$

where $\eta(i)$ is the corresponding position in S according to the position i in S'; $\gamma(t_1, t_2)$ and $\delta(x_1, y_1, x_2, y_2)$ where $t_1, t_2 \in \{A, C, G, U, '\_'\}$ and $x_1, x_2, y_1, y_2 \in \{A, C, G, U\}$, are scores for character similarity and for base pair similarity, respectively. The problem is to find an alignment to maximize the score.

3.2   Substructure of Simple Non-standard pseudoknot

We solve the problem using dynamic programming. The key is to define a substructure to enable us to find the solution recursively. For ease understanding of what a substructure is, we draw the pseudoknot structure using another approach (see Fig. 3).

We use simple non-standard pseudoknots with degree 4 for illustration. The result can be easily extended to general k. Fig. 3(b) shows the same pseudoknot structure as in Fig. 3(a). By drawing the pseudoknot structure this way, the base pairs can be drawn without crossing and can be ordered from the top to bottom. According to this ordering, we can define a substructure based on four points on the sequence (see Fig. 3(c) in which the substructure is highlighted in bold) such that all base pairs are either with both end points inside or outside the substructure. Note that in Fig. 3(c), $t = 1$ (t is odd), if $t = 2$ (t is even), we have to use a slightly different definition for substructures, otherwise base pairs cannot be ordered from top to bottom without crossing each other (see Fig. 3(d) and (e)). Note that the two base pairs that cross in Fig. 3(d) is due to the way

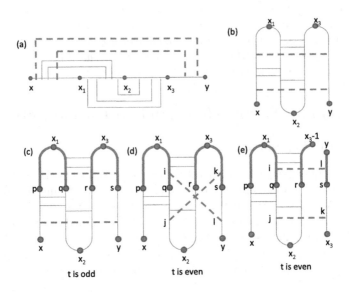

Fig. 3. Substructure of a simple non-standard pseudoknot

we draw the pseudoknot, they do not actually cross each other. i.e. basepairs (i,l) and (j,k) do not form a pseudoknot.). These are the only cases we need to consider.

Now, we formally define what a substructure is. Let $S[x..y]$ be an RNA sequence with known simple non-standard pseudoknot structure M of degree 4. Note that x_1, x_2, x_3 and t are known. Let $v = (p, q, r, s)$ be a quadruple with $x \leq p < x_1 \leq q < x_2 \leq r \leq x_3 < s \leq y$. If t is odd, define the subregion $R_{odd}(S, v) = [p, q] \cup [r, s]$. Otherwise, define the subregion $R_{even}(S, x_3, v) = [p, q] \cup [r, x_3 - 1] \cup [s, y]$. Note that x_3 is not a parameter, but a fixed value for S. Let $Struct(R_x) = \{(i, j) \in M | i, j \in R_x\}$ where R_x is a subregion.

We say that a subregion R_x defines a valid substructure ($Struct(R_x)$) of M if there does not exist $(i, j) \in M$ such that one endpoint of (i, j) is in R_x and the other is outside the region. Obviously, $Struct(R_x)$ is also a simple non-standard pseudoknot structure.

3.3 Dynamic Programming

Let $S[1, m]$ be the query sequence with known structure M and $T[1, n]$ be the target sequence with unknown structure. Note that the pivot points x_1, x_2, x_3 and t for S is known. We can apply the definitions of R_{odd} and R_{even} to T. If t is odd, for any $v' = (e, f, g, h)$ such that $1 \leq e < f < g < h \leq n$, we define the subregion $R_{odd}(T, v') = [e, f] \cup [g, h]$. If t is even, for any $v' = (e, f, g, h)$ and x_3' such that $1 \leq e < f < g < x_3' \leq h \leq n$, we define the subregion $R_{even}(T, x_3', v') = [e, f] \cup [g, x_3' - 1] \cup [h, n]$. Note that since the structure of T is unknown, x_3' is a parameter.

Define $C(R_x, R_y)$ be the score of the optimal alignment between a subregion R_x in S with substructure $Struct(R_x)$ and a subregion R_y in T. The score of the optimal alignment between S and T can be obtained as follows. If t is odd, setting $v^* = (1, x_2 - 1, x_2, m)$ includes the whole query sequence S, the entry $\max_{x_2'}\{C(R_{odd}(S, v^*), R_{odd}(T, v' = (1, x_2' - 1, x_2', n)))\}$ provides the answer. On the other hand, if t is even, setting $v^* = (1, x_2 - 1, x_2, x_3)$, the entry $\max_{x_2'} \max_{x_3' > x_2'}\{C(R_{even}(S, x_3, v^*), R_{even}(T, x_3', v' = (1, x_2' - 1, x_2', x_3')))\}$ provides the optimal score.

The value of $C(R_x, R_y)$ can be computed recursively. Assume that t is odd. Let $R_x = R_{odd}(S, (p, q, r, s))$ and $R_y = R_{odd}(T, (e, f, g, h))$. If (p, q) is a base pair in $Struct(R_x)$, there are four cases to consider. Case 1: MATCH$_{both}$ - aligning the base pair (p, q) of S with (e, f) of T; Case 2: MATCH$_{single}$ - aligning only one of the bases in (p, q) with the corresponding base in (e, f); Case 3: INSERT - insert a space on S; Case 4: DELETE - delete the base-pair (p, q) from S. Lemma 1 summarizes these cases.

The other cases, (q, r) is a base pair or (p, s) is a base pair, are similar. Note that if more than one such base pair exists (e.g. both (q, r) and (p, s) are base pairs), we only need to follow the recursion on one of the pairs. However, you cannot pick any of them in an arbitrary manner, otherwise, when we fill the dynamic programming table, we need to fill all entries for all possible subregions of S. We will address this issue in the later part of this section.

Lemma 1. *Let* $v = (p, q, r, s)$ *and* $v' = (e, f, g, h)$. *Let* t *be odd. And* $R_x = R_{odd}(S, v)$, $R_y = R_{odd}(T, v')$. *If* (p, q) *is a base pair, then* $C(R_x, R_y) = \max$

$$
\begin{cases}
//MATCH_{both} \\
C(R_{odd}(S, (p+1, q-1, r, s)), R_{odd}(T, (e+1, f-1, g, h))) \\
\quad + \gamma(S[p], T[e]) + \gamma(S[q], T[f]) + \delta(S[p], S[q], T[e], T[f]); \\
//MATCH_{single} \\
C(R_{odd}(S, (p+1, q-1, r, s)), R_{odd}(T, (e+1, f, g, h))) + \gamma(S[p], T[e]) + \gamma(S[q], `\_`), \\
C(R_{odd}(S, (p+1, q-1, r, s)), R_{odd}(T, (e, f-1, g, h))) + \gamma(S[p], `\_`) + \gamma(S[q], T[f]); \\
//INSERT \\
C(R_{odd}(S, (p, q, r, s)), R_{odd}(T, (e+1, f, g, h))) + \gamma(`\_`, T[e]), \\
C(R_{odd}(S, (p, q, r, s)), R_{odd}(T, (e, f-1, g, h))) + \gamma(`\_`, T[f]), \\
C(R_{odd}(S, (p, q, r, s)), R_{odd}(T, (e, f, g+1, h))) + \gamma(`\_`, T[g]), \\
C(R_{odd}(S, (p, q, r, s)), R_{odd}(T, (e, f, g, h-1))) + \gamma(`\_`, T[h]), \\
//DELETE \\
C(R_{odd}(S, (p+1, q-1, r, s)), R_{odd}(T, (e, f, g, h))) + \gamma(S[p], `\_`) + \gamma(S[q], `\_`)
\end{cases}
$$

On the other hand, if none of these are base pairs, assume that $p + 1 < x_1$ and $S[p]$ is a single base, then we can compute $C(R_x, R_y)$ recursively according to another three cases. Case 1: Match - aligning $S[p]$ with $T[e]$; Case 2: INSERT - insert a space on S; Case 3: Delete - delete $S[p]$.

Lemma 2. *Let $v = (p, q, r, s)$ and $v' = (e, f, g, h)$. Let t be odd. And $R_x = R_{odd}(S, v)$, $R_y = R_{odd}(T, v')$. If $p + 1 < x_1$ and $S[p]$ is a single base, then $C(R_x, R_y) = \max$*

$$
\begin{cases}
C(R_{odd}(S, (p+1, q, r, s)), R_{odd}(T, (e+1, f, g, h))) + \gamma(S[p], T[e]) \ //MATCH \\
//INSERT: \text{same as the one defined in Lemma 1} \\
C(R_{odd}(S, (p+1, q, r, s)), R_{odd}(T, (e, f, g, h))) + \gamma(S[p], \text{`-'}) \ //DELETE
\end{cases}
$$

For t is even, we consider whether $(p, q), (q, r)$, and (q, s) are base pairs in $Struct(R_x)$ and we need to consider all possible cases for x'_3 since the structure of T is unknown (i.e.,the pivot points are unknown).

To fill the dynamic programming table, not all entries for all possible subranges of S needs to be filled. For any given subregion $v = (p, q, r, s)$ in S, we first define $pair_{min}(v)$ and $single_{min}(v)$ as follow. If there exists a set of base pairs, say $\{(i_1, j_1), ..., (i_d, j_d)\}$, such that all $i_k, j_k (1 \le k \le d)$ equals to p (if $x \le p < x_1$), q (if $x_1 \le q < x_2$), r (if $x_2 \le r < x_3$) or s (if $x_3 \le s \le y$), then $pair_{min}(v)$ is the pair with minimum value of i. Also, if there exists a set of single bases (i.e. the positions which do not belong to any base pair), say $\{u_1, ..., u_d\}$, such that all $u_k (1 \le k \le d)$ equals to p (if $x \le p < x_1$), q (if $x_1 \le q < x_2$), r (if $x_2 \le r < x_3$) or s (if $x_3 \le s \le y$), then $single_{min}(v)$ is the one with minimum value.

Now, we define a function $\zeta(v)$ to determine subregions in S, for which we need to fill the corresponding C entires.

Case 1. If $(i, j) = pair_{min}(v)$ exists, then

$$
\zeta(v) = \begin{cases}
(p+1, q-1, r, s), & \text{if } (i, j) = (p, q) \\
(p, q-1, r+1, s), & \text{if } (i, j) = (q, r) \\
(p+1, q, r, s-1), & \text{if } (i, j) = (p, s) \text{ i.e. } t \text{ is odd} \\
(p, q-1, r, s+1), & \text{if } (i, j) = (q, s) \text{ i.e. } t \text{ is even}
\end{cases}
\tag{2}
$$

Case 2. If $pair_{min}(v)$ does not exist, then $u = single_{min}(v)$ should exist and

$$
\zeta(v) = \begin{cases}
(p+1, q, r, s), & \text{if } u = p \\
(p, q-1, r, s), & \text{if } u = q \\
(p, q, r+1, s), & \text{if } u = r \\
(p, q, r, s-1), & \text{if } u = s \text{ and } t \text{ is odd} \\
(p, q, r, s+1), & \text{if } u = s \text{ and } t \text{ is even}
\end{cases}
\tag{3}
$$

It is obvious that if v defines a subregion with a valid substructure, $\zeta(v)$ also defines a valid substructure. For t is odd, let $v* = (1, x_2 - 1, x_2, m)$. We only need to fill in the entries for C provided v can be obtained from $v*$ by applying ζ function repeatedly. If t is even, let $v* = (1, x_2 - 1, x_2, x_3)$. Intuitively, ζ guides which recursion formula to use. And there are only $O(m)$ such v values. The following lemma summarizes the time complexity for this algorithm.

Lemma 3. *For any sequence $S[1..m]$ with simple non-standard pseudoknot of degree 4 and any sequence $T[1..n]$, let c be the max length of $[x'_3, n]$, the optimal alignment score between $S[1..m]$ and $T[1..n]$ can be computed in $O(cmn^4)$.*

Note that the factor c is only needed when t is even due to the extra parameter x'_3. We examined all sequences in Rfam and PseudoBase, we found that usually the length of the final segment c is short (< 15) and the average length is only 5.4 with most of the cases having lengths from 5 to 7. So, we can assume that c is a constant. The algorithm can be extended to simple non-standard pseudoknot of degree k easily.

Theorem 1. *For any sequence $S[1..m]$ with simple non-standard pseudoknot of degree k and any sequence $T[1...n]$, the optimal alignment score between $S[1..m]$ and $T[1..n]$ can be computed in $O(kmn^k)$.*

4   Algorithm for Recursive Pseudoknot

We use the recursive pseudoknot of degree 4 to illustrate the algorithm. The approach can be easily extended to general k. Let $S[1..m]$ be the query sequence with recursive pseudoknot structure M. Recall the definition of a recursive pseudoknot. There can be disjoint recursive regions, namely $M_{a_1,b_1}, \ldots, M_{a_s,b_s}$, in M. By removing all these recursive regions, the remaining structure $M - (M_{a_1,b_1} \cup \cdots \cup M_{a_s,b_s})$ together with the remaining sequence $S[1..a_1 - 1]S[b_1 + 1..a_2 - 1]\ldots S[b_s+1..m]$ are referred as level-0. For each removed recursive region M_{a_i,b_i}, we can apply the same procedure to define level-1, level-2, ..., level-ℓ structures (see Fig. 4 for an example). In this section, we assume all the recursive regions M_{a_i,b_i} in all levels including $M_{1,m}$ after removing the next-level substructure are simple non-standard pseudoknots or regular structure. Among all these recursive regions with degree-4, if there exists one of them for which the value of t is even, then the whole recursive pseudoknot structure is referred as an *even* structure, otherwise, it is an *odd* structure.

Let $T[1..n]$ be the target sequence. Define $H[a_i, b_i, x', y']$ be the score of the optimal alignment between the recursive region $S[a_i, b_i]$ with structure M_{a_i,b_i}

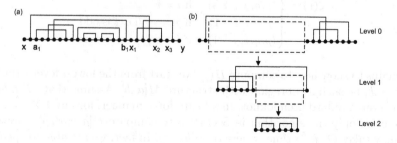

Fig. 4. An example showing that recursive pseudoknot of degree 4 can be divided into levels: Level 0 is a simple non-standard pseudoknot of degree 4; Level 1 is another simple non-standard pseudoknot; Level 2 is a regular structure

and $T[x'..y']$, where $1 \le x' < y' \le n$. We now show how to compute the score of the optimal alignment between S and T recursively. We assume that level-0 structure is a simple non-standard pseudoknot and consider t is odd. Let $v = (p, q, r, s)$ be a quadruple that defines a substructure of S. Let $S[p..y_p]$ be a recursive region. The following lemma shows how to compute $C(R_x, R_y)$ where $R_x = R_{odd}(S, v)$ and $R_y = R_{odd}(T, v')$.

Lemma 4. *Let $v = (p, q, r, s)$ and $v' = (e, f, g, h)$. Assume that t is odd. $R_x = R_{odd}(S, v)$ and $R_y = R_{odd}(T, v')$. If $S[p..y_p]$ is a recursive region, then*

$$C(R_x, R_y) = \max \begin{cases} //MATCH \\ \max_{e \le w \le f}\{C(R_{odd}(S, (y_p + 1, q, r, s)), R_{odd}(T, (w + 1, f, g, h))) + H(p, y_p, e, w)\} \\ //INSERT \\ \text{same as INSERT defined in Lemma 1} \\ //DELETE \\ C(R_{odd}(S, (y_p + 1, q, r, s)), R_{odd}(T, (e, f, g, h))) + \sum_{p \le w \le y_p} \gamma(S[w], \text{'-'}) \end{cases}$$

Other cases, such as $S[x_q..q]$ or $S[r..y_r]$ or $S[x_s..s]$ is a recursive region, are handled in a similar way. Again, we need to determine for which subregions in S, we need to fill in the corresponding C entries. So, we enhance ζ function as follows.

Consider a quadruple $v = (p, q, r, s)$ in a region $S[x...y]$ where the structure is a simple non-standard pseudoknot of degree 4 if all the next-level subregions inside are excluded. Let us define $\text{subregion}_{min}(v)$ as follows: if there exists a set of next-level subregions, say $\{[i_1, j_1], ..., [i_d, j_d]\}$ where $x \le i_k < j_k \le y$ for all $1 \le k \le d$ such that either i_k or j_k equals to p (if $x \le p < x_1$), q (if $x_1 \le q < x_2$), r (if $x_2 \le r < x_3$) or s (if $x_3 \le s \le y$), then let $\text{subregion}_{min}(v)$ be the region with minimum value of i. We add the following case to ζ function. Note that the t value refers to the structure for $S[x..y]$ excluding all next-level subregions.

Case 0 of $\zeta(v)$: If $[i, j] = \text{subregion}_{min}(v)$ exists, then

$$\zeta(v) = \begin{cases} (j + 1, q, r, s), & \text{if } i = p \\ (p, i - 1, r, s), & \text{if } j = q \\ (p, q, j + 1, s), & \text{if } i = r \\ (p, q, r, i - 1), & \text{if } j = s \text{ //i.e. } t \text{ is odd} \\ (p, q, r, j + 1), & \text{if } i = s \text{ //i.e. } t \text{ is even} \end{cases} \tag{4}$$

It remains to show how to compute $H()$. We start from the lowest level structure. Let $S[a..b]$ be such a subregion with structure $M[a..b]$. Assume that $M[a..b]$ is a simple non-standard pseudoknot. In a brute-force manner, for any $1 \le x' < y' \le n$, we can apply the algorithm in Section 3 to compute $H[a, b, x', y']$. However, this may takes $O(\alpha n^6)$ time, where $\alpha = (b - a)$. In fact, we are able to speed up the computation so that $H()$ can be computed in $O(\alpha n^4)$ time if t is odd and $O(\alpha n^5)$ if t is even. Note that once a subregion is considered to compute $H()$ function, in subsequent steps, we do not need to process that subregion again.

The following theorem shows the main result of this section (the full details will be given in the full paper) assuming that inside the recursive pseudoknot, there are some simple non-standard pseudoknots.

Theorem 2. *To compute the optimal alignment score between a query sequence* $S[1...m]$ *with recursive pseudoknot of degree* k (≥ 4) *and a target sequence* $T[1...n]$, *it can be done with the following time complexity.*

$$time\ complexity = \begin{cases} O(kmn^{k+1}), & \textit{if it is an odd structure;} \\ O(kmn^{k+2}), & \textit{if it is an even structure;} \end{cases}$$

Our algorithm for recursive pseudoknot can be easily adapted to cases in which the recursive pseudoknot can have a mix of regular structure, standard pseudoknot of degree k, and simple non-standard pseudoknot of degree k.

5   Preliminary Evaluation

We selected two families in Rfam 9.1 database to test the effectiveness of our algorithms. RF00094 is a family with members having recursive pseudoknots and RF00140 is another with members having simple non-standard pseudoknots. Both structures of RF00094 and RF00140 are of degree 4 (see Fig. 1(a) and (b)). In the experiment, we select a known member of the family, scan the genome of another species and check whether the known members of that species can be identified. Following the evaluation method of [15], since there is no existing software which can perform structural alignment for complex pseudoknot structure, we compared the performance of our programs with BLAST. We use default parameters for BLAST except the wordsize is set to 7. Fig. 5 shows the result for family RF00094. The two regions of known members have the highest scores from our program. Although one of their BLAST score is the highest, another one is not quite distinguishable from others. For RF00140, there are two known members in this genome region and our program identifies 4 possible candidates

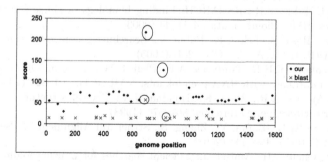

Fig. 5. Comparison of resulting scores from our program and BLAST when scanning the whole genome (AB118824) for family RF00094 using AB037948/685-775 as the query sequence. The circled points are the known members of the same family in genome AB118824.

including these two known members. On the other hand, BLAST resulted in three regions but none of them matches with any of known members.

6 Conclusions

In the paper, we provided the first algorithms to handle structural alignment of RNA with two complex pseudoknot structures, recursive pseudoknots and simple non-standard pseudoknots. Further directions include speeding up these algorithms and considering other more complicated pseudoknot structures.

References

1. Frank, D.N., Pace, N.R.: Ribonuclease P: unity and diversity in a tRNA processing ribozyme. Annu. Rev. Biochem. 67, 153–180 (1998)
2. Nguyen, V.T., et al.: 7SK small nuclear RNA blinds to and inhibits the activity of CDK9/cyclin T complexes. Nature 414, 322–325 (2001)
3. Yang, Z., et al.: The 7SK small nuclear RNA inhibits the CDK9/cyclin T1 kinase to control transcription. Nature 414, 317–322 (2001)
4. Liu, C., et al.: NONCODE: an integrated knowledge database of non-coding RNAs. NAR 33(Database issue), D112–D115 (2005)
5. Griffiths-Jones, S., et al.: Rfam: an RNA family database. NAR 31(1), 439–441 (2003), http://www.sanger.ac.uk/Software/Rfam/
6. Eddy, S.: Non-coding RNA genes and the modern RNA world. Nature Reviews in Genetics 2, 919–929 (2001)
7. Hen, J., Greider, C.W.: Functional analysis of the pseudoknot structure in human telomerase RNA. PNAS 102(23), 8080–8085 (2005)
8. Dam, E., Pleij, K., Draper, D.: Structural and functional aspects of RNA pseudoknots. Biochemistry 31(47), 11665–11676 (1992)
9. Adams, P.L., Stahley, M.R., Kosek, A.B., Wang, J., Strobel, S.A.: Crystal structure of a self-splicing group I intron with both exons. Nature 430, 45–50 (2004)
10. Rivas, E., Eddy, S.: Secondary structure alone is generally not statistically significant for the detection of noncoding RNAs. Bioinformatics 16(7), 583–605 (2000)
11. Klei, R.J., Eddy, S.R.: RSEARCH: Finding homologs of single structured RNA sequences. BMC Bioinformatics 4(1), 44 (2003)
12. Zhang, S., Hass, B., Eskin, E., Bafna, V.: Searching genomes for noncoding RNA using FastR. IEEE/ACM TCBB 2(4) (2005)
13. van Batenburg, F.H.D., Gultyaev, A.P., Pleij, C.W.A., Ng, J., Oliehoek, J.: Pseudobase: a database with RNA pseudoknots. NAR 28(1), 201–204 (2000)
14. Ferre-D'Amare, A.R., Zhou, K., Doudna, J.A.: Crystal structure of a hepatitis delta virus ribozyme. Nature 395, 567–574 (1998)
15. Han, B., Dost, B., Bafna, V., Zhang, S.: Structural Alignment of Pseudoknotted RNA. JCB 15(5), 489–504 (2008)
16. Le, S.Y., Chen, J.H., Maizel, J.: Efficient searches for unusual folding regions in RNA sequences. In: Structure and Methods: Human Genome Initiative and DNA Recombination, vol. 1, pp. 127–130 (1990)

Improving Inference of Transcriptional Regulatory Networks Based on Network Evolutionary Models

Xiuwei Zhang and Bernard M.E. Moret

Laboratory for Computational Biology and Bioinformatics
EPFL (Ecole Polytechnique Fédérale de Lausanne)
Lausanne, Switzerland
and Swiss Institute of Bioinformatics
{xiuwei.zhang,bernard.moret}@epfl.ch

Abstract. Computational inference of transcriptional regulatory networks remains a challenging problem, in part due to the lack of strong network models. In this paper we present evolutionary approaches to improve the inference of regulatory networks for a family of organisms by developing an evolutionary model for these networks and taking advantage of established phylogenetic relationships among these organisms. In previous work, we used a simple evolutionary model for regulatory networks and provided extensive simulation results showing that phylogenetic information, combined with such a model, could be used to gain significant improvements on the performance of current inference algorithms.

In this paper, we extend the evolutionary model so as to take into account gene duplications and losses, which are viewed as major drivers in the evolution of regulatory networks. We show how to adapt our evolutionary approach to this new model and provide detailed simulation results, which show significant improvement on the reference network inference algorithms. We also provide results on biological data (*cis*-regulatory modules for 12 species of *Drosophila*), confirming our simulation results.

1 Introduction

Transcriptional regulatory networks are models of the cellular regulatory system that governs transcription. Because establishing the topology of the network from bench experiments is very difficult and time-consuming, regulatory networks are commonly inferred from gene-expression data. Various computational models, such as Boolean networks [1], Bayesian networks [11], dynamic Bayesian networks (DBNs) [17], and differential equations [8,27], have been proposed for this purpose, along with associated inference algorithms. Results, however, have proved mixed: the high noise level in the data, the paucity of well studied networks, and the many simplifications made in the models all combine to make inference difficult.

Bioinformatics has long used comparative and, more generally, evolutionary approaches to improve the accuracy of computational analyses. Work by Babu's group [3,4,26] on the evolution of regulatory networks in *E. coli* and *S. cerevisiae* has demonstrated the applicability of such approaches to regulatory networks. They posit a simple evolutionary model, under which network edges are simply

S.L. Salzberg and T. Warnow (Eds.): WABI 2009, LNBI 5724, pp. 415–428, 2009.

added or removed; they study how well such a model accounts for the dynamic evolution of the two most studied regulatory networks; they then investigate the evolution
of regulatory networks with gene duplications [26], concluding that gene duplication
plays a major role, in agreement with other work [23].

Phylogenetic relationships are well established for many groups of organisms; as the
regulatory networks evolved along the same lineages, the phylogenetic relationships
informed this evolution and so can be used to improve the inference of regulatory networks. Indeed, Bourque and Sankoff [7] developed an integrated algorithm to infer regulatory networks across a group of species whose phylogenetic relationships are known;
they used the phylogeny to reconstruct networks under a simple parsimony criterion. In
previous work [29], we presented two refinement algorithms, both based on phylogenetic information and using a likelihood framework, that boost the performance of any
chosen network inference method. On simulated data, the *receiver-operator characteristic (ROC)* curves for our algorithms consistently dominated those of the standard approaches used alone; under comparable conditions, they also dominated the results from
Bourque and Sankoff. Both our previous approach and that of Bourque and Sankoff are
based on an evolutionary model that considers only edge gains and losses, so that the
networks must all have the same number of genes (orthologous across all species).
Moreover, the gain or loss of an edge in that model is independent of any other events.
However, this process accounts for only a small part of regulatory network evolution;
in particular, gene duplication is known to be a crucial source of new genetic function
and a mechanism of evolutionary novelty [23,26].

In this paper we present a model of network evolution that takes into account gene
duplications and losses and their effect on regulatory network structures. Such a model
provides a direct evolutionary mechanism for edge gains and losses, while also enabling
broader application and more flexible parameterization. To refine networks using phylogenetic information within this broader framework, we use the reconciliation of gene
trees with the species tree [2,10,21] to process gene duplication and loss events. We then
extend our refinement algorithms [29]; thanks to the more complex model, these refinement algorithms are more broadly applicable, while also returning better solutions. We
provide experimental results confirming that our new algorithms provide significant
improvements over the base inference algorithms.

2   Background

Our refinement algorithms [29] work iteratively in two phases after an initialization
step. First, we obtain the regulatory networks for the family of organisms; typically,
these networks are inferred from gene-expression data for these organisms, using standard inference methods. We place these networks at the corresponding leaves of the
phylogeny of the family of organisms and encode them into binary strings by simply
concatenating the rows of their adjacency matrix. We then enter the iterative refinement cycle. In the first phase, we infer ancestral networks for the phylogeny (strings
labelling internal nodes), using our own adaptation of the FastML [22] algorithm; in the
second phase, these ancestral networks are used to refine the leaf networks. These two
phases are then repeated as needed. Our refinement algorithms are formulated within a

maximum likelihood (ML) framework and focused solely on refinement—they are algorithmic boosters for one's preferred network inference method. Our new algorithms retain the same general approach, but include many changes to use the duplication/loss data and handle the new model.

2.1 Base Network Inference Methods

We use both DBN and differential equation models as base inference methods in our experiments. When DBNs are used to model regulatory networks, an associated structure learning algorithm is used to infer the networks from gene-expression data [12,17,18]; so as to avoid overly complex networks, a penalty on graph structure complexity is usually added to the ML score, thereby reducing the number of false positive edges. In [29] we used a coefficient k_p to adjust the weight of this penalty and study different tradeoffs between sensitivity and specificity, yielding the optimization criterion $\log Pr(D|G, \hat{\Theta}_G) - k_p \#G \log N$, where D denotes the dataset used in learning, G is the (structure of the) network, $\hat{\Theta}_G$ is the ML estimate of parameters for G, $\#G$ is the number of free parameters of G, and N is the number of samples in D.

In models based on differential equations [8,27], a regulatory network is represented by the equation system $dx/dt = f(x(t)) - Kx(t)$, where $x(t) = (x_1(t), \cdots, x_n(t))$ denotes the expression levels of the n genes and K (a matrix) denotes the degradation rates of the genes. The regulatory relationships among genes are then characterized by $f(\cdot)$. To get networks with different levels of sparseness, we applied different thresholds to the connection matrix to get final edges.

In our experiments we use Murphy's Bayesian Network Toolbox [20] for the DBN approach and TRNinfer [27] for the differential equation approach; we refer to them as *DBI* and *DEI*, respectively.

2.2 Refinement Algorithms in Our Previous Work

The principle of our phylogenetic approach is that phylogenetically close organisms are likely to have similarly close regulatory networks; thus independent network inference errors at the leaves get corrected in the ancestral reconstruction process along the phylogeny. We gave two refinement algorithms, *RefineFast* and *RefineML*. Each uses the globally optimized parents of the leaves to refine the leaves, but the first simply picks a new leaf network from the inferred distribution (given the parent network, the evolutionary model parameters, and the phylogeny), while the second combines the inferred distribution with a prior, the existing leaf network, using a precomputed *belief coefficient* that indicates our confidence in the current leaf network, and returns the most likely network under these parameters.

2.3 Reconciliation of Species Tree and Gene Trees

To infer ancestral networks with the extended network evolution model, we need a full history of gene duplications and losses. We reconstruct this history by reconciling the gene trees and the species tree. The species tree is the phylogenetic tree whose leaves correspond to the modern organisms; gene duplications and losses occur along the branches of this tree. A gene tree is a phylogenetic tree whose leaves correspond to genes in orthologous gene families across the organisms of interest; in such a tree, gene duplications and speciations are associated with internal nodes.

When gene duplications and losses occur, the species trees and the gene trees may legitimately differ in topology. Reconciling these superficially conflicting topologies— that is, explaining the differences through a history of gene duplications and losses—is known as *lineage sorting* or *reconciliation* and produces a list of gene duplications and losses along each edge in the species tree. While reconciliation is a hard computational problem, algorithms have been devised for it in a Bayesian framework [2] or using a simple parsimony criterion [10].

3   The New Network Evolution Model

Although transcriptional regulatory networks produced from bench experiments are available for only a few model organisms, other types of data have been used to assist in the comparative study of regulatory mechanisms across organisms [9,24,25]. For example, gene-expression data [25], sequence data like transcriptional factor binding site (TFBS) data [9,24], and *cis*-regulatory elements [25] have all been used in this context. Moreover, a broad range of model organisms have been studied, including bacteria [4], yeast [9,25], and fruit fly [24], thus spanning a broad evolutionary range. However, while these studies are promising, they have not to date sufficed to establish a clear model for regulatory networks or their evolution.

Our new model remains simple, but can easily be generalized or associated with other models, such as the evolutionary model of TFBSs [16]. In this new model, the networks are represented by binary adjacency matrices. The evolutionary operations are:

- *Gene duplication*: a gene is duplicated with probability p_d. After a duplication, edges for the newly generated copy can be assigned as follows:

  Neutral initialization: Create connections between the new copy and other genes randomly according to the proportion π_1 of edges in the background network independently of the original copy.

  Inheritance initialization: Connections of the duplicated copy are reported to correlate with those of the original copy [4,23,26]. This observation suggests letting the new copy inherit the connections of the original, then lose some of them or gain new ones at some fixed rate [6].

  Preferential attachment: The new copy gets preferentially connected to genes with high connectivity [5,6].

- *Gene loss*: a gene is deleted along with all its connections with probability p_l.
- *Edge gain*: an edge between two genes is generated with probability p_{01}.
- *Edge loss*: an existing edge is deleted with probability p_{10}.

The model parameters are thus p_d, p_l, the proportions of 0s and 1s in the networks $\Pi = \begin{pmatrix} \pi_0 & \pi_1 \end{pmatrix}$, the substitution matrix of 0s and 1s, $P = \begin{pmatrix} p_{00} & p_{01} \\ p_{10} & p_{11} \end{pmatrix}$, plus parameters suitable to the initialization model.

4 The New Refinement Methods

We begin by collecting the regulatory networks to be refined. These networks may have already been inferred or they can be inferred from gene-expression data at this point using any of the standard network inference methods introduced in Sec. 2.1. The genes in these networks are not required to be orthologous across all species, as the duplication/loss model allows for gene families of various sizes. Refinement proceeds in the two-phase iterative manner already described, but adding a step for lineage reconciliation and suitably modified algorithms for ancestral reconstruction and leaf refinement:

1. Reconstruct the gene trees, one for each gene family.
2. Reconstruct the history of gene duplications and losses by reconciling these gene trees and the given species tree. From this history, determine gene contents for the ancestral regulatory networks (at each internal node of the species tree).
3. Infer the edges in the ancestral networks. We do this using a revised version of FastML, described in Sec. 4.1.
4. Refine the leaf networks with new versions of *RefineFast* and *RefineML*, described in Secs. 4.2 and 4.3.
5. Repeat steps 3 and 4 as needed.

4.1 Inferring Ancestral Networks

FastML [22] assumes independence among the entries of the adjacency matrices and reconstructs ancestral characters one site at a time. When the gene content is the same in all networks, FastML can be used nearly as such. In our new model, however, the gene content varies across networks. We solve this problem by embedding all networks into one that includes every gene that appears in any network, taking the union of all gene sets. We then represent a network with a ternary adjacency matrix, where the rows and columns of the missing genes are filled with a special character x. All networks are thus represented with adjacency matrices of the same size. Since the gene contents of ancestral networks are known thanks to reconciliation, the entries with x are already identified in their matrices; other entries are reconstructed by our revised version of FastML, with a new character set $S' = \{0, 1, x\}$. We modify the substitution matrix and take special measures for x during calculation. The substitution matrix P' for S' can be derived from the model parameters in Sec. 3, without introducing new parameters.

$$P' = \begin{pmatrix} p'_{00} & p'_{01} & p'_{0x} \\ p'_{10} & p'_{11} & p'_{1x} \\ p'_{x0} & p'_{x1} & p'_{xx} \end{pmatrix} = \begin{pmatrix} (1-p_l)\cdot p_{00} & (1-p_l)\cdot p_{01} & p_l \\ (1-p_l)\cdot p_{10} & (1-p_l)\cdot p_{11} & p_l \\ p_d\cdot\pi_0 & p_d\cdot\pi_1 & 1-p_d \end{pmatrix}$$

Given P', let i, j, k denote a tree node, and $a, b, c \in S'$ possible values of a character at some node. For each character a at each node i, we maintain two variables:

- $L_i(a)$: the likelihood of the best reconstruction of the subtree with root i, given that the parent of i is assigned character a.
- $C_i(a)$: the optimal character for i, given that its parent is assigned character a.

On a binary phylogenetic tree, for each site, the revised FastML then works as follows:

1. If leaf i has character b, then, for each $a \in S'$, set $C_i(a) = b$ and $L_i(a) = p'_{ab}$.
2. If i is an internal node and not the root, its children are j and k, and it has not yet been processed, then
   - if i has character x, for each $a \in S'$, set $L_i(a) = p'_{ax} \cdot L_j(x) \cdot L_k(x)$ and $C_i(a) = x$;
   - otherwise, for each $a \in S'$, set $L_i(a) = \max_{c \in \{0,1\}} p'_{ac} \cdot L_j(c) \cdot L_k(c)$ and $C_i(a) = \arg\max_{c \in \{0,1\}} p'_{ac} \cdot L_j(c) \cdot L_k(c)$.
3. If there remain unvisited nonroot nodes, return to Step 2.
4. If i is the root node, with children j and k, assign it the value $a \in \{0,1\}$ that maximizes $\pi_a \cdot L_j(a) \cdot L_k(a)$, if the character of i is not already identified as x.
5. Traverse the tree from the root, assigning to each node its character by $C_i(a)$.

4.2   Refining Leaf Networks: RefineFast

RefineFast uses the parent networks inferred by FastML to evolve new sample leaf networks. Because the strategy is just one of sampling, we do not alter the gene contents of the original leaves—duplication and loss are not taken into account in this refinement step. Let A_l and A_p be the adjacency matrices of a leaf network and its parent network, respectively, and let A'_l stand for the refined network for A_l; then the revised *RefineFast* algorithm carries out the following steps:

1. For each entry (i, j) of each leaf network A_l,
   - if $A_l(i, j) \neq x$ and $A_p(i, j) \neq x$, evolve $A_p(i, j)$ by P to get $A'_l(i, j)$;
   - otherwise, assign $A'_l(i, j) = A_l(i, j)$.
2. Use the $A'_l(i, j)$ to replace $A_l(i, j)$.

In this algorithm, the original leaf networks are used only in the first round of ancestral reconstruction, after which they are replaced with the sample networks drawn from the distribution of possible children of the parents.

4.3   Refining Leaf Networks: RefineML

To make use of the prior information (in the original leaf networks), *RefineML* uses a *belief coefficient* k_b for each edge of these networks. In the DBN framework, these coefficients can be calculated from the *conditional probability table* (CPT) parameters of the predicted networks.

As in *RefineFast*, the refinement procedure does not alter the gene contents of the leaves. Using the same notations as in Sec. 4.1 and 4.2, *RefineML* aims to find the A'_l which maximizes the likelihood of the subtree between A_p and A'_l. The revised *RefineML* algorithm thus works as follows:

1. Learn the CPT parameters for the leaf networks reconstructed by the base inference algorithm and calculate the *belief coefficient* k_b for every site.
2. For each entry (i, j) of each leaf network A_l, do:
   - If $A_l(i, j) \neq x$ and $A_p(i, j) \neq x$, let $a = A_p(i, j)$, $b = A_l(i, j)$,
     (a) let $Q(c) = k_b$ if $b = c$, $1 - k_b$ otherwise;
     (b) calculate the likelihood $L(a) = \max_{c \in \{0,1\}} p_{ac} \cdot Q(c)$;
     (c) assign $A'_l(i, j) = \arg\max_{c \in \{0,1\}} p_{ac} \cdot Q(c)$.
   - Otherwise, assign $A'_l(i, j) = A_l(i, j)$.
3. Use $A'_l(i, j)$ to replace $A_l(i, j)$.

5   Experimental Design

To test the performance of our approach, we need regulatory networks as the input to our refinement algorithms. In our simulation experiments, we evolve networks along a given tree from a chosen ancestral network to obtain the "true" leaf networks. Then, in order to reduce the correlation between generation and reconstruction of networks, we use the leaf networks to create simulated expression data and use our preferred network inference method to reconstruct networks from the expression data. These inferred networks are the true starting point of our refinement procedure—we use the simulated gene expression data only to achieve better separation between the generation of networks and their refinement, and also to provide a glimpse of a full analysis pipeline for biological data. We then compare against the "true" networks (generated in the first step) the inferred networks after and before refinement.

Despite of the advantages of such simulation experiments (which allow an exact assessment of the performance of the inference and refinement algorithms), results on biological data are highly desirable, as such data may prove quite different from what was generated in our simulations. TFBS data is used to study regulatory networks, assuming that the regulatory interactions determined by transcription factor (TF) binding share many properties with the real interactions [9,13,24]. Given this close relationship between regulatory networks and TFBSs and given the large amount of available data on TFBSs, we chose to use TFBS data to derive regulatory networks for the organisms as their "true" networks—rather than generate these networks through simulation. In this fashion, we produce datasets for the *cis*-regulatory modules (CRMs) for 12 species of *Drosophila*.

With the extended evolutionary model, conducting experiments with real data involves several extra steps besides the refinement step, each of which is a source of potential errors. For example, assuming we have identified gene families of interest, we need to build gene trees for these genes to be able to reconstruct a history of duplications and losses. Any error in gene tree reconstruction leads to magnified errors in the history of duplications and losses. Assessing the results under such circumstances (no knowledge of the true networks and many complex sources of error) is not possible, so we turned to simulation for this part of the testing. This decision does not prejudice our ability to apply our approach to real data and to infer high-quality networks: it only reflects our inability to compute precise accuracy scores on biological data.

5.1   Experiments on Biological Data with the Basic Evolutionary Model

We use regulatory networks derived from TFBS data as the "true" networks for the organisms rather than generating these networks through simulations. Such data is available for the *Drosophila* family (whose phylogeny is well studied) with 12 organisms: *D. simulans, D. sechellia, D. melanogaster, D. yakuba, D. erecta, D. ananassae, D. pseudoobscura, D. persimilis, D. willistoni, D. mojavensis, D. virilis* and *D. grimshawi*. The TFBS data is drawn from the work of Kim *et al.* [16], where the TFBSs are annotated for all the 12 organisms on 51 CRMs.

We conduct separate experiments on different CRMs. For each CRM, we choose orthologous TFBS sequences of 6 transcription factors (TFs): *Dstat, Bicoid, Caudal,*

Hunchback, Kruppel and *Tailless*, across the 12 organisms. Then for each organism, we can get a network with these 6 TFs and the target genes indicated by the TFBS sequences, where the arcs are determined by the annotation of TFBSs, and the weights of arcs are calculated from the binding scores provided in [16]. (In this paper we do not distinguish TFs and target genes and call them all "genes.") These networks are regarded as the "true" regulatory networks for the organisms.

Gene-expression data is then generated from these "true" regulatory networks; data is generated independently for each organism, using procedure *DBNSim*, based on the DBN model [29]. Following [18], *DBNSim* uses binary gene-expression levels, where 1 and 0 indicate that the gene is, respectively, *on* and *off*. Denote the expression level of gene g_i by x_i, $x_i \in \{0,1\}$; if m_i nodes have arcs directed to g_i in the network, let the expression levels of these nodes be denoted by the vector $y = y_1 y_2 \cdots y_{m_i}$ and the weights of their arcs by the vector $w = w_1 w_2 \cdots w_{m_i}$. From y and w, we can get the conditional probability $Pr(x_i|y)$. Once we have the full parameters of the leaf networks, we generate simulated time-series gene-expression data. At the initial time point, the value of x_i is generated by the initial distribution $Pr(x_i)$; x_i at time t is generated based on y at time $t-1$ and the conditional probability $Pr(x_i|y)$.

We generate 100 time points of gene-expression data for each network in this manner. With this data we can apply the approach in Sec. 4. *DBI* is applied to infer regulatory networks from the gene-expression data. The inferred networks are then refined by *RefineFast* and *RefineML*. The whole procedure is run 10 times to provide smoothing and we report average performance over these runs.

5.2   Experiments on Simulated Data with the Extended Model

In these experiments, the "true" networks for the organisms and their gene-expression data are both generated, starting from three pieces of input information: the phylogenetic tree, the network at the root, and the evolutionary model. While simulated data allows us to get absolute evaluation of our refinement algorithms, specific precautions need to be taken against systematic bias during data simulation and result analysis. We use a wide variety of phylogenetic trees from the literature (of modest sizes: between 20 and 60 taxa) and several choices of root networks, the latter variations on part of the *yeast* network from the KEGG database [15], as also used by Kim *et al.* [17]; we also explore a wide range of evolutionary rates, especially different rates of gene duplication and loss. The root network is of modest size, between 14 and 17 genes, a relatively easy case for inference algorithms and thus also a more challenging case for a boosting algorithm.

We first generate the leaf networks that are used as the "true" regulatory networks for the chosen organisms. Since we need quantitative relationships in the networks in order to generate gene-expression data from each network, in the data generation process, we use adjacency matrices with signed weights. Weight values are assigned to the root network, yielding a weighted adjacency matrix A_p. To get the adjacency matrix for its child A_c, according to the extended network evolution model, we follow two steps: evolve the gene contents and evolve the regulatory connections. First, genes are duplicated or lost by p_d and p_l. If a duplication happens, a row and column for this new copy will be added to A_p, the values initialized either according to the *neutral initialization* model or the *inheritance initialization* model. (We conducted experiments under both models.) We denote the current adjacency matrix as A'_c. Secondly, edges in A'_c are mutated according to p_{01} and p_{10} to get A_c. We repeat this process as we traverse down the tree to obtain weighted adjacency matrices at the leaves, which is standard practice in the study of phylogenetic reconstruction [14,19].

Simulation experiments allow us to record the real gene duplication and loss history during data generation, so that we can test the pure accuracy of the refinement algorithms, without mixing their performance with that of gene tree reconstruction or reconciliation.

To test our refinement algorithms on different kinds of data, besides *DBNSim*, we also use Yu's GeneSim [28] to generate continuous gene-expression data from the weighted networks. Denoting the gene-expression levels of the genes at time t by the vector $x(t)$, the values at time $t+1$ are calculated according to $x(t+1) = x(t) + (x(t) - z)C + \varepsilon$, where C is the weighted adjacency matrix of the network, the vector z represents *constitutive expression values* for each gene, and ε models noise in the data. The values of $x(0)$ and $x_i(t)$ for those genes without parents are chosen uniformly at random from the range $[0,100]$, while the values of z are all set to 50. The term $(x(t) - z)C$ represents the effect of the regulators on the genes; this term needs to be amplified for the use of *DBI*, because of the required discretization. We use a factor k_e with the regulation term (set to 7 in our experiments), yielding the new equation $x(t+1) = x(t) + k_e(x(t) - z)C + \varepsilon$.

With two data generation methods, *DBNSim* and GeneSim, and two base inference algorithms, *DBI* and *DEI*, we can conduct experiments with different combinations of data generation methods and inference algorithms to verify that our boosting algorithms work under all circumstances. First, we use *DBNSim* to generate data for *DBI*. $13 \times n$ time points are generated for a network with n genes, since larger networks generally need more samples to gain comparable inference accuracy to smaller ones. Second, we apply *DEI* to datasets generated by GeneSim to infer the networks. Since the *DEI* tool TRNinfer does not accept large datasets (with many time points), here we use smaller datasets than the previous group of experiments with at most 75 time points. For each setup, experiments with different gene duplication and loss rates are conducted, and each experiment is run 10 times to obtain average performance.

5.3 Measurements

We want to examine the predicted networks at different levels of sensitivity and specificity. For *DBI*, on each dataset, we apply different penalty coefficients to predict regulatory networks, from 0 to 0.5, with an interval of 0.05, which results in 11 discrete coefficients. For each penalty coefficient, we apply *RefineFast* and *RefineML* on the predicted networks. For *DEI*, we also choose 11 thresholds for each predicted weighted connection matrix to get networks on various sparseness levels. For each threshold, we apply *RefineFast* on the predicted networks. We measure specificity and sensitivity to evaluate the performance of the algorithms and plot the values, as measured on the results for various penalty coefficients (for *DBI*) and thresholds (for *DEI*) to yield ROC curves. In such plots, the larger the area under the curve, the better the results.

6   Results and Analysis

6.1   Results on Biological Data with Basic Evolutionary Model

Experiments were conducted on different CRMs of the 12 *Drosophila* species; here we
show results on two of them. In both experiments, regulatory networks have 6 TFs and
12 target genes, forming networks with 18 nodes. Average performance for the base inference algorithm (*DBI*) and for the two refinement algorithms over 8 runs for these two
experiments is shown in Fig. 1 using ROC curves. In the two plots, the points on each

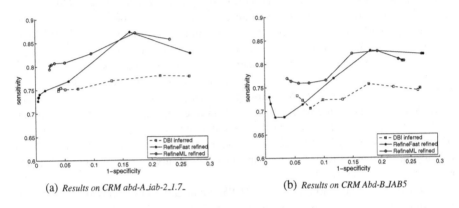

(a) *Results on CRM abd-A_iab-2_1.7_*   (b) *Results on CRM Abd-B_IAB5*

Fig. 1. Performance of refinement algorithms on *Drosophila* data

curve are obtained with different structure complexity penalty coefficients. From Fig. 1
we can see the improvement of our refinement algorithms over the base algorithm is
significant: *RefineML* improves significantly both sensitivity and specificity, while *RefineFast* loses a little sensitivity while gaining more specificity for sparse networks.[1]
The dominance of *RefineML* over *RefineFast* shows the advantage of reusing the leaf
networks inferred by base algorithms, especially when the error rate in these leaf networks is low. Results on other CRMs also show similar improvement of our refinement
algorithms.

Besides the obvious improvement, we can also observe the fluctuation of the curves:
theoretically sensitivity can be traded for specificity and vice versa, so that the ROC
curves should be in the shapes similar to those in Fig. 2, but the curves in Fig. 1 are not
as well shaped. Various factors can account for this: the shortage of gene-expression
data to infer the network, uncertainty and noise of biological data, the special structure
of the networks to be inferred, or the relatively small amount of data involved, leading
to higher variability. (We have excluded the first possibility by generating larger gene-
expression datasets for inference algorithms, where similar fluctuations still occur.)

We analyze the difference level between the "true" networks of the 12 organisms,
to obtain a view of the evolutionary rate of regulatory networks in our datasets. For

---

[1] In both CRMs, the standard deviation on sensitivity is around 0.05 and that on specificity
around 0.005.

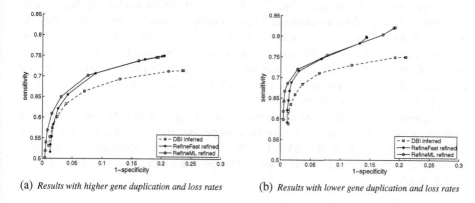

(a) *Results with higher gene duplication and loss rates*   (b) *Results with lower gene duplication and loss rates*

Fig. 2. Performance with extended evolution model and *DBI* inference method

Table 1. The proportion of edges shared by different numbers of species

Number of species	1	2	3	4	5	6	7	8	9	10	11	12
Proportion of edges	0.19	0.18	0.03	0.07	0.03	0.09	0.03	0.09	0.02	0.07	0.07	0.13

each CRM, we take the union of the edges in all 12 networks, classify these edges by the number of networks in which they are present, and calculate the proportion of each category. The overall proportions on all CRMs are shown in Table 1; a large fraction of edges are shared by less than half of the organisms, meaning that the networks are quite diverse. Therefore, the improvement brought by the refinement algorithms is due to proper use of phylogenetic information rather than the averaging effect of trees.

6.2   Results with Extended Evolutionary Model

Simulation experiments on various combinations of gene-expression data generation methods and network inference methods were conducted. All results we show below are on one representative phylogenetic tree with 37 nodes on 7 levels. Since the results of using *neutral initialization* and *inheritance initialization* in data generation are very similar, we only show results with the *neutral initialization* model.

We do not directly compare the extended model with the basic, as the two do not lend themselves to a fair comparison — for instance, the basic model requires equal gene contents across all leaves, something that can only be achieved by restricting the data to a common intersection, thereby catastrophically reducing sensitivity.

With *DBI* as the Base Inference Algorithm

In this series of experiments, *DBNSim* is used to generate gene-expression data, and *DBI* as base inference algorithm. The root network has 16 genes. Fig. 2 shows the average performance of these algorithms over 10 runs. In [29] we tested different evolutionary rates, which were mainly edge gain or loss rates; here we focus on testing different gene duplication and loss rates. Fig. 2(a) shows results under a relatively high rate of gene duplication and loss (resulting in 32 gene duplication and 25 loss events along the tree), while Fig 2(b) has a lower rate (with 15 gene duplication events and 7 gene loss events), again averaged over 10 runs. Given the size of the tree, these are high rates of gene duplication and loss, yet, as we can see from Fig. 2, the improvement gained by our refinement algorithms remains clear, with Fig. 2(b) showing slightly more improvement than Fig. 2(a), especially for sensitivity.

Comparing Fig. 1 and 2, one can observe that in Fig. 1 *RefineML* does better than *RefineFast*, but in Fig. 2 they have comparable performance. The main reason is that, as described in Section 4, *RefineML* benefits from the useful information in the *DBI* inferred networks, and bad performance of *DBI* provides little advantage for *RefineML* over *RefineFast*.

With *DEI* as the Base Inference Algorithm

Here GeneSim is used to generate continuous gene-expression data for network inference with *DEI*. The root network has 14 genes. We also show results for 2 experiments: Fig. 3(a) has higher gene duplication and loss rates, resulting in 15 gene duplications and 7 gene losses, while datasets in Fig. 3(b) have an average of 8 gene duplications and 3 losses. The *DEI* tool that we use, aims to infer networks with small gene-expression datasets. Fig. 3 shows the average performance of *DEI* and *RefineFast* for both experiments over 10 runs. *RefineFast* significantly improves the performance of the base algorithm, especially the sensitivity. (Sensitivity for *DEI* is poor in these experiments, because of inherent lower sensitivity of TRNinfer, as seen in [29] and also because of the reduced size of the gene-expression datasets.) Again more improvement is observed with the experiment which has less gene duplications and losses. This is because high duplication and loss rates give rise to a large overall gene population, yet many of them exist only in a few leaves, so that there is not much phylogenetic information to be used to correct the prediction of the connections for these genes.

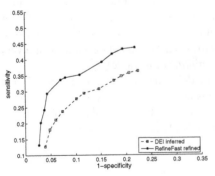

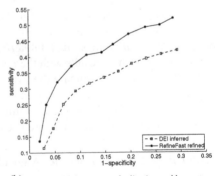

(a) *Results with higher gene duplication and loss rates*   (b) *Results with lower gene duplication and loss rates*

Fig. 3. Performance with extended evolution model and *DEI* inference method

7   Conclusions and Future Work

We presented a model, associated algorithms, and experimental results to test the hypothesis that a more refined model of transcriptional regulatory network evolution would support additional refinements in accuracy. Specifically, we presented a new version of our evolutionary approach to refine the accuracy of transcriptional regulatory networks for phylogenetically related organisms, based on an extended network evolution model, which takes into account gene duplication and loss. As these events are thought to play a crucial role in evolving new functions and interactions [23,26], integrating them into the model both extends the range of applicability of our refinement algorithms and enhances their accuracy. Results of experiments under various settings show the effectiveness of our refinement algorithms with the new model throughout a broad range of gene duplications and losses.

We also collected regulatory networks from the TFBS data of 12 *Drosophila* species and applied our approach (using the basic model), with very encouraging results. These results confirm that phylogenetic relationships carry over to the regulatory networks of a family of organisms and can be used to improve the network inference and to help with further analysis of regulatory systems and their dynamics and evolution.

Our positive results with the extended network evolution model show that refined models can be used in inference to good effect. Our current model can itself be refined, by using the widely studied evolution of TFBS [16,24].

References

1. Akutsu, T., Miyano, S., Kuhara, S.: Identification of genetic networks from a small number of gene expression patterns under the Boolean network model. In: Proc. 4th Pacific Symp. on Biocomputing PSB 1999, pp. 17–28. World Scientific, Singapore (1999)
2. Arvestad, L., Berglund, A.-C., Lagergren, J., Sennblad, B.: Gene tree reconstruction and orthology analysis based on an integrated model for duplications and sequence evolution. In: Proc. 8th Conf. Research in Comput. Mol. Bio. RECOMB 2004, pp. 326–335 (2004)
3. Babu, M.M., Luscombe, N.M., Aravind, L., Gerstein, M., Teichmann, S.A.: Structure and evolution of transcriptional regulatory networks. Curr. Opinion in Struct. Bio. 14(3), 283–291 (2004)
4. Babu, M.M., Teichmann, S.A., Aravind, L.: Evolutionary dynamics of prokaryotic transcriptional regulatory networks. J. Mol. Bio. 358(2), 614–633 (2006)
5. Barabási, A.-L., Oltvai, Z.N.: Network biology: understanding the cell's functional organization. Nat. Rev. Genet. 5, 101–113 (2004)
6. Bhan, A., Galas, D.J., Dewey, T.G.: A duplication growth model of gene expression networks. Bioinformatics 18(11), 1486–1493 (2002)
7. Bourque, G., Sankoff, D.: Improving gene network inference by comparing expression time-series across species, developmental stages or tissues. J. Bioinform. Comput. Bio. 2(4), 765–783 (2004)
8. Chen, T., He, H.L., Church, G.M.: Modeling gene expression with differential equations. In: Proc. 4th Pacific Symp. on Biocomputing PSB 1999, pp. 29–40. World Scientific, Singapore (1999)
9. Crombach, A., Hogeweg, P.: Evolution of evolvability in gene regulatory networks. PLoS Comput. Bio. 4(7), e1000112 (2008)

10. Durand, D., Halldórsson, B.V., Vernot, B.: A hybrid micro-macroevolutionary approach to gene tree reconstruction. J. Comput. Bio. 13(2), 320–335 (2006)
11. Friedman, N., Linial, M., Nachman, I., Pe'er, D.: Using Bayesian networks to analyze expression data. J. Comput. Bio. 7(3-4), 601–620 (2000)
12. Friedman, N., Murphy, K.P., Russell, S.: Learning the structure of dynamic probabilistic networks. In: Proc. 14th Conf. on Uncertainty in Art. Intell. UAI 1998, pp. 139–147 (1998)
13. Harbison, C.T., Gordon, D.B., Lee, T.I., et al.: Transcriptional regulatory code of a eukaryotic genome. Nature 431, 99–104 (2004)
14. Hillis, D.M.: Approaches for assessing phylogenetic accuracy. Syst. Bio. 44, 3–16 (1995)
15. Kanehisa, M., et al.: From genomics to chemical genomics: new developments in KEGG. Nucleic Acids Res. 34, D354–D357 (2006)
16. Kim, J., He, X., Sinha, S.: Evolution of regulatory sequences in 12 *Drosophila* species. PLoS Genet. 5(1), e1000330 (2009)
17. Kim, S.Y., Imoto, S., Miyano, S.: Inferring gene networks from time series microarray data using dynamic Bayesian networks. Briefings in Bioinf. 4(3), 228–235 (2003)
18. Liang, S., Fuhrman, S., Somogyi, R.: REVEAL, a general reverse engineering algorithm for inference of genetic network architectures. In: Proc. 3rd Pacific Symp. on Biocomputing, PSB 1998, pp. 18–29. World Scientific, Singapore (1998)
19. Moret, B.M.E., Warnow, T.: Reconstructing optimal phylogenetic trees: A challenge in experimental algorithmics. In: Fleischer, R., Moret, B.M.E., Schmidt, E.M. (eds.) Experimental Algorithmics. LNCS, vol. 2547, pp. 163–180. Springer, Heidelberg (2002)
20. Murphy, K.P.: The Bayes net toolbox for MATLAB. Comput. Sci. Stat. 33, 331–351 (2001)
21. Page, R.D.M., Charleston, M.A.: From gene to organismal phylogeny: Reconciled trees and the gene tree/species tree problem. Mol. Phyl. Evol. 7(2), 231–240 (1997)
22. Pupko, T., Pe'er, I., Shamir, R., Graur, D.: A fast algorithm for joint reconstruction of ancestral amino acid sequences. Mol. Bio. Evol. 17(6), 890–896 (2000)
23. Roth, C., et al.: Evolution after gene duplication: models, mechanisms, sequences, systems, and organisms. J. Exp. Zool. Part B 308B(1), 58–73 (2007)
24. Stark, A., Kheradpour, P., Roy, S., Kellis, M.: Reliable prediction of regulator targets using 12 *Drosophila* genomes. Genome Res. 17, 1919–1931 (2007)
25. Tanay, A., Regev, A., Shamir, R.: Conservation and evolvability in regulatory networks: The evolution of ribosomal regulation in yeast. Proc. Nat'l Acad. Sci. USA 102(20), 7203–7208 (2005)
26. Teichmann, S.A., Babu, M.M.: Gene regulatory network growth by duplication. Nature Genetics 36(5), 492–496 (2004)
27. Wang, R., Wang, Y., Zhang, X., Chen, L.: Inferring transcriptional regulatory networks from high-throughput data. Bioinformatics 23(22), 3056–3064 (2007)
28. Yu, J., Smith, V.A., Wang, P.P., Hartemink, A.J., Jarvis, E.D.: Advances to Bayesian network inference for generating causal networks from observational biological data. Bioinformatics 20(18), 3594–3603 (2004)
29. Zhang, X., Moret, B.M.E.: Boosting the performance of inference algorithms for transcriptional regulatory networks using a phylogenetic approach. In: Crandall, K.A., Lagergren, J. (eds.) WABI 2008. LNCS, vol. 5251, pp. 245–258. Springer, Heidelberg (2008)

Author Index